SUNFLOWER SCIENCE AND TECHNOLOGY

AGRONOMY

A Series of Monographs

The American Society of Agronomy (ASA) and Academic Press published the first six books in this series. Subsequent books were published by ASA alone, but in 1978 the associated societies, ASA, Crop Science Society of America (CSSA), and Soil Science Society of America (SSSA), published Agronomy 19. The books numbered 1 to 6 on the list below are available from Academic Press, Inc., 111 Fifth Avenue, New York, NY 10003; those numbered 7 to 19 are available from ASA, 677 S. Segoe Road, Madison, WI 53711.

General Editor Monographs 1 to 6, A. G. NORMAN

1. C. EDMUND MARSHALL: The Colloid Chemistry of the Silicate Minerals, 1949
2. BYRON T. SHAW, *Editor:* Soil Physical Conditions and Plant Growth, 1952
3. K. D. JACOB: Fertilizer Technology and Resources in the United States, 1953
4. W. H. PIERRE and A. G. NORMAN, *Editor:* Soil and Fertilizer Phosphate in Crop Nutrition, 1953
5. GEORGE F. SPRAGUE, *Editor:* Corn and Corn Improvement, 1955
6. J. LEVITT: The Hardiness of Plants, 1956

7. JAMES N. LUTHIN, *Editor:* Drainage of Agricultural Lands, 1957

General Editor, D. E. Gregg

8. FRANKLIN A. COFFMAN, *Editor:* Oats and Oat Improvement

Managing Editor, H. L. Hamilton

9. C. A. BLACK, *Editor-in-Chief,* and D. D. EVANS, J. L. WHITE, L. E. ENSMINGER, and F. E. CLARK, *Associate Editors:* Methods of Soil Analysis, 1965.
Part 1—Physical and Mineralogical Properties, Including Statistics of Measurement and Sampling
Part 2—Chemical and Microbiological Properties

Managing Editor, R. C. Dinauer

10. W. V. BARTHOLOMEW and F. E. CLARK, *Editor:* Soil Nitrogen, 1965

Managing Editor, H. L. Hamilton

11. R. M. HAGAN, H. R. HAISE, and T. W. EDMINSTER, *Editors:* Irrigation of Agricultural Lands, 1967

Managing Editor, R. C. Dinauer

12. R. W. PEARSON and FRED ADAMS, *Editors:* Soil Acidity and Liming, 1967

Managing Editor, R. C. Dinauer

13. K. S. QUISENBERRY and L. P. REITZ, *Editors:* Wheat and Wheat Improvement, 1967

Managing Editor, H. L. Hamilton

14. A. A. HANSON and F. V. JUSKA, *Editors:* Turfgrass Science, 1969

Managing Editor, H. L. Hamilton

15. CLARENCE H. HANSON, *Editor:* Alfalfa Science and Technology, 1972

Managing Editor, H. L. Hamilton

16. B. E. CALDWELL, *Editor:* Soybeans: Improvement, Production, and Use, 1973

Managing Editor, H. L. Hamilton

17. JAN VAN SCHILFGAARDE, *Editor:* Drainage for Agriculture, 1974

Managing Editor, R. C. Dinauer

18. GEORGE F. SPRAGUE, *Editor:* Corn and Corn Improvement, 1977

Managing Editor, D. A. Fuccillo

19. JACK F. CARTER, *Editor:* Sunflower Science and Technology, 1978

Managing Editor, D. A. Fuccillo

SUNFLOWER SCIENCE AND TECHNOLOGY

EDITOR:

JACK F. CARTER

Editorial Committee
Gerhardt N. Fick   Eric D. Putt
Donald L. Smith

Managing Editor
D. A. Fuccillo

Number 19 in the series
AGRONOMY

American Society of Agronomy,
Crop Science Society of America,
Soil Science Society of America, Inc., Publishers
Madison, Wisconsin, USA
1978

Library of Congress Cataloging in Publication Data

Sunflower Science and Technology.

(Agronomy, a series of monographs; no. 19)
Includes bibliographies and index.
1. Sunflower.   I. Carter, Jack F. 1919–
II. Series.
SB299.S9S86 338.1′7′593355 78-16192
ISBN 0-89118-054-0

The American Society of Agronomy, Inc.
677 S. Segoe Road, Madison, Wisconsin, USA 53711

Printed in the United States of America

Frontispiece. Top, a field of hybrid sunflower in North Dakota; bottom, a sunflower head visited by a bumblebee pollinator.

Sunflower Seeds

A. Black
B. Black with grey stripe
C. Black with white stripe
D. Dark grey with white stripe
E. Grey with white stripe
F. White with grey stripe
G. White

FOREWORD

When I was a small boy in Tennessee, my grandmother planted a few sunflower seeds in the corner of the garden. She said they were planted for the three B's: beauty, bees, and birds. How those plants did grow. They held dominion over all other plants in the garden because of their size and beauty.

My next remembered encounter with sunflower came in an excellent restaurant in Shanghai, China, at the end of World War II. There I discovered that sunflower seeds were not just for birds, but a delicious toasted and salted hors d'oeuvre with the English translation of "time killer." One needed a tough fingernail and patience to open them.

During graduate school I learned from Kansans that the sunflower was a state flower, originated in America, and occasionally was planted in large fields for commercial purposes. I was surprised in the 1960's suddenly to start hearing and reading about the high oil sunflower cultivars being developed and grown in the USSR and their introduction to Canada and the USA.

The sunflower might be described as the Golden Girl of American Agriculture. For the plant was native born, acclaimed first in foreign lands, and then returned home to a tumultuous welcome by both growers and researchers. This amazing saga of the sunflower has not yet ended.

The American Society of Agronomy, the Crop Science Society of America, and the Soil Science Society of America are pleased to bring to you this monograph on an increasingly important world food crop. We express our deepest appreciation to Editor Jack Carter and the many authors and reviewers for contributing their expertise and time.

Madison, Wisconsin
February 1978

JOHNNY W. PENDLETON
President
American Society of Agronomy

GENERAL FOREWORD

Publication of this Sunflower Monograph is evidence of the growing importance of this crop to agronomy and crop science in the USA. This book has been evolving for several years, from the inception of the idea for a monograph, its suggestion to the Society's Monograph Committee, and now to its materialization into book form by the industrious members of the editorial committee and their able chairman, Dr. Jack Carter.

The editors and authors of this book contributed vast efforts to produce it. Together with their coauthors, and with the assistance of reviewers, a splendid text was written to describe the current status of sunflower research, and the historical significance, taxonomy, production, pests, and various other aspects of this important crop.

Sunflower Science and Technology is the 19th monograph in the series prepared by the American Society of Agronomy since 1949. The first six volumes were published by Academic Press, Inc., of New York, but since 1957 the society has become the publisher. A complete list of the titles in the series may be found among these first pages. The monographs represent a significant and continuing activity of the American Society of Agronomy, its officers, and its approximately 10,000 members located in 100 countries around the world.

The American Society of Agronomy is associated with the Crop Science Society of America and the Soil Science Society of America. The societies share many objectives and activities in promoting these branches of agriculture and scientific disciplines. Members of the societies contribute generously of their time and talents in producing various publications, including monographs, and in pursuing other activities in the interest of human welfare.

Sunflower Science and Technology is the first monograph cosponsored by the three societies under a new agreement by these organizations. The book should be of great interest and benefit to researchers, teachers, students, producers, and other users. The societies consider it as one of their major contributions to mankind because of the worldwide adaptation of sunflower as a field crop and its growing importance in the USA. Through the presentation of up-to-date scientific and practical material on this subject, the societies aim to make sunflower an even more useful and widely grown crop for the benefit of all people.

In behalf of the society members and myself in particular, I sincerely thank Dr. Carter for his successful performance as editor, the editorial committee members, the many authors, Domenic Fuccillo, managing editor, and all others who have contributed directly or indirectly to the accomplishments of this worthy project.

Madison, Wisconsin
April 1978

MATTHIAS STELLY
Editor-in-Chief
ASA Publications

PREFACE

The colorful, single-headed sunflower of the *Frontispiece* is a valuable and useful cousin of the native sunflower of North America. The native sunflower was distributed widely across the Central Plains from north to south and occurred rarely in much of the rest of the continent. The wide genetic diversity of sunflower is still available in the indigenous populations, although population density and dispersion is decreased with increased pressure of cultivation and urbanization, grazing by animals, etc. The native sunflower was used by natives of North America, and archaeological evidence indicates large headed types with large seeds existed many centuries ago. Sunflower was used for food in mixtures of cooked vegetables and in food "concentrates" by some native Americans of the Plains.

Sunflower was taken to Europe and Asia in the 16th century and spread widely from west to east as an ornamental plant and as kitchen garden food. It was selected first for large head and large seed types as a food crop. Later, plant breeding techniques were used in the USSR and adjacent countries to produce types with high oil percentage. Sunflower has become the major annual oilseed crop of the USSR and other countries, including Argentina.

The nonoilseed, confectionery, or food type sunflower was brought back to North America by immigrants and cultivated in the Northern Plains of North America. Sunflower has been grown in the Northern Great Plains of the USA and in the Prairie Provinces of Canada for the last century. Production areas fluctuated, but they were economically nonsignificant until recent years. The introduction of the high oil types from the USSR, e.g., 'Peredovik,' into Canada and then on to the USA, and favorable prices for the oilseed sunflower for export mainly has led to sunflower production of about 910,000 ha in the USA and 67,000 ha in Canada in 1977. Interest for domestic use of sunflower oil as an edible salad and cooking oil, and in manufactured products, because of its high unsaturation, is increasing. The production first of improved, disease resistant, open-pollinated types, then true hybrids from cytoplasmic male sterility/fertility restorer techniques in the last 5 to 10 years has made sunflower economically competitive with other crops. Recent surpluses of other crops in the Northern Plains of North America have increased the competitiveness of sunflower in 1977 and apparently for 1978.

The Monograph Committee of the American Society of Agronomy, the Crop Science Society of America, and the Soil Science Society of America and the Executive Committee of the societies considered the preparation of a Sunflower Monograph for several years prior to authorizing its writing. Sunflower has become much more important as a crop during the period of planning and writing of the monograph, reaching approximately 910,000 ha in 1977 in the USA, most of which is oilseed sunflower. The production of sunflower is expected to increase further in 1978.

Since sunflower was an extremely minor crop in North America until a few years ago, only a few plant and soil scientists had conducted research on the crop. Botanists such as Charles B. Heiser, Jr., and taxonomists had conducted the major studies about sunflower and provided the only detailed publications, except for the pioneering research of Eric D. Putt and associates in Canada. Only a few scientists were available to write about sunflower in this monograph. Authors from public and private agencies, who were distributed geographically as much as possible, were chosen to represent the various disciplines.

This first Sunflower Monograph in English is designed to cover the historical aspects of the crop and the species, the economic and food function of the crop, the usefulness of the products from sunflower, problems of production and efforts made to improve sunflower, and its expected position in commerce in North America and the world in the future. The monograph is written to appeal to a wide audience in North America and to those involved in sunflower production, merchandizing, processing, and sunflower improvement throughout the world.

The great contribution of writing and personal time by authors, the editorial committee, and outside reviewers is acknowledged. Also acknowledged is the special secretarial and other assistance of Mrs. Elaine Dobrinz, Agronomy Department, North Dakota State University, Fargo. The Editorial Committee and the societies also recognize the willingness of the various public and private agencies to permit the authors and editorial committee to devote many hours of time to literature search, writing, and editorial activities to produce this Sunflower Monograph. Lastly, the counsel and advice of Domenic Fuccillo, managing editor, and of the staff at Society Headquarters at Madison, Wisconsin, is acknowledged, as well as the financial assistance of ASA, CSSA, and SSSA for various costs of the Editor.

Fargo, North Dakota
January 1978

JACK F. CARTER
Editor

JACK F. CARTER: B.S., University of Nebraska; M.S., Washington State University; Ph.D., University of Wisconsin; Professor and Chairman, Agronomy Department, North Dakota State University (NDSU), Fargo, editor of publications on sunflower and administrator in sunflower research programs by USDA-NDSU in North Dakota.

CONTRIBUTORS

JEROME F. BESSER, Fish and Wildlife Service, U.S. Department of Interior, Building 16, Denver Federal Center, Denver, CO 80225

JACK F. CARTER, Agronomy Department, North Dakota State University, Fargo, ND 58102

DAVID W. COBIA, Agricultural Economics, North Dakota State University, Fargo, ND 58102

D. GORDON DORRELL, Agriculture Canada, Research Station, Box 3301, Morden, Manitoba ROG 1JO

HARRY O. DOTY, JR., Economics, Statistics, and Cooperatives Service, U.S. Department of Agriculture, Washington, D.C. 20250

GERHARDT N. FICK, SIGCO Sunflower Products, Box 150, Breckenridge, MN 56520

CHARLES B. HEISER, JR., Botany Department, Indiana University, Bloomington, IN 47401

HARVEY J. HIRNING, Agricultural Engineering Department, North Dakota State University, Fargo, ND 58102

JOHN A. HOES, Agriculture Canada, Research Station, Box 3301, Morden, Manitoba ROG 1JO

VERNON L. HOFMAN, Agricultural Engineering Department, North Dakota State University, Fargo, ND 58102

PAULDEN F. KNOWLES, Department of Agronomy, University of California, Davis, CA 95616

JAMES R. LOFGREN, Dahlgren, Inc., Crookston, MN 56716

DARNELL R. LUNDSTROM, Agricultural Engineering Department, North Dakota State University, Fargo, ND 58102

ERIC D. PUTT, Agriculture Canada, Research Station, Box 3301, Morden, Manitoba ROG 1JO

ROBERT G. ROBINSON, Department of Agronomy and Plant Genetics, University of Minnesota, St. Paul, MN 55108

RONALD T. SCHULER, Agricultural Engineering Department, University of Minnesota, St. Paul, MN 55108

J. T. SCHULZ, Department of Entomology, North Dakota State University, Fargo, ND 58102

DONALD L. SMITH, Cal/West Seeds, P. O. Box 1428, Woodland, CA 95696

ERNEST D. P. WHELAN, Agriculture Canada, Research Station, Lethbridge, Alberta T1J 4B1

DAVID E. ZIMMER, Science and Education Administration, U.S. Department of Agriculture, Coastal Plains, Research Station, Tifton, GA 31794

Conversion Factors for English and Metric Units

To convert column 1 into column 2, multiply by	Column 1	Column 2	To convert column 2 into column 1, multiply by
		Length	
0.621	kilometer, km	mile, mi	1.609
1.094	meter, m	yard, yd	0.914
0.394	centimeter, cm	inch, in.	2.540
		Area	
0.386	kilometer², km²	mile², mi²	2.590
257.1	kilometer², km²	acre, acre	0.00405
2.471	hectare, ha [0.01 km²]	acre, acre	0.405
		Volume	
0.00973	meter³, m³	acre-inch	102.8
3.532	hectoliter, hl	cubic foot, ft³	0.2832
2.838	hectoliter, hl	bushel, bu	0.352
1.057	liter	quart (liquid), qt	0.946
		Mass	
1.102	ton (metric)	ton (English)	0.9072
220.5	quintal, q	pound, lb	0.00454
2.205	kilogram, kg	pound, lb	0.454
		Yield or Rate	
0.446	ton (metric)/hectare	ton (English)/acre	2.242
0.892	kg/ha	lb/acre	1.121
0.892	quintal/hectare	hundredweight/acre	1.121
		Pressure	
14.22	kg/cm²	lb/inch², psi	0.0703
14.50	bar	lb/in.², psi	0.06895
0.9869	bar	atmosphere, atm*	1.013
0.9678	kg/cm²	atmosphere, atm*	1.033
14.70	atmosphere, atm*	lb/in.², psi	0.06805
		* An "atmosphere" may be specified in metric or English units.	
		Temperature	
1.80C + 32	Celsius, C	Fahrenheit, F	0.555(F − 32)

Note From the Editorial Committee:
Some Standard Sunflower Terminology for the Future

SUNFLOWER: the crop, not sunflowers, that is, discard the "s". Compare to wheat. corn, soybean, cotton, etc.

OILSEED: sunflower grown primarily for extraction of edible oil from the seeds, plus other supplementary uses.

NONOILSEED: sunflower grown primarily for human food, i.e., so-called confectionery uses, or birdfood, other petfood, etc.

ACHENE: the ripened sunflower ovary.

"SEED": equivalent of achene in colloquial use.

KERNEL: the seed with hull or ovary wall removed (commonly called nutmeat, a term to be discouraged).

TO HULL OR DEHULL: means removal of the hull or ovary wall from the seed or "kernel"

CULTIVAR: equals variety in this monograph.

CONTENTS

Chapter 3   Morphology and Anatomy

P. F. KNOWLES

Chapter 4   Production and Culture

R. G. ROBINSON

HARRY O. DOTY, JR.

Chapter 1

History and Present World Status

ERIC D. PUTT

The cultivated sunflower [*Helianthus annuus* var. *marcocarpus* (DC.) Ckll.] (38) ranks with soybean [*Glycine max* (L.) Merr.], rapeseed (*Brassica campestris* L. and *B. napus* L.), and peanut (groundnut) (*Arachis hypogaea* L.) as one of the four most important annual crops in the world grown for edible oil. It has been the main source of edible vegetable oil in Russia and other eastern European countries for decades. In the last 10 to 15 years, production has increased greatly in other countries, and since 1966 increased tenfold in Canada and the USA. The oilseed cultivars of today, which consistently contain more than 40% oil, offer attractive raw material to the processors. Food manufacturing industries readily accept the high quality oil.

Significant amounts of sunflower seed are consumed in other than oil markets. Some large seed is used as whole, roasted seed, much like peanut. Some seed is dehulled and the kernels are sold as confectionery "nuts." Smaller, whole seed is used in the rations for pet birds and small animals, as well as in home feeders for wild birds.

This chapter examines the origins of the cultivated sunflower in North America, the early history of the plant, and its introduction, distribution, and adoption as a commercial agricultural crop in Europe. The chapter also describes the return of the sunflower to North America and its spread throughout other parts of the world. The chapter closes with a brief, statistical account of the crop in the past three decades and its place alongside the other major annual oilseed crops.

SUNFLOWER AMONG NORTH AMERICAN INDIANS

Early Evidence

Archaeological evidence reveals the use of sunflower among American Indians (36). At least one reference indicates cultivation of sunflower began in Arizona and New Mexico about 3000 B.C. (83). The geographical range

From: Carter, Jack F. (ed.). 1978. *Sunflower Science and Technology.* Agronomy 19. Copyright © 1978 by the American Society of Agronomy, Crop Science Society of America, and Soil Science Society of America, 677 South Segoe Road, Madison, WI 53711 USA.

widened, the crop reaching from Arkansas to the Dakotas and eastward to Ontario, Canada, and Pennsylvania. Much of the evidence indicates that a type of monocephalic, domesticated sunflower existed in the prehistoric North American Indian culture and Whiting (101) credits V. N. Jones with the opinion that sunflower may have been domesticated before corn (*Zea mays* L.) was introduced to North America. Lees (57) states that archaeologists, using carbon-14 dating, have established evidence of sunflower in the Mississippi-Missouri Basin 2,800 years ago. C. B. Heiser gives further information on the archaeology of the crop in another chapter of this monograph.

Within recorded history, after the early discoveries in the Americas by European explorers, frequent reference is made to the use of the sunflower plant by the native people. Heiser (35) has prepared one of the most comprehensive reviews on this topic. The wild sunflower, *Helianthus annuus* ssp. *lenticularis,* was used in the western U.S. for food, in medicine, and in ceremonies. Most frequently the seed was ground or pounded into flour for cakes or mush. Heiser reports frequent records of use of the cultivated sunflower over most of the remainder of the USA, except the Southeast. Champlain, the explorer, found sunflower growing in eastern Canada, and in 1588 a type grown in Virginia was clearly the cultivated sunflower. The plant also occurred in the northeastern part of Texas.

The Mandans cultivated sunflower in North Dakota. Wilson (103) gives an account of the culture of the crop by a Hidsata woman born about 1839. The description of plants with heads up to 28 cm across clearly suggests that the cultivated type was grown. The account refers to different colors of seed which were "well-fixed," implying pure breeding. These seed ranged through black, white, red, and striped, and were named for the color. There also was evidence of variation in plant type.

The southwest U.S. rendered the most extensively documented account of sunflower among the Indians (35). Castetter (16) states the Pueblo Indians of the Rio Grande Valley grew the cultivated sunflower for its edible seeds. Whiting (101) suggests the Hopi Indians possibly had the cultivated sunflower by the middle of the 13th century. Further south in Mexico, F. Hernandez, in 1570–75, described the cultivated sunflower, and Heiser (34) gives an account of an unknown cultivated form which he collected about 1944.

Thus there is both archaeological and written evidence of a form of sunflower, similar to the present cultivated types, having been part of the culture of North American Indians for 30 centuries or more. The plant extended over a wide area of the USA and into eastern Canada. Some of the types seem to have continued as distinct, biological entities into the modern era. Examples are the Hopi sunflower (33) and others described by Heiser (35).

Uses of Sunflower by the Indians

Frequent references occur to the North American Indians using sunflower for food; they may have cultivated the crop in the eastern U.S. before they had maize as a food plant (36). Whiting (101) suggests that the

Hopi Indians had corn and sunflower in pre-Spanish times or by the middle of the 13th century. Harvard (32), in a review of food plants of the North American natives, stated that *H. annuus* was a staple food from the Arctic Circle to the Tropics and from the Missouri to the Pacific. Seeds were ground and made into bread.

The Hopi Indians in the Southwest used the seed for a variety of purposes including the production of *piki*, a wafer-like bread cooked on a hot piki-stone (101). They also cracked and ate the seed like nuts (33, 101). Castetter (16) indicates several of the Indian tribes in the Southwest grew sunflower for edible seed, ground it, and made the flour into cakes. Wilson's account (103) shows the seed was used for food by the Indians of North Dakota. They roasted and pounded it into meal in a mortar. The meal then became a component of "four vegetables mixed," which consisted of a mixture of sunflower, bean, squash, and cornmeal, which was cooked by boiling for a few minutes. They considered it their best item of food. No meat was added to the mixture because sunflower gave enough fat. In another form, the meal was made into balls by rolling and shaping in the palms of the hands. The warriors carried these balls on the road for an instant source of food. Jenness (45) states the Iroquoian tribes grew sunflower with beans and maize and that Indians in Virginia used the oil from it in making bread. This reference is the only one found of the use of the oil as food, a practice which occurred as early as 1590 (83).

The Indians of North America also had several non-food uses for the sunflower. The seed provided a purple dye for basketry and textiles as well as for body paint in the Southwest (101). The stems were used to construct a ventilated hood for the piki-stone. The plant was credited with medicinal value. Jenness (45) and Harvard (32) both mention use of the oil to anoint the hair and skin. The extensive review of Heiser (35) mentions some tribes using sunflower for ceremonial purposes.

EUROPEAN ADOPTION

First Introductions

The first social contacts between North America and Europe were through Spain. It is reasonable to assume therefore, that the early Spanish explorers made the original introductions of sunflower to Europe. Zukovsky (105) states that the earliest records were of sunflower seed from New Mexico obtained by a Spanish expedition in 1510. This seed was sown in a botanic garden in Madrid. Heiser's and other reviews (29, 35, 83) credit the herbalist Dodonaeus with the first published description of the sunflower in 1568. He indicated the plant existed in Spain at this date, and records of other herbalists show it gradually moved eastward and northward in the European continent.

Spread through Europe

The original description by Dodonaeus in 1568 showed a monocephalic type similar to the commercial sunflower of today. Evidence from other botanical investigations (83), however, shows that many variants existed in Europe in the 16th century. This evidence indicates that several types arrived in Europe at different times. The movement of sunflower over Europe can be divided into two phases. The first was for use as an ornamental horticultural plant and the second as a plant to provide foods (38).

After the introduction to Spain, the sunflower plant moved to Italy and France. Semelczi-Kovacs (83) credits the Italian botanists Cortuso and Matthiolus, who published detailed descriptions, with the dissemination of the plant. At about the same time, the French botanist Dalechamps described specimens with a single stem and others with a branched stem in Lyons. He elaborated on the size and ornamental aspects of the specimens. Other botanists, and physicians searching for new medicinal plants, encouraged interest in the sunflower so that by the late 16th century it was grown in gardens in Belgium, Holland, Switzerland, Germany, and England. The Nuremburg physician Camerarius was prominent among those describing the sunflower in Germany. The botanist John Gerade noted its presence in England in 1597 (83) and there are records of it in Belgium in 1576 (35).

The migration of the sunflower eastward in Europe occurred during the 17th century. Some writers credit Germany as the point of departure (83). Lippac, cited by Semelczi-Kovacs (83), records the plant as an ornamental in Hungary in 1664, and other authors mentioned in his review continued to describe it as a garden plant as late as 1798. The height and size of leaves and flowers were impressive in these descriptions. One record, which was probably an exaggeration, speaks of a plant 12.2 m high and several studies refer to plants 6.1 to 7.3 m tall.

Coincident with the spread and use of sunflower as an ornamental garden plant, the practice of eating sunflower seeds became common and was noted as early as 1740. Zukovsky, as cited by Semelczi-Kovacs (83), indicates they replaced nuts and hazelnuts as a delicacy. This custom also led to selection for plants with single large heads and large seeds, or a type which might be considered an early cultivar. Prior to the use of the seed as a food, parts of the plant notably the petioles and young flowers, had been used as vegetable delicacies. In fact, Dodonaeus noted the value of the petioles for use as a food similar to asparagus when he wrote his original description.

In addition to the ornamental and food uses of the plant, several medicinal properties were attributed to sunflower. These included the healing of wounds from the gum of broken stalks, a diuretic capability, and the healing of kidneys with the stem turpentine (83). Also, it was credited with aiding in control of malaria when grown in marshy areas.

The first European record of using sunflower seed as a source of oil is an English patent No. 408, in 1716 granted to Arthur Bunyan (13), for inventing:

> *How from a certaine English seed might be expressed a good sweet oyle of great use to all persons concerned in the woollen manufacture, painters, leatherdressers, etc. . . . such oyle so to be made is to be expressed from the seed of the flowers commonly called & knowne by the name of sunflowers of all sorts, both double & single.*

Two implications are evident in this patent. Firstly, the oil was intended for industrial rather than edible use, and secondly, extensive variation in the plant type existed.

Evolution as a Crop in Russia

Zukovsky (105) credits Peter the Great with introducing the sunflower into Russia in the 18th century. The original seed came from the Netherlands (83). Again, as in other parts of Europe, it was grown mainly as an ornamental plant, but some literature mentions cultivation for oil occurred as early as 1769 (83). The first suggestion of extracting oil from the seed was recorded in the proceedings of the Russian Academy in 1779 (29). Several accounts, probably stemming from a single source, indicate 1830 to 1840 as the period when the manufacture of oil began on a commercial scale in Russia (10, 19, 80). Inadvertently the Holy Orthodox Church of Russia may have encouraged sunflower as an oilseed. Strict Lenten regulations prohibited many oil foods, but they omitted specific mention of sunflower. Consequently sunflower oil became popular as a food (36).

Following the initial identification and use of sunflower for oil in Russia, its cultivation expanded rapidly. By 1854, in the area of Voronezh, 84 small mills were producing about 2,000 metric tons of oil per year. Migrating peasants were credited with spreading the crop to the Ukraine and the Kuban Region, and by 1880 the crop occupied 150,000 ha in Russia (83). By the beginning of the 20th century, it was becoming a major crop in Russia. This conclusion can be reached from knowledge that the stalks were used for the production of potash. The dry stems contain nearly 5% of their weight as potash (2) and yield 45 to 56 kg/ha in the Caucasus Region of Russia. Twenty-four factories in this Region were producing 12,600 to 16,200 tons/year of potash in 1905. Assuming this to be the British long ton of 1,017 kg, this amount translates to annual plantings exceeding 0.4 million ha. About 75% of this production was exported. The average yearly production in the period 1911 to 1916 has been reported at 21.5 million hl (39). Severin (84) shows the area planted in Russia had increased to 0.9 million ha in 1915 with a yield of 377 kg/ha, and states that distinct types of seed were grown by this date in the country. One type was a small, well-filled, round seed with thin hull which was used for the extraction of edible oil. The oil percentage in this seed ranged between 20 and 30. The other type of seed was for direct human consumption. It was a large, long seed with thick, heavy hull and oil percentage ranging from only 15 to 20.

When the sunflower became a major agricultural crop in Russia, breeding commenced. Gundaev (29) credits the peasants of the country with the development of the first local cultivars. They included both types mentioned by Severin (84). Their small garden plots contained only a few plants and usually were grown in effective spatial isolation. This feature plus natural selection by the limits of climate, led to a larger number of local cultivars by 1880. One of the mean advances was the effective selection for material maturing earlier than the original introductions, which required as much as 170 days to ripen. By 1915 some cultivars had oil concentration of 38%. Zukovsky (105) states selection for oil content commenced in 1860, but Gundaev's account (29) indicates concerted breeding work started in the decade of 1890 with search for resistance to the moth *Homoeosoma nebullela* Hb., and by about 1925 breeding programs were underway at several agricultural institutes and experimental stations. V. S. Pustovoit developed the most successful program at Krasnodar (69). He achieved outstanding results in raising the percentage of oil while at the same time maintaining or improving yield of the seed. In 1940 the average percentage of oil of the main cultivar in the USSR was 33, and by 1965 Pustovoit was testing strains with 55% oil.

Evolution as a Crop in Other European Countries

Semelczi-Kovacs (83) shows interest in sunflower in Hungary for oil paralleling the early developments in Russia. Several authors which he cites mention the plant in the last decade of the 18th century. Baron Lilien in Ercsi made the first effort to cultivate it commercially in 1812. The same review shows early use of sunflower for oil in 1794 in Transylvania, now a portion of Romania, and the presence of oil mills there by 1814. Semelczi-Kovacs follows the evolution of the crop in some detail, primarily in competition with rapeseed, and notes that it proved superior to rapeseed on sandy soil. By the end of the century sunflower was the primary oil crop in Hungary and was being exported to western and northern Europe as well as the USA (83). Russia was the only other country with greater production at the beginning of the 20th century.

France had some significant production in the middle of the 19th century based on exports of oil cake to Britain. These exports were 120,000 metric tons between 1836 and 1840 and 234,000 metric tons by 1847 (83). Other countries, including Germany, also encouraged production of the crop but without any great success. One of the major deterrents was lack of cultivars suited to the cool climate common in much of the northern European areas (83).

In Europe outside of Russia, the periods of World War I and II coincided with development of the crop and gave it a major impetus. For example, Romania planted 24,000 ha in 1916. By 1938, this had increased to 200,000 ha and in 1948 to 326,000 ha (83). In 1931, the country began exporting oil and seed to northern and western Europe (25).

During World War I vegetable oil imports to Bulgaria were interrupted and the government intervened to start sunflower growing. The plantings increased from 4,800 ha in 1920 or shortly after the end of the war, to

Table 1—Sunflower seed production in European countries (23).

Country	1955	1974
	— metric tons × 1,000 —	
Bulgaria	204	260
France	8	40
Hungary	164	114
Romania	300	511
Spain	2	437
Yugoslavia	104	201
USSR	4,240	6,761

78,000 ha in 1928 and 156,000 ha in 1936 (83). Sunflower production followed a similar trend in Hungary (83), where small farmers grew much of their sunflower in mixed crops and depended upon it for edible oil and oil for illumination. Data for 1923 show the mixed crop containing sunflower at 100,000 ha and pure crop at only 2,600 ha. By 1939 the respective figures were 135,000 and 5,500 ha but in the next decade the pure crop increased rapidly. By 1948 there were 160,000 ha of pure sunflower crop and it was the nation's most important source of fat (83).

Yugoslavia adopted sunflower production more slowly. It had reached only 19,500 ha by 1939 (83) and gradually moved to 104,000 ha by 1955.

Present Importance in Europe

The USSR clearly has been much the largest producer of sunflower in Europe in this century and continues to occupy that position today. The area harvested in 1934 to 1938 was 3,276,000 ha (23) and only 545,000 ha in the remainder of Europe. Although the proportions have changed since 1938, the USSR is still the leading producer in Europe. The figures for 1974 were 4,650,000:1,608,000 ha, respectively.

Sunflower production has increased gradually in the last two decades in the USSR. The rate of increase in other parts of Europe has been gradual also, but at a somewhat faster pace (23). Dramatic expansion in production has occurred in some countries (Table 1). The expansion outside of the USSR is believed to be due in large measure to the impact of the breeding program in the USSR on the yield and percentage oil in the seed. The improved cultivars spread from the USSR into other parts of Europe and stimulated the crop.

Sunflower now occupies a larger area than any other annual oilseed crop in the countries listed in Table 1, with the exception of France, where rape is the main oilseed crop at 342,000 ha in 1974. Romania grew 248,000 ha of soybean in 1974 and was the only other country in the group that had major production of oilseed other than sunflower. The reader must recognize that other than France, these are southern European countries. In Germany, the countries of Scandinavia, and Finland, rapeseed remains the predominant oilseed crop.

SUNFLOWER IN NORTH AMERICA

Reintroduction from Europe

Most sunflower historians agree that the present cultivated sunflower in North America stems from materials reintroduced from Russia after the crop became widely grown there. The precise date or route of these introductions is not clear. Most references indicate the date as the latter part of the 19th century. Semelczi-Kovacs (83) in his review found evidence that seed was being ordered from Russia by American farmers in the decade beginning 1880. Heiser (37) reports two seed firms were offering the 'Mammoth Russian' cultivar in their catalogues in the same period, and that the U.S. consul in St. Petersburg (now Leningrad) sent cultivars to the USA in 1893. There are clues showing awareness of the plant in the USA somewhat earlier for its esthetic value. Lees (57), for example, notes a sunflower motif was used in a wrought iron screen at the Centennial International Exposition in Philadelphia in 1876.

Another highly likely route for introduction of the sunflower from Russia to the USA must have been via immigrants bringing small quantities of sunflower seed in their personal possessions. Descendents of Mennonite immigrants to Canada have told the author that their parents brought seed from Russia because it was a popular item in private kitchen gardens for producing seed for roasting and eating whole. The first of these migrations occurred about 1875.

Early Uses

Although the first introductions of the sunflower crop appear to have been stimulated by observations on the use of it as an oilseed in Russia, the preponderance of early use in North America was for silage, both in Canada and the USA. In 1901, Wiley of the USDA published a comprehensive bulletin on the plant (102). He describes some seed production in the 1890 decade. It was used mainly in scratch feeds for poultry. The publication gives more attention to the use of the crop for silage and describes studies for this use conducted in Vermont, New York, and Maine as well as at Ottawa, Canada. In the succeeding 20 to 30 years, many investigations on methods of growing the crop and its value for silage were conducted. These ranged from Quebec (42) to Saskatchewan (52) in Canada and from West Virginia to the State of Washington in the USA (98).

The crop produced high yields of silage. In many instances the nutritive value compared favorably with corn silage. Those with strong interest in sunflower considered it useful for silage on the fringe corn-growing areas because sunflower possessed frost and drought tolerance superior to corn. Today, little or none of the crop is grown for silage in North America, but some interest continues in it for this purpose in other countries (9, 12), and some studies have been conducted recently to reevaluate the crop and the newer cultivars for silage production in double-cropping systems in the USA (86, 87).

In the early decades of this century, some sunflower seed was produced in the USA. The earliest account, aside from that of Wiley (102), is a comment by Mell (63) in 1918 on the potential value of the crop for the oilseed industry. The same item indicated that seed was grown extensively in Missouri and Arkansas in 1914. A release by USDA in 1948 shows the mean annual production in the country from 1919 to 1947 was 2.7 million ka, ranging from only 404,000 kg in 1930 to 7.3 million kg 2 years previously in 1928. The average annual area harvested for seed from 1932 to 1947 was slightly over 2,350 ha, varying from 1,200 to over 4,000 ha. The major producing states over this time span were Illinois, Missouri, and California. Until 1931, Missouri, and occasionally Illinois, was the largest producer. Thereafter, California maintained continuous ascendancy to 1947 (5). The seed was used mainly in poultry scratch feeds (39, 88).

Some interest and investigation on crushing of the sunflower seed and use of the oil occurred in the early decades of this century. None led to continuing acceptance of the crop by the oilseed processors of the USA. The largest effort was a crush of 91 metric tons by a cotton oil mill in Tennessee in 1920 (88), which is presumably the same lot mentioned by Hensley (39). The oil from this trial was used by the paint and varnish trade. In 1926, French and Humphrey (25) reported trial crushing tests of seed, and refining investigations and trial uses of the oil from seed supplied by the Southeast Missouri Sunflower Growers' Association. They noted that the USDA officially classified sunflower as an edible oil, but their studies did not rank it equal to corn oil, primarily because of flavor problems. They also considered it inferior to linseed oil in drying qualities and found the resulting films had no weathering qualities superior to linseed oil.

The only other comment noted on crushing was in 1936 by Pickett (67). He saw a potential expansion of the oilseed industry in California using sunflower seed and peanuts as raw material.

In summary, the period 1900 to 1940 shows a major interest in the sunflower as a silage crop, but also some production of sunflower seed and interest in crushing it for oil. The vagaries of yield and price however, some doubts about the value of the oil for both edible and industrial purposes, and competition from other U.S. oilseed crops as well as imported oils prevented production of sunflower seed from evolving into a significant agricultural crop.

Evolution of Sunflower as an Oilseed Crop

Canadian Pioneering

In the middle of the decade, 1930 to 1939, the Government of Canada recognized the dependency of the country on imported sources of edible oil and requested the Canada Department of Agriculture to examine crops, including sunflower, which might have potential for oilseed production in Canada. Because it had been produced as a forage and silage crop in Western Canada, the sunflower was assigned to the Dominion Forage Crops Laboratory, Saskatoon, Saskatchewan (now part of the Saskatoon Research Station, Agriculture Canada). Obviously the tall, late-maturing cultivars, such as Mammoth Russian, which had been grown for silage, were not suited for oilseed production under Canadian conditions. Earlier and shorter material more amenable to mechanical harvest was essential. Fortunately, such material was available in gardens of Mennonite farmers, who had grown and matured sunflower seed for their own use continuously since settling in several communities near Saskatoon at the beginning of the century.

In the fall of 1936, approximately 400 single heads, representing a cross-section of the types available within the 'Mennonite' cultivar, were collected from gardens of three farmers in the community of Rosthern. This germplasm was planted in progeny rows to form the major portion of the first breeding nursery in 1937. Additional material included about 20 introductions from Russia and a similar number of inbred lines from a small project at the Central Experimental Farm in Ottawa. In addition to the breeding work, studies commenced in 1937 on the effects of plant populations and dates of seeding on the yield of seed.

Inbreeding was started in the Mennonite material, workers selecting among the various phenotypes for desirable agronomic characters such as strength of stem, early maturity, vigor, and high seed yield per plant as well as high oil percentage in the seed. Studies of the associations of several characters were undertaken by Putt (72) along with observations on flowering processes and morphological characters, development of a crossing technique, and some preliminary genetic studies (70, 71).

It was observed quickly that some of the Russian introductions were quite uniform. One of these, designated S-490, had a distinctly dwarf stature, small seed, and high oil percentage compared to the Mennonite material. Cultural trials by Putt and Unrau (79) with the Mennonite cultivar and open-pollinated seed of S-490 produced some of the first information that was used as a guide to farmers in growing the crop for seed.

Remnant seed of S-490, and some seed produced by self-pollination, was increased in space isolation and licensed as the cultivar 'Sunrise' in 1942. By this time, the edible oil supply had become critical because of World War II. Accordingly, the Canadian government promoted sunflower production. About 2,000 ha were grown in Saskatchewan and Alberta in 1943. In September of the same year the Minister of Trade and Commerce obtained an Order-in-Council establishing official grades for sunflower seed. The grades included Mennonite and Sunrise as part of the grade name; thus the difference in oil content of the two cultivars was recognized (40). The most recent Canadian grades make cultivar designation an optional notation at the request of the owner or shipper in the "remarks" section of the grade certificate (15).

In the seasons immediately following 1943, the sunflower crop shifted almost entirely to the Red River Valley of south central Manitoba. This area is populated by Mennonite farmers who had grown the seed in their gardens for generations. No strong inducements were needed for them to expand cultivation to a field scale. They also possessed experience and equipment for handling row crops. The area was the prime producer of sugarbeet (*Beta vulgaris* L.) in Manitoba and the sole producer of grain corn in western Canada. This knowledge gave them an advantage over the farmers of Saskatchewan and Alberta. Furthermore, the area had the benefit of a somewhat longer growing season than Saskatchewan and Alberta.

Yields and cash returns from the crop were good in Manitoba. Mean annual gross return per ha from sunflower was $71.02 for the 6 years 1943 to 1948, and $102.92 for the 2 years 1947 and 1948. Respective returns for the two periods from other crops grown in Manitoba were as follows: wheat (*Triticum aestivum* L.)—$79.04, $81.62; oats (*Avena sativa* L.)—$52.06, $59.59; barley (*Hordeum vulgare* L.)—$48.85, $59.89; grain corn—$59.45, $87.84 (6). Consequently the area planted to sunflower increased rapidly to over 24,000 ha in 1949. There also was interest in the crop in the Red River Valley of North Dakota and Minnesota, and about 8,000 ha were grown in these states in 1948 (73). Three poor crops followed the good returns of 1947 and 1948. Severe storms during harvest caused high losses due to lodging in 1949. Severe spring flooding in 1950 delayed planting until June and resulted in low yields of immature, poor quality seed. A severe epidemic of sunflower rust accentuated by 19 days of precipitation in August reduced yields to only 365 kg/ha in 1951. Consequently plantings in Manitoba dropped to only 1,215 ha in 1952, the low point since the crop first was promoted for oilseed.

Discovery of simply inherited rust resistance by Putt and Sackston (77, 78) and the knowledge among growers that breeding for improved cultivars was underway, plus more favorable weather conditions led to gradual recovery of the crop to a mean of almost 11,000 ha in Manitoba for the remainder of the decade, 1953 to 1959.

The early breeding work at Saskatoon occurred soon after the commercial acceptance and exploitation of heterosis in hybrid corn. Putt (70) noted natural crossing up to 78% in some lines early in the program. Later observations showed it varied from 5 to 95% (95) among lines. This led to recognition and subsequent exploitation of the phenomenon of heterosis in sunflower. Hybrids of inbred lines outyielded the parental mean by almost 250% and the best open-pollinated cultivar by over 60%. One hybrid was named 'Advance' and released to the public. The female parent was an inbred line descending from one of the original collections from the farm gardens in the Rosthern region; the male parent was the cultivar Sunrise (96). The female parent showed a higher propensity for cross fertilization than the male. Dominant monogenic rust resistance was added to it by backcrossing, and under the designation 'S-37-388RR' it became the female parent of 'Advent' with Sunrise as the continuing male parent. Later a single cross replaced Sunrise to produce 'Admiral,' which is slightly earlier than Advent.

Although this method to utilize heterosis showed promise under experimental conditions, in commercial practice the proportion of actual hybrids in the "hybrid" seed averaged only 39 to 45% with some samples as low as 19%. Thus the maximum potential of heterosis was not obtained. Putt (74) later concluded that use of synthetics would be a better means of using inbred lines.

The 1943 and 1944 Canadian crops were shipped to Hamilton, Ontario, and processed there. The growers recognized that shipment of the whole seed such long distances was a needless expense because the hull contained no extractable oil and was of little or no value for other purposes. Therefore, a group of 800 Manitoba farmers and local businessmen formed a producers' cooperative which constructed a small crushing plant. It commenced operation with the 1946 crop (24) and has processed sunflower seed every year since then. Initially, this crushing plant produced only crude oil using the expeller process and mechanical filtering. Later alkali refining, bleaching, deodorizing, and packaging equipment was installed, and the oil was introduced under a registered brand name to the grocery shelves. The organization also pioneered in using the sunflower hulls to produce under license a fireplace specialty fuel known as Pres-to-Logs. The process had been developed earlier to utilize waste sawdust.

In the decade commencing in 1960, two events had a marked effect on the sunflower industry in North America. The first was the introduction of cultivars from the USSR that had maturity levels comparable to Advent and Admiral, but much higher oil content. The second was the discovery of cytoplasmic male sterility.

Russian Cultivars and U.S. Interest

A cultivar, 'Peredovik', was licensed in 1964 in Canada. It yielded the same amount of seed as Advent but had 43.6% oil compared to only 32.8% in Advent (75). Consequently, all sunflower germplasm in the Canadian breeding program except for some lines with specific genetic characters, such as disease resistance and recessive branching, became obsolete. The program was reoriented to the Soviet cultivars and inbreeding and hybridization work commenced to develop disease-resistant lines with the high oil percentage. This high oil percentage of Peredovik greatly improved efficiency of the processing operation and immediately was reflected in higher prices to the grower and increased interest in the crop.

At the same time, the U.S. oilseed processing industry, which had been conscious of the Canadian ventures with the crop, recognized that the high oil percentage of the Soviet cultivars made them economically attractive for processing and marketing. By 1967, over 39,000 ha (V. Erlandson, USDA Statistical Reporting Service, St. Paul, Minn., private communication, 1976) of these cultivars were grown in North Dakota and Minnesota, and commercial crushing began. The USDA and the State of Minnesota, which had both carried out small programs, expanded research on the crop. In addition, members of the corporate sector with expertise in production of seed for planting or interests in oilseed crushing gave increasing attention to the crop by promotion and offering firm production contracts to growers.

The Sunflower Association of America (Box 2051, Fargo, ND 58102) was organized in March, 1975, to promote the crop. It is comprised of individuals and agencies that choose to support the crop, including farm organizations, processors, growers, researchers, commission companies, and exporters, as examples. The organization publishes six issues annually of a magazine called *The Sunflower*. This magazine disseminates information on all facets of the crop from cultural practices to market information on sunflower products. The publication also serves as an advertising medium for those firms participating in the sunflower seed industry as well as agribusiness.

Cytoplasmic Male Sterility

The second major event that affected the sunflower industry in North America was the discovery by Leclercq in France (56) of cytoplasmic male sterility, followed closely by the identification of fertility restoring genes (7, 22, 51). These two discoveries provided a mechanism for overcoming the problems of utilizing heterosis that were encountered with hybrids such as Advance. Hybrid sunflower seed now can be produced that contains 100% hybrids, making it comparable to hybrid corn seed. The system was incorporated rapidly into the breeding program of public and corporate establishments and added greatly to the interest which already had been stimulated by the introduction of the Soviet cultivars. In 1977, over 800,000 ha of the crop are being grown in the USA (55). Details of the use of this important tool of cytoplasmic male sterility in combination with genetic fertility restoration appear in another chapter of this monograph.

Current Importance of Nonoilseed Types

Semelczi-Kovacs (83), in his review, mentions the use of seed for food early in the growth of sunflower in Europe. Wealthy people were the first to use the seed and later farmers grew it in their kitchen gardens. He notes one reference in 1744 to the practice of eating seeds and another in 1789, "Its seeds are a favourite food of children and birds." Similar observations were reported by Hooper who commented that the streets of St. Petersburg were littered with hulls in some seasons (41). This practice of eating whole, large seeds migrated to North America with the immigrants from Europe. A notable example is the culture of the plant in farm gardens of the Mennonite people, who used seed they brought with them, and ultimately formed the base for the Canadian breeding program. Other writers also refer to the use of the whole seed for food in Russia (80), or by Russian settlers in the USA (46). It is notable that seed for this purpose was recognized to be of sufficient importance in Russia to establish a separate category for it in the official grades (97).

In Canada and the USA the affluent society following World War II fostered snack or confectionery foods and an expanded market for rations for pet birds and animals. The large seed or nonoilseed type of sunflower moved out of the kitchen garden to become a commercial crop. In 1955, one-fourth or over 2,000 ha of the crop in Manitoba was grown for large seed. By 1963 this area had increased to 6,500 ha. Much of this seed was exported to the USA and stimulated interest in production of the crop there, so that by 1967, total plantings of the nonoilseed cultivars in the USA exceeded 52,600 ha (93). Some fluctuation occurred, but by 1974 the area had risen to 97,200 ha (93). Prior to these developments, small plantings of 1,000 to 2,000 ha had been the main production of large sunflower seed in the USA (53).

Thomason (93) indicates that about 50% of the nonoilseed production is used for confectionery, primarily in two forms. The largest seed is separated, roasted, packaged whole, and sold in the same way as roasted peanuts in the shell. The remainder is dehulled, and the kernels used for confectionery or in confectionery products, often appearing in roasted form in competition with packaged shelled nuts of other species. The fraction of smaller seed from the nonoilseed crop is used mainly in the rations for pet birds and seed-eating animal pets, and also in outdoor feeding stations for wild birds. There is some preference in the trade in pet rations for a seed with prominent black and white stripes.

The expanding market for this specialty type of confectionery seed led to limited breeding programs in the public sector. These have produced the cultivars 'Commander' (76), 'Mingren' (82), and 'Sundak' (104), and there is increasing interest in the breeding of these types in the corporate sector in the USA.

SUNFLOWER IN OTHER CONTINENTS

Sunflower historians tend to associate the legend and the modern development of the sunflower crop largely with Russia and the eastern countries of Europe. The literature quickly reveals Argentina as a second important area. In North America the thinking tends to center on the early portion of this century, as the period when the sunflower became established outside of Europe. Historians also tend to regard North America as the first continent to make use of the crop beyond the European region. Significant interest existed in the crop on all continents, however, often in the middle part of the 19th century or even earlier.

Sunflower in Asia

Hooper (41) describes experiments with the crop in India. The earliest mention of oil extraction was in 1875, but the oil was disappointing commercially for use as a lubricant and for paints. Other observations on the oil were more positive and noted its edibility. Cultural trials dated back to 1855 and appeared spasmodically through to 1905, the last year covered by Hooper's report. He mentions a cultivar named 'Giant Russian' in 1905, probably the same cultivar that was introduced to the USA as Mammoth Russian. The experimenters concluded that yields were too low to make sunflower a regular oilseed crop in India. Jamieson and Baughman (44) in 1922 stated sunflower oil was of commercial importance in India and China. Pieraerts (68) made a similar comment, but no supporting documentation was cited by these authors and the writer has found none. Since the cultivars with high percentage oil have become available over the world, India has shown renewed interest in the crop. The country expected to plant 180,000 ha in 1972 and to increase plantings to 1 million ha by 1978–79 (61).

Along with the plans for increased production, an extensive promotion and research campaign is underway on the crop in India. Both Jensma (47) and Hooper (41) comment on poor seed set. Clearly, poor seed set is one of the most serious deterrents to production in India. Putt observed, during a year in India sponsored by the Canadian International Development Agency, that poor seed set is much worse than any seen in several countries of the temperate zone. Research on the problem is being carried out at five centers under the All-India Co-ordinated Research Project for Oilseeds, and additional work on the crop is incorporated in the All-India Co-ordinated Research Project for Dryland Agriculture. Two issues of the 1973 volume of the *Oilseeds Journal,* published by the Directorate of Oilseeds Development, Hyderabad, were devoted entirely to the sunflower crop. Advice and information is being obtained through international aid agencies. Personnel from countries where the crop is grown are being sent to India to provide advice, and Indian nationals are being sent abroad for training.

Hancock (30) has noted that adjoining Pakistan is looking for increased sunflower production, and Malik (60) reports promising experimental yields and field research.

Several other countries in Asia, including Iran, Iraq, and Turkey, have made notable strides in expanding planting for oilseed. In Iran, Karami (49) reported that the crop increased from 1,800 ha in 1967 to over 100,000 ha in 1970, and at the Seventh International Sunflower Conference, Cherafat (17) indicated that the crop represented 85 to 90% of the oilseed production in the country. Iraq has investigations underway which show the crop has promise in that country (90). In Turkey, Semelczi-Kovacs (83) states the first experiments were conducted after World War I and the crop reached 100,000 ha by 1946. Other personal informants from Turkey indicate significant culture of the crop commenced about 1943 and by 1956 it had reached 168,000 ha. This high point was followed by a decline to 81,000 ha in 1962, largely because of the disease caused by broom-rape (*Orobanche cumana* Wallr.), which is a seed-producing root parasite. Resistance to this parasite was introduced with the Russian cultivar 'VNIIMK 1646' in 1963. This cultivar also had much higher oil content than those previously grown. It and other Soviet cultivars led to rapid recovery and almost explosive increase in planting to 495,000 ha in 1972 (30, 91). Tasan (92) stated the crop provided 48% of the edible oil in Turkey in 1974.

The Philippines and China are the only other countries of Asia in which an interest in the sunflower crop is evident. Merrill in 1912 listed *Helianthus annuus* L. as a cultivated species with single heads in the Philippines (64). Webster, in 1920, indicated that the crop had been imported about 1900 and grown on a small scale for seed to be roasted and eaten like peanuts or to be used as a poultry feed. He was also aware that a good oil could be obtained from the seed (99). In 1971, sunflower as a commercial crop was revived in the Philippines and a substantial research program on development of cultural practices and genotypes suited to the environment was set in motion. A potential is seen of growing two crops annually after the regular rice crop. Experimental data show cultivars such as 'Krasnodarets' and Peredovik require a growing period 25 to 30 days less than in the temperate zones [(14); and personal communication, F. F. Campos, 1976]. Thus the prospect of two crops per year is feasible in the Philippines, providing the practice does not cause an increase of soil-borne diseases. Data from FAO (23) indicate production of sunflower in China in the range of 50,000 to 55,000 ha in the last decade, but precise information is not readily available.

Sunflower in Australia

Australia shows much the same pattern for the sunflower crop as North America. First, significant interest occurred for silage production in the early decades of this century. An almost completely senescent period followed, succeeded by renewed interest in the decade starting in 1940. A strong upsurge in interest has occurred since the high oil cultivars from the USSR became available in 1963.

In New South Wales the first indication of production of seed is suggested by comments on cultivation in an extension article in 1896 (1). Although the article mentioned the value of the seed for oil, evidently it was used only for poultry feed. Subsequent publications from the same and other states 20 to 30 years later refer to the use and value of the crop for silage. In part, this interest appears to have followed work in Canada and the USA on sunflower as a fodder and silage crop (100). Observations and trials on the value of the crop for silage are reported in 1922 from New South Wales (58). Some seed production and interest in silage production were recorded in the state of Victoria in 1918 (11). The information on silage draws heavily on work carried out previously in Montana. While the value of the oil was recognized, reference is made only to experimental production giving 18 to 21 hl/ha.

The Australian literature states that the crop continued as an "unimportant sideline" in the eastern states until 1940. Then, because of wartime effects of international trade, officials of New South Wales proposed sunflower as a useful oilseed crop. They extolled the value of sunflower seed for oil and drew attention to the fact that over 1 million pounds of Australian currency was being spent annually on the import of edible oils (50).

Some interest in seed production appears to have continued in Queensland, and again the wartime conditions stimulated this interest because of the tripling in the price of imported seed. Staughan (89) published information on production practices for the crop in 1945. This early publication of information in the 1940 decade from both New South Wales and Queensland dealt primarily with the tall-growing Mammoth Russian type of cultivar. At the same time, on the other side of the continent in the state of Western Australia, similar attention was being given to sunflower and information about the crop was being published (21).

By 1948 in Australia the transfer to the dwarf Canadian cultivars, notably Mennonite, Advance, and Sunrise, had begun and evolution of harvesting equipment virtually identical to that used today was evident (19). The Soviet cultivars, with their high oil percentage, made their first impact in the middle of the 1960 decade in Victoria (48). They were released in 1966 in New South Wales (20), and specific reference is made to the release of Peredovik in Queensland in 1968–69. Peredovik had 40% oil compared to 30% for 'Polestar,' which had previously been grown to meet the birdseed market (62). This period was the beginning of a rapid increase in plantings, production, and release of information in Australia. Production was only 500 metric tons in 1967–68, and reached 143,600 metric tons in 1971–72 (27). Plantings increased sharply to 239,000 ha in 1972–72, but have since dropped to 78,000 ha in 1974–75. Although graphic fluctuations in area planted and production have occurred in the last few years, apparently sunflower will continue as an established crop in Australia. Growers with experience are expected to continue in the crop. The crushing industry recognizes the value of sunflower seed (47). The most recent positive development is the formation of the Australian Sunflower Association, composed of growers, processors, researchers, and personnel in the extension and marketing fields. The organization appears to be an exact counterpart of the Sunflower Association of America, even to publishing a periodical under the same title, *The Sunflower*. The first issue was released in January, 1976. It contained items dealing with topics ranging from the development of hybrids to the quality of oil and carried advertising from all segments of the industry. (The address of the secretary of the Australian Sunflower Association is P. O. Box 337, Toowoomba, Queensland 4350, and enquiries regarding *The Sunflower* should be addressed to P. O. Box 360, Toowoomba, Queensland 4350.)

Sunflower in Africa

In Africa in the last decade, FAO reports show a range of 200,000 ha to over 400,000 ha (23). Much of this production occurs in South Africa where 32,000 ha were recorded in 1946. The history of the development of the sunflower crop on this continent, however, and notably South Africa itself, appears limited. Some items show interest in the crop in Central Africa in the early part of this century. Inference can be taken that the crop was fostered by colonial powers in this period for possible export to England and other countries as a source of oil. Some sunflower production is reported in Rhodesia. Timson (94) noted the areas planted exceeded 2,400 ha in 1924–25, and mentioned the first official grades for sunflower seed observed in the literature. These grades set up standards and required all seed for export to be examined by a Government Grain Inspector. Earlier Mainwaring (59) commented on the crop for seed as well as for silage and mentioned two firms in the market for seed for oil in 1922. He indicates, however, that local manufacture of the oil was insignificant. The main use of the seed was for poultry and other livestock feed. Rindl (81) in 1920 commented on production practices for Rhodesia as well as quality of seed and prices for it as viewed by the market in London. He also commented on prices offered in Johannesburg vs. those by a local oil factory and the results of 26 cooperative experiments in Cape Province, which is now a portion of the Republic of South Africa. He was optimistic about the crop and mentions a prophecy of the Rhodesian Department of Agriculture which stated that, "sunflowers are destined to occupy an important place in the agricultural economy of that country". At the same date, 1920, he indicated South Africa was importing both seed and oil in undetermined quantities. The seed was being used for poultry and livestock feed and the oil for soap manufacture. He stated little was grown in the Union of South Africa itself.

In Kenya, persons knowledgeable about sunflower are of the opinion that the crop was introduced by European farmers in the early part of the 1920 decade (H. N. G. Selly, personal communication, 1976) and seed produced primarily for birdfeed. The Government of Kenya is now engaged in a breeding program and an oil-crushing industry based on sunflower is developing. Also in East Africa, Childs (18), working in Tanganyika (Tanzania), recommended cultural practices for the crop in 1948 and reported on trial plantings of the previous year. The crop appears to be adapted over a range of 900 to 2,130 m elevation. He believed that sunflower seed could be a desirable cash crop for farmers on certain soil types in the Iringa District (18).

In the central regions of the African continent, Hannay (31) cites observations by J. Pieraerts (68) on the crop in the Belgian Congo (Zaire). It was introduced as early as 1906, though by 1925, when Pieraerts's article was published, the crop was not cultivated to any large extent (68). The comments indicate that cultivation was directed toward production for oilseed purposes. In North Africa, reference to cultural trials in the early part of this century has not been noted. Closer to the present day, Imani (43) reviewed the use of the crop in Morocco in 1972 and the plans to increase production there. He noted in the period prior to 1960 that the cultivars 'Jupiter' and 'Rudorf,' with only 28% to 30% oil, had been used. With these cultivars, domestic production of edible oils in the country was stationary at slightly under 1,000 metric tons annually. In 1959, a program of promotion and investigation, based on the Soviet cultivars '6540' and Peredovik, was initiated. By 1971, total domestic production of edible oils, primarily from sunflower and cotton, had risen to 10,500 metric tons. The program had a target of 200,000 ha of sunflower. FAO releases indicate a rising trend in plantings for this country, but the area is far short of this target at present (23).

Sunflower in South America

Argentina is synonymous with the thought of sunflower production in South America. As in North America, it is a reasonable speculation that the crop was introduced to Argentina by European immigrants. The country commenced to export sunflower seed early in this century, the amount varying from 63 metric tons to nearly 5,000 metric tons annually during the 1920 decade and reaching almost 10,000 metric tons in 1934 (26). In the same period, the crop gained prominence as a contributor to production of edible vegetable oils within the country. The first recorded production was 53 metric tons, or less than 1% of the total, but by 1938 it had risen to over 53,000 metric tons or 66% of the total edible vegetable oil produced in Argentina (4). Some of the increase, as a proportion of the total production in the country, was due to decrease in peanut oil, but the major factor was the contribution of sunflower to the rapidly expanding oilseed industry (4).

The first data noted on the area planted in Argentina was 84,000 ha in 1934, doubling to 176,000 ha in 1936 (85). The area varied somewhat but moved steadily upward to almost 1.6 million ha in 1946. Later it dropped back to 1 million ha in 1956 and now occupies approximately 1.2 million ha (23).

Uruguay is the only other country with production of consequence in South America. This country grew an average of 5,000 ha in the 3 years 1935 to 1938 and since then has had significant amounts, ranging from 28,000 ha in 1942 to 180,000 ha in 1952 to 1956. Since 1956 it has declined to somewhat less than 100,000 ha today (23). The prominence of the crop in this small country may be attributed to its success in neighboring Argentina, and particularly in the adjacent provinces of Buenos Aires and Entre Rios (3).

Chile and Brazil are other countries of South America that presently produce, or have produced, sunflower seed. Awareness of the crop was evident in both Chile and Brazil early in this century. In Chile in 1917 Opazo (66) recognized sunflower as a promising oilseed that could reduce the need for importation of edible oil into the country. He described experimental work carried out as early as 1905 and 1909 which gave encouraging yields and high quality oil. In 1941, the FAO reported 7,000 ha in Chile and by 1954 this had risen to 54,000 ha. The crop declined to 9,000 ha in 1974, giving way to rapeseed (23).

The Secretary of Agriculture for Brazil expressed interest in sunflower in 1924 (28), but apparently the crop has not been grown to any extent in that country, either immediately subsequent to 1924 or in recent years. FAO statistics indicate 21,000 ha grown there in 1944, but otherwise make no mention of the crop in Brazil (23). This sudden isolated surge may have resulted from the unusual conditions existing in world production and international trade of edible oils during World War II.

SUNFLOWER IN THE WORLD OILSEED ECONOMY

World Production of Annual Oilseeds

Sunflower is one of four major annual crops grown for edible oil. The others are soybean, peanut, and rapeseed. Cotton (*Gossypium hirsutum* L.) is a fiber crop, but its seed byproduct contributes significantly to world edible oil supply. Comparative mean annual production of oil for the decade 1965 to 1974 in millions of metric tons was as follows: cottonseed—2.7; peanut—3.3; soybean—6.1; sunflower—3.6; and rapeseed—2.0 (8). Several minor annual oilseed crops including safflower (*Carthamus tinctorius* L.), and sesame (*Sesamum indicum* L.), and other species, do not make the contributions of peanut, soybean, sunflower, and rapeseed to world oil supplies. Corn oil is another minor component of the world total. Flax (*Linum usitatissimum* L.) and castorbean (*Ricinus communis* L.) are also annual oilseed crops but they produce inedible oil used mainly in the paint and coatings industry so they are not classified with the previously mentioned group. Therefore this section of the chapter considers sunflower in relation to peanut, soybean, and rapeseed.

In the last 40 years, data published by FAO (23) show production of soybean and sunflower seed has increased at a faster rate than that of peanut and rapeseed (Table 2). The mean annual production of soybean in the 1972–75 period was 5.0 times greater than in the 1934–38 period. In the same periods, the mean annual production for sunflower increased 4.3-fold, whereas it was only 3.3 and 1.8 for peanut and rapeseed, respectively. Over the period of 40 years, five-year time blocks show sunflower production, as a portion of the total production of the four crops, has fluctuated between 10.4 and 12.6%, but the variation has shown no clear pattern of an increasing or decreasing trend.

Continental Production of Sunflower

Europe is clearly the largest producer of sunflower among the individual continents. This relationship is true both in quantity and as a proportion of the combined production of sunflower, rapeseed, soybean, and peanut (Table 2). Production ranged from almost 2.3 million metric tons annually in the middle of the decade starting in 1930 to over 8.1 million metric tons at present. Most of this production is from the USSR and the Eastern European countries. In the last 5 to 10 years, the sunflower has declined noticeably as a proportion of the total pool generated by the four crops. This decline is due to the disproportionate increase in rapeseed, which doubled from 1.2 to 2.4 million metric tons, compared to sunflower, which increased from only 6.6 to 8.1 million metric tons.

Compared with Europe, the production of sunflower seed on the other continents is almost insignificant. The one exception is South America

Table 2—Mean annual production of four major oilseed crops and production of sunflower expressed as percent of total of the four. Compiled from FAO Yearbook of Food and Agricultural Statistics (23).

Area & period	Soybean	Peanut	Rapeseed	Sunflower	Sunflower as % of total
	metric tons × 1,000				
World					
1934–38	11,900	5,300	3,910	2,448	10.4
1948–52	15,952	9,512	2,820	3,912	12.2
1952–56	20,625	11,329	3,151	4,605	11.6
1957–61	27,904	13,591	3,720	5,834	11.4
1962–66	34,071	15,952	4,449	7,800	12.5
1967–71	44,974	17,711	6,128	9,886	12.6
1972–75	60,066	17,388	7,311	10,540	11.1
Europe					
1934–38	123	25	299	2,295	83.7
1948–52	192	27	781	2,636	72.5
1952–56	143	31	682	3,712	81.2
1957–61	281	25	796	4,872	81.6
1962–66	460	23	1,212	6,660	79.7
1967–71	617	21	1,895	8,010	80.0
1972–75	799	21	2,458	8,145	71.3
South America					
1934–38	--	107	46	157	50.6
1948–52	55	263	--	1,018	76.2
1952–56	106	364	--	643	57.7
1957–61	217	732	29	753	43.5
1962–66	486	1,101	61	782	32.2
1967–71	1,346	1,171	67	1,070	29.3
1972–75	7,227	1,006	55	926	10.0
Oceania					
1934–38	--	6	--	--	--
1948–52	--	9	--	1	10.0
1952–56	--	11	--	1	8.3
1957–61	--	24	--	2	7.6
1962–66	1	19	--	2	9.5
1967–71	4	38	18	17	22.0
1972–75	50	39	16	112	51.6
Asia					
1934–38	10,600	3,050	3,560	--	--
1948–52	8,299	5,874	2,001	130	0.8
1952–56	10,784	7,154	2,401	174	0.8
1957–61	11,817	7,999	2,681	124	0.5
1962–66	11,895	8,412	2,822	207	0.8
1967–71	12,339	9,592	2,938	429	1.7
1972–75	13,217	9,678	3,434	647	2.4
Africa					
1934–38	--	1,300	--	1	--
1948–52	17	2,376	20	68	2.7
1952–56	23	2,909	20	79	2.6
1957–61	27	3,860	17	126	3.1
1962–66	59	5,216	5	116	2.1
1967–71	69	5,490	6	132	2.3
1972–75	86	4,853	17	270	5.2
North America					
1934–38	11,695	555	1	2	--
1948–52	7,397	922	16	19	0.2
1952–56	9,561	765	48	7	0.1
1957–61	15,586	908	201	11	0.1
1962–66	21,172	1,118	349	41	0.2
1967–71	30,559	1,396	1,112	134	0.4
1972–75	38,640	1,790	1,332	441	1.0

(Table 2). The trend that the data show for the sunflower crop in relation to the total for the four major annual oilseeds is noteworthy. The proportion has declined drastically from 76.2% of the total in the 1948 to 1952 period to only 10.0% in recent years. This decline is due largely to the huge increase in soybean production in South America which has occurred primarily in Brazil (23). If this crop is omitted, then the change from the 1948 to 1952 period to the present day has been from 80% down to 47%. The decrease of sunflower to 10.0% also is due, in part, to the increasing peanut production in South America.

All other continents except Europe and South America show an increased sunflower production in relation to the other crops (Table 2). This increase is most noticeable in Oceania, or more precisely, Australia. It had virtually no production of the annual oilseeds except for some peanut or groundnut until the beginning of the 1970 decade. In the 4-year period 1972–75, annual sunflower production reached 112,000 metric tons compared with only 2,000 metric tons in the 1962–66 period. Respectively, these quantities represent 51.6 and 9.5% of the total oilseed production in the country. Equally striking is the indication of a surge of soybean production in the country. It and sunflower seem destined to form the base for an edible oilseed industry in Australia. The relative future of the two crops depends on too many factors to hazard forecasts. If the red meat industry and intensive livestock feeding expands on the continent, the greater protein component in soybean could favor increased production of it rather than sunflower. Increasing production of nearby off-shore palm oil could compete strongly with future production and use of sunflower oil in Australia.

In Asia, sunflower production increased fivefold from the period 1948–52 to 1972–75. At a mean of 647,000 metric tons annually in the latter period, however, it still represents only 2.4% of the total production of the four major annual oilseed crops on the continent. Most of the increase is due to the rising production in Turkey. In 1974, as an example, the Turkish production comprised 80% of the Asian total. All other Asian countries for which records were given (23) have remained virtually constant in their production in the last 5 years. Unless they, or some other country, turn toward the crop, it is unlikely that sunflower will become a major oilseed crop in Asia. The agricultural potential of Turkey alone is insufficient to make sunflower a major crop on the continent. India could change the Asian picture. If the present 5-year national plan for this country becomes a fact and the crop becomes firmly established there, then Asian production will show substantial increases. Within India at present, peanut is the major oilseed crop. It can be suggested that sunflower may displace a portion of this crop as it did in Argentina. At the same time, it must be recognized that land tenure systems, agricultural practices, and culture in the two countries differ greatly. For example, the Indian farmer relies heavily on the forage from groundnuts (peanuts) to feed his bullocks, which are his prime source of power. Consequently, he is reluctant to move to other crops, even though the salable produce from them may be of higher value.

Sunflower production in Africa has increased fourfold since the 1948–52 period (Table 2). Peanut, the other major oilseed crop in Africa, has doubled since the 1948–52 period. Nevertheless, the production of peanut is so large that the output of sunflower seed remained between 2 and 3% of the total production of the four annual oilseed crops through to 1971. In the 1972–75 period, it did exceed 5%. The annual statistics released by FAO (23) show most of the increase occurred in South Africa and the same source shows a strong upward trend in peanut production in this country as well. Peanut is the major oilseed crop of many of the central African countries. Nigeria follows India, for example, as the largest peanut-producing country in the world. Accordingly, at this time it is difficult to foresee any large increase in the relative importance of sunflower as an oilseed crop in the African continent.

Soybean far outweighs the other annual oilseed crops in North America (Table 2). It is one of the largest farm crops in the USA (23). At present, sunflower is insignificant in relation to it and small in relation to peanut and rapeseed in North America. Sunflower, however, did increase by 10-fold from the early portion of the 1960 decade to the time frame 1972–75. The most interesting aspect of the sunflower in North America is the U.S. rediscovery of the crop in the past 10 years. Some of the increase occurred in Canada, but primarily it took place in the tristate area of North Dakota, Minnesota, and South Dakota in the USA. The crop increased from only 29,000 metric tons in 1963 to 291,000 and 625,000 metric tons, respectively, in 1974 and 1975. Even though the percentage contribution of sunflower to the total annual oilseed production in North America has been and still remains low, the proportion has moved from 0.1% to 1.0% in a geometric progression in the 5-year periods starting in 1957 (Table 2). If this progression continues in the USA, the crop is destined for substantial increases in the next decade. The planting of over 800,000 ha in 1977 (55), which projects a production of 800,000 to over 1 million metric tons of seed, based on recent yields, indicates the increase is occurring.

Within the USA, the proportion of the sunflower crop devoted to oilseed and nonoilseed cultivars has changed markedly. In the first 3 years of the 6-year span 1969–75, total area planted increased from 71,700 ha to 174,500 ha. Oilseed types accounted for 28 to 43% of this area. The plantings rose to 845,800 ha in 1977, and the oilseed types comprised 88% of this much larger area (54, 55, 93). These data show the long-term large markets for sunflower seed are in the oilseed type.

Several features in the North American or U.S. and Canadian agriculture could cause expanding production of sunflower. 1) The crop should be well-adaped to the large central plains area of the continent from Texas to the Prairie Provinces of Canada. After all, the cultivated species, *Helianthus annuus,* or close relatives of it, are indigenous over this entire area. 2) The agriculture is highly mechanized and much of the present equipment is suited in the existing form, or with minor modifications, for use in sunflower seed production. 3) The arrival of hybrid sunflower is expected to

Table 3—Rank of the 10 countries producing the largest quantity of sunflower seed, 1934 to 1975. Compiled from FAO Yearbook of Food and Agricultural Statistics (23).

Country	1934–38	1948–52	1952–56	1957	1958	1959	1960	1961	1962	1963	1964	1965	1966	1967	1968	1969	1970	1971	1972	1973	1974	1975
USSR	1	1	1	1	1	1	1	1	1	1	1	1	1	1	1	1	1	1	1	1	1	1
Romania	2	4	3	3	3	3	3	3	3	2	2	3	3	3	3	3	3	3	2	3	3	3
Argentina	3	2	2	2	2	2	2	2	2	3	3	2	2	2	2	2	2	2	3	2	2	2
Bulgaria	4	5	4	4	4	4	4	4	4	4	4	4	4	4	4	4	4	4	5	5	5	6
Hungary	5	3	5	8	5	6	9	7	6	6	7	7	9	9	7	7	8	9	—	10	10	10
Yugoslavia	6	7	7	7	8	7	7	5	5	5	5	5	5	5	5	5	6	6	7	6	7	8
Czechoslovakia	7	—	8	—	—	—	—	—	—	—	—	—	—	10	—	—	—	—	—	—	—	—
Uruguay	8	8	—	5	10	10	—	10	8	9	10	10	7	7	9	9	9	8	6	7	8	4
United States	9	—	—	—	—	—	—	—	—	—	—	—	—	—	—	—	—	—	—	—	—	—
Rhodesia	10	—	—	—	—	—	5	8	10	8	6	6	6	6	6	6	5	5	4	4	4	5
Turkey	—	6	6	6	7	5	6	6	7	7	8	8	8	8	8	8	—	10	—	9	9	9
Chile	—	9	10	10	9	8	8	9	9	10	9	9	10	—	10	10	—	—	—	—	—	—
South Africa	—	—	—	9	6	9	10	—	—	—	—	—	—	—	—	8	7	7	9	—	—	—
China Mainland	—	10	9	—	—	—	—	—	—	—	—	—	—	—	—	—	10	7	—	—	—	7
Spain	—	—	—	—	—	—	—	—	—	—	—	—	—	—	—	—	—	—	8	8	6	—
Australia	—	—	—	—	—	—	—	—	—	—	—	—	—	—	—	—	—	—	10	—	—	—

improve yield and disease control in the crop. 4) The American farmer is well-conditioned to the use of hybrids through his experience with hybrid corn and hybrid sorghum (*Sorghum bicolor* L.) and will accept the sunflower hybrids more readily than farmers in other countries. Considering these factors, the outlook for increasing importance of the sunflower crop on the North American continent is positive. The events of the next decade will determine whether the sunflower becomes a significant, permanent part of the agriculture on the continent.

Sunflower Production Ranked by Country

Production statistics are one means of assessing the importance and trends of a crop in the world. A quick view of the total situation also can be obtained by knowing which are the major producing countries. The 10 in the top rank based on FAO data (23) are shown for the past four decades in Table 3, which clearly illustrates the ascendency of Russia and the eastern European countries. Romania vies with Argentina for second position and has reached this rank occasionally. The Romanian production reflects a concerted breeding and development effort and the early development of hybrids using a seedling marker closely linked to the genetic male sterility (65). FAO data show the effectiveness of this program (23). Mean yield in the period 1963-75 has been 1,356 kg/ha in Romania vs. only 738 kg/ha in Argentina. The effectiveness of research and development work also is reflected in the yield figures in the USSR where the average annual yield has been 1,282 kg/ha.

The other main feature evident in Table 2 is the changing pattern among the producers below rank 4 or Bulgaria. Note, for example, that Bulgaria has fallen to sixth position in the years since 1972, and that the USA has moved into the fourth position in these years from being unranked in the early part of the decade beginning 1960. Hungary shows an opposite trend dropping from ranks 3 and 5 in the early years to rank 10 in 1973 and 1974, and below rank 10 in 1972. Both Spain and the USA have appeared in the ranking in the 4 years 1971 to 1973, and have reached sixth and fourth places, respectively. These events suggest that these two countries will play a continuing, significant role in world sunflower production.

Sunflower in World Trade

Hancock (30) has reviewed sunflower seed, sunflower oil, and meal, giving attention to significant production shifts and the role of these products in international trade in vegetable oils. In the late part of the decade commencing 1960, sunflower oil trading surged to third place and 16% of the trade, due to exports from eastern European countries. Hancock concludes that, at present, sunflower occupies sixth place among the edible soap-vegetable oils and accounts for about 8% of the total world trade in this commodity. The drop in sunflower oil is attributed to reduced crops and increased domestic consumption in eastern Europe. Present world trade in sunflower is comparable to rapeseed and groundnut (peanut) oils, but is much less than soybean oil, which is the major oil in this commodity group. Traditionally, sunflower oil on world markets has commanded some premium over soybean oil, but it is discounted in relation to peanut oil. Hancock is of the opinion this discount may reflect a traditional consumer preference rather than a true difference in the quality of the two oils.

Considering the medium-term outlook, Hancock confirms the obvious from the foregoing review. The world output of sunflower seed and its products will be dominated by developments in the USSR. Although the relative share of world production from this country has declined in recent years, it still contributes over 50% of the total. Furthermore, in the present tenth 5-year plan terminating in 1980, the target is for a 26% increase or 1.6 million metric tons over the previous 5-year plan. Hancock anticipates, however, that most of this increased production will be consumed domestically because of increased internal demands for vegetable oils. He visualizes the same situation for most other countries which may embark on expanded production (30). The U.S. may be one exception to this generalization. It has a large export surplus of soybean oil and about 90% of the recent expanded production of oilseed sunflower has been exported (54).

LITERATURE CITED

1. Anonymous. 1896. Notes on the cultivation of the sunflower. Agric. Gaz. New South Wales 7:643.
2. ————. 1916. Cultivation and utilization of sunflower, niger and safflower seed. Bull. Imp. Inst. London, 14:89–101.
3. ————. 1939. Los oleaginosos en la Argentina. Rev. Econ. Argent. 38:224–226.
4. ————. 1940. The Argentine vegetable oil industry. Rev. River Plate.88(2532):21–23.
5. ————. 1948. Sunflower seed—United States. Bur. Agric. Econ., Agric. Estimates, USDA.
6. ————. 1969. Yearbook of Manitoba Agriculture 1968. Manitoba Dep. Agric.
7. ————. 1970. Key to hybrid sunflowers found by USDA scientist. Crops Soils 23(3): 21.
8. ————. 1977. Oilseeds in Canada, an oilseed system reference. Food Systems Branch, Agric. Canada, Ottawa, Canada.
9. Andrieu, J., et C. Demarquilly. 1972. Valeur nutritive et quantité ingérée de la plante de tournesol sur pied et après ensilage. p. 463–467. In Expo. et Discuss. 5e Conf. Int. sur le Tournesol. Clermont-Ferrand, France.
10. Atkinson, A., J. B. Nelson, C. N. Arnett, W. E. Joseph, and Oscar Tretsven. 1919. Growing and feeding sunflowers in Montana. Mont. Agric. Exp. Stn. Bull. 131.
11. Audas, J. W. 1918. The sunflower, its cultivation, and utilization. J. Dep. Agric. Victoria 16:620–626.
12. Bengtsson, A. 1960. Trials with sunflower as a fodder plant. Medd. Stat. JordborForsok Uppsala. No. 104. (Seen in Field Crop Abstr. 14:1394.)
13. Bunyan, A. 1716. Extracting oil from the seed of the sunflower. Patent No. 408. (Printed by G. E. Eyre and W. Spotteswoode, Printers to the Queen's Most Excellent Majesty. 1856.)
14. Campos, F. F. 1972. Research on sunflower production and breeding under tropical conditions. p. 503–507. In Expo. et Discuss. 5e Conf. Int. sur le Tournesol. Clermont-Ferrand, France.

15. Canadian Grain Commission. 1975. Grain grading handbook for Western Canada effective Aug. 1, 1975. Winnipeg, Manitoba.

16. Castetter, E. F. 1935. Ethnobiological studies in the American Southwest. I. Uncultivated native plants used as sources of food. Univ. N. Mex. Bull. 266, Biol. Ser. 4(1).

17. Cherafat, F. 1976. Economical survey and prospects of sunflower culture in Iran. p. 9–10. *In* Abstr. of Papers 7th Int. Sunflower Conf., Krasnodar, USSR.

18. Childs, A. H. B. 1948. Sunflower production in the Iringa District. East Afr. Agric. J. 14:77–78.

19. Clydesdale, C. S., and J. Hart. 1948. Sunflowers for seed on the Darling Downs. Queensland Agric. J. 67:318–331.

20. Cutting, F. W. 1975. Sunflowers, an important source of vegetable oil. Agric. Gaz. New South Wales 86(3):21–23.

21. Elliott, H. G. 1949. The sunflower (*Helianthus annuus*). J. Agric. West. Australia 26: 44–49.

22. Enns, H., D. G. Dorrell, J. A. Hoes, and W. O. Chubb. 1970. Sunflower research, a progress report. p. 162–167. *In* Proc. 4th Int. Sunflower Conf., Memphis, Tenn.

23. FAO. 1947–75. Yearbook of food and agricultural statistics. Vols. 1–29 inclusive. Food and Agric. Organ. of the U.N., Rome, Italy.

24. Francis, E. K. 1955. In search of Utopia. The Mennonites of Manitoba. D. W. Friesen & Sons Ltd., Altona, Manitoba.

25. French, H. E., and H. O. Humphrey. 1926. Experiments on sunflower seed oil. Missouri Univ. Bull. 27(7). Eng. Exp. Stn. Ser. 25.

26. Garcia Mata, Carlos. 1936. Estudio economico de la produccion y consumo de acutes comestibles en la Argentina. Republ. Argent. Minist. de Agric. de la Nacion, Buenos Aires Ed. 2.

27. Garside, A. L. 1975. Sunflowers—more than a pretty flower. Tasmanian J. Agric. 46(2): 75–79.

28. Granato, L. 1924. O girasol; sua cultura e exploracao industrial. Bol. de Agr. 24:47–61. Sec. da Agr., Commer. e Obras Publica de S. Paulo.

29. Gundaev, A. I. 1971. Basic principles of sunflower selections. p. 417–465. *In* Genetic principles of plant selection. Nauka, Moscow. (In Russian.)

30. Hancock, R. F. 1976. Sunflower seed, oil and meal. Recent developments and prospects. *In* 7th Int. Sunflower Conf., Krasnodar, USSR.

31. Hannay, A. M. 1971. The sunflower, its cultivation and uses. A selected list of references. Econ. Libr. List No. 20, USDA.

32. Harvard, V. 1895. Food plants of the North American Indians. Bull. Torrey Bot. Club 22:98–123.

33. Heiser, C. B. 1945. The Hopi sunflower. Missouri Bot. Gard. Bull. 33:163–166.

34. ———. 1946. A "new" cultivated sunflower from Mexico. Madrono 8:226–229.

35. ———. 1951. The sunflower among the North American Indians. Proc. Am. Phil. Soc. 95:432–448.

36. ———. 1955. Origin and development of the cultivated sunflower. Am. Biol. Teach. 17:161–167.

37. ———. 1976. The sunflower. Univ. Okla. Press, Norman, Okla.

38. ———, D. M. Smith, S. B. Clevenger, and W. C. Martin. 1969. The North American sunflowers (*Helianthus*). Mem. Torrey Bot. Club 22(2):1–218.

39. Hensley, H. C. 1924. Production of sunflower seed in Missouri. Missouri Univ. Coll. Agric., Agric. Ext. Circ. 140.

40. Heoney, A. D. P. 1943. Order-in-Council establishing grades for processed sunflower and rapeseed PC 7301. Governor-General in Council, Government House, Ottawa, Canada.

41. Hooper, D. 1907. The sunflower in India. Agric. Ledger 1:323–333 (Veg. Prod. Ser. 100).

42. Hopper, W. C. 1930. Sunflowers as a silage crop. J. Agric. Hort. (Quebec) 33(12):178.

43. Imani, A. 1972. L'evolution des cultures de tournesol au Maroc. p. 497–502. *In* Expo. et Discuss. 5e Conf. Int. sur le Tournesol. Clermont-Ferrand, France.

44. Jamieson, G. S., and W. F. Baughman. 1922. The chemical composition of sunflower seed oil. J. Am. Chem. Soc. 44:2952–2957.

45. Jenness, D. 1958. The Indians of Canada. 4th ed., Natl. Museum Can. Bull. 65.

46. Jensen, I. J. 1923. Sunflowers under irrigation in Montana. Montana Agric. Exp. Stn. Bull. 162.

47. Jensma, J. R. 1973. Sunflower—alive and well in the tropics. World Farming 15(7):8–10.
48. Jones, G. O., and F. J. Barkla. 1968. Sunflowers—a developing oilseed crop. J. Agric., Dep. Agric. Victoria 68:298–299.
49. Karami, E. 1972. Sunflower production and problems in Iran. p. 486–489. *In* Expo. et Discuss. 5e Conf. Int. sur le Tournesol. Clermont-Ferrand, France.
50. Kerle, W. D. 1940. Sunflowers—a possible sideline. Agric. Gaz. New South Wales 51: 479–481.
51. Kinman, M. L. 1970. New developments in the USDA and state experiment breeding programs. p. 181–183. *In* Proc. 4th Int. Sunflower Conf., Memphis, Tenn.
52. Kirk, L. E. 1924. Sunflower production in Saskatchewan. Univ. Saskatchewan Coll. Agric. Bull. 22.
53. Knowles, P. F., and L. L. Davis. 1955. Sunflowers as a field crop. Univ. California Agric. Ext. Serv. Leafl. 2/55.
54. Kromer, G. W., and S. A. Gazelle. 1977. Situation and outlook—Sunflower seed. Econ. Res. Serv. FOS-286:18. USDA.
55. ————, and ————. 1977. Situation and outlook—Sunflower seed. Econ. Res. Serv. FOS-288:22–24. USDA.
56. Leclercq, P. 1968. Une stérilité mâle cytoplasmique chez le tournesol. Ann. Amélior. Plantes 19:99–106.
57. Lees, C. B. 1965. Sunflower Hortic. 43:24–25. (August.)
58. Little, L. G. 1922. Field experiments with sunflowers. A comparison with maize for silage. Agric. Gaz. New South Wales 23:622–624.
59. Mainwaring, C. 1922. The common sunflower (*Helianthus annuus*). Rhodesia Agric. J. 19:295–301.
60. Malik, S. J. 1972. Sunflower (*Helianthus annuus* L.) in West Pakistan. p. 528–531. *In* Expo. et Discuss. 5e Conf. Int. sur le Tournesol. Clermont-Ferrand, France.
61. Mani, N. S. 1974. Recent advances in sunflower development in India. p. 91–94. *In* Proc. 6th Int. Sunflower Conf., Bucharest, Romania.
62. McAllister, J. E., and I. F. Swan. 1970. Sunflower on the Darling Downs. Queensland Agric. J. 96:381–384.
63. Mell, C. D. 1918. The sunflower. Sci. Am. Suppl. 85:411.
64. Merrill, E. D. 1914. A flora of Manila. Manila Bur. Printing, Philippines.
65. Muresan, T., et V. A. Vranceanu. 1972. Aspects de la culture du tournesol en Roumanie. p. 508–512. *In* Expo. et Discuss. 5e Conf. Int. sur le Tournesol. Clermont-Ferrand, France.
66. Opazo G., R. 1917. Cultivo de la maravilla. Serv. de Agron. Reg. y de Ensenanza Agr. Ambulante Bol. 29.
67. Pickett, J. E. 1936. A new oil and feed industry. Pac. Rural Press 132(9):210–211.
68. Pieraerts, J. 1925. Le grand soleil ou tournesol. Bull. Agric. du Congo Belge 16:393–406.
69. Pustovoit, V. S. 1964. Conclusions of work on the selection and seed production of sunflowers. Agrobiology 5:662–697. (In Russian.)
70. Putt, E. D. 1940. Observations on morphological characters and flowering processes in the sunflower (*Helianthus annuus* L.). Sci. Agric. 21:167–179.
71. ————. 1941. Investigations of breeding technique for the sunflower (*Helianthus annuus* L.). Sci. Agric. 21:689–702.
72. ————. 1943. Association of seed yield and oil content with other characters in the sunflower. Sci. Agric. 23:377–383.
73. ————. 1949. Sunflower seed production. Co-op Vegetable Oils Ltd., Altona, Manitoba.
74. ————. 1962. The value of hybrids and synthetics in sunflower seed production. Can. J. Plant Sci. 42:488–500.
75. ————. 1965. Sunflower variety Peredovik. Can. J. Plant Sci. 45:207.
76. ————. 1965. Breeding for large sunflower seed. Res. Farmers 10(2):10–11.
77. ————, and E. Rojas M. 1955. Field studies on the inheritance of resistance to rust in the cultivated sunflower (*Helianthus annuus* L.). Can. J. Agric. Sci. 35:557–563.
78. ————, and W. E. Sackston. 1955. Rust resistance in sunflowers (*Helianthus annuus* L.). Nature 176:77.

79. ————, and J. Unrau. 1943. Influence of various cultural practices and seed and plant characters in the sunflower. Sci. Agric. 23:384–398.

80. Quesenberry, G. R., O. C. Cunningham, and L. Foster. 1921. The culture and feeding of Russian sunflowers. New Mexico Agric. Exp. Stn. Bull. 126.

81. Rindl, M. 1920. Sunflower seed and oil. S. Afr. J. Ind. 3:256–264.

82. Robinson, R. G. 1967. Registration of Mingren sunflower. Crop Sci. 7:404.

83. Semelczi-Kovacs, A. 1975. Acclimatization and dissemination of the sunflower in Europe. Acta Ethnogr. Acad. Sci. Hung. 24:47–88. (In German.)

84. Severin, G. 1935. The trade in sunflower seed. Int. Rev. Agric. 26:281S–284S.

85. ————. 1939. Sunflower seed and sunflower oil. p. 77–89. In Studies of principal agricultural products of the world market No. 4. Oils and Fats: Production and international trade, Pt. I, Int. Inst. Agric., Rome, Italy.

86. Sheaffer, C. C., J. H. McNeman, and N. A. Clark. 1976. The sunflower as a silage crop. Maryland Agric. Exp. Stn. Pub. 893.

87. ————, ————, and ————. 1977. Potential of sunflower for silage in double cropping system. Agron. J. 69:543–546.

88. Sievers, A. F. 1940. The sunflower: Its culture and uses. Bur. Plant Ind., USDA.

89. Straughan, W. R. 1945. Sunflowers for seed. Queensland Agric. J. 61:5–7. (July).

90. Tabrah, T. A. J., and G. S. Grewal. 1974. Sunflower cultivation in Iraq. p. 85–89. In Proc. 6th Int. Sunflower Conf., Bucharest, Romania.

91. Tasan, Recai. 1968. A glance at the sunflower culture and improvement activities in Turkey. Processed 3 pp. Agric. Res. Inst. Yesilkoy—Istanbul, Turkey.

92. Tashan, R. 1974. Sunflower study memorandum in Turkey. p. 22–24. In Proc. 6th Int. Sunflower Conf., Bucharest, Romania.

93. Thomason, F. G. 1974. The U.S. sunflower seed situation. Econ. Res. Serv. FOS-275:27–36. USDA.

94. Timson, S. D. 1928. The sunflower (Helianthus annuus). Rhodesia Agric. J. 25:281–296.

95. Unrau, J. 1947. Heterosis in relation to sunflower breeding. Sci. Agric. 27:414–427.

96. ————, and W. J. White. 1944. The yield and other characters of inbred lines and single crosses of sunflowers. Sci. Agric. 24:516–525.

97. Vehov, G. K. 1949. The sunflower. The science of the staple commodity, its delivery and storage. Zagitizdat, Moscow. p. 1–44. (Extended summary, Imp. Bur. of Plant Breeding and Genet., Cambridge, England.)

98. Vinall, H. N. 1922. The sunflower as a silage crop. USDA Bull. 1045.

99. Webster, P. J. 1920. Cultural directions for field crops and vegetables. Phillipp. Agric. Rev. 13:80–88.

100. Wenholz, H. 1920. Sunflowers as silage. Agric. Gaz. of New South Wales 31:721–723.

101. Whiting, A. E. 1939. Ethnobotany of the Hopi. Mus. North. Ariz. Bull. 15.

102. Wiley, H. W. 1901. The sunflower plant. Its cultivation, composition and uses. Div. Chem. USDA Bull. 60.

103. Wilson, G. L. 1917. Agriculture of the Hidsata Indians, an Indian interpretation. Univ. Minnesota Stud. Soc. Sci. 9.

104. Zimmer, D. E., and G. N. Fick. 1973. Registration of Sundak sunflower. Crop Sci. 13:584.

105. Zukovsky, P. M. 1950. Cultivated plants and their wild relatives. (Abstr. transl. by P. S. Hudson, 1962). Commonw. Agric. Bur., Farnham Royal, England.

ERIC D. PUTT: B.S.A., M.Sc., Ph.D.; Director, Research Station, Agriculture Canada, Morden, Manitoba, Canada. Sunflower breeder and agronomist; major contribution to early breeding, promotion, and establishment of sunflower as an oilseed crop in Canada with both Agriculture Canada and private industry; author and coauthor of over more than 40 scientific publications, mainly on genetics, resistance to disease, and breeding methods of sunflower; advisor on sunflower breeding and production to Government of Chile, 1950–60 through FAO, and to Government of India, 1974–75 through Canadian International Development Agency.

Chapter 2

Taxonomy of *Helianthus* and Origin of Domesticated Sunflower

CHARLES B. HEISER, JR.

The sunflowers of the genus *Helianthus* comprise some 67 species, all native to the Americas; the majority of them are found in the USA. A few of the species are rather rare, some are conspicuous elements of the natural vegetation, and a number are quite weedy, growing in areas largely disturbed by man. Several of the species have been brought into cultivation. Two of these, *H. annuus* L., the common sunflower, and *H. tuberosus* L., the Jerusalem artichoke, are food plants, the former cultivated for its oily seeds and the latter for its tubers. These two plants perhaps deserve the distinction of being the only important food plants domesticated in prehistoric times in what became the USA. Several sunflowers have been cultivated as ornamentals, the principal ones being *H. annuus, H. argophyllus* T. & G., and *H. debilis* Nutt. among the annuals, and *H. decapetalus* L., *H.* × *laetiflorus* Pers., *H. maximiliani* Schrad., *H.* × *multiflorus* L., and *H. salicifolius* A. Dietr. among the perennials (12). The beauty of the sunflower has inspired both the poet and the artist, albeit more in Europe than in the homeland. One species, *H. annuus*, is the state flower of Kansas.

From the taxonomic standpoint *Helianthus* has been considered a difficult genus, and as Mason (17) has pointed out it contains "many species of bewildering variability." Natural hybridization occurs between many of the species contributing to the taxonomic complexity. Polyploidy in the perennial species also has been shown to contribute to the taxonomic problems. In addition to these phenomena, many of the species are, genetically speaking, extremely variable. Moreover, many of the species are phenotypically plastic, the large number of environmental modifications often increasing our difficulty of placing a given plant in its proper species.

From: Carter, Jack F. (ed.). 1978. *Sunflower Science and Technology.* Agronomy 19. Copyright © 1978 by the American Society of Agronomy, Crop Science Society of America, and Soil Science Society of America, 677 South Segoe Road, Madison, WI 53711 USA.

One of the early students of the genus was Asa Gray (5), who recognized 42 species in North America. The first attempt at a comprehensive taxonomic reversion was that of E. E. Watson (21), who included 108 species, 15 of them from South America. The recent treatment by Heiser and his students accepts 17 species for South America (9) and 50 for North America (14). Keys, descriptions, maps, and drawings are provided in the latter work to facilitate identification of the species. This work will serve as the basis for systematic treatment adopted in the present account. The most recent revision by Anaschenko (1) represents a radical departure from pre-

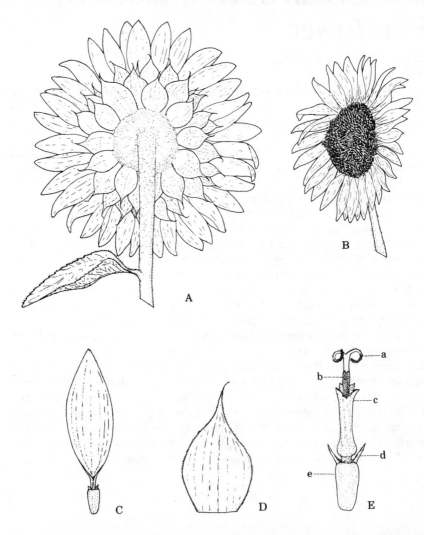

Fig. 1—*Helianthus annuus*. A. View of head from back, showing rays and phyllaries (reduced). B. View of head from size, showing ray and disk flowers (reduced). C. Ray flower ×3/5. D. Phyllary, ×4/5. E. Disk flower, ×1 3/5. a. branch of the style; b. anther; c. corolla; d. pappus; e. ovary. (Drawing by Lewis Johnson).

vious treatments and can hardly be justified on either morphological or cytogenetic grounds.

GENERIC DESCRIPTION

Helianthus is placed in the tribe Heliantheae of the family Compositae. The species for the most part are tall and coarse annual or perennial herbs, or rarely shrubby. Some of the species have a perennial root, but most of the perennial species arise from rhizomes, rarely from tubers or crown buds. The stems are simple or ramose, and along with most other parts of the plant, vary from glabrous to very densely pubescent. The leaves are opposite or alternate in arrangement. The first true leaves are always opposite, but in some species soon become alternate. The leaves usually are petiolate and three nerved, vary in shape from linear to ovate, and are entire or serrate, very rarely somewhat lobed. The heads (Fig. 1) are radiate, and the ray flowers are neutral or pistillate, but always sterile; the ligules usually are rather large, and with few exceptions, some shade of yellow. The phyllaries, or bracts of the involucre, are green and subequal to imbricate. The receptacle is nearly flat to low conic and bears chaffy bracts which embrace the achenes. The rather numerous disk flowers are five lobed, perfect and fertile, and yellow, red or purple. The two branches of the style are somewhat flattened, usually hispidulous, and with the marginal stigmatic lines poorly developed. The achenes are obovate to linear-obovate, slightly compressed or sometimes slightly quadrangular, glabrous or pubescent, and range in length from about 2 to 9 mm, to much larger in the domesticated forms of *H. annuus*. The pappus is of two scales or awns, occasionally with small supplementary scales or squamellae, all readily deciduous. The species are found in a variety of habitats, ranging from extremely dry to rather wet. Most species grow in full sunlight, although a few are adapted to shaded areas. The majority of the species flower in late summer in the USA. The basic chromosome number is 17, and diploid, tetraploid, and hexaploid species are known.

GENERIC RELATIONS

The closest relative of the genus *Helianthus* is *Viguiera,* a large genus whose species range from the southwestern United States to South America. Probably *Helianthus* developed from *Viguiera* or *Viguiera*-like ancestors. The separation of the two genera is based on the pappus. In *Helianthus* the pappus generally consists of two deciduous awns, whereas in *Viguiera* it is of awns and squamellae, both of which are persistent, or it is lacking entirely. There are a few species that are somewhat intermediate between the two genera, so that on morphological grounds there could be some justification for combining the two into a single genus, in which case the older name *Helianthus* would take preference.

The genus *Tithonia,* comprising ten or so species of Mexico and Central America, also is very closely related to *Helianthus.* Species of *Tithonia* may be distinguished readily from those of *Helianthus* by the persistent pappus, and in addition usually have more strongly four-angled achenes, and fistulose, often somewhat enlarged, peduncles. Two species of *Tithonia* sometimes are cultivated as garden ornamentals, one under the name of orange sunflower. In addition to the characters noted above, the cultivated species often are distinguished readily from the true sunflowers by the orange color of their ligules.

Other genera closely related to *Helianthus* are *Phoebanthus* and *Heliomeris. Phoebanthus* is limited to Florida, and its two species are separated from *Helianthus* most readily by the possession of small horizontal tubers. *Heliomeris,* which is combined frequently with *Viguiera,* comprises about a half dozen species of the Southwest, and one rather anomalous species, *Heliomeris porteri* (Gray) Ckll. of Georgia, all of which may be distinguished from *Helianthus* by the lack of the pappus. The southwestern species of *Heliomeris* so far reported all have a chromosome number of $n = 8$, whereas *Heliomeris porteri* is $n = 17$. *Heliomeris (Viguiera) porteri* has been hybridized with four species of *Helianthus* (11), and these are the only intergeneric hybrids known that involve *Helianthus.* Although very distinct morphologically from any species of *Helianthus, Heliomeris porteri* may be related more closely to *Helianthus* than it is to the other species of *Heliomeris.* Its proper taxonomic placement requires more study.

KEY TO SECTIONS OF *HELIANTHUS*

Perennials; leaves opposite throughout or becoming alternate above
    Plants shrubby; South American........................ I. Fruticosi
    Plants herbaceous; North American
        Plants from tap roots or long creeping roots;
        western United States and Mexico...................... II Ciliares
        Plants from rhizomes, tubers or crown buds;
        mostly eastern and central North America............. III Divaricati
Annuals, or rarely tap rooted perennials; middle and
upper leaves usually alternate............................. IV Annui

Names for the sections of *Helianthus* have never been formally adopted by taxonomists, but it seems useful to recognize four major groups, although the assignment of a few species to one or the other of the groups is somewhat arbitrary.

Section I. Fruticosi. The South American species, ranging down the Andes from southern Colombia into Peru, are not related closely to the North American species. Although it is likely that they also arose from *Viguiera,* it was probably from a different stock than that giving rise to the North American species. Thus if they are included in *Helianthus,* the genus is most likely biphyletic. On the technical character of the pappus, the

South American species should be assigned to *Helianthus,* but in other characters they are nearer to *Viguiera.* Thus they might equally well be placed into *Viguiera* or made into a separate genus. The two species for which chromosome numbers have been reported are $n = 17$.

Section II. Ciliares. Two subgroups or series may be recognized in this section: 1) Pumuli: plants with stout roots, leaves usually petiolate and harshly pubescent; Rocky Mountains and western U.S.—*H. gracilentus* A. Gray, *H. pumilus* Nutt., and *H. cusickii* A. Gray. 2) Ciliares: plants with slender roots, leaves sessile to short petiolate, glabrous to hispid; southwestern U.S. and Northern Mexico—*H. arizonensis* R. Jackson, *H. laciniatus* A. Gray, and *H. ciliaris* DC. All of the species in this section are diploid with the exception of *H. ciliaris* in which both tetraploid and hexaploid counts are reported. Species of this section appear to be well isolated genetically from members of the other sections, although only a limited number of crosses have been attempted. *Helianthus ciliaris,* blueweed, has often been a troublesome weed in agricultural lands.

III. Divaricati. This large section is well represented in the eastern and central U.S. Two species, *H. nuttallii* T. & G. and *H. californicus* DC. are found in the Western States. Species boundaries are not well marked in some members of this section, and this has probably been most responsible for the genus being considered a difficult one taxonomically. Both interspecific hybridization and polyploidy have contributed to the "species problem" in this group. The diploid members of this section probably all have a common origin, but it is possible that some of the polyploid species contain one genome derived from species belonging to another section. Five, admittedly somewhat artificial, series have been recognized. These are, with synopses, species included, and chromosome numbers (in parentheses), as follows:

1. Divaricati—Roots fibrous to coarse, not tuberous; rhizomes usually long and slender, sometimes becoming tuberous; leaves mostly or all opposite with stem leaves well developed except in *H. occidentalis* and some races of *H. rigidus*; lobes of the disk corollas yellow, except in *H. rigidus.* Wide ranging species except for *H. eggertii* which is limited to Kentucky and Tennessee. *H. mollis* L. (17), *H. occidentalis* Riddell (17), *H. divaricatus* L. (17), *H. hirsutus* L. (34), *H. decapetalus* L. (17 and 34), *H. eggertii* Small (51), *H. strumosus* L. (34 and 51), *H. tuberosus* L. (51), *H. rigidus* (Cass.) Desf. (51).

2. Gigantei—Roots coarse, often enlarged; rhizomes usually short and thick; stem leaves mostly alternate, usually lanceolate; lobes of the disk corolla yellow except in *H. salicifolius.* Wide ranging species except for *H. salicifolius,* chiefly of the Ozarks, *H. californicus* of California and northern Baja California, and *H. schweinitzii* of North and South Carolina. *H. giganteus* L. (17), *H. grosseserratus* Martens (17), *H. nuttallii* T. & G. (17), *H. maximiliani* Schrad. (17), *H. salicifolius* A. Deitr. (17), *H. californicus* DC. (51), *H. resinosus* Small (51), *H. schweinitzii* T. & G. (51).

3. Microcephali—Roots fibrous to coarse; rhizomes lacking or poorly developed; stem leaves well developed, except in *H. longifolius,* becoming

alternate above; disk corollas yellow. One species, *H. microcephalus,* wide ranging, others with rather restricted distributions, mostly southeastern. *H. microcephalus* T. & G. (17), *H. glaucophyllus* D. M. Smith (17), *H. laevigatus* T. & G. (34), *H. smithii* Heiser (34), *H. longifolius* Pursh (17).

4. Angustifolii—Roots fibrous to coarse; rhizomes lacking to well developed; basal leaves quite different in shape from stem leaves, latter narrow and usually alternate and somewhat revolute or undulate on margins; disk corollas yellow or purple. Species of southeastern U.S. *H. angustifolius* L. (17), *H. simulans* E. E. Wats. (17), *H. floridanus* A. Gray (17).

5. Atrorubentes—Roots fibrous or cord-like; rhizomes usually lacking; basal rosettes present; stem leaves usually few and much smaller than basal leaves; disk corollas purple except in *H. carnosus.* Mostly southeastern United States. *H. silphioides* Nutt. (17), *H. atrorubens* L. (17), *H. heterophyllus* Nutt. (17), *H. radula* (Pursh) T. & G. (17), *H. carnosus* Small (17). On the basis of morphology *H. carnosus* and *H. radula* are the most unusual species in the genus.

Attempts at securing hybrids between the diploid species of this section with the annual species have been largely unsuccessful, although E. D. P. Whelan (22) recently has reported securing hybrids of both *H. giganteus* and *H. maximiliani* with *H. annuus.* A number of hybrids have been secured between the tetraploid species with various annuals, and hybrids also have been produced between some of the hexaploids and *H. annuus.* The hybrid of *H. annuus* and *H. tuberosus* has been used in Russia in the improvement of the cultivated sunflower (18). Other perennials have thus far been little used in breeding programs, but it is possible that some of the other hexaploid species, and perhaps even the tetraploids and some of the diploids, also might serve as a source of genes for use in the improvement of the cultivated sunflower.

IV. Annui—Thirteen species are now placed in this section. One species included here in the earlier treatment (14) has been transferred to *Viguiera,* where it becomes *V. ludens* (Shinners) M. C. Johnst. The chromosome number for *H. similis* has not yet been reported; all of the other species are diploid ($n = 17$). The species of this section, with the possible exception of *H. similis* and *H. agrestis,* appear to be a phylogenetic unit. The species for the most part are well marked, although natural hybridization is rather common. Unfortunately, there are no good morphological characters that will always serve to distinguish the wild annual species from the perennials. The following characteristics, however, in combination are often useful: leaves mostly alternate, commonly ovate and long petiolate, disk flowers red or purple and rather numerous, disk nearly flat, pales rather elongate. The majority of the species are found in the southwestern United States and are well adapted to dry or extremely dry conditions and sandy soils.

In as much as this section includes the cultivated sunflower and is of special interest to the plant breeder, brief descriptions of all of the species are given.

KEY TO ANNUAL SPECIES

Phyllaries usually over 4 mm wide, ovate or oblong-ovate, frequently abruptly attenuate; larger leaves over 10 cm broad, cordate or subcordate; plants usually over 1.5 m tall
  Leaves, phyllaries and stems rough hairy; leaves
  usually serrate . . . . 13. *H. annuus*
  Leaves, phyllaries and stems with dense silvery white pubescence;
  leaves entire or serrulate . . . . 12. *H. argophyllus*
Phyllaries usually 4 mm or less broad, lanceolate to ovate-lanceolate, usually gradually attenuate; larger leaves usually less than 12 cm broad, cuneate, truncate or cordate; plants usually less than 2.0 m tall
  Tips of pales (chaff) in center of head densely white pubescent
    Leaves, phyllaries, and stems densely canescent;
    leaves whitish to grayish. . . . . 5. *H. niveus*
    Leaves, phyllaries and stems strigulose to hispid, rarely somewhat hirsute or subglabrous; leaves green or bluish-green
      Larger leaves usually twice or more as long as broad, usually deltoid-lanceolate to lanceolate, truncate to cuneate;
      phyllaries short attenuate . . . . 8. *H. petiolaris*
      Larger leaves usually less than twice as long as broad, lanceolate to deltoid-ovate to ovate, truncate to cordate
        Phyllaries long attenuate, exceeding disk; leaves subentire to serrulate; stems hispid . . . . 9. *H. neglectus*
        Phyllaries short attenuate, scarcely exceeding disk; leaves serrate to serrulate; stems hispid to hirsute . . . . 7. *H. praecox*
  Tips of pales in center of head glabrous to hispid at tips but not densely white pubescent
    Pales conspicuously exceeding disk-flowers; phyllaries
    hirsute. . . . . 10. *H. bolanderi*
    Pales equalling or only slightly exceeding disk-flowers; phyllaries variously pubescent to glabrous
      Leaves very rough hispid with conspicuous hairs; phyllaries usually exceeding disk; achenes densely pubescent
        Phyllaries greatly exceeding disk; trichomes of leaves and stems scattered; achenes over 5 mm long; scales of the pappus linear or linear-lanceolate . . . . 4. *H. anomalus*
        Phyllaries slightly exceeding disk; trichomes of leaves and stems dense; achenes to 5 mm long; some of pappus
        scales ovate. . . . . 3. *H. deserticola*
      Leaves variously pubescent to glabrous, if hispid, hairs not conspicuous; phyllaries equalling or exceeding disk; achenes glabrous or densely pubescent
        Leaves densely canescent, tomentose, or villous, at least beneath; pappus of two awns and usually several smaller scales

Leaves mostly alternate, canescent or tomentose, nearly equally
so on both surfaces, smooth to the touch, usually truncate to
cuneate; disk corollas usually red or
purple tipped.............................. 5. *H. niveus*
Leaves mostly opposite, lower surface densely villous, upper
surface less so and somewhat rough, usually cordate to subcordate; disk corollas yellow................. 2. *H. similis*
Leaves hispid or strigose to glabrous; pappus usually of 2 awns
only
Leaves lanceolate to ovate lanceolate, cuneate; plants usually
over 1.5 m tall; of wet, not sandy soil
Stems nearly glabrous, usually glaucous; base of leaf and
petiole ciliate with prominent white
trichomes.............................. 1. *H. agrestis*
Stems usually somewhat hispid, not glaucous; base of leaf
and petiole not prominently ciliate........ 11. *H. paradoxus*
Leaves deltoid-ovate, rarely ovate-lanceolate, cuneate, truncate
or cordate; plants usually of sandy soil, usually less than 1.5 m
tall, sometimes decumbent; usually in sandy soils
Leaves gradually tapering to tip; phyllaries short acuminate
to long attenuate, usually not more than
3 mm wide.............................. 6. *H. debilis*
Leaves often with constriction near middle; phyllaries short
acuminate, sometimes more than
3 mm broad........................... 7. *H. praecox*

1. *H. agrestis* Pollard. Annual from shallow root system, 1 rarely 2 m in
height, branches slender; stem and branches glaucous and glabrous or with
a few scattered hairs; leaves lanceolate, acuminate, mostly opposite, becoming alternate near heads, 8.5 to 18.0 cm long, 1.5 to 4.0 cm wide, thin, light
green both sides, shiny, irregularly and usually sparingly serrate, teeth
mucronate, petiole quite short, 5 to 12 mm long, prominent scattered long,
stiff white hairs principally along lower margin and petiole; peduncles
slender, short tomentose near summit; disk conical, about 1.0 cm in diam.
or slightly larger; phyllaries lanceolate, acuminate, usually somewhat longer
than the disk, loose but not reflexed, hispidulose along the margins, otherwise glabrous, 1 to 2 mm broad, rays 12 to 18 mm long; lobes and throat of
disk florets purple; style branches yellow; anthers purple; pales glabrous;
achenes 2 to 3 mm long, glabrous and inconspicuously tuberculate.
Distribution: Central Florida and Thomas, Co., Ga.
    This species does not appear to be closely related to any of the other
annual species and perhaps it could be derived from the ancestral group of
the Section Divaricati. All attempts at hybridizing it with other annual
species have failed. This species also differs from most others in that it is
self-compatible.

2. *H. similis* (Brandegee) Blake (Synonym: *Viguiera similis* Brandegee).
Spreading perennial shrub, root unknown, stem slightly tomentose,
glabrate, to 1.2 m tall; leaves opposite or uppermost subopposite, ovate,

long acuminate, cordate at base, shallowly and regularly serrate, villous to tomentose and greenish above, white-villous to tomentose below, to 16 cm long by 9 cm wide, petiole to 2 cm long; heads few, disk up to 1 cm broad; peduncles 2 to 8 cm long; phyllaries linear-lanceolate to lanceolate, ca. 1 cm long and 1.6 mm wide, acuminate; pales entire, somewhat keeled, hispidulous on mid-line; disk flowers yellow, sparingly pubescent on tube and densely so on lobes; pappus of two awns and several squamellae; achenes 2.5 mm long, nearly glabrous.

Distribution: known only from southernmost Baja Calif.

This species is somewhat transitional between *Viguiera* and *Helianthus* and might equally well be placed in *Viguiera*. Its nearest relative is probably *H. niveus*, but it does not appear to be a close relationship. This species has not been grown under artificial conditions, and neither its chromosome number nor crossing relationships is known.

3. *H. deserticola* Heiser. Annual; 0.2 to 0.6 m tall, stem green or red, little branched, densely hispid; leaves lanceolate to lanceolate-ovate, cuneate at base, densely hispid-hirsute, entire, 2.5 to 5.0 cm long, 1.0 to 2.0 cm wide, petiolate; heads several, disk 1.3 to 2.5 cm in diam.; phyllaries lanceolate to linear-lanceolate, short-acuminate, slightly exceeding disk, mostly 1.4 cm long by 2 mm wide; rays 9 to 12, 1.0 to 2.0 cm long; pales hirsute at apex, rarely glabrous; achenes 4 to 5 mm long, slender, pilose with hairs 0.5 to 1.0 mm long; pappus of 2 awns, about 2.5 mm long, and several ovate squamellae, 0.5 to 1.0 mm long.

Distribution: Utah, Nevada and northwestern Arizona; rare.

This species appears related to *H. anomalus, H. petiolaris* subsp. *fallax* and *H. niveus*. Little experimental work has been done with it, as great difficulty has been encountered in germination of seeds. Filled achenes have been obtained in crosses with *H. annuus,* but these failed to germinate.

4. *H. anomalus* Blake. Annual 0.2 to 0.8 m tall, stems whitish or greenish, somewhat glaucous, nearly glabrous or tuberculate-hispid, sparingly branched, leaf blades ovate or ovate-lanceolate, sparingly tuberculate-hispid, entire, cuneate at base, 5.0 to 10.0 cm long, 3 to 6 cm wide, petiolate; phyllaries linear-lanceolate, conspicuously long ciliate on lower margins, sparsely so on backs, 2.0 to 4.0 mm wide, 1.5 to 2.5 cm long, usually greatly exceeding the disk; heads few, disk 2.0 to 2.7 cm diam.; rays few light yellow, 2.0 to 3.0 cm long; pales long attenuate, hispid to near summit; achenes 6 to 9 mm long, slender, buff-colored, densely villous with hairs 1 to 2 mm long; pappus of several linear scales, generally of two long ones and several short ones.

As with the previous species great difficulty has been experienced in germinating seeds of this species. Filled achenes have been obtained in crosses with *H. annuus, H. deserticola, H. debilis* and *H. petiolaris*; only those from the last two named species germinated. Next to *H. annuus,* this species has the longest achenes known in the genus. Perhaps it deserves investigation as a potential oil crop for arid regions. One herbarium specimen collected by H. R. Voth in Arizona, exact locality nor date given, originally identified as *H. annuus* and now deposited in the Chicago Museum of Natural History, is labeled "Plants of the Hopi," but details on its uses, if any, and whether it was cultivated are not indicated. The large head of this specimen suggests the possibility that it may have been a cultivated plant.

5. *H. niveus* (Benth.) Brandegee (synonyms: *Encelia nivea* Benth., *Viguiera tephrodes* A. Gray, *V. nivea* A. Gray, *V. sonorae* Rose and Standl. *Gymnolomia encelioides* A. Gray, *H. dealbatus* A. Gray, *H. tephrodes* A. Gray, *H. petiolaris* var. *canescens* A. Gray, *H. petiolaris* var. *canus* Britton, *H. canus* Woot. & Standl.). Perennial from long tap root, somewhat shrubby, erect or decumbent, 0.3 to 1.0 m tall; stem appressed white to grayish-sericeous-villous, glabrate and woody below; leaves nearly all opposite or upper alternate, spathulate, ovate or ovate-lanceolate, entire or undulate to crisped, 2.0 to 5.0 cm long, 1.5 to 3.0 cm wide, appressed white to grayish-sericeous-villous; petioles scarcely distinct to 3.0 cm long; heads generally solitary on short to long peduncles; phyllaries 2 to 3 mm wide; lobes of disk-corolla from reddish-purple to nearly yellow; pales entire or shallowly 3-cuspidate; achenes (3-)4(-5) mm long, sparingly pubescent; pappus extremely varible, usually of two awns, sometimes with one to several squamellae less than 0.5 mm long.

Distribution: western Texas to southern California to northern Mexico.

This species comprises three major races: subsp. *niveus* of Baja California; subsp. *tephrodes* (A. Gray) Heiser of Yuma Co., Ariz., Imperial Co., Calif. and western Sonora, Mexico; and subsp. *canescens* (A. Gray) Heiser. Although showing some connection with *Viguiera*, this species is clearly a *Helianthus*, and perhaps may be regarded as the most "primitive" species of the genus. It, or something very similar to it, may be the nearest to the ancestral stock that gave rise to all the North American sunflowers. This species has been hybridized with species belonging to all the three sections of the North American sunflowers.

6. *H. debilis* Nutt. (Synonyms: *H. cucumerifolius* T. & G., *H. vestitus* E. E. Wats.). Annual or short-lived perennial; stem decumbent or erect, rarely to 2 m tall, glabrous, hispid or hirsute, sometimes strongly purple mottled; leaves ovate, deltoid-ovate, to lance-ovate, usually dark green, cuneate to cordate, acute to acuminate at apex, nearly entire to deeply and irregularly serrate, nearly glabrous to hispid, or somewhat hirsute beneath, to 14 cm long and 13 cm wide, usually much smaller, petiolate; disk 1.0 to 2.2 cm in diam.; phyllaries lanceolate, short acute to long attenuate, 1.5 to 3.0 mm wide (sometimes wider in cultivated forms), glabrous or sparingly hispid to hispidulous; rays pale yellow to orange yellow, 11 to 20, 1.2 to 2.3 cm long, 0.5 to 1.2 cm wide (larger in cultivated forms); lobes of disk-corollas purple (rarely yellow in cultivated forms of ssp. *cucumerifolius*); pales 3-cuspidate, middle cusps acuminate, glabrous to hispid, rarely slightly villous; achenes 2.5 to 3.0 mm long, sparingly pubescent to glabrous.

Distribution: southeastern United States.

This species comprises five subspecies: subsp. *debilis* of east coastal Florida; subsp. *vestitus* (E. E. Wats.) Heiser, keys of central west Florida;

subsp. *tardiflorus* Heiser, western and northern Florida to Alabama; subsp. *silvestris* Heiser, northeastern Texas; and subsp. *cucumerifolius* (T. & G.) Heiser, native to southeastern Texas, occasional as a weed elsewhere, and widely cultivated as an ornamental. Members of this species have given hybrids in crosses with most of the other annual species. Natural hybrids with *H. annuus* are known from Texas. Its closest relative appears to be *H. praecox*.

7. *H. praecox* Engelm. and Gray. Annual; stem erect or becoming somewhat procumbent to 1.5 m tall, hispid to hirsute, purple-mottled or straw colored; leaves deltoid-ovate, usually with slight constriction near middle, cuneate to subcordate, acuminate, frequently grayish green, hispid to sparingly hirsute, serrulate or shallowly serrate with sharp teeth, blade to 9 cm long and 7 cm wide, petiolate; peduncles 15 to 40 cm long; disk 1.3 to 1.8 cm in diam.; phyllaries broadly lanceolate, short acute, 3 to 4 mm wide, hispidulous to sparingly hispid-hirsute; rays 11 to 16, 1.7 to 2.7 cm long, 0.7 to 1.2 cm wide; disk-flowers purple; pales 3-cuspidate, middle cusps short acuminate, hispid to short-villous; achene 2.5 to 3.0 mm long, lightly villous to nearly glabrous.
Distribution: southeastern Texas.

Three subspecies are recognized: subsp. *praecox,* subsp. *runyonii* Heiser, and subsp, *hirtus* Heiser. Until fairly recently this species was included within *H. debilis*. Its closest relatives are *H. debilis* and *H. petiolaris* and it stands somewhat intermediate between them. Hybrids have been secured with four other annual species, and natural hybridization occurs with *H. annuus*.

8. *H. petiolaris* Nutt. (Synonyms: *H. patens* Lehm., *H. integrifolius* Nutt.). Annual 0.4 to 2.0 m tall, usually much branched, stems hispid or strigose to hispidulous or nearly glabrous; leaves deltoid-ovate, deltoid-lanceolate to lanceolate, entire to shallowly serrate or obscurely serrate, cuneate to truncate at base, 4.0 to 15.0 cm long, 1.0 to 8.0 cm wide, strigose on both surfaces, frequently bluish-green in color, petiolate; peduncles 5 to 40 cm long; disks 1.0 to 2.4 cm in diam.; phyllaries lanceolate to ovate-lanceolate, hispidulous on backs, rarely slightly hirsute to nearly glabrous, short ciliate or glabrous on margins, 2.0 to 4.5 (to 5.5) mm wide; middle cusps of chaff slightly exceeding disk-flowers, those in center of head densely long white ciliate or hirsute at apex; disk flowers reddish-purple, very rarely yellow; achenes lightly villous, 3.5 to 4.5 mm long; pappus of 2 awns.
Distribution: west central North America, occasional as a weed elsewhere.

This widespread species comprises two morphological races: subsp. *petiolaris* of the Great Plains and subsp. *fallax* Heiser of the Southwest. The species also includes three cytological races which differ from one another by one or two translocations (9). Hybrids of *H. petiolaris* have been secured with most of the other annual species, and natural hybridization is common with *H. annuus*. Chromosomes of *H. annuus* transferred into cytoplasm of *H. petiolaris* have given rise to cytoplasmic male sterility in the latter species (15).

9. *H. neglectus* Heiser. Annual, 0.8 to 2.0 m tall, usually much branched;

stem densely hispid to hispid-hirsute below, hispid to nearly glabrous above; lower leaves ovate, cordate or rarely truncate, subentire or remotely serrulate, 7 to 14 cm long, 7.5 to 12.3 cm wide, petioles nearly as long as leaves; peduncles 10 to 40 cm long, phyllaries 25 to 35, lanceolate, long attenuate, 2.5 to 4.0 cm broad; disk ca. 2.3 to 2.8 cm in diam; rays 21 to 31, 3.0 to 3.9 cm long, 1.0 to 1.4 cm wide; pales 3-cuspidate, middle cusps in center of the disk white hirsute; achene ca. 4 mm long, with short appressed hairs, becoming glabrous.

Distribution: Southeastern New Mexico and adjacent Texas.

Although this species has a very limited range, it is found in some abundance within this area. Artificial hybrids have been produced with several of the other annual species, and natural hybridization probably occurs with *H. petiolaris,* which appears to be its closest relative. Some of its features also suggest a fairly close relation to *H. annuus.*

10. *H. bolanderi* A. Gray (Synonyms: *H. scaberrimus* Benth., *H. exilis* A. Gray). Annual to 1.5 m tall; stem hispid to hirsute; leaves linear-lanceolate to ovate, cuneate to truncate at base, entire to prominently serrate, hispid to sparingly hirsute on both surfaces, blade to 15 cm long, 12 cm wide, usually much smaller, petiolate; disk 1.5 to 2.5 cm in diam; phyllaries oblong to lanceolate, gradually attenuate, hirsute to hirsute-villous, generally conspicuously exceeding disk in length, 3.0 to 4.5 mm broad; rays 10 to 17, usually bifid at apex; lobes of disk-corollas red-purple or yellow; pales 3-cuspidate, middle cusps subulate, conspicuously exceeding disk flowers, glabrous at tip; achenes 3.0 to 4.5 mm long.

Distribution: southern Oregon to central California.

This rather distinctive species does not appear particularly closely related to any other species. Hybrids have been secured with five other annual species. Natural hybrids are known with *H. annuus,* and introgression from *H. annuus* appears to have led to the development of a weedy race of *H. bolanderi.*

11. *H. paradoxus* Heiser. Annual 1.3 to 2.0 m tall, stem nearly glabrous to hispid, branched above, branches rather short; leaves lanceolate to ovate-lanceolate, cuneate at base, to 17.5 cm long by 8.5 cm wide, usually much smaller, scabrous, entire or lowermost remotely serrate or with a few prominent teeth, upper leaves much reduced, those of the inflorescence bract-like, petiolate; phyllaries 15 to 25, spreading or slightly recurved at tips, oblong-lanceolate, acuminate, sparingly scabrous or hispid, margin ciliate with gland tipped hairs or hispid, about equally disk, 3 to 4 mm wide; disk 1.4 to 2.0 cm in diam; rays 12 to 20; pales glabrous at tips; achenes 3 to 4 mm long, glabrous.

Distribution: western Texas, very rare.

When first described in 1958 this species was known only from one locality in Pecos County, Texas. Another population has since been found in Reeves County, Texas by Barton Warnock of Sul Ross State Univ. This is certainly one of the rarest of the sunflowers. The species is related to *H. annuus,* but it is not a particularly close relationship. Artificial hybrids have been secured with four other species, and natural hybrids are known with *H. annuus.*

12. *H. argophyllus* T. & G. Annual; 1.0 to 3.0 m tall; stem densely white tomentose to floccose, much branched; leaves ovate to ovate-lanceolate, entire or shallowly serrate, truncate to subcordate, 15 to 25 cm long, nearly as wide, densely pubescent with long silky hairs; petiolate; heads 2.0 to 3.0 cm in diam; phyllaries ovate to ovate-lanceolate, abruptly long attenuate, densely white villous, tomentose or floccose, 5 to 7 mm broad; rays yellow to orange-yellow, 15 or more; disk flowers usually deep purple; pales 3 cuspidate, middle cusps attenuate, glabrous or lightly villous; achenes obovate, somewhat flattened, 4 to 6 mm long.
Distribution: southern and eastern Texas.

This species is native to Texas and has been reported as a weed in Florida and Australia, perhaps representing escapes from cultivation. The dense silvery pubescence of this species has made it desirable as a garden ornamental. It seems quite clearly to be the nearest relative of *H. annuus,* and natural hybrids are known with this species. Artificial hybrids have been produced with several other annual species.

Some years ago when examining the writer's sunflowers, G. Ledyard Stebbins called attention to the long silky hairs on the peduncles of some of the cultivars of the domesticated sunflower (*H. annuus*), quite unlike the pubescence on the wild and weedy types of this species. He asked if introgression from *H. argophyllus* could be responsible. There is no record of the Indians having cultivated the sunflower in areas where *H. argophyllus* occurs today, but the sunflower was cultivated in other parts of Texas and Louisiana. Although it is possible that there was natural introgression of *H. argophyllus* into the cultivated sunflower before its introduction into Europe in the sixteenth century, it seems much more likely that the introgression is more recent and artificial, for the Russians are known to have crossed *H. argophyllus* with the cultivated sunflowers in the early part of this century in an attempt to secure disease resistance in the latter (19).

13. *H. annuus* L. (Synonyms: *H. indicus* L., *H. tubaeformis* Nutt., *H. platycephalus* Cass., *H. macrocarpus* DC., *H. ovatus* Lehm., *H. lenticularis* Dougl., *H. colossus* Kunze, *H. erthrocarpus* Barth., *H. multiflorus* Hook., *H. grandiflorus* Wender, *H. lindheimerianus* Scheele, *H. cirrhoides* Lehm., *H. aridus* Rydb., *H. jaegeri* Heiser). Annual, 1.0 to 3.0 m tall, unbranched to much branched, stem usually hispid; leaves mostly alternate, ovate-lanceolate to ovate, conspicuously serrate, lower-most leaves cordate to subcordate, usually hispid both surfaces, 10 to 40 cm long, 5 to 35 cm wide, long petiolate; disk (1.5 to) 2.0 cm in diam or usually much larger; phyllaries ovate or ovate-lanceolate, abruptly attenuate, hirsute, hispid, or rarely glabrous on backs, usually ciliate on margin, (3 to) 5 mm or more broad; rays 17 or more, rarely fewer, usually 2.5 cm or more long; lobes of disk-corollas reddish or purplish, occasionally yellow; pales deeply 3-cuspidate, middle cusps attenuate, hispid or rarely glabrous near apex; achenes (3.0) to 5.0 (15.0) mm long, glabrate, variously colored; pappus scales usually two, extra scales sometimes present, particularly in cultivated forms.

Distribution: southern Canada to northern Mexico, most common in the western U.S.

This species is quite weedy, nearly always being found in areas in some way disturbed by man, and, not surprisingly, it is the most widely distributed species of *Helianthus*. It grows from sea level to 2,500 m and is found in areas of both low and moderate rainfall. This species is extremely variable morphologically, and as yet no set of formal names has been proposed that is adequate to classify all the variants. Attempts to describe some of the variation have been made (7, 14), but clearly this species requires more study, particularly in the western United States. The key presented below will indicate some of the major kinds that may be found, although the extensive intergradation of the various types perhaps makes it undesirable to attempt to give Latin names to the principal variants. There is a large reservoir of genes in the wild and weedy forms of this species, some of which may prove of value in attempting to improve the cultivated sunflower.

KEY TO TYPES OF *H. ANNUUS*

I. Naturally occurring plants (wild or weedy)

Phyllaries 3 to 4 mm wide, glabrous to hispidulous, disks small (less than 2.0 cm in diam) . . . . A

Phyllaries over 3.5 mm broad, conspicuously ciliate on margins, glabrous to densely pubescent on surface

Disks medium-sized (ca. 2.0 to 3.0 cm in diam); rays few, ca. 15 to 25.

Phyllaries usually densely hirsute . . . . B

Phyllaries hispid to slightly hirsute

Rays yellow; leaves often deeply and irregularly serrate C

Rays orange-yellow; leaves usually more shallowly
and regularly serrate. . . . . D

Disks large (ca. 3.0 to 5.0 cm in diam); rays on terminal heads usually more than 25

Plants quite tall (to 3 m). . . . . E

Plants shorter (1.7 to 2.2 m). . . . . F

II. Cultivated Plants

Plants usually branched; phyllaries 6 to 9 mm broad; heads single or double; rays variously colored . . . . G

Plants usually unbranched; phyllaries over 8.5 mm broad; heads single; rays orange-yellow. . . . . H

I. Wild and weedy sunflowers. A. This sunflower, which is found in the southwestern United States, may well represent a primitive form of the species and the nearest approach to a truly wild type. It has been recognized as *H. annuus* subsp. *jaegeri* Heiser. B. A distinct ecotype or altitudinal race is found in the Rockies which is characterized by a rather dense pubescence.

This race has not been given a formal name but is sometimes included within *H. annuus* subsp. *lenticularis* (Dougl.) Ckll. C. The sunflower of eastern Texas has been designated as *H. annuus* subsp. *texanus* Heiser, although it is doubtful that it deserves subspecific recognition. This race may owe its distinctive features to introgression from *H. debilis* subsp. *cucumerifolius*. D. Throughout much of the western United States extending into Canada and Mexico is found a highly variable group of sunflowers that have sometimes been designated as *H. annuus* subsp. *lenticularis*. E. The common weed sunflower, often a serious pest in cultivated fields and also common along roadsides, in the central United States is usually a tall, much branched plant. This race has not been formally designated with a name. F. The weed sunflower of waste places around cities in the central, and occasionally in the eastern, states seems to be somewhat different from the previous one. This form seems to be most similar to the nomenclatural type of the species and must be designated *H. annuus* subsp. *annuus*. This sunflower appears to be the one most closely related to the giant cultivated sunflower and may be similar to the types that gave rise to the domesticated plant. It also is possible that some of its similarities to the domesticated plant are the result of introgression from the domesticated plant. Individuals with yellow disks are occasionally found in populations of this weed. Yellow disked forms are extremely rare in the other wild and weed races.

II. Cultivated sunflower. G. The common sunflower is occasionally grown as an ornamental. The ornamentals include the so-called double forms which arose in Europe, and the red sunflower which was developed by T. D. A. Cockerell from wild plants in Colorado (12). A number of these are offered by seed companies in the USA, but they appear to be more popular in Europe than in North America. These ornamentals also may be referred to *H. annuus* subsp. *annuus* and designated by cultivar names. H. The giant, generally monocephalic, sunflower is placed in *H. annuus* var. *macrocarpus* (DC.) Ckll. and further divided into cultivars. These in turn may be grouped for convenience into the following categories: those cultivars known only from the archaeological record; the old cultivars, including the surviving varieties of the North American Indians (6) and the Old World cultivars, the majority of which arose in Russia (20); and the modern cultivars.

ARCHAEOLOGICAL SUNFLOWER

Archaeological material of the sunflower has now been recovered from several sites in North America. Thus far only remains of wild sunflower have been found in the southwestern United States. Presumably, the seeds were used for food, but this is not definitely known. The material that has been recovered in other areas, Arkansas, the Dakotas, Kentucky, Missouri, Nebraska, Ohio, Ontario, and Pennsylvania, however, most probably is remains of cultivated sunflower. There are also reports from Illinois and Tennessee that most likely represent the weedy type of *H. annuus* (25).

Fig. 3—Achenes of *maiz de teja* from mexico.

In archaeological sites generally only achenes are found, but sunflower heads or fragments of heads have been recovered from two sites. Of particular interest is the material from Newt Kash Hollow, Kentucky, which probably dates to sometime in the first millenium B.C. (24, 25). In this material there is one head 7.5 cm, two heads each about 5.0 cm, and two very small heads that are 2.8 and 1.5 cm in diam (Fig. 2). While it is possible that these heads come from two different kinds of plants, the larger heads from monocephalic domesticated types (the size rules out weed sunflowers) and the smaller heads from branched weedy types, it probably is more likely that they all come from branched domesticated sunflowers with the larger heads being the terminal ones and the smaller ones from the lateral branches. If so, such a sunflower would be a transitional type between the weed sunflower and the large headed monocephalic types, and thus might represent a stage in the domestication. Sunflowers probably similar to these archaeological ones were found among the Arikara Indians of North Dakota in recent times. The Arikara sunflower, however, may not represent a primitive type but may result from the hybridization of large headed monocephalic plants with weed sunflower (6).

The other heads come from an Ozark Bluff Dweller site in northwestern Arkansas. These could not be located for study, but M. R. Gilmore in his notes deposited at the Univ. of Michigan has recorded that two heads had a diameter of 17 cm and that there were also smaller heads present. The size of the larger heads indicates that they were little inferior to post-Columbian types. Exact dates for these sunflowers are not known but Yarnell (25) considers that the site represents a relatively late occupation.

Fig. 2—Heads of sunflowers from Newt Kash Hollow site, Kentucky.

In the archaeological sunflower material from sites in Kentucky, Missouri, Nebraska, and North Dakota, some or all of the achenes bear a prominent beak. Such a beak is not found on wild species of *Helianthus*, and among the races of extant Indian and modern domesticated sunflower a prominent beak is found only in the variety *maiz de teja* (Fig. 3) of Mexico (6). It thus seems that the beak developed in early domesticated sunflower only to be replaced later by the unbeaked types except in Mexico. Thus it might be argued that *maiz de teja* represents a very early introduction into Mexico.

Among the archaeological achenes, the extremely long and narrow ones from North Dakota are the most unusual. These achenes, now on deposit at the Ethnobotanical Museum of the Univ. of Michigan, were collected by W. D. Strong at the Old Fort Abraham Lincoln site, apparently an old Mandan village occupied as late as 1750. The exact age of the achenes is unknown, however. These achenes, like most achenes of archaeological sunflowers, are charred and extremely brittle so only a few were measured because of possible damage in handling. Three collections are present, and from a random sampling the following measurements were secured.

	Accession number	Number measured	Length in mm (range and mean)	Width in mm (range and mean)
a.	2048I	8	11.0–13.4–16.9	2.5–3.1–3.5
b.	2050A	6	13.8–15.7–17.3	3.5–4.4–5.0
c.	2048H	9	14.7–16.2–19.1	3.5–4.1–5.1

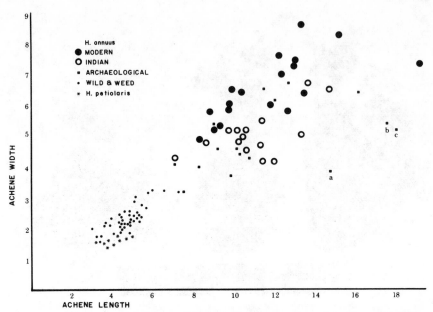

Fig. 4—Length and width (in mm) of achenes of sunflowers. *Helianthus annuus*: modern domesticated (from 6 with addition of one cultivar from Turkey), Indian domesticated (6), archaeological domesticated (6 with addition of two samples from 25), wild and weed types (7), and *H. petiolaris*. The archaeological samples from North Dakota are labeled a, b and c (see text). Each dot or symbol represents the means of a sample from a different geographical source or from a different cultivar.

Since the achenes are charred they are somewhat smaller than their original size. To determine the loss in charring so that comparisons could be made with uncharred achenes, 10 achenes each from seven different varieties were artificially charred, and it was found that on the average the charred achenes were 90% as long and 85% as wide as the uncharred ones. Therefore, the appropriate extrapolations for them, as well as for the other charred archaeological samples, were made before plotting them on the graph (Fig. 4) in order to arrive at the approximate original size.

As can be seen from Fig. 4, the achenes from North Dakota are quite outside the range of the other samples of domesticated sunflower. In length they are exceeded only by one sample of modern achenes which comes from one of Jack Harlan's collections in Turkey. It is the extreme narrowness of the North Dakota achenes rather than their length, however, that is of greater interest. Although it is probable that the achenes represent nothing more than an extreme development within *H. annuus*, nevertheless their uniqueness invites some speculation. These achenes had been identified previously as *H. annuus* by Volney Jones, an authority on American archaeological plants. These achenes also possess the beaks referred to above. It seems most unlikely that they could belong to anything other than *Helianthus*, but it might be asked whether or not they are from *Helianthus annuus*. Several other annual species, including *H. petiolaris* (Fig. 4), possess much narrower achenes than do *H. annuus*. Since characters diagnostic of the species other than size are not present in the achenes, all archaeological material over a certain size has been assumed to be referable to *H. annuus* as it was the only annual sunflower found in cultivation at the time of European contact. Thus the possibility that another species was at one time domesticated and the domesticate had become extinct before it was known by Europeans cannot be entirely ruled out. In this connection, it is of interest to point out that one plant, *Iva annua,* apparently had been domesticated in central North America and had become extinct as a domesticated plant in prehistoric times (23).

ORIGIN AND DEVELOPMENT OF THE
DOMESTICATED SUNFLOWER

Over 20 years ago the hypothesis was advanced that the wild sunflower (*Helianthus annuus*) of the western U.S., whose seed was used by man for food, became a camp-following weed and was introduced into the central U.S. where it became domesticated (7, 8, 13). The reason for considering the place of domestication as being in the central part of the country rather than the Southwest or Mexico, where agriculture was considered to be much earlier, was that the only archaeological sunflowers referable to the domesticated variety came from the central and eastern area. In the past 20 years there has been increased archaeological work in North America and Mexico, but none of the recent discoveries requires any alteration of the original hypothesis.

Donald Lathrap (oral communication, 1973), while accepting an origin of domestication in the central area, has suggested that the sunflower may have been cultivated in the Southwest and spread eastward as a cultivated plant. Unfortunately the archaeological record has not revealed—and probably will never reveal—whether wild sunflower was actually cultivated. Nevertheless, it is interesting to consider why, if a wild sunflower were cultivated in the Southwest, it did not become domesticated there. 1) Possibly since the sunflower is an outcrossing plant it would have been difficult to fix genes for a domesticated, larger seeded type because of a continuous influx of genes from the wild plants. On the other hand, it is difficult to imagine that the sunflower could not have been cultivated in some places sufficiently isolated from wild types to eliminate crossing, or that disruptive selection could not have led to the development of a domesticated form 2) In the original hypothesis of the origin it was postulated that in prehistoric times only wild sunflowers occurred in the Southwest and that as the sunflower spread eastward the weedy type developed. Thus, man would have cultivated wild types in the Southwest whereas weedy types could have originally been cultivated in the central area. Perhaps the weedy types were more amenable to improvement than were wild types, and thus they, and not the wild type, eventually gave rise to the domesticated form.

Whether the sunflower was domesticated in only one place or whether it was independently domesticated in several places is not known, nor is it likely that evidence will be forthcoming that will allow us to choose between these two alternatives.

Anderson (2, 3) has suggested that the weed sunflower (*H. annuus*) that gave rise to the domesticated sunflower was modified by introgression from *H. petiolaris.* The two species hybridize in many places today (10) and although hybrid swarms are produced there is no clear-cut evidence of introgression of *H. petiolaris* into *H. annuus.*[1] In fact, none of the differences between wild and weed races of *H. annuus* is in the direction of *H. petiolaris.* On the other hand, the extremely narrow achenes in the North Dakota archaeological sunflowers might conceivably be the result of introgression. Although one can not definitely point to hybridization with *H. petiolaris* as having a role in the origin of the domesticated sunflower, it is clear that hybridization between the two species has been of great significance recently as will be discussed later.

From the early historical accounts in North America and the descriptions left to us by the herbalists in the sixteenth century we know that the Indians had developed a very tall (2 m or more) sunflower with a single massive head, not unlike some of the older varieties still grown today. The early European visitors to America observed the sunflower in cultivation by the Indians in widely scattered places from southern Canada and Virginia to Arizona and northern Mexico, but nowhere did it appear to be a primary crop. This sunflower was introduced to Europe and eventually spread around most of the world. The first introductions were probably from Mexico to Spain, and subsequently from the eastern area of North America to other parts of Europe. The first published description and illustration is that of Dodonaeus in 1568 (4). The sunflower, while attracting considerable curiosity, appears to have been little used for food in Europe until it reached Russia where it soon became appreciated and large areas were found suitable for its cultivation. The use of the sunflower as a source of oil in Russia was suggested in 1779 and selection for high oil content was started in 1860 (26). The oil content of the seed has since been increased from around 28% to 50%.

Another major change has been the development of dwarf (1 to 1.5 m tall) and semidwarf types with rather small heads. Selection for smaller size was carried out to create types that could be more readily harvested mechanically than was possible with the taller varieties. Thus, in a sense, the modern breeder has reversed the direction of selection practiced by the Indians and returned to varieties with heads often little larger than that of the Newt Kash Hollow archaeological sunflowers.

Interspecific hybridization has figured prominently in sunflower improvement in recent years. The Russians have had notable success in securing resistance to several diseases in the sunflower by using the Jerusalem

---

[1] Some introgression may occur in the reverse direction, however. Occasional populations of *H. petiolaris* not in contact with *H. annuus* have some individuals with unusually broad phyllaries with hispid margins. These characters could be derived from *H. annuus.*

artichoke (*H. tuberosus*) as a source of genes. The Jerusalem artichoke is a hexaploid, whereas the sunflower is diploid, and the two species are placed in different sections of the genus. On this point, Pustovoit (18) said: "In spite of the great difficulties encountered in the breeding process, 'distant hybridization' of the cultivated sunflower with wild species of *Helianthus* has permitted us to obtain plants of the cultivated sunflower type, with 'group immunity' to broomrape, rust, downy mildew, and sunflower moth; with great possibilities of incorporating resistance to other diseases; and with sharply increased seed yields, and yield of oil per hectare." Thus far *H. tuberosus* is the only perennial species that has been used to any extent in sunflower breeding but as indicated earlier several other perennials would seem to hold promise.

A second example of an important advance through the use of interspecific hybridization is Leclercq's transfer of the genome of *H. annuus* to the cytoplasm of *H. petiolaris* to secure cytoplasmic male sterile sunflowers. In his crosses of *H. petiolaris* with *H. annuus* Leclercq secured F_1 plants with normal pollen production. These plants in turn were backcrossed to *H. annuus* and in the backcrosses he secured plants showing male sterility which proved to be cytoplasmically controlled. These cytoplasmic male sterile sunflower plants have made practical the production of the commercial hybrid sunflower which has shown dramatic increases in yield over the older varieties. It would, of course, be desirable to have other sources of cytoplasm for the production of hybrids. The most likely source of cytoplasms for the production of other male sterile sunflower would be the other annual species of the genus, particularly *H. argophyllus, H. bolanderi, H. debilis, H. neglectus, H. paradoxus,* and *H. praecox* which are known to cross readily with *H. annuus.*

In 1971 the author had occasion to grow a few plants of *H. petiolaris* of the same accession that had been supplied to Leclercq. One of the seven plants grown proved to be male sterile. Both *H. annuus* 'Mammoth Russian' and a sister plant were crossed to the male sterile plant. The seven F_1 hybrids with *H. annuus* were all male sterile, and this condition has been maintained through three backcrosses. That these cytoplasmic male steriles and those of Leclercq are the same seems evident since they are derived from the same original population. Moreover, these cytoplasmic male sterile plants of *H. annuus* have pollen production restored by the same gene that is known to restore it in male steriles derived from Leclercq's crosses.

Crosses of a sister plant of *H. petiolaris* to the male sterile gave eight F_1 hybrids with normal pollen production. From a sib cross of these F_1's, an F_2 was secured that gave 28 normal plants and 9 male steriles, a very good fit for a 3:1 segregation. The next season a larger F_2 generation was grown but several plants died before flowering because of a stem rot. Of those that did flower, 27 were normal, 8 were male sterile, and 3 plants had terminal heads that were male sterile but flowers of some or all of the lateral heads produced pollen. Leclercq (15, 16) also noted unstable male steriles in some of his cultures. Although more work on the genetics is called for to explain the unstable condition, these results suggest that male sterility within *H. petiolaris* is controlled by a single gene ms, with MsMs and Msms individuals being pollen producers and msms individuals being male steriles.

SUMMARY

The genus *Helianthus* comprises 67 species native to the Americas. Two species, *H. annuus* and *H. tuberosus,* are cultivated as food plants and several species are grown as ornamentals. *Helianthus* is most closely related to *Viguiera; Thithonia* and *Phoebanthus* also are close relatives. The nature of the pappus, generally of two deciduous awns, serves most readily to separate *Helianthus* from the related genera. The genus has a base chromosome number of $n = 17$, and diploid, tetraploid, and hexaploid species are found. *Helianthus* is divided into four sections: I. Fruticosi—South American shrubby species, which appear to have had an origin separate from that of North American species; II. Ciliares—six perennial species of western North America; III. Divaricati—30 perennial species mostly of eastern North America which are placed in five series; IV. Annui—13 species, mostly annuals, of North America. A key to species, descriptions, distributions, and notes are provided for all species of Section Annui. *H. annuus,* one of the most variable and widespread species, includes wild, weed, and cultivated forms. The hypothesis advanced over 20 years ago that the wild sunflower (*Helianthus annuus*) of the western U.S. became a camp-following weed and was introduced into the central part of the country where it was domesticated is still considered most plausible. Such an hypothesis is supported by the archaeological evidence. The possibility exists however, that the wild sunflower was first cultivated in the Southwest and moved eastward as a cultivated plant. The sunflower known archaeologically from Newt Kash Hollow, Kentucky is thought to represent an early domesticated sunflower, transitional between the weed sunflower and later domesticated ones. The long, narrow achenes from an archaeological site in North Dakota are of particular interest; and while they most probably represent an extreme development of *H. annuus,* the possibility is raised that they might come from another species. Interspecific hybridization is not definitely known to be involved in the origin of the domesticated sunflower, but artificial interspecific hybridization has contributed to the improvement of the sunflower in this century. The increase in oil content and the creation of dwarf varieties, types resistant to diseases, and cytoplasmic male sterile sunflowers have been notable among the recent developments in the sunflower. The cytoplasmic male sterile sunflower is of particular significance in that it has given a practical method for the production of hybrid sunflowers.

LITERATURE CITED

1. Anashchenko, A. 1974. On the taxonomy of the genus *Helianthus* L. Bot. Zhurn. 59: 1472-1481.
2. Anderson, Edgar. 1952. Plants, Man and Life. Little Brown, Boston.
3. ————. 1956. Man as a maker of new plants and new plant communities. *In* William L. Thomas, Jr. (ed.) Man's role in changing the face of the earth. Univ. of Chicago, Chicago, Ill.
4. Dodonaeus, Rembrantus. 1568. Florum et coronarium odoratorumque nonullarum herbarium historia. . .Antverpiae. Ex officina Christophore Planteni.
5. Gray, Asa. 1889. Synoptical Flora of North America. Smithsonian Institution, Washington, D.C.
6. Heiser, C. B. 1951. The sunflower among the North American Indians. Proc. Am. Phil. Soc. 95:432-448.
7. ————. 1954. Variation and subspeciation in the common sunflower, *Helianthus annuus*. Am. Midl. Nat. 51:387-405.
8. ————. 1955. The origin and development of the cultivated sunflower. Am. Biol. Teacher 17:161-167.
9. ————. 1957. A revision of the South American species of *Helianthus*. Brittonia 8: 283-295.
10. ————. 1961. Morphological and cytological variation in *Helianthus petiolaris* with notes on related species. Evolution 15:247-258.
11. ————. 1963. Artificial hybrids of *Helianthus* and *Viguiera*. Madroño 17:118-127.
12. ————. 1976. The Sunflower. Univ. of Oklahoma Press, Norman.
13. ————. 1976. Sunflowers. p. 36-38. *In* N. W. Simmonds (ed.) Evolution of crop plants. Longman, London and New York.
14. ————, D. M. Smith, S. Clevenger, and W. C. Martin. 1969. The North American sunflowers. Mem. Torrey Club 22:1-218.
15. Leclercq, P. 1969. Une stérilité mâle cytoplasmique chez le Tournesol. Ann. Amelior. Plantes 19:99-106.
16. ————. 1971. La stérilité mâle cytoplasmique der Tournesol. Ann. Amelior. Plantes 21:45-54.
17. Mason, H. L. 1957. A flora of the marshes of California. Univ. of California Press, Berkeley.
18. Pustovoit, G. 1966. Distant (interspecific) hybridization in sunflowers. p. 82-99. *In* Proc. 2nd Inter. Sunflower Conference. Morden, Manitoba.
19. Satsyperov, F. A. 1916. *Helianthus annuus* × *H. argophyllus* A. Gray. Bull. Appl. Bot., Genet., Pl. Breed. 9:207-244.
20. Ventslavovich, F. S. 1941. *Helianthus*. *In* Wulff, E. V. Flora of cultivated plants USSR 7:349-436.
21. Watson, E. E. 1929. Contributions to a monograph of the genus *Helianthus*. Papers Mich. Acad. Sci. 9:305-475.
22. Whelan, E. D. P. 1976. Sterility problems in interspecific hybridization of *Helianthus* species. Proc. Sunflower Forum. Sunflower Assoc. of America.
23. Yarnell, R. A. 1972. *Iva annua* var. *macrocarpa*: extinct American cultigen? Am. Anthropol. 74:335-341.
24. Yarnell, R. A. 1976. The origin of agriculture: Native plant husbandry north of Mexico. *In* Charles A. Reed (ed.) The origins of agriculture, Mouton, The Hague (in press).
25. ————. 1976. Sunflower, sumpweed, and amaranth and semi-cultivated plants. *In* The handbook of North American Indians. Washington, D.C. (in press).
26. Zhukovski, P. M. 1950. Cultivated plants and their wild relatives. Abr. transl. by P. S. Hudson. 1962. Commonwealth Agric. Bureaux, Farnham Royal, England.

CHARLES B. HEISER, JR.: A.B., M.A., Ph.D., Professor of Botany, Indiana University, Bloomington, Indiana; plant taxonomist and cytogeneticist; major contribution to systematics, cytogenetics and origin of *Helianthus*; author or coauthor of 25 papers, a monograph, and a book on sunflower.

Chapter 3

Morphology and Anatomy

P. F. KNOWLES

Cultivated sunflower (*Helianthus annuus* L.) is an unusual plant. It is distinguished from all other cultivated plants by its single stem and conspicuous, large inflorescence. Quantitative characteristics, such as height, head size, achene size, and time to mature, vary greatly. Those characteristics determine somewhat the use of the plant—as a source of edible oil, as food for people and animals, or as forage. Floriculturists are interested in the variability in flower color, petal structure, and petal size.

Descriptive classes are proposed herein for some morphological characters to improve the uniformity of descriptions of inbred lines and cultivars. Many descriptive characters are affected by the environment, so measurements should be made of plants grown under optimum field conditions, which include: spacing of 0.5 m²/plant, planting date of 15 April to 15 May in the Northern Hemisphere and 15 October to 15 November in the Southern Hemisphere; adequate soil water and nutrients for vigorous growth; and freedom from pests that affect morphology. The inbred line, HA 89, released by the USDA, is suggested as a standard type for comparison because of its wide use as a breeding line.

STEM

The stem of cultivated sunflower is typically unbranched, although branched types frequently appear in commercial fields and often are used as male parents in producing hybrid cultivars. Stem dimensions and development, including branching, are influenced strongly by the environment.

Dimensions

Stem length of commercial sunflower cultivars varies from 50 to over 500 cm, and stem diameter from 1 to about 10 cm. Rare types 12 m high were reported by Dodonaeus, and cited by Cockerell (2). In the USA there appears to be no association between stem length and utilization. In Cali-

From: Carter, Jack F. (ed.). 1978. *Sunflower Science and Technology*. Agronomy 19. Copyright © 1978 by the American Society of Agronomy, Crop Science Society of America, and Soil Science Society of America, 677 South Segoe Road, Madison, WI 53711 USA.

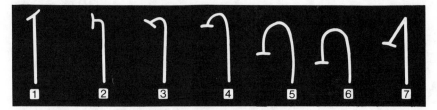

Fig. 1—Stem curvature classes: 1 = straight stem; 2 = sharply curved stem, but not drooping; 3 = stem curved to about 15% of the total stem length of the plant; 4 = same as 3 but curvature 16 to 35%; 5 = same as 3, but curvature 36 to 65%; 6 = same as 3, but curvature over 65%; and 7 = broken stem with the head adhering to, or falling from, the stem. From E. S. Shein, Northrup, King & Company, Woodland, Calif.

fornia, however, the nonoilseed cultivars 'Black Manchurian' and 'Russian Graystripe' are much taller than cultivars used for oil. Cultivars used for forage are usually tall and late.

Skrdla et al. (38), at the North Central Regional Plant Introduction Station, at Ames, Iowa, have recorded the range of plant height as a six-digit number, the first three giving the height of the shortest plant, and the last three that of the tallest plant. For example, 180225 indicates a range of 180 to 225 cm. The length of the curved portion of the stem often is recorded as calculated by substracting the distance of the head from the soil surface from the stem length.

Classes for stem curvature are illustrated and described in Fig. 1. Curvature can be determined at any stage of development, although the preferred time is just before, or at, maturity. The hanging positions of the head in classes 3 and 4 are preferred because the flowers are not exposed to the sun and the seeds are protected from birds. Curvatures of classes 5, 6, and 7 suggest stem weakness and complicate mechanical harvest.

Stem length is determined by number and length of internodes. Both tall and short plants with many internodes will have thick stems because of a positive association between internode number and stem thickness. If stems are short because of fewer internodes, however, stems will be thinner.

Internode Elongation

Garrison (6) found that the first internode of 'Giant Russian' began to elongate when the first true leaves were 20 to 30 mm long. Elongation, following a sigmoid pattern, was completed in 2 to 3 weeks at which time the leaves subtended by the internode completed their elongation. Succeeding nodes and leaves showed a similar pattern of development, with the elongation of successive internodes overlapping. Elongation occurred over a distance of 15 to 40 cm from the stem apex. Initially elongation of an internode was more rapid at the base, then shifted progressively up the internode, and ceased at the base when the internode reached 60% of its final length. Elongation resulted from both cell division and cell elongation, as

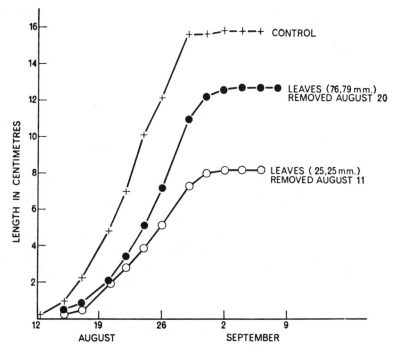

Fig. 2—How removal of the second pair of leaves affects the growth of subadjacent second internode of sunflower. From Wetmore and Garrison (42).

measured in the pith. From an initial internode length of 1 mm with a column of 30 to 40 cells (average cell length 20 to 30 μm), the mature internode was 200 to 250 mm long and had columns of 700 to 800 cells (average cell length 280 to 350 μm). Cell division occurred throughout the elongating internode, but the percentage of dividing cells was greater in earlier stages of the elongation of any one section. Increased internode length was associated with increased cell numbers, but no constant relationship existed between internode length and cell length. Cells elongated in nodal regions also, and were only slightly shorter than cells of adjacent internodes. The transition in cell length was gradual from internode to node.

Jones and Phillips (14) found internode elongation to be associated with the parts of the plant that produce gibberellic acid. The petiolate leaves of the apical bud and the stem leaves were most active in production of gibberellic acid. Excision experiments showed that young leaves, particularly those in the apical buds, were necessary for continued internode extension. Removing the second pair of leaves affects the second internode immediately below, as graphed in Fig. 2 (42). Internode elongation was reduced about half that of the control when leaves were excised just before early growth of the internode below. When half-grown leaves were removed after the internode was 3 mm long, the reduction was about 20%. Removing the second pair of leaves at an early stage did not affect elongation of the first and third internodes.

Putt (28) reported that long internodes and long leaf petioles appeared to be associated with weak stems.

Stems, leaves, and involucral bracts usually are covered with hairs, which may be delicate to coarse depending on genotype. Occasional genotypes may be glabrous.

Anatomy

A cross-section of a sunflower stem is shown and described in Fig. 3. The stem consists of three tissue systems, the dermal, the fundamental, and vascular. The dermal system (also termed the epidermis) consists of one layer of cells with cutinized walls and a cuticle on the outer wall. Epidermal cells divide radially and expand tangentially to accommodate expansion in stem diameter.

The fundamental tissue includes the cortex, pith, and interfascicular parenchyma. The cortex includes the collenchyma, which is comprised of thick-walled cells, often of two layers, under the epidermis and parenchyma. An endodermis, the innermost layer of cells of the cortex, is identified sometimes (26).

The pith, sometimes referred to as the medulla, is inside the ring of vascular bundles. It consists of spongy parenchyma. Small phloem bundles may be present in the pith and the interfascicular regions. Interfascicular parenchyma lies between the vascular bundles and connects the pith and cortex, and sometimes is termed a pith or medullary ray.

The vascular system consists of a ring of bundles in the periphery of the stem. In young stems or at the top of the plant, the bundles are distinct and form strands, whereas at lower levels they merge to form a continuous ring (5).

A vascular bundle is shown in cross-section in Fig. 4, and in longitudinal section in Fig. 5. It consists of phloem and xylem separated by cambium. Cambium that extends into the medullary ray is called interfascicular cambium. Phloem and xylem cells are generated by meristematic cells of the cambium dividing periclinally. Sclerenchyma caps are adjacent to, and outside of, the phloem of each bundle.

Phloem cells differentiate into sieve tubes and companion cells, which conduct food for the plant. Sieve tubes consist of cells arranged end to end, with perforations in the ends of the cells. The parenchymal companion cells are contiguous to sieve tube cells and appear to play a major role in the movement of sugars into and out of the sieve tubes (5).

The xylem has many cells, called tracheary cells, that are modified for conducting water. Because of their thick walls, tracheary cells strengthen the stem. Secondary wall thickenings of the tracheary cells give the cells different appearances. Differentiation of tracheary cells in the bundle begins on the inside and proceeds outward, with cells having in sequence the following wall thickenings: annular; helical or spiral; scalariform (helical with

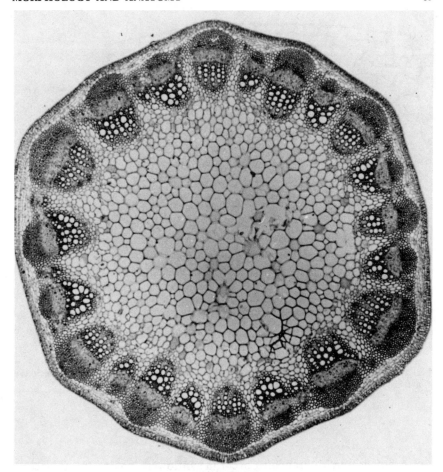

Fig. 3—Cross-section of the stem of sunflower in the upper part of the plant. The collenchyma of two layers of cells is under the epidermis of one layer of cells. Parenchyma is between the collenchyma and the ring of vascular bundles. Interfasicular parenchyma lies between the bundles. Inside the ring of vascular bundles is the pith, almost entirely parenchyma cells. 16×. From Steeves and Sussex (40). Reprinted with permission.

coils interconnected); netlike or reticulate; and pitted. Categories of tracheids are not distinct, and cells occur with intermediate types of wall thickenings. Libriform fibers occur which have thicker walls than those of tracheids. Many parenchymous cells also are present.

The bundle of cells outside the phloem is made up of fibrous cells with thick secondary walls. They are knit together to form strands. They are sometimes called pericyclic fibers, with the bundle called the pericycle. The fibrous strands strengthen the stem.

One or more of the vascular bundles leave the central cylinder at each node to enter the petiole of a leaf.

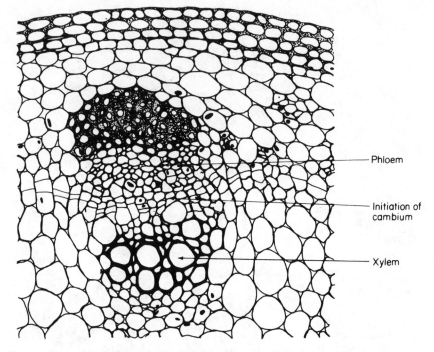

Fig. 4—Cross-section of the vascular bundle of the stem of sunflower. 215×. From Steeves
and Sussex (40). Reprinted with permission.

Branching

Many degrees of branching occur in sunflower, ranging from the cultivated type with a single head to the many branched wild type. Branch length
varies from a few centimeters to distances longer than the main stem.
Branches may be concentrated at the base, at the top, or over the entire
plant. Although branch heads usually are smaller than the main head, some
first-order branches may have a terminal head almost as large as the main
head. Ross (31) reports a high negative correlation (r = −0.709) between
number of branches and yield.

Hockett and Knowles (12) used the following categories to classify hybrid progeny of branched and unbranched types: 0 = no branching; 1 =
basal branching; 2 = top branching; 3 = fully branched with the head of
the main stem larger than the other heads; and 4 = fully branched without a
large central head (wild type). Types 1 to 4 were subdivided further by
branch length: a = short, or peglike, and up to 15 cm long; b = intermediate, or over 15 cm and less than half the length of the main stem; and c =
long, or longer than half the length of the main stem.

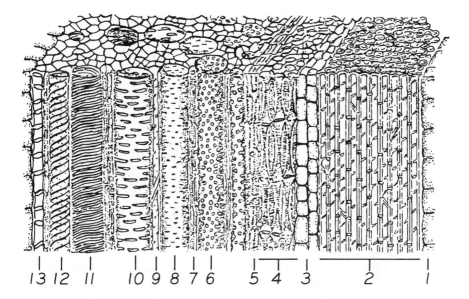

Fig. 5—Longitudinal section of a vascular bundle of the stem of sunflower. 1 = parenchyma cell of the cortex; 2 = fiber cells of the pericycle; 3 = parenchyma cells separating the pericycle and phloem; 4 = sieve tubes and companion cells; 5 = cambium; 6 = tracheid with pits; 7 = parenchyma; 8 = tracheids with reticulate or netlike thickenings; 9 = libriform fibers; 10 = tracheids with scalariform thickenings; 11 & 12 = tracheids with helical or spiral thickenings; 13 = tracheids with annular or ringlike thickenings. Redrawn from Pustovoit (26).

ROOT

Dimensions

Recent studies of root development of sunflower are scarce. Weaver (41), studying the cultivar, 'Russian,' at Lincoln, Nebraska, found that a strong central taproot penetrated the soil to a depth of 150 to 270 cm, and that plant height ranged from 75 to 125 cm. Early in plant development, numerous strong lateral roots originated from the enlarged taproot in the top 10 to 15 cm of the soil. The laterals spread widely, from 60 to 150 cm, and developed in the upper 30 cm of soil. If laterals turned down, they did not penetrate beyond 90 to 120 cm. The roots diffused almost completely through the top 60 cm of the soil. Pustovoit (27) reported root depths of 4 to 5 m in cultivars with a growth period of 100 to 110 days.

The root is simpler than the stem morphologically, mostly because it has no nodes and leaves. Like the stem, the root has both a primary and a secondary stage of development.

Primary Growth

The primary root, or taproot, develops from the radicle of the seed. It remains in a primary stage of growth until secondary thickening takes place. Branch roots go through a primary stage also, with some never entering a secondary stage.

In a primary stage, the root is similar in anatomy at different depths. It has the same three tissue systems as the stem: the dermal system, or epidermis; ground system or cortex; and the vascular system (Fig. 6).

The epidermis consists of one layer of cells. In young roots, epidermal cells develop root hairs a few millimeters back from the root tip. Epidermal cells with or without root hairs absorb water and nutrients. Where the epidermis persists, it become cutinized.

The cortex consists only of parenchyma cells. The innermost layer is the endodermis. A band of suberin, referred to as the Casparian strip, surrounds each endodermal cell on the radial and transverse walls. Casparian strips extend across the middle lamella, and provide a barrier against diffusion of the soil solution from the root hairs and other epidermal cells to tracheal elements. As a result, the soil solution must move through the cytoplasm of endodermal cells.

The pericycle, which lies inside the endodermis, is the first layer of cells of the vascular cylinder. The xylem occupies the center of the vascular cylinder and has four projections extending to the pericycle. Phloem alternates with the xylem projections.

The root cap is a specialized structure at the tip of the primary root. It protects the apical meristem and facilitates the movement of the root through the soil. As cells slough off from the tip of the root cap, the cap is renewed by the meristematic cells in the apical meristem contributing cells forward of the meristem. The root cap appears to contribute much of the mucilage that causes soil to adhere to root tips. If the root cap is removed, the root will lose its geotropic response, but will continue to grow (5).

Lateral Roots

Lateral roots originate in the pericycle (Fig. 6), which acquires dense cytoplasm and divides periclinally and anticlinally (5). Further divisions of pericyclic cells produce a protrusion, called the root primordium. As cells continue to divide, the root primordium penetrates and emerges from the cortex. En route through the cortex the apical meristem and root cap initiate and develop.

Secondary Growth

During secondary growth the root expands in diameter, secondary vascular tissues develop, the cortex sloughs off, and periderm (corky tissue) replaces the cortex.

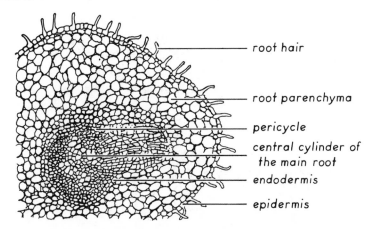

root hair

root parenchyma

pericycle

central cylinder of
the main root

endodermis

epidermis

Fig. 6—Cross-section of the primary root of sunflower, showing the beginning of a lateral root. Redrawn from Pustovoit (26).

The vascular tissue, with the xylem inside and the phloem outside, develops from a cambium that is initiated in both pericyclic and undifferentiated procambial cells between the primary phloem and the primary xylem (5). Eventually the cambium becomes circular in shape. It has the same function as stem cambium.

The pericycle expands by both periclinal and anticlinal divisions during the development of a secondary vascular system (5). As a consequence, the cortex is forced outward, becomes ruptured, and is cast off along with the epidermis and endodermis. The cork cambium or phellogen arises in the outer part of the pericycle. It forms phellum, or cork, toward the outside and phelloderm toward the inside. Phellum cells are dead at maturity and are covered on the outside with suberin, a fatty substance.

LEAVES

When the seedling emerges from the soil, the cotyledons unfold, and the first pair of true leaves is visible at the tip of the shoot axis. Leaves are produced in opposite alternate pairs, and after five opposite pairs a whorled form of alternate phyllotaxy develops (25). Leaves on single-stemmed plants may vary in number from 8 to 70. There appears to be some association between leaf number and time to maturity, for plants with numerous leaves are usually late in maturing.

Morphology

Leaves vary in size, shape of the entire leaf, shape of the tip and base, shape of the margin, shape of the surface, hairiness, and petiolar characteristics. Figure 7 gives some descriptive terms used for leaves. Because leaves

Outline Shape

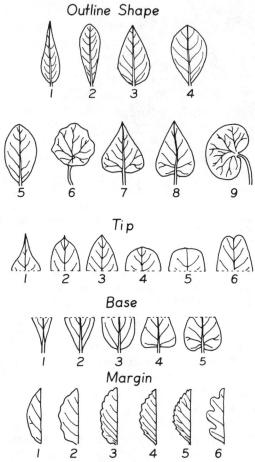

Fig. 7—Terms used in describing general shapes of leaves, leaf tips, leaf bases, and leaf margins. Outline shape: 1 = lanceolate; 2 = oblanceolate; 3 = ovate; 4 = obovate; 5 = elliptic; 6 = orbicular; 7 = deltoid; 8 = cordate; and 9 = reniform. Tip: 1 = acuminate; 2 = cuspidate; 3 = acute; 4 = obtuse; 5 = truncate; and 6 = emarginate. Base: 1 = attentuate; 2 = acute; 3 = obtuse; 4 = truncate; and 5 = cordate. Margin: 1 = entire; 2 = entire; 3 = crenate; 4 = dentate; 5 = serrate; and 6 = lobed.

on different parts of the plant differ in size, cultivar descriptions should designate leaves arising from a specific node at a specific stage of development.

Leaf blades are largest in the midregion of the plant. Length is measured from the tip to the base of the blade, and width at the widest part of the leaf. Length/width ratios may be mentioned in cultivar descriptions. The most common shape is cordate, or heart-shaped.

The surface of leaves may be flat, concave ("cupped"), or convex. The leaf surface may be scabrous (having short projections or barbs on the outer wall of epidermal cells that give a rough texture), hispid (having short stiff hairs), or hirsute (having long silky hairs). Occasionally the leaf surface is glabrous.

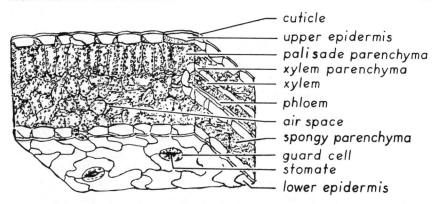

cuticle
upper epidermis
palisade parenchyma
xylem parenchyma
xylem
phloem
air space
spongy parenchyma
guard cell
stomate
lower epidermis

Fig. 8—Structure of the leaf blade of sunflower. Redrawn from Pustovoit (26).

Leaves vary in intensity of green color. Cultivar descriptions use the terms light green, green, and dark green.

Palmer and Phillips (25) studied the orientation of sunflower leaves and petioles. On emerging from the terminal bud, the petiole had almost a vertical position until it was 1 cm long. As elongation continued, the petiole curved away from the stem, and the distal portion curved downward until at leaf senescence the petiole appeared as a shallow inverted "U." The leaf blade held a horizontal position throughout early development and rapid elongation, but at maturity hung more or less vertically.

The length of the petiole and its shape in cross-section is variable and may be used in cultivar descriptions. The angle of the petiole to the main stem has been recorded in cultivar descriptions as semierect, horizontal, and hanging down. The petiole may be enlarged at its juncture with the main stem in some cultivars, although more particularly in some inbred lines.

Anatomy

Like the stem and root, the leaf consists of dermal, ground, and vascular systems. The leaf is typical in cross-section (Fig. 8) of many dicotyledons.

The epidermis appears as a single layer of cells on both sides of the leaf. Stomates are more abundant on the lower than on the upper surface.

The ground tissue is mainly mesophyll consisting of palisade parenchyma next to the upper leaf surface, and spongy parenchyma in the balance of the leaf not occupied by the vascular system. Air spaces occur between palisade cells, and more so between spongy parenchyma cells. The spaces facilitate photosynthesis. The mesophyll cells contain many chloroplasts.

A prominent feature of the leaf is the vascular system, or venation. It consists of a large midvein, running the length of the leaf, and strong lateral veins with extensive branching.

Involucral Bracts

Involucral bracts, also termed phyllaries, are modified leaves (22), with the modification increasing from the outer to the inner bracts. Modifications include a reduction in the leaf blade, a change of the petiole into a leaflike sheath, a scattering of the vascular bundles, and a reduction in mesophyll tissue. Dorsoventral relationships are changed in that the upper or inner surfaces of the bracts have more stomates than the lower or outer surfaces. Also, palisade parenchyma develops next to the lower surface of the involucral bract, rather than next to the upper surface, as in the leaf.

Involucral bracts vary in length, width, and shape of the tip. The descriptive terms of Fig. 7 may be used to describe cultivars.

Flower Bracts

Flower bracts are modified leaves that subtend each disk flower. They are boat-shaped and partially enclose the developing ovaries. A bract has two shoulders and a sharp tip. Cultivar descriptions may record both the actual length of the bract and the projections of the bract beyond the achene.

Heliotropism

Schaffner (34) studied the heliotropism or nutation of cultivated sunflower. The developing heads faced east (50 to 70° from north) in the morning, followed the movement of the sun, and in the evening faced west (60 to 90° from north). Both leaves and involucral bracts were involved in the process, being oriented for greater exposure of the upper leaf surfaces to the sun. Shell and Lang (36) reported that leaf movement lagged behind the azimuthal movement of the sun by 12° or 48 min in time. Shell et al. (37) found that the upper leaf area normal (perpendicular) to the sun's beam was 0.67, whereas in a random distribution of leaves it would be 0.50.

Schaffner (34) believed that heliotropism was an adaptation of the plant to get increased exposure to the sun for photosynthesis. Schell and Lang (36) calculated that photosynthesis would be 9.5% greater with heliotropism than with an optimum arrangement of fixed leaves, and 23% greater than with a spherical distribution of leaves.

Only young leaves manifest heliotropism. After anthesis the heads generally will be tilted toward the east, northeast (in the northern hemisphere), or southeast (in the southern hemisphere), although some plants do not follow the general pattern.

Heliotropism was not manifested when plants were grown in a cabinet with overhead lighting, a result indicating that the field response depended on the sun's movement and not on an endogenous rhythm (36). Also, heliotropism is not expressed on cloudy days.

Lam and Leopold (16) found in sunflower seedlings decapitated above the cotyledons that: darkening of one cotyledon will cause curvature of the stem toward the lighted cotyledon; the darkened cotyledon sustains an enhanced growth rate in the stem below it; light suppresses the growth-stimulating effects of a single cotyledon; and more diffusible auxin is obtained from the stem below darkened cotyledons than below lighted ones.

INFLORESCENCE

The inflorescence is of major interest to the agronomist and plant breeder because seed yield is determined largely by the size of the inflorescence and the percentage of fertile flowers. Sunflower is one of the most photogenic crops at flowering because of its large inflorescence with yellow-orange ray flowers.

Floral Initiation

Schuster and Boye (35) found conflicting reports on the responses of sunflower to different day lengths. Some reports classified sunflower as day-neutral, whereas other reports classified it as a short-day plant. Using ten different cultivars originating from different areas and planted under field conditions in Germany, they showed that seven cultivars were day-neutral and that three late-maturing cultivars were short-day types. The time from seeding to flowering decreased for all cultivars by about 10 days with a delay of seeding from mid-April to mid-June. The time to flowering increased, however, by about 6 days in only seven cultivars when planting date was delayed further to mid-July, and was reduced by about 1 day in three late cultivars. It was assumed that the latter three cultivars were responding to shorter day lengths.

When grown in a growth chamber at 22.0 and 12.5 C and day lengths of 13 and 17 hours, the 10 cultivars differed in responses, as illustrated in Fig. 9 for 'VNIIMK 8931' and 'Kiswardai' (35). At 22.0 C the time to flowering for VNIIMK 8931 was 1 day less with 17-hour than with 13-hour days, whereas for Kiswardai it was 34 days longer. At 12.5 C, however, both cultivars responded similarly; VNIIMK 8931 and Kiswardai were respectively 12 and 6 days later with 17-hour than with 13-hour days. The numbers of leaves decreased for VNIIMK 8931 and increased for Kiswardai with long days at both temperatures. Long days increased plant height for both cultivars at both temperatures.

Dyer et al. (4) found that the cultivar, 'Mammoth Russian,' behaved as a short-day plant. With both 8-hour and 12-hour days, 80% or more of the plants had macroscopic flowers or flower buds 55 days after seeding, whereas with 16-hour days no floral development occurred. Stem length was distinctly shorter with 8-hour days than with 12-hour and 16-hour days.

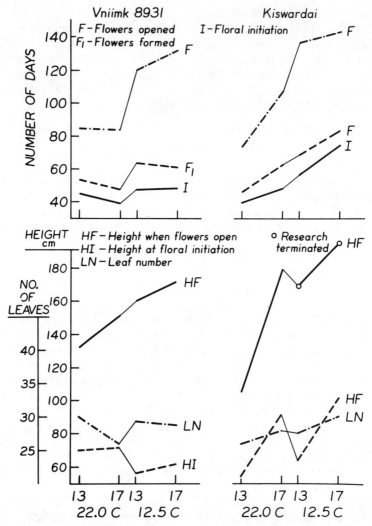

Fig. 9—Responses of two sunflower cultivars. VNIIMK 8931 and Kiswardai, to day lengths of 13 and 17 hours, and temperatures of 22.0 and 12.5 C. Days are from emergence. Redrawn from Schuster and Boye (35).

An interruption in the dark period of 12-hour days with 1 hour of light caused long day effects. The plants eventually flowered at all day lengths up to 20 hours.

Goyne et al. (8) reported that the cultivar, 'Sunfola 68-2,' showed a day-neutral to short-day response and the hybrid cultivar, 'Hysun-30,' a long-day response. They were able to predict the time of 50% flowering using growing degree day summations, but found that the base temperatures for computations were −1.3 C for Sunfola 68-2, and −5.9 C for Hysun-30. Doyle (3) found for four cultivars, 'Voronezskij,' 'Stepniak,'

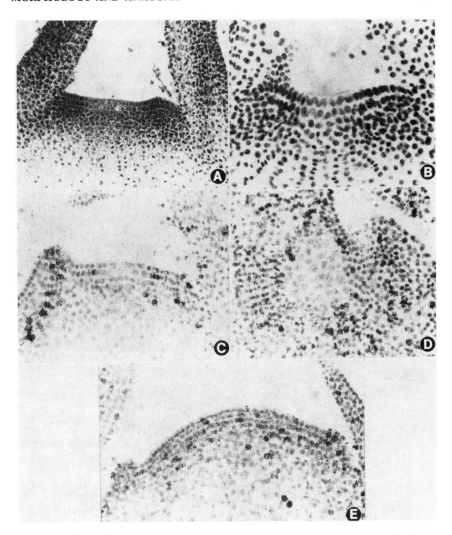

Fig. 10—Shoot apices of *Helianthus annuus*. A–D. Vegetative apices. A. Median longitudinal section, safranin and hematoxylin stain. 138×. B. Median longitudinal section. 204×. C. Autoradiograph of median longitudinal section of apex fed for 24 hours with thymidine-H³. 204×. D. Similar prepration of transverse section. 204×. E. Similar prepration of median longitudinal section of reproductive apex. 204×. B, C, and D Feulgen stain. From Steeves et al. (39).

'Peredovik,' and 'Smena,' that the time from seeding to the appearance of the first anther was influenced much less by photoperiod than by temperature.

The information summarized above and other literature indicate that the day-length responses of sunflower are complex and strongly influenced by temperature. It also is apparent that short-day types are not prevented from flowering by long days, but only delayed in flowering.

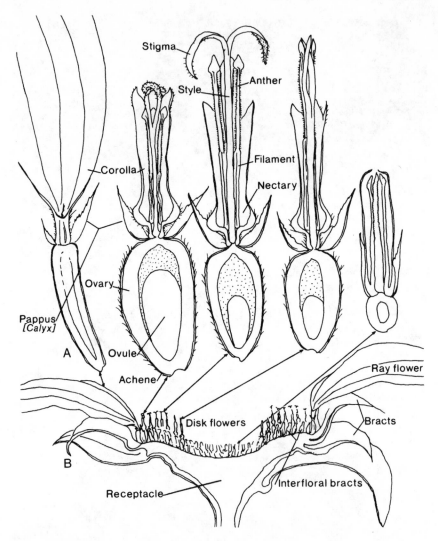

Fig. 11—Longitudinal section of a sunflower head, with individual flowers. A. Left, single ray flower; right, four disk flowers in different stages of development. 3.5×. B. Head. 0.35×. From McGregor (19).

Steeves et al. (39) found that the shoot apex of the cultivar, 'Peredovik,' was still in a vegetative stage 2 weeks after planting. The flat growing point had a distinct central area that stained less intensely than the peripheral area (Fig. 10, A, B). Nuclei were larger in cells in the central area than in surrounding cells, and less intensely stained. Excised apices provided with a nutrient labeled with thymidine-H^3 showed less DNA synthesis in the central area (Fig. 10, C, D), as measured by autoradiography. The intense staining and mitotic activity of the periphery of the growing point in the vegetative stage was associated with the initiation of leaves, 20

Fig. 12—Head shapes of sunflower at maturity: 1 = flat, the back sloping enough to drain water; 2 = concave; 3 = convex; 4 = flat, but with the periphery of the head rolled slightly, enough to collect water when the head is hanging down; 5 = irregular; and 6 = trumpet-shaped, usually concave, and having a coarse, heavy receptacle, often associated with a hole in the center of the head. From E. S. Shein, Northrup, King & Company, Woodland, Calif.

of them in the period 7 to 21 days after planting. Onset of the flowering phase occurred 21 days after planting when there was a conspicuous broadening and bulging of the apex. At that stage all cells of the flat apex stained the same, and cells of the central area were active mitotically (Fig. 10, E). Schuster and Boye (35) found that the time of floral initiation varied from 30 to 110 days after emergence, depending on cultivar and temperature. Figure 9 gives the results for two cultivars. Steeves et al. (39) believed that the area of less intense staining in the center of the apex during the vegetative stage may be meristematic cells that remain quiescent until onset of the reproductive phase.

The time of floral initiation is a critical stage in development. It merits more study with an objective to measure the effects of environmental factors at that stage on the final development of the sunflower inflorescence.

Inflorescence

The inflorescence is a capitulum or head (Fig. 11), characteristic of the Compositae family. It consists of 700 to 3,000 flowers in oilseed cultivars, and occasionally up to 8,000 flowers in nonoilseed cultivars (27). The outer whorl of flowers has a display role, and those over the balance of the head produce seed. Involucral bracts, or phyllaries, which may vary in form and size, surround the head.

Head diameter is obtained by measuring the part of the head that includes the disk flowers. It may vary from 6 to 75 cm (11). The head may vary from concave to convex. Six classes are given and described in Fig. 12.

The attitude of the head is variable, although during flowering it usually is vertical to the soil surface and faces east. As the seeds develop, the heads usually nod, and at maturity hang facing downward. The face of a head which is vertical to the ground will show sun damage at the top of the head exposed to the sun during seed development. Exposed flowers and achenes turn brown and seeds fail to develop inside the hulls. Heads facing upward suffer severe sun damage in California. Rudorf (33) has reported that hanging heads are less attractive to some species of birds. The angle of the head from horizontal in cultivar descriptions should be given as follows: 0 = 0° angle; 1 = 10° angle; and so on to 18, which is an inverted head.

Ray Flowers

The flowers of the outer whorl of the head have five elongated petals united to form straplike structures which give them the name ray or ligulate flowers. The ray flowers are usually a golden yellow, but may be pale yel-

Fig. 13—Variations in flowers of sunflower. A. Typical head. B. More than the usual number of ray flowers. C. Ray flowers reduced to two. D. Ray flowers in the center of the head. E. Chrysanthemum type. F. Intermediate between chrysanthemum type and typical type.

low, orange yellow, or reddish. Rudorf (33) reports the rare occurrence of white ray flowers. Ray flowers have vestigial styles and stigmas but no anthers, hence are sterile. Mutant types occur with more than the normal numbers of ray flowers, and occasionally none (Fig. 13).

Disk Flowers

The flowers over the remainder of the head are called disk flowers (Fig. 14). These are arranged in arcs radiating from the center of the head. Each is subtended by a sharp-pointed, chaffy bract and consists of a basal in-

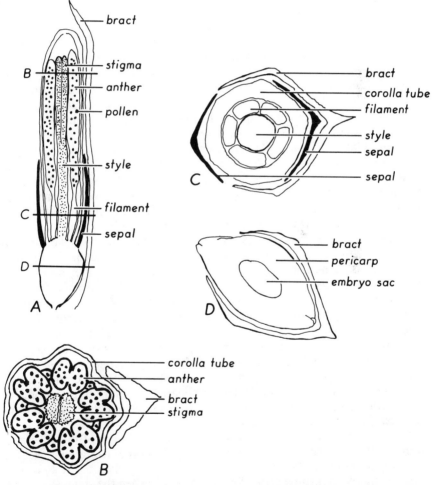

Fig. 14—Schematic drawing of a disk flower at the bud stage. A. Longitudinal section. The locations of sections B, C, and D are indicated. 6×. B. Cross-section through the upper part of the flower. The pollen is emptied inward and carried upward by the bristles of the stigma as the style elongates. 15×. C. Cross-section through the lower portion of the flower. 15×. D. Cross-section through the ovary. 15×. Adapted from Krauter (15).

ferior ovary, two pappus scales (often considered to be modified sepals), and a tubular corolla of five petals united except for the tips. Five anthers are united to form a tube with separate filaments attached to the base of the corolla tube. Inside the anther tube is the style, terminating in a divided stigma with receptive surfaces in close contact in the bud stage before the flower opens. When the flower is fully developed the style is elongated and the divided stigma curls outward, the two lobes often making full circles.

In the ornamental chrysanthemum type the disk flowers are much elongated and somewhat quilled (Fig. 13). Other types are intermediate between the chrysanthemum and normal type.

Anthesis

The outer whorl of disk flowers opens first, at about the time that the ray flowers spread out from their folded position against the buds of disk flowers. Successive whorls of one to four rows of disk flowers open daily for 5 to 10 days. The flowering period will be longer if heads are large or the weather is cool and cloudy. The ray flowers begin to fall about 1 day after the disk flowers at the center of the head finish flowering.

Putt (28) recorded the flowering process in disk flowers. At about 0700 hours the anther tube is exserted fully by elongation of the filaments. Immediately thereafter, pollen is discharged inside the anther tube. Next, the basal portion of the style elongates and pushes the stigma through the anther tube. The stigma appears at about 1700 hours of the same day, and by the following morning is fully emerged, with receptive surfaces exposed. At the same time there is a recession in the anther tube due to loss of turgidity of the filaments. During these processes the pollen is caught by the hairy surface of the emerging stigma and pushed out of the anther tube. After exposure of the receptive surfaces of the stigma, pollination and fertilization soon occur. Thereafter the stigma withers and recedes. On the morning of the third day, the anthers and stigma have shriveled and retreated back into the corolla tube.

Putt (28) found that pollen grains germinated only on the inner surfaces of the stigma lobes. A few pollen grains had germinated by 1900 hours on the first day, before the stigmas had fully emerged. By 0800 hours of the following day, sufficient pollen had germinated to ensure fertilization. He found considerable variation from plant to plant, and higher temperatures favored earlier pollen germination.

By removing the style and stigma at different stages of the flowering process, Putt (28) showed that most of the pollen tubes had grown past the base of the style by 0800 hours of the second day, when the stigmas had divided and were turning back. Two hours later, when the stigmas were receding, all pollen tubes had grown past the base of the style.

Putt (28) showed that pollen is moved little by wind; any movement occurs as the pollen falls from the head. The pollen is spiny (Fig. 15, F), adapted to being transported by insects.

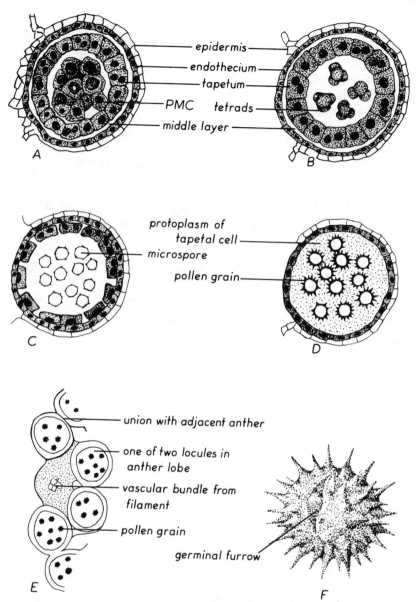

Fig. 15—Development of anther and pollen, A-D. Development of one locule of a lobe in cross-section, the second locule being to the left. 65×. A. The locule just prior to meiosis of the PMCs (or microsporocytes). The middle layer is not well defined, having degenerated and having been crushed between endothecium and tapetum. B. Meiosis complete, and tetrads of the tetrahedral type formed. C. Microspores released and in early stage of development. Tapetal cells shrinking. D. Tapetal protoplasm extruded into the locule, and exine of pollen grains well-developed. E. Diagram of anther showing arrangement of locules. 19×. F. Pollen grain showing one of three germinal furrows. 530×. A-D adapted from Nakashima and Hosokawa (20), and F redrawn from Wodehouse (43).

Anther and Pollen Development

The anther has two lobes, each producing two pollen sacs (Fig. 15, E). The lobes are united to lobes of adjacent anthers. Nanda and Gupta (21) showed that anthers are united by a cementing substance secreted by the epidermis of adjacent anthers. A hyaline membrane is formed prior to meiosis of microsporocytes. By meiosis I, the membrane begins to disorganize, and by dehiscence the anthers are almost free again.

At the microsporocyte (pollen mother cell) stage the anther walls consist of four layers: the epidermis; the endothecium; the middle layer, which has degenerated and been squeezed by adjacent cell layers; and the tapetum (Fig. 15, A) (20). The tapetal cells are large and often binucleate. Pollen mother cells are packed closely. After meiosis, tetrahedral tetrads are formed, each of four microspores (Fig. 15, B). When the microspores are released, they begin to form thick walls (Fig. 15, C). Before and during the release of the microspores the tapetal cells begin to shrink, and their nuclei migrate to the outer wall (Fig. 15, C). Later, while the spiny exine of the pollen grain is forming, the tapetal cells extrude their protoplasm into the locule (Fig. 15, D); this is referred to as the plasmodial or amoeboid type of tapetal behavior. The plasmodium is associated with the developing pollen grains until they are mature. Horner (13) has divided microsporogenesis into 11 stages, beginning with sporogenous cells and continuing to the fully developed pollen.

The pollen grain (Fig. 15, F) has well-developed spines and three germinal furrows (43). Walls consist of an outer layer called the exine, and an inner layer, termed the intine. The exine is well-developed except in furrows, where the intine is thicker. Pollen walls carry the proteins that produce "hay fever" reactions to sunflower pollen in some people.

The two pollen sacs of an anther lobe merge their contents before dehiscence of the pollen (Fig. 14, B). At dehiscence, the pollen is shed inward toward the pistil, where it is picked up by, or pushed ahead of, the emerging stigma.

Before the pollen is shed, the nucleus of the pollen divides mitotically to form a vegetative or tube nucleus, and a generative nucleus. The generative nucleus then divides to form two sperms.

A pollen grain germinates when it reaches the stigma, and the pollen tube emerges from a pore under one of the furrows. The pollen tube penetrates the stigma, grows down the style and into the ovule, and penetrates the embryo sac at the micropylar end. The vegetative nucleus and two sperms are a short distance behind the advancing tip of the pollen tube.

Male Sterility

Two types of male sterility, genetic and cytoplasmic, have been found in sunflower. In a study of genetic male sterility, Nakashima and Hosokawa

(20) reported that the difference in the development of fertile and male-sterile types was first apparent when tetrads were released into the anther locule and grew as microspores. Tapetal cells remained intact, instead of shrinking, as in the normal type (Fig. 15, C), and their nuclei were spherical in shape and were located in the center of the cells. Tapetal cells of sterile types grew larger, instead of disintegrating, and microspores were irregular in appearance. At the stage when there were spinous pollen grains in the normal type, the tapetal cells of male-sterile types grew larger, developed vacuoles, and then suddenly disintegrated, whereupon the anther became flat in shape. At that stage, microspores would not stain. Leclercq (17) found that pollen of genetic male-sterile plants stained much less than normal pollen with both acetocarmine and methylene blue in lactophenol. Pollen produced by male-sterile plants in a wet year lacked the spines on the exine, and contained oily yellow granules suggestive of a carotenoid. Genetically male-sterile plants produced no pollen in a dry year.

Putt and Heiser (30) reported the occurrence of two types of genetic male sterility. In one, discovered at Bloomington, Indiana, anthers were empty and split apart before exsertion, and pollen grains were not present or were uneven in size, clumped, and failed to stain. The second, found at Morden, Manitoba, produced pollen grains that were smaller than normal, nonstaining, and not clumped. They also found two inbred lines which showed variable amounts of partial genetic male sterility. One line had a mean of 75% normal pollen, and the other a mean of 32%.

Cytoplasmically male-sterile (CMS) plants usually have anthers one-half the normal length which do not project out of the corolla (18). Anthers are fused only at their bases, not at the tips. In some stocks, anthers are normal in length and project from the corolla, although empty of pollen. Female flower parts appear normal and are fertile when pollinated.

Horner (13) found that microsporogenesis of CMS types is similar to that of normal types through to meiosis I (the first three stages). From the early to late tetrad stage (stages 4 and 5), the tapetal cells of CMS anthers enlarge much more than in the normal type. At the end of stage 5 the tapetum degenerates and along with the tetrads becomes completely disorganized. Microspores of CMS anthers produce only a rudimentary exine and become flattened and crushed. No further development occurs in CMS anthers.

Embryogenesis

The development of the embryo sac before and after fertilization has been detailed by Newcomb (23, 24). After two meiotic divisions of the megaspore mother cell, a linear tetrad of four megaspores is formed (Fig. 16, A). The megaspore at the chalazal end enlarges to become the functional megaspore, and the other megaspores and the nucellus degenerate (Fig. 16, B). Enlargement of the chalazal megaspore includes an increase in the amount of cytoplasm and vacuolation. After the first mitotic division of the

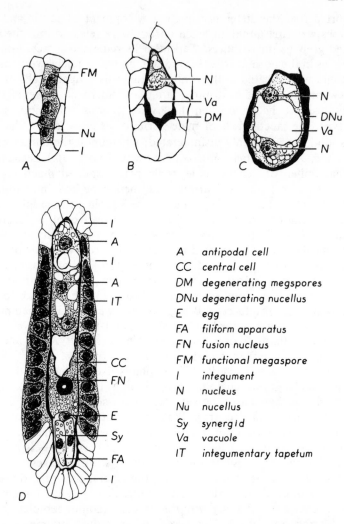

A   antipodal cell
CC   central cell
DM   degenerating megspores
DNu degenerating nucellus
E   egg
FA   filiform apparatus
FN   fusion nucleus
FM   functional megaspore
I   integument
N   nucleus
Nu   nucellus
Sy   synergid
Va   vacuole
IT   integumentary tapetum

Fig. 16—Development of the embryo sac of sunflower. A. Linear tetrad of four megaspores. The upper chalazal megaspores are larger than the lower micropylar megaspores. The upper of the two chalazal megaspores will become the functional megaspore. 290×. B. Functional megaspore which has increased in size and vacuolation. The darkly stained areas at the base are degenerating megaspores. Nucellar cells are beginning to degenerate. 465×. C. Two-nucleate embryo sac, with nuclei at the chalazal and micropylar ends, separated by a large vacuole. 440×. D. Early stage of a cellular embryo sac following two mitotic divisions of the nuclei of C. One antipodal (chalazal cell) has one nucleus and the other has two, and both are highly vacuolate. At the micropylar end, two synergid cells and an egg cell are each somewhat pear-shaped and are squeezed together with some of the contents of the central cell around them; the egg in the drawing is behind the synergids. The nucleus of the egg cell is at the chalazal end of the cell and is often close to the fusion nucleus. Nuclei of the synergids are closer to the midregion. A filiform apparatus between the synergids is developing. The central cell with the fusion nucleus, formed by fusion of two polar nuclei, occupies the rest of the embryo sac. The integumentary tapetum is well-developed, with the cells having thick wells. 245×. Adapted from Newcomb (23).

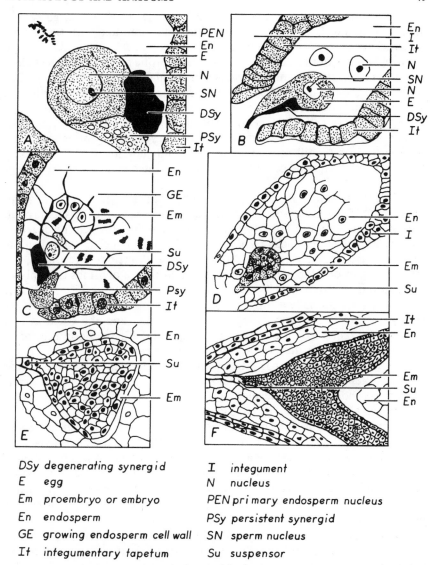

DSy degenerating synergid
E   egg
Em   proembryo or embryo
En   endosperm
GE   growing endosperm cell wall
It   integumentary tapetum

I   integument
N   nucleus
PEN primary endosperm nucleus
PSy persistent synergid
SN   sperm nucleus
Su   suspensor

Fig. 17—Development of the embryo sac after fertilization. A. Embryo sac during fertilization. The primary endosperm nucleus is in the metaphase of the first endosperm nuclear division. A sperm nucleus is present in the nucleus of the egg. The degenerating synergid stains
darkly. 510×. B. Later stage. There are two endosperm nuclei. A sperm nucleus is present
in the egg nucleus. The degenerating synergid is partially collapsed. The integumentary tapetum surrounds the embryo sac. 205×. C. Micropylar region of an embryo sac containing a
proembryo and a growing endosperm. The ends of the walls of the endosperm cells are growing toward the chalazal end. The basal cell of the suspensor is adjacent to the synergids.
170×. D. Embryo sac containing a globular proembryo and endosperm still growing toward
the chalazal end. Some of the integumentary tapetum near the micropylar end is degenerating. 100×. E. Early heart-stage of the embryo. There are large areas free of endosperm next
to the embryo. 80×. F. Late heart-stage of the embryo, with large spaces separating it from
the endosperm. 50×. Adapted from Newcomb (24).

functional megaspore, the nuclei lie on opposite sides of a large central vacuole (Fig. 16, C). With a second mitotic division, and with further enlargement of the vacuole, two nuclei are located at each of the chalazal and micropylar ends of the embryo sac. Another mitotic division produces eight nuclei. Meanwhile, the embryo sac has been enlarging and elongating. Cell walls form around nuclei (Fig. 16, D), resulting in two antipodal cells, one with a single nucleus and the other with two. Other cells include two synergids, an egg cell, and a central cell with a fusion nucleus, or rarely the two polar nuclei. At the micropylar ends of the synergids, contiguous walls thicken to form the filiform apparatus. The fully developed embryo sac extends beyond the tapetal cells at both ends and is in contact with integument cells.

Before fertilization the pollen tube appears to grow through integument tissue along the micropyle, and enters the embryo sac through the filiform apparatus (24). After presumably passing through the degenerating synergid, sperm nuclei fertilize, first, the fusion nucleus, and then the egg-cell nucleus (Fig. 17, A, B). After the egg-cell nucleus is fertilized, the embryo begins development. After several mitotic cell divisions the embryo has a globular form with a short stalk at the micropylar end (Fig. 17, C, D). The short stalk, initially a single cell, becomes the suspensor. The globular embryo grows larger with more mitotic divisions, and the chalazal end becomes flattened to initiate the heart stage (Fig. 17, E). Soon thereafter two cotyledon primordia project beyond the apical meristem (Fig. 17, F). At this stage the suspensor is much elongated, after which it begins to degenerate. Newcomb (24) believes that the suspensor is involved in translocation of metabolites to the developing embryo. The suspensor and developing embryo are surrounded by endosperm.

Until six to eight endosperm nuclei are formed, they are all contained in the central cell. Endosperm wall formation begins at the micropylar end of the embryo sac. Eventually all of the endosperm micropylar to the antipodals becomes filled with cellular endosperm. Newcomb (24) proposed that the embryo at the heart stage begins to draw on the endosperm for nutrients. The persistent synergid slowly degenerates, although it is often still present at the heart stage of embryogenesis.

When the embryo is developed fully the cotyledons surround two embryonic leaves and a small growing point. Guttenberg (9) gave details of the development of the embryo, with particular reference to the root.

ACHENES

The achene, or fruit, of the sunflower consists of a seed, often called the kernel, and adhering pericarp, usually called the hull (Fig. 18). As the achenes mature over the head, all parts of the flowers above the ovary drop away. The achenes usually grade in size from the largest at the periphery of the head to the smallest at the center.

All achenes develop a hull whether a seed develops or not. Empty achenes are frequently somewhat pinched in appearance. Disk flowers in

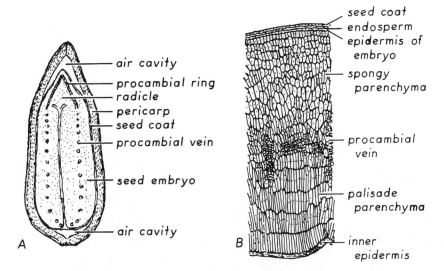

Fig. 18—Diagrams of: A. Longitudinal section of an achene of sunflower showing pericarp or hull and seed. 4×. B. Cross-section of a cotyledon of the seed. The inner epidermis will be the upper surface of the cotyledon when the seed germinates. 28×. Redrawn from Pustovoit (26).

the center of the head fail to produce seed in some plants or breeding lines and the mature achenes appear chaff-like. Both genotype and adverse environment appear to be involved. Cultivar descriptions should record the diameter of the central sterile area.

Achenes vary from 7 to 25 mm long and 4 to 13 mm wide. Weight per 100 achenes will vary from 4 to 20 g. Lengthwise, the achenes may be linear, oval, or almost round, and crosswise they may be flat to almost round. Large achenes usually have thick hulls and are not well filled. Small achenes, in contrast, usually have tightly fitting thin hulls.

Pericarp

Hanausek (10) and Roth (32) have described the structure of the pericarp (Fig. 19). The epidermal cells have thick, outer walls covered with a thin cuticle. Longitudinally they are elongated and four-sided, and rectangular in cross-section. Radial and lower walls are thin. The surface of the developing achene has straight hairs either inclined outward or appressed. Most hairs are double, although some are single, and three occasionally occur together. The double hairs are united for most of their length, with only the tips being separate. Both are attached to a basal cell that projects slightly above the surrounding epidermal cells. Many of the hairs are removed in harvest and handling of the seeds and contribute a fire hazard in artificial drying.

As the pericarp develops there is initially a single layer of cells below the epidermis, termed the hypodermis. The cells are thin-walled with large

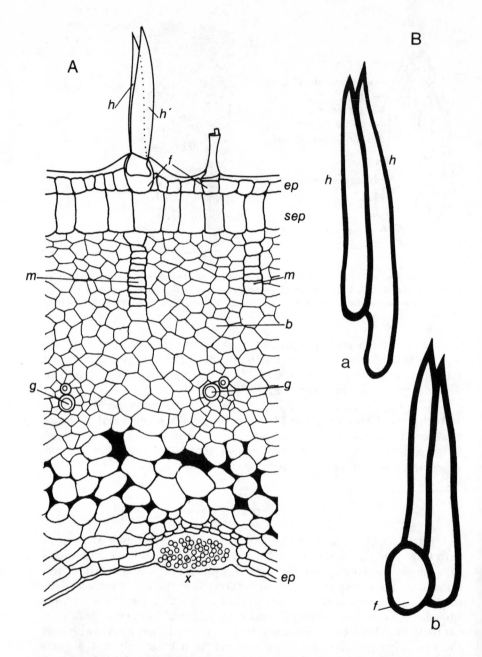

Fig. 19—Pericarp of sunflower. A. Cross-section through the ovary wall at an early stage: ep = epidermis; h-h′ = double hairs, with h = outer and h′ = inner; f = basal cell; sep = subepidermal layer of cells; m = medullary raylike row of cells; b = middle layer; g = xylem; ep′ = inner epidermis; and x = fiber bundle. B. Double hairs: a = without the basal cell, and b = with basal cell (f). C. Some days later, with notation as in A. hb = the first

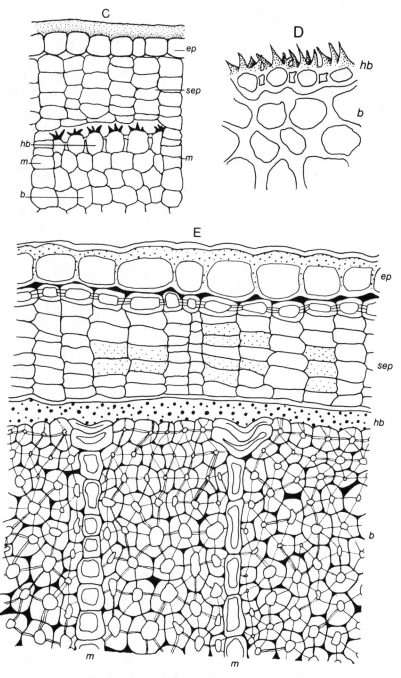

row of cells of the future sclereid layer, with small dark-colored conelike projections. D.
Cross-section of some cells of the first row in association with sclereid cells. E. Cross-section
of outer portion of mature pericarp, with notation as in A. Redrawn from Hanausek (10).

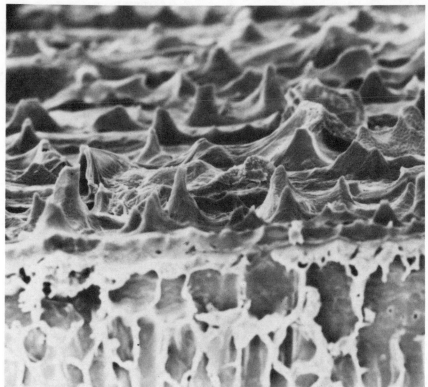

Fig. 20—Photograph of the armor layer of sunflower pericarp as revealed by the scanning electron microscope. 490 ×. From Carlson and Witt (1).

nuclei. After a few days they divide radially to form four to six layers of cells arranged in radial rows, the innermost cells being the oldest. The cell walls are thick in the outer layer and thin in other layers. The layers of cells have some of the characteristics of a periderm.

Below the hypodermal layer is the middle layer, made up of axially oriented, sharp-cornered, polygonal cells. The cells of this layer are initially thin-walled and, except for the outer layer, develop secondary thickening of the walls to become sclereids.

Single layers of parenchymal cells occur at regular intervals in the middle layer, and run in a radial direction. They separate the middle layer into bundles running longitudinally. At maturity the outer cells are curved inward and folded on one another as if pressed from the outside.

In cultivars possessing an "armor layer," the outer layer of cells of the middle layer develops in an unusual way. These, on their outer sides against the hypodermis, develop small, pointed cones and hooklets, which mechanically separate that row of cells from the hypodermis. Later, similar outgrowths appear on the radial walls of the cells, leading to the development of spaces between them. The walls turn brown, and finally black. Eventually the cellular nature of the layer disappears, and there is left a black mass of disorganized material between the hypodermis and the middle layer. The conelike projections persist to maturity of the achene. Figure 20 shows the armor layer. This dark-colored layer has been reported to provide protection from larvae of the sunflower moth [*Homoeosoma electellum* Hulst in the USA (1) and *H. nebulella* Hbn. in Russia].

The inner layer of cells is parenchyma. These cells are thin-walled and loosely packed. A small vascular bundle is located in the parenchyma between each of the "rays." The inner epidermis and occasional bundles of fibers are on the inside of the pericarp.

Putt (29) reported that the striping of the achene is produced in epidermal cells. A purple color may be produced under the epidermis in cells of the hypodermis. If present, the purple is usually so heavy that it entirely masks the color of other coats, so that the seed appears purple. Finally there is the armor layer. When present, it shows through other layers as a gray tint.

Achenes may vary from completely white through different shades of brown and gray to black. They are often gray or white, with dark-gray, brown, or black striping in variable amounts. The following color code is used by the North Central Regional Plant Introduction Station, Ames, Iowa, USA (38); BK = black; BK-WH = black with white stripes; BK-GY = black with gray stripes; BK-BR = black with brown stripes; BR = solid brown; BR-WH = brown with white stripes; BR-BK = brown with black stripes; WH = solid white; WH-BK = white with black stripes; WH-GY = white with gray stripes; GY = solid gray; GY-WH = gray with white stripes.

Seed

The seed consists of a seed coat, endosperm, and embryo (Fig. 18). The seed coat is thin and has three layers, with the inner and outer layers parenchymal, and a middle layer of spongy parenchyma. The endosperm consists mostly of a single layer of aleurone cells coalesced with the seed coat. The embryo is made up mostly of cotyledons. It consists largely of palisade parenchyma, with cells containing oil-rich large aleurone particles and protein crystals (7).

ACKNOWLEDGMENTS

I am indebted to Mr. E. Steven Shein, Northrup, King and Company, Woodland, California, who provided a major input into developing morphological classes, reviewed the manuscript, and provided literature. I thank Kate Tague and Margaret Voldal who prepared the figures.

LITERATURE CITED

1. Carlson, E. C., and Robert Witt. 1974. Moth resistance of armored-layer sunflower. Calif. Agric. 28(11):12–14.

2. Cockerell, T. D. A. 1915. Specific and varietal characters in annual sunflowers. Am. Nat. 49:609–622.

3. Doyle, A. D. 1975. Influence of temperature and daylength on phenology of sunflowers in the field. Aust. J. Exp. Agric. Anim. Husb. 15:88–92.

4. Dyer, H. J., John Skok, and N. J. Scully. 1959. Photoperiodic behavior of sunflower. Bot. Gaz. 121:50–55.

5. Esau, Katherine. 1977. Anatomy of seed plants. 2nd ed. John Wiley and Sons, New York.

6. Garrison, Rhoda. 1973. The growth and development of internodes in Helianthus. Bot. Gaz. 134:246–255.

7. Gassner, G. 1973. Mikroscopische Untersuchung pflanzicher Lebensmittel. (Microscopic Investigation of Plant Food Products). Gustav Fischer, Publisher, Stuttgart. p. 128–130.

8. Goyne, P. F., D. R. Woodruff, and F. D. Churchett. 1977. Prediction of flowering in sunflowers. Aust. J. Exp. Agric. Anim. Husb. 17:475–482.

9. Guttenberg, H. v., J. Burmeister, and H. J. Brosell. 1955. Studien über die Entwicklung des Wurzelvegetationspunktes der Dikotyledonen. II. (Studies on the development of the root tip of dicotyledons. II.). Planta 46:179–222.

10. Hanausek, T. F. 1902. Zur Entwicklungsgeschichte des Perikarps von Helianthus annuus. (Development of the pericarp of Helianthus annuus.). Ber. Dtsch. Bot. Ges. 20:449–454.

11. Heiser, C. B. 1976. The sunflower. University of Oklahoma Press, Norman.

12. Hockett, E. A., and P. F. Knowles. 1970. Inheritance of branching in sunflowers, Helianthus annuus L. Crop Sci. 10:432–436.

13. Horner, H. T., Jr. 1977. Comparative light- and electron-microscopic study of microsporogenesis in male-fertile and cytoplasmic male-sterile sunflower (Helianthus annuus). Am. J. Bot. 64:745–759.

14. Jones, J. L., and I. D. J. Phillips. 1966. Organs of gibberellin synthesis in light-grown sunflower plants. Plant Physiol. 41:1381–1386.

15. Krauter, Dieter. 1975. Zum Unschlagbild: Langschnitt durch den Blutenkorb einer Sonnenblume (An illustration: longitudinal section through the head of a sunflower). Microkosmos 64:255–256.

16. Lam, S. L., and A. C. Leopold. 1966. Role of leaves in phototropism. Plant Physiol. 41:847–851.

17. Leclercq, P. 1966. Une stérility mâle utilisable pour la production d'hybrides simples de tournesol. (A male sterility utilizable for the production of simple hybrids of sunflower.) Ann. Amelior. Plant. 16:135–144.

18. ————. 1969. Une stérility mâle cytoplasmique chez le tournesol. (Cytoplasmic male sterility in sunflower.) Ann. Amelior. Plant. 19:99–106.

19. McGregor, S. E. 1976. Insect pollination of cultivated crop plants. Agricultural Handb. 496. USDA, ARS.

20. Nakashima, H., and S. Hosokawa. 1974. Studies on histological features of male-sterility in sunflower (Helianthus annuus L.). Proc. Crop Sci. Soc. Japan 43:475–481.

21. Nanda, Kanan, and S. C. Gupta. 1975. Syngenesious anthers of Helianthus annuus—a histochemical study. Bot. Not. 128:450–454.

22. Napp-Zinn, K. 1951. Anatomische und morphologische Untersuchungen an den Involucral- und Spreublättern von Compositen. (Anatomical and morphological studies of the involucral and achene bracts of composites.) Oesterr. Bot. Z. 98:142–170.

23. Newcomb, William. 1973. The development of the embryo sac of sunflower Helianthus annuus before fertilization. Can. J. Bot. 51:863–878.

24. ————. 1973. The development of the embryo sac of sunflower Helianthus annuus after fertilization. Can. J. Bot. 51:879–890.

25. Palmer, J. H., and I. D. J. Phillips. 1963. The effect of the terminal bud, indoleacetic acid, and nitrogen supply on the growth and orientation of the petiole of Helianthus annuus. Physiol. Plant. 16:572–584.

26. Pustovoit, V. S. 1975. The sunflower. (In Russian). Kolos, Moscow.

27. ————. 1967. Handbook of selection and seed growing of oil plants. (In Russian). English translation available from U.S. Department of Commerce, Springfield, Virginia.

28. Putt, E. D. 1940. Observations on morphological characters and flowering processes in the sunflower (*Helianthus annuus* L.). Sci. Agric. 21:167–179.

29. ————. 1944. Histological observations on locations of pigments in the akene wall of the sunflower (*Helianthus annuus* L.). Sci. Agric. 25:185–188.

30. ————, and C. B. Heiser. 1966. Male-sterility and partial male-sterility in sunflowers. Crop Sci. 6:165–168.

31. Ross, A. M. 1939. Some morphological characters of *Helianthus annuus* L., and their relationship to the yield of seed and oil. Sci. Agric. 19:372–379.

32. Roth, Ingrid. 1977. Fruits of angiosperms. Encyc. Plant Anat. 10:258–290. Gebrüder Borntraeger, Berlin and Stuttgart.

33. Rudorf, Wilhelm. 1961. Die Sonnenblume, *Helianthus annuus* L. (The sunflower, *Helianthus annuus* L.). Handb. Pflanzenzuecht. 5:89–114. Paul Parey, Berlin and Hamburg.

34. Schaffner, J. H. 1900. The nutation of *Helianthus*. Bot. Gaz. 29:197–200.

35. Schuster, W., and R. Boye. 1971. Der Einfluss von Temperatur und Tageslange auf verschiedene Sonnenblumensorten unter kontrollierten Klimabedingungen und in Freiland. (The influence of temperature and day length on different sunflower cultivars under controlled climatic conditions and in the field). Z. Pflanzenzuecht. 65:151–176.

36. Shell, G. S. G., and A. R. G. Lang. 1976. Movements of sunflower leaves over a 24-H period. Agric. Meteorol. 16:161–170.

37. ————, ————, and P. J. M. Sale. 1974. Quantitative measures of leaf orientation and heliotropic response in sunflower, bean, pepper and cucumber. Agric. Meteorol. 13: 25–37.

38. Skrdla, W. H., R. L. Clark, and J. L. Jarvis. 1976. List of seed available at the North Central Regional Plant Introduction Station, Ames, Iowa. USDA mimeo. 33 p.

39. Steeves, T. A., M. Anne Hicks, J. M. Naylor, and Patricia Rennie. 1969. Analytical studies on the shoot apex of *Helianthus annuus*. Can. J. Bot. 47:1367–1375.

40. ————, and I. M. Sussex. 1972. Patterns in plant development. Prentice-Hall, Englewood Cliffs, N.J.

41. Weaver, J. E. 1926. Root development of field crops. McGraw-Hill Book Co., New York.

42. Wetmore, R. H., and R. Garrison. 1966. The morphological ontogeny of the leafy shoot. p. 187–199. *In* E. G. Cutter (ed.) Trends in plant morphogenesis. Longmans, Green, and Co., London.

43. Wodehouse, R. P. 1935. Pollen Grains. McGraw-Hill Book Co., New York.

PAULDEN F. KNOWLES, B.S.A., M.Sc., Ph.D.; Professor of Agronomy, University of California, Davis. Geneticist-agronomist working on several oilseed crops including safflower, sunflower, soybeans, and rapeseed; contributed to the development of safflower in California; coauthor of a textbook on plant breeding; author of over 80 publications and reports on oilseed developmental projects; advisor on research and development of oilseed crops to the USDA in Turkey and Pakistan, to FAO in Egypt, to the Ford Foundation in Thailand, and to the Development and Resources Corporation in Iran.

Chapter 4

Production and Culture

R. G. ROBINSON

Heredity and environment determine the yield of sunflower (*Helianthus annuus* L.). Heredity is more controllable than environment because it is fixed when the cultivar is chosen. Environment is only partially controllable. Control of the environment to maintain optimum conditions for modern sunflower production is the topic of this chapter.

In many areas of the temperate zones of the world, sunflower produces more oil per hectare than any other species. Most cultivars have a potential seed yield exceeding 3,000 kg/ha, but average yields in most states or provinces in North America are less than 1,500 kg/ha. Obviously, under normal production practices the environment is not controlled sufficiently to allow yields to reach the cultivars' potentials.

ADAPTATION AND PHYSIOLOGIC CHARACTERISTICS OF THE SUNFLOWER PLANT

Sunflower performs well in most temperate zones. The same cultivars often are grown in Europe, Asia, Africa, Australia, North America, and South America. Cultivars of few other crops show this wide range of adaptation. Various morphological, physiological, and economic characteristics of sunflower account for its wide adaptation.

Light and Photosynthesis

Sunflower requires adequate light. Sunflower grown in shade that excluded 40% of the natural light suffered a 64% reduction in yield. Yields were reduced 36% if shading was limited to only 27 days during flowering and seed development (135). Most crops are inefficient users of light and become light-saturated at considerably less than full sunlight (108 klux). Sun-

From: Carter, Jack F. (ed.). 1978. *Sunflower Science and Technology.* Agronomy 19. Copyright © 1978 by the American Society of Agronomy, Crop Science Society of America, and Soil Science Society of America, 677 South Segoe Road, Madison, WI 53711 USA.

flower and corn (*Zea mays* L.) did not become light-saturated at relatively high levels of light (71).

Sunflower leaves are phototropic with a lag of 12°, or 48 min in time, behind the sun's azimuth. Average daily interception of sunlight was 9% greater than the maximum obtainable from a fixed position of leaves. Shell and Lang (173) calculated that phototropism could increase average daily photosynthesis by 10 to 23% compared with static leaf arrangements. Klimov et al. (91) found that leaves of a dwarf cultivar received photosynthetically active radiation of 464 cal min^{-1} cm^{-2} at 0800 hour compared with only 296 cal min^{-1} cm^{-2} for 'Majak.' The difference was attributed to greater petiole mobility of the dwarf which changed leaf orientation to intercept light. Reception of energy was about the same (555 cal min^{-1} cm^{-2}) for both cultivars at 1200 hour.

Crops like corn, sorghum (*Sorghum bicolor* (L.) Moench), and *Amaranthus* spp. have photosynthetic rates exceeding 50 mg CO_2dm^{-2} hour^{-1} under good conditions. But sunflower leaves maintain a medium-high photosynthetic rate through a wide range of temperatures. El-Sharkawy and Hesketh (45) found that 30 to 35C was optimum, and net photosynthetic rate exceeded 30 mg CO_2dm^{-2} hour^{-1} from 23C to 39C in a glasshouse study. Photosynthesis continued at a medium-high rate even when leaves wilted, and it did not decline to zero until temperature exceeded 45C.

Oxygen released by respiration inhibits photosynthesis of sunflower (104) and most other crops (Warburg effect). But photosynthesis of corn, sorghum, and sugarcane (*Saccharum officinarum* L.) is not affected by oxygen (63). CO_2 compensation concentration is the minimum level of CO_2 needed for plant survival. Normal air contains 300 μl/liter. Krenzer et al. (94) classified sunflower as high (>40 μl/liter) and corn as low (<10 μl/liter) in CO_2 compensation concentration. Both the Warburg effect and the CO_2 compensation concentration tend to reduce the photosynthetic efficiency of sunflower.

Respiration

Respiration occurring during the night, called dark respiration, supplies the biological energy for sunflower growth. The substrates consumed are carbohydrates, lipids, and proteins. Photorespiration occurs during daylight in sunflower and other C_3 species (66, 119, 214). Glycolic acid is the substrate, and the energy released is of no use to sunflower. Ludwig and Canvin (104) reported that photorespiration in sunflower was over three times the rate of dark respiration and was independent of CO_2 concentration from 0 to 300 μl/liter. Photorespiration does not occur in C_4 species like corn, sorghum, sugarcane, and some species of *Amaranthus* (66, 119, 214).

Development of C_4 metabolism in sunflower would greatly increase the crop's photosynthetic efficiency. Photorespiration of sunflower can be inhibited by reducing oxygen concentration in the air from its normal 21 to 2%, and this reduction increased net photosynthesis by 43% (70).

Net Assimilation Rate

Sunflower has a relatively high net assimilation rate (NAR). Warren Wilson (207) reported a common range of 10 to 18 g m^{-2} leaf area^{-1} day^{-1}, but under good conditions the rates exceeded 28 g m^{-2} leaf area^{-1} day^{-1}. He calculated that this high NAR would require maximum rates of photosynthesis of 50 to 65 mg CO_2dm^{-2} hour^{-1} in active leaves. These rates equal those of C_4 species and suggest a higher potential for sunflower than previously reported. Togari et al. (192) reported NAR for rice (*Oryza sativa* L.), corn, sunflower, and soybean [*Glycine max* (L.) Merr.] as 30, 24, 21, and 18 g m^{-2} leaf area^{-1} day^{-1}, respectively.

Temperature

Tolerance to both cold and high temperatures contributes to sunflower's adaptation in different environments. Sunflower seeds germinate at 4 C, but temperatures of at least 8 to 10 C are required for satisfactory germination (35). Emergence was faster at 15 C than at 10 C (179). Seed vernalization treatments of 0.5 C for 7, 14, 21, 28, or 35 days after germination at 30 C for 24 hours did not affect yield, seed weight, height, or number of days to maturity (107).

Young plants resist frost. Seedlings in the cotyledon stage have survived temperatures of -5 C. This resistance gradually declines until by the six to eight-leaf stage, temperatures slightly below freezing may injure the crop. Freezing of young plants may injure the terminal bud and result in branched plants of low yield. During flowering, freezing affects both sunflower and its pollinators. Much of the 1965 crop in Minnesota was exposed to a mid-August freeze. Some heads had a circular zone of empty achenes, and the achenes inside and outside the zone were normal. Crops like corn and soybean are killed by slight frosts in the fall, but temperatures must be less than -2 C to kill maturing sunflower plants.

NAR was maximum at 28 C, but temperature effects between 18 and 33 C were small and overshadowed by variation in light (208). The optimum range for seed production was 21 to 24 C (41). In a controlled environment of 18 to 20 C night temperature, day temperatures of 24 to 26 C gave higher yields, more seeds/plant, and heavier seeds of higher oil percentage than did day temperatures of 38 to 40 C (92).

Although temperatures during seed development strongly affect fatty acid composition of sunflower oil, they had less effect on oil percentage of seed developed at constant temperatures of 10, 16, 21, or 26 C (29). Oil concentrations were about 40% at 21 C compared with about 37% at the other temperatures. Protein concentrations increased from 14 to 20% as temperatures increased (29). Higher temperatures, 35 C, lowered oil percentage (38).

Photoperiod

Sunflower often is classified as insensitive to photoperiod because it will flower through a wide range of daylengths. Although night length is the important factor in plant response, photoperiod is expressed in length of day. Photoperiod influences changes in the terminal bud. Foliage leaf differentiation is replaced by differentiation of head and reproductive organs. Consequently, sunflower response to different photoperiods is indicated by leaf number per stem. Unfortunately much photoperiod research is reported in terms of days to flowering or in length of other growth periods. These intervals may be affected by photoperiod but more often are affected by temperature, soil moisture, or soil fertility.

A 12 to 14-hour photoperiod sometimes is considered a classification division between long and short-day plants. Sunflower production in North America is mostly between 30° and 50° N Lat. Within the range of practical planting dates for these latitudes, the daylength duration ranges from 13 to 16 hours at the normal times of head differentiation (101). Robinson (152, 157) reported that daylight variation within this range had little effect on several important cultivars. Three inbreds (157) and one cultivar (170), however, responded to differing photoperiods like short-day plants. In comparisons of 13 with 17-hour days and 15 with 21-hour days, sunflower cultivars gave short-day reactions (42, 169).

Photoperiod is not important in choosing planting date or production area within the temperate zones of North America. It can be very important in hybrid seed production if only one parent has a short-day reaction and the crossing fields are grown in different latitudes or in different seasons of the year.

Latitude Effects

Area of production does not appreciably affect appearance of a cultivar, but it does affect days to flowering and oil composition. Data from plantings made 14 May at nine locations from Texas (31°N. Lat) to Manitoba (49°N. Lat) showed that days from planting to ray flower stage increased from south to north. Average increase was 1.9 days per degree latitude (157).

Oil from sunflower seed grown in northern USA and Canada contains about 70% linoleic acid and has a high ratio of polyunsaturated to saturated fatty acids (89, 146). In contrast, oil from seed produced in southern U.S. contains only 40 to 50% linoleic acid if the sunflower was planted in the spring. The northern oil has an iodine value of 130 to 138 and is used for an edible or a semidrying industrial oil. The southern oil has an iodine value of 105 to 121 and is used as an edible oil. Higher temperatures during seed development were associated with the lower linoleic and higher oleic fatty acid percentages of the southern seed oil (87, 146). Oleic acid concentration in sunflower oil was 26% in seed grown in Minnesota and 51% in seed grown in Texas (118).

Water Requirement

The water requirement, or transpiration ratio, is defined as the grams of water transpired/g of dry matter produced above soil level. Transpiration ratios based on seed, rather than on total dry matter, are rarely used because factors other than transpiration determine the proportions of seed and stover. Water requirement varies among years and locations because transpiration rate is affected by the aerial environment—humidity, temperature, wind, and light. It is a measure of the crop's efficiency in using water when soil moisture is at an optimum level. It is not a measure of drought resistance. Sunflower is an inefficient user of water as indicated by a water requirement of 577 compared with 349 for corn, 304 for sorghum, 267 for proso millet (*Panicum miliaceum* L.), and 377 for sugarbeet (*Beta vulgaris* L.) at Akron, Colo. (172). Other inefficient crops include soybean, cotton (*Gossypium hirsutum* L.), oilseed rape (*Brassica napus* L.), field bean (*Phaseolus vulgaris* L.), wheat (*Triticum aestivum* L.), and oat (*Avena sativa* L.) with water requirements of 646, 568, 714, 700, 557, and 583, respectively. Water requirements in Saskatchewan were about 68% of those in Colorado; sunflower, corn, and wheat had water requirements of 386, 240, and 375, respectively (12). In Europe, Mihalyfalvy (115) reported a water requirement of 600 for sunflower in field tests and about 1,000 in pot tests.

Drought Tolerance

Sunflower is not highly drought tolerant but often produces satisfactorily when other crops are seriously damaged. Characteristics accounting for this response include an extensive and heavily branched tap root system with a potential lateral spread and depth exceeding 2 m. Barnes reported that sunflower roots extracted more soil moisture than did corn roots (12). Although resistance to water movement from the soil to sunflower leaves was only half that of soybean (23), injury to leaves from low water potentials was greater in sunflower than in soybean (180).

Photosynthesis continued at high levels of moisture stress even though leaf growth did not (22). Consequently, short periods of drought may not greatly reduce seed yield because growth can proceed at night when transpiration is low. The critical period for seed yield starts about 20 days before and ends about 20 days after flowering (145, 154, 162, 163). Rollier and

Pierre (163) reported that stress must continue for nearly 5 weeks of this 40-day critical period to reduce yields. Stress during the 10 days including anthesis was most damaging to seed yield (132). Extreme drought stopped anthesis, but it resumed after rain 1 month later (154). Oil yield was affected most by stress during the 20 days after flowering (162).

In extreme drought, some stalks break from 10 to 60 cm above the soil at heading time (154). This is a natural way of reducing the population. Extreme drought also causes lower leaves to dry prematurely. This leaf loss is not of great importance if it occurs after anthesis, but Johnson found that artificial defoliation 1 or 2 days before flowering reduced yields greatly (80). When the lower four to eight leaves were removed, yields were 30% below those of the undefoliated control. If all 16 or 20 leaves were removed, yields were reduced 93%.

Sunflower and *Cruciferae* spp. germination percentages were higher than many vegetable crops in dry soil (39). In a sandy loam with a permanent wilting percentage of 8.6%, sunflower emerged 73% at 8% and 89% at 9% soil moisture. In most fields, however, a good supply of moisture is needed for emergence of a satisfactory stand.

In comparison with some other field crops, sunflower production costs are lower and price of harvested seed is higher. Consequently, farmers' losses may be less with sunflower than with other crops in drought years.

Soil

Sunflower grows well in soils ranging in texture from sand to clay. It does not require as high fertility as crops like corn, wheat, or potato (*Solanum tuberosum* L.) to produce satisfactory yields. It does not use nutrients as efficiently as grain crops, however, because of the high concentration of chemical elements in sunflower seed and stover (153).

The salinity of a soil is often expressed in terms of the electrical conductivity (mmhos/cm) of water from a saturated soil. Based on its salt tolerance of 2 to 4 mmhos/cm, sunflower is classed as low in salt tolerance but somewhat better than field bean or soybean. Corn, wheat, rye (*Secale cereale* L.), and sorghum were rated medium (5 to 10 mmhos/cm) while barley (*Hordeum vulgare* L. and *H. distichum* L.) and sugarbeet were rated high (over 10 mmhos/cm) in salt tolerance (50).

At 12 mmhos/cm salinity, oil percentage was 21% lower than that of sunflower seed grown in a 2.5 mmhos/cm soil. Fatty acid composition of the oil did not change with change in salinity of the soil (120). Germination was not affected by NaCl concentrations up to 0.7%, but 1% NaCl reduced germination (61). Sunflower germinations ranged from 0 to 23% in 2% NaCl concentrations.

Soils should have good drainage for sunflower production, but the crop does not differ greatly from many other field crops in this respect. In sunflower flooded to the first leaves, Kawase (85) found that ethylene increased in the roots and stems below the water. Later, chlorophyll breakdown and epinasty of the leaves developed. Plants flooded longer than 3 days did not recover. The increased ethylene concentration in flooded plants was largely, although not exclusively, responsible for the damage symptoms. Cause of the increase in ethylene was attributed to blockage of ethylene escape by water (86).

Phenology

Lengths of five periods of sunflower growth amounted to an average of 11 days from planting to emergence, 33 days from emergence to head visible, 27 days from head visible to first anther, 8 days from first to last anther, and 30 days from last anther to maturity over a 3-year period with six cultivars (152). But temperature is a major factor in the phenological development of sunflower. Consequently, growing degree-day (GDD) summations were much better than day summations for prediction of phenological stages.

Robinson et al. (152, 157) used 7 C as a temperature base for GDD summations and indicated that any reasonable base was satisfactory for cultivar comparisons or descriptive data at a location. For each degree of latitude northward, however, days from planting to first anther stage increased nearly 2 days and GDD increased, decreased, or remained the same depending on the temperature base chosen. Keefer et al. (87) and Doyle (42) found that base temperatures of 0 or 1 C gave the most consistent GDD summations over a wide range of planting dates.

Cultivar differences in relative maturity were shown by the length of the planting to head visible or emergence to head visible stages. The commonly observed decrease in day summations for these periods from early to late planting was caused by increased GDD/day (152).

Anderson (6) defined physiologic maturity of sunflower as the time when achene dry weight, oil percentage, and linoleic acid percentage are at their maximums. Oleic acid percentage also was maximum at this time. The backs of most heads were yellow but 10% were brown. The heads contained 70% water and the achenes 40% water. GDD summation from first anther was 1052 (1 C base). Fenelonova (49) found that respiration of seeds/unit of dry matter decreased with age, but respiration/seed increased to a maximum 18 days after flowering. After 29 days, CO_2 emission exceeded O_2 absorption.

PLANTING PRACTICES

Dates and Depth of Planting

Dates of Planting

A wide range of planting dates can be used for sunflower, because the growing season in most areas is longer than needed for commonly-grown cultivars. Highest yields and oil percentages are obtained by planting early—soon after spring-sown small grain crops in the northern hemisphere. Planting from 1 to 20 May (138, 149) is optimum for northern USA and Canada, 20 April to 20 May for northern California (14), and 15 March through April in southern USA (81, 194). Planting by mid-March gave higher yields of seed and oil than did later plantings in Spain (100, 102). In Argentina, planting dates vary from July to November depending on the area and latitude. Planting in October gave highest yields at Pergamino, Argentina (103). In Cordoba, Argentina, the optimum planting time for the best cultivar was from 15 December to 15 January. 'Smena' and 'Peredovik' cultivars performed well however, when planted from September through November indicating their potential for southern Argentina (13). Seed yields and oil percentages decreased with delay in planting from 26 September to 23 October in New Zealand (96).

Days from planting to emergence and emergence to flowering decreased from early to late planting because of increased GDD/day (152).

Planting in April is hazardous in parts of northern USA and Canada where frosts may occur in late May or early June. Warm weather following April planting can result in plants that have lost their early resistance and are in the frost-susceptible, six-leaf stage at the time of a late frost. Another hazard from April planting is cold weather and slow growth when plants are emerging. This lengthening of the susceptible period for systemic infection by *Plasmopara halstedii* (Farl.) Berl. & de T. may increase losses from downy mildew disease. The rain and wet soil conducive to spread of the fungus also can occur with late plantings, but the plants are susceptible for fewer days.

Sunflower can be grown as a winter crop in Florida, Hawaii, and other areas with mild winters. Attempts in other areas to plant sunflower in the fall for emergence and growth the next spring have usually resulted in unsatisfactory stands. A new approach to fall planting still in the developmental stage is to coat the seed with an inner coating of talc, binder, plasticizer, and hydrogen peroxide; a cover of methycellulose; and an outer coating of plastic (168). The outer coating is impermeable to moisture but is ruptured by frost during the winter. The procedure is patented and involves the use of the Wurster air suspension equipment.

Planting date strongly affects the proportions of linoleic and oleic acids in sunflower oil. Robinson (159) suggested that planting sunflower late might increase the low linoleic acid concentrations in sunflower seed grown in southern USA. The predicted increase occurred in 1969 in Georgia but not in 1970–1971 trials (81). Australian research supports the idea of using planting dates to control oil quality. Linoleic acid in oil increased from less than 50% from September plantings to over 70% from March plantings. Mid-December planting gave linoleic acid levels of 48 to 57%, and Keefer et al. (87) projected that the desired 60% minimum could be achieved by planting in late December. They found a high correlation between linoleic acid concentrations in the seed and temperatures during the 0 to 35 or the 21 to 35-day periods after the beginning of flowering. Low temperatures were associated with high concentrations. These field results are supported by data from growth chambers where oil from seed developed at 10, 16, 21, or 26 C had about 76, 52, 36, and 24% linoleic acid, respectively (29). In California, linoleic acid concentrations increased from 61 to 72% as planting was delayed from 22 April to 15 July (14).

Canvin (29) concluded that temperature had little effect on oil percentage of sunflower grown at constant temperatures of 10, 16, 21, or 26 C during seed development. Although some date of planting trials suggest a relationship between temperature and oil percentage, prediction of oil percentage based on temperature data remains uncertain.

Despite potentially lower yields from late planting, sunflower is often planted in July after small grain harvest in southern USA. Thus, it may be possible in areas with long growing seasons to produce an early crop with high oleic-low linoleic and a late crop with low oleic-high linoleic oil for different markets.

Chlorogenic acid causes sunflower meal to turn yellow-green under alkaline conditions, and this limits its use in food products. Dorrell (40) found that the chlorogenic acid concentration in sunflower flours declined steadily from 4.2 to 3.3% as date of planting was delayed from 14 May to 15 June. Maximum synthesis of the acid occurred during 15 to 34 days after flowering, and mean temperature during this period for seven dates of planting declined from 19.3 C for the 14 May planting to 12.2 C for the 15 June planting.

Damage by sunflower head moth larvae (*Homoeosoma electellum* Hulst) may be affected by date of planting. Early plantings have been the most severely injured in Minnesota, but in California early planting was suggested as a method of reducing losses from head moth (31). The increasing area of land devoted to sunflower in the United States will be an important factor in developing recommendations for plantings that will flower when moth flights are minimum.

Depth of Planting

A planting depth of 3 cm is excellent in moist soil or if rain is imminent. Seed can be planted to a maximum depth of 10 cm if necessary to reach moisture. Although the seed will germinate when soil moisture is near the wilting point (39), imbibition of water through the hulls is slow. Nonoilseed cultivars often emerge more slowly than oilseed cultivars which have thinner hulls. Emergence of oilseed cultivars was slightly greater from large than from small seeds (82, 141). If planting depth exceeded 7 cm, however, emergence of small seeds was only 69% that of large (141).

Sunflower planted 9 to 15 cm deep yielded 10 to 28% less than that planted 7 cm deep (183). On a silt loam soil packed by heavy rains, emergence from depths of 3, 8, 13, and 18 cm was 97%, 68%, 42%, and 5%, respectively (154). Plants from the 3-cm depth emerged and flowered 4 days earlier than those from 8 cm and 7 to 8 days earlier than those from 13 cm deep. Depth increases of 3-cm increments from 3 to 15 cm decreased emergence, root and shoot length, and dry matter in an experiment with seeds planted in pots (179).

On sandy soil or in environments where the topsoil dries quickly, planting should be slightly deeper than on fine-textured soils. Emergence from 3, 4, 5, and 7 cm was 85, 95, 92, and 84%, respectively, in the southern U.S. (82).

Shallow plantings emerge sooner than deep, because the seed is closer to the surface. Soil temperatures near the surface are strongly affected by air temperatures during the day when temperatures are usually high enough for growth. The temperature effect is especially important at early dates of planting when low temperatures retard germination and elongation.

Use of lister planters permits deep planting in trenches with shallow coverage of the seed. A major disadvantage of lister planting in many areas is the cold environment in the furrow for the seed and seedling.

Plant Populations and Planting Rates

Sunflower yield is the product of three components: (a) number of heads/ha, (b) number of seeds/head, and (c) average weight/seed. Since most cultivars produce one head/plant, component (a) is determined by plant population. The other two components are affected by the first component and by cultivar, weather, soil, and sunflower pests.

The relationship among the components of yield and sunflower performance shown in Table 1 is an average of 12 trials with oilseed and 2 trials with nonoilseed cultivars at 4 locations in Minnesota (160). Sunflower adjusts to low populations by increasing weight/seed and seeds/head and to high populations by decreasing weight/seed and seeds/head. So, yield which is the product of the three components remains relatively constant through a wide range of populations. But, in environments where yields approach the yield capacity of the cultivar, adjustments among components of yield are not sufficient. For example, identical trials on similar soil at two locations 148 km apart resulted in yields of about 2,200 kg/ha from all populations at the drier location. At the other location, yields were 3,100, 3,500, 3,800, 4,100, and 4,600 kg/ha at populations of 37,000, 49,000, 62,000, 74,000, and 86,000 plants/ha, respectively (160). Thus yield at the dry location was limited by factors other than population. Consideration of data from only one of the two locations could lead to an erroneous recommendation.

Disagreement on the optimum plant population is common. Differences within a region are as great as those between different countries of the world. In many trials where population differences had no effect, yields were medium or low indicating that factors other than population were limiting yield. Consequently, even the lowest population was capable of producing the highest yield.

High populations are needed for highest yields according to results of seven trials with oilseed cultivars and five trials with nonoilseed cultivars in

Table 1—Effect of components of yield on seed size and oil percentage and on stalk lodging of sunflower. From Robinson et al. (160).

Components of yield			Large	Seed	
Heads/ha	Seeds/head	Weight/seed	seed†	oil	Lodging
		g	———— % ————		score‡
37,000	831	0.073	52	42.6	1.5
49,000	727	0.067	44	43.2	1.8
62,000	632	0.062	33	43.2	2.1
74,000	548	0.060	31	43.4	2.4
86,000	501	0.058	26	43.8	2.5
L.S.D. 5%	--	0.002	8	0.6	0.2

† Nonoilseed cultivars held on an 0.8 cm round-hole screen.
‡ 1 erect, 9 prostrate.

eastern North Dakota from 1971 to 1973 (215). Oilseed cultivars yielded 2,258, 2,630, and 3,190 kg/ha, respectively at 36,000, 48,000, and 72,000 plants/ha. Nonoilseed cultivars yielded 2,303, 2,625, and 3,201 kg/ha at 29,000, 36,000, and 48,000 plants/ha, respectively. But plant populations had little effect on yield in other trials in eastern North Dakota and the adjoining areas of Minnesota (160, 186). In western North Dakota, yields were higher from populations of 25,000 than from 50,000 plants/ha (3).

In Manitoba, populations of about 62,000 plants/ha were optimum (46). Similar populations were recommended in Saskatchewan although yields from populations of 25,000 and 75,000 did not differ (200). Depth of root system was greatest at 25,000 with 23% of the roots below 38 cm compared with 15 and 20%, respectively at 75,000 and 125,000 plants/ha. All available moisture, however, was used to a depth of 120 cm by all populations.

Yields from populations of 33,000, 44,000, and 65,000 plants/ha did not differ in California (14). Similar results were obtained from some research in Georgia (82), but other research showed much higher yields from 60,000 than from 20,000 or 30,000 plants/ha (110).

Considerable research in Europe indicates that about 40,000 plants/ha is a good population (44, 90, 99, 102, 106, 121, 140). Insect damage was greatest in low populations and was 3, 6, and 22% damaged seeds in populations of 40,000, 20,000, and 10,000 plants/ha, respectively (44). A common theory that higher populations are needed for short than for tall cultivars is supported by a comparison of Peredovik with dwarf cultivars (127).

In India, yields increased as population increased from 20,000 to 67,000 plants/ha and then decreased with higher populations. Yields were highest at populations from 43,000 to 54,000 plants/ha in Pakistan and Bangladesh (88, 114). Populations of 40,000 to 50,000 plants/ha were recommended in Argentina (103).

Contrary to experience with corn, published research does not show a pattern of high populations being best for high rainfall areas and low populations for arid areas. However, populations of 50,000 to 60,000 plants/ha were suggested for irrigated fields compared to 40,000 to 50,000 without irrigation in Romania (2). Optimum populations in Texas were 35,000 for dryland and 62,000 for irrigated fields (194). Total water use (23 cm) was not affected by plant population in western North Dakota where sunflower extracted water to a depth of 150 cm (3).

The strong effect of plant population on seed size is very important to packagers of nonoilseed sunflower for food. Only the 37,000 and 49,000 populations in Table 1 had a sufficiently high percentage of large seeds to meet food market requirements. Medium and high populations produced seed of higher oil percentage than did low populations (160, 184). High populations also produce small heads that remain upright and dry faster than large heads (90, 139). The use of preharvest desiccant-sprays, however, may reduce that advantage of small heads.

Flowering date often is not affected by populations adapted to the environment (160). Excessive populations delayed flowering 1 to 4 days (3). Heights often increase slightly as population increases (90, 139). Increased lodging accompanies increased populations (Table 1). Such increases are commonly attributed to shading of the stem and to a reduced stalk diameter. Another important cause is the rapid spread of basal stem rot caused by *Sclerotinia sclerotiorum* (Lib.) de Bary in dense populations (74, 213). The disease organism spreads through root contact, and wide spacings between plants delay this contact.

In summary, plant populations for oilseed production in North America should be 35,000 to 50,000 plants/ha on sands or on soils with sandy subsoils within 50 cm of the surface; 50,000 to 60,000 on silt loam, clay, or irrigated sandy soils; and 60,000 to 70,000 on fields where expected yields exceed 3,500 kg/ha. Where large seed is required from nonoilseed cultivars, a plant population of 35,000 is recommended, but this should be increased to 40,000 in fields where yield expectations exceed 2,500 kg/ha.

Plant population/ha is determined by the planting rate, which is often expressed in kg/ha. Seed weight, viability, and expected mortality are used to calculate the planting rate for a desired plant population.

$$\text{Planting rate} = \frac{\text{Plants/ha} \times \text{g/100 seeds}}{1,000 \times \text{germination percentage}}$$

The rate calculated from the equation should be increased by 5% on sandy soils and up to 20% on other soils to allow for seed and young plant mortality (160).

Planting rate recommendations often have been higher than necessary because seed costs were low compared to most crops. Emergence failures and plant mortality are less likely to reduce yield in high than in low populations. A plant in a population of 70,000 plants/ha with a missing plant on each side is still at a population of 35,000, but a plant with a gap on each side in a population of 40,000 plants/ha is at only a population of 20,000. The high cost of hybrid seed and precision planting equipment encourages use of less seed. Nonoilseed cultivars often have less than half as many seeds/kg as oilseed cultivars so seed cost is especially important in nonoilseed production.

Seed cost/ha can be reduced by using the smallest *good* seed grade of the desired cultivar. Crops produced on plots planted with large, medium, or small seed sieved from ungraded seed lots did not differ in yield, oil percentage, or test weight (155). A "trash" grade of small, light weight seed comprising 2% of the ungraded seed lot produced plants of low seedling vigor and 22% lower yield than plants from ungraded seed at the same population. Use of small and medium seed resulted in savings over large seed of 36 and 16%, respectively. Complete removal of dehulled seed from seed lots of small achenes is not always achieved. Such seed lots are undesirable for precision planting because dehulled seed may germinate poorly. Small seed should not be planted deeper than 7 cm because of poor emergence (141).

Arrangement of the Population

Distribution or arrangement of the population is altered by varying distance between rows, by planting seeds in groups, or by changing row direction.

Direction of Rows

Sunflower is phototropic from emergence to flowering. The head and leaves face east in the morning and west in the evening. About 1 day before the ray flowers open, phototropic movement ceases and most heads face the east. When sunflower was harvested by hand, some growers of tall cultivars preferred rows directed north-south (NS) because heads overhanging a wagon pulled along the east side of the row could be gathered easily.

Calculations from a statistical model of light interception show that grain sorghum in north-south (NS) rows would intercept 44% and in east-west (EW) rows 37% of direct sunlight in August in Colorado (5). Total light interception by corn between 0700 and 1500 hours in Wisconsin did not differ with row direction (190). But in both crops, EW rows intercepted more light at noon and less at 0900 and 1400 hours than did NS rows. Evapotranspiration did not differ appreciably between EW and NS rows (190). Consequently, NS rows had a potential yield advantage over EW rows. But direction of prevailing winds could easily offset this advantage. Furthermore, statistical models and measurements on nonheliotropes may not apply to sunflower.

Few scientific attempts have been made to measure the effect of row direction on crop yields because of the inconvenient plot arrangements and the large area required. The large plots needed to avoid border effects usually require large turning areas for farm machinery, and replications are needed.

Robinson (156) reported that sunflower grown in EW and NS rows did not differ in yield, oil percentage, large seed percentage, seed weight, or test weight, but significantly more lodging occurred in EW rows. Other directions were studied by using a circular planting technique. A "compass" was used to draw circles of 6-m radii on the seedbed. Straight lines drawn from the center of a circle to points 10 degrees apart on its circumference marked 36 "wheel-spoke" plots. Thus each circle consisted of 18 row-direction spokes replicated twice. Spacings of plants along these spokes decreased from the center to the outside so that all plants occupied equal areas.

Sunflower yields from the 36 spokes/circle showed random variation not fitting any directional pattern. Yields from the NS plus its 10 and 20-degree spokes on both sides were compared with yields from the EW plus its 10 and 20-degree spokes on both sides, and again no significant difference occurred between predominantly NS and EW rows. Wheel-spoke alignments in circular plantings are an effective way of comparing yields of many row directions in minimum space. However, lodging cannot be evaluated in such plantings because of lack of border plants.

Harvesting losses are sometimes slightly greater when combines approach EW rows from the west. Some growers with combine pans 25 cm wide (suitable for all row spacings in contrast with wider pans for specific row spacings) found that such losses were reduced by driving diagonally across the rows. Diagonal travel, though, is usually precluded by the practice of hilling the rows at the last cultivation. Hilling the rows reduces lodging (154).

NS rows may be slightly preferable to EW, but effect of row direction on sunflower is not important in commercial production. Seed production fields that require examination of heads to detect pollen production should be planted in NS rows for efficient roguing. For research plots, EW rows with plot labels on the east end often are preferred, because it is easier to evaluate most plots when all heads face the viewer.

Rows at the end of the field usually are planted at right angles to the main rows. The advantage of planting end rows first is that the inner row serves as a mark to raise and lower the planter. This helps prevent gaps or excess population. If the soil is so loose that machinery tires will move the seed, end rows should be planted last. A greater rate of planting for end rows will offset loss of stand from machinery tires when the main rows are cultivated.

Check-row or Hill Planting

Check-row or hill planting has not been a common practice for sunflower in North America, but it has been used in Europe and other continents. It usually involves clumps of two or more plants/hill rather than single-plant spacing. The checkerboard plant arrangement permits cultivation in both directions. Consequently, control of quackgrass (*Agropyron repens* (L.) Beauv.) and other weeds is better than from cultivation in one direction.

Plants equally spaced in the row yielded more than clumps of two or more plants at both 32,000 and 64,000 plants/ha in Minnesota (159). Single-plant spacing also was best in Manitoba in rows 46 cm apart. But in 91 by 91-cm spacings to allow cross cultivation, 5 to 9-plant hills yielded more than 1 to 3-plant hills because they gave a larger population (139). At populations of 42,000 plants/ha in Romania, plantings at one or two seeds/hill did not differ appreciably in yield (143).

Annual weeds can be controlled with herbicides or by postemergence harrowing. Consequently, check-row planting is an obsolete practice except for some fields severely infested with perennial weeds.

Spacing Between Rows

The space between rows is largely determined by the machinery available and the needs of all crops grown by each farmer. A 56-cm spacing often is used by sunflower growers who also raise sugarbeet and 76 or 97-cm spacings by those who grow corn, sorghum, or soybean. Small grain farmers who do not have row crop planters sow sunflower with a grain drill either in widely spaced rows for intertillage or in noncultivated rows 15 to 46 cm apart. Bed width determines row spacing when furrow irrigation is practiced. Bed widths may vary from 76 to 122 cm. Two sunflower rows about 25 cm apart are sometimes planted on the wider beds.

In dryland areas where a year of fallow is used to store soil moisture, some sunflower is grown in single rows 3 to 6 m apart instead of fallow. Rows 3.7 m apart produced slightly over half as much seed as did rows 0.9 m apart (138). Plant spacings of 8 to 10 cm apart are recommended for these widely spaced rows. Other sunflower is grown in single or multiple rows 6 to 21 m apart to reduce wind damage to interseeded crops, to reduce soil erosion losses, or to hold and increase uniformity of snow cover.

Under nonarid conditions and at optimum populations, sunflower should produce highest yields when interrow and intrarow plant spacings are equal. This equidistant spacing produces an earlier and more complete soil cover than other spacings. As a result, more sunlight is intercepted by the foliage. The greater interception increases photosynthesis and reduces evaporation of water from the soil. The more complete soil cover also intercepts more rainfall and may reduce runoff and soil erosion. Equidistant spacing is approached more closely by rows 50 to 76 cm apart than by wider spacings.

Maximum evapotranspiration occurs when water is readily available to the plant and at the soil surface. Under these conditions, neither space between rows nor plant population has much effect on water loss. Closely spaced rows intercept more solar energy. Consequently, transpiration is high relative to evaporation. In widely spaced rows, more solar energy reaches the ground and causes more evaporation. As a result, the total water loss is approximately equal in widely and closely spaced rows. But when the soil surface dries, evaporation is limited by the moisture supply in the upper 10 to 15 cm of soil. However, transpiration is not affected by the dry surface soil. So, more water is lost from narrow than from widely spaced rows. These effects of row spacing on light interception and evapotranspiration may be of lower magnitude in sunflower that spreads its foliage by leaning toward the east after flowering than in erect crops.

Row spacing also affects the time when stored soil moisture between the rows is used. Closely spaced rows are more likely to deplete this moisture before flowering because of earlier root penetration throughout the interrow area. In widely spaced rows, sunflower used soil moisture from at least 1.2 m on both sides of the row (72). This result indicates that in arid areas, a row spacing of about 2.5 m would provide maximum interrow moisture storage for use during the flowering and seed development stages.

When adoption of closely spaced rows results in appreciably greater distances between plants, losses from basal stem rot disease may be reduced (74).

Sunflower grown in rows 56, 76, or 97 cm apart did not differ in seed yield, oil percentage, large seed percentage, seed weight, seed test weight, height, or flowering date. Results were consistent among five populations and ten trials in Minnesota (160). Consequently the recommendations for populations apply to all row spacings from 50 to 100 cm. Stalk lodging increased as row spacing increased and plant spacing in the rows decreased.

The circular planting technique described under "Direction of Rows" was used to compare 16 row spacings ranging from 36 to 102 cm on sand (Udorthentic Haploborolls, sandy, mixed) and silt loam (Typic Hapludolls, fine-silty over sandy or sandy-skeletal, mixed, mesic) soils. A row spacing plot consisted of 18 plants at the same distance from the center of the circle in a 180° arc. Neither yield, oil percentage, nor seed size differed among row spacings (160). The circular planting technique permits evaluation of many row spacings in a minimum area. Consequently, trends in yield and other characteristics can be detected that might not be apparent in the usual experiment involving only a few spacings.

Sunflower in rows 36, 53, and 89 cm apart did not differ in yield in Saskatchewan nor was there any difference in yields from rows 51 and 102 cm apart in Tennessee (67, 200). Sunflower in rows 36 cm apart had both the shallowest and deepest roots—32% in the upper 13 cm and 25% below 38 cm (200). The 53-cm spacing had a root distribution of 30% in the upper 13 cm and 15% below 38 cm compared with 22 and 18%, respectively, in rows 89 cm apart. Root systems had a greater lateral spread and were deeper in a rectangular (135 × 10 cm) than in an equidistant (37 × 37 cm) plant arrangement.

Some fields in western Canada are planted with a grain drill or disker in noncultivated rows 15 to 41 cm apart. Recommended populations are 62,000 (138) to 74,000 (43) plants/ha. In weed-free plots with rows 30 cm apart, populations of 25,000 to 50,000 plants/ha gave higher yields than did populations of 75 to 100,000 plants/ha in western North Dakota (3).

CROP SEQUENCE AND ROTATION

Crop rotation is a recurring succession of crops on the same field. Crop sequence is the order in which crops are grown on a field. Rotations were of most common use in Europe and on farms producing crops and livestock in

North America. Their use has declined greatly with increased specialization of crop and livestock farms and greater use of commercial fertilizer and pesticides. Sunflower became an important crop in North America during the time that rotations became less important. Rotations unsupported by fertilizer cannot maintain soil fertility, and the best means of maintaining soil fertility is to produce high yields with large crop residues. Other than diversification for the individual farmer, the major advantages of crop rotation are weed, insect, and disease control. These pests can be controlled more effectively by changing crop sequences than by fixed rotations.

Major considerations in determining crop sequence are control of volunteer sunflower, control of sunflower pests, soil moisture and fertility depletion by sunflower, effect of preceding crop on sunflower, and effect of sunflower on succeeding crops.

Volunteer Sunflower and Its Control

Many seeds fall to the ground before and during sunflower harvest. A small loss of 45 kg/ha equals 5 to 20 times a normal rate of seeding. Enough of this seed survives the winter to create a weed problem in the following crop. Seed dormancy of most cultivars is very short in cool temperatures and does not exceed 10 weeks at warm temperatures so sunflower will not become a weed except through neglect. Nevertheless considerable seed, probably buried too deeply or trapped in dry clods the first year, germinates the second year after a crop of sunflower. Consequently, control of volunteer sunflower is needed in two crops following sunflower unless late-planted crops are used.

Damage from volunteer sunflower is not confined to reduced yield of the infested crop. Crop sequence cannot be highly effective in disease and insect control unless volunteer sunflower is killed at an early stage of growth. Volunteer sunflower may interfere with pure seed production that requires fields isolated from other sunflower. Uncontrolled volunteer sunflower on the seed field in previous years or in neighboring fields the same year may result in seed that fails to meet quality standards.

Research in Minnesota (R. G. Robinson, unpublished data) devoted to the volunteer problem has shown that less herbicide is needed to control sunflower growing in crops than growing alone. Average control by crops alone without help from herbicides amounted to 63, 48, 30, and 70%, respectively for oat, canarygrass (*Phalaris canariensis* L.), flax (*Linum usitatissimum* L.), and pea (*Pisum sativum* L.). The best time to spray herbicides postemergence on these crops is when sunflower is in the 4 to 5-leaf stage but spraying up to the 12-leaf stage also gave good control. Earlier spraying was less effective because some sunflower had not emerged. Selective (no crop injury) control of sunflower in all four crops was achieved with 2,4-D [(2,4-dichlorophenoxy)acetic acid], MCPA [[(4-chloro-*o*-tolyl)oxy]acetic acid], 2,4-DB [4-(2,4-dichlorophenoxy)butyric acid], MCPB [4-[(4-chloro-*o*-tolyl)oxy]butyric acid], and bentazon [3-isopropyl-1H-2,1,3-benzothiadiazin-(4)3H-one 2,2-dioxide]. Dinoseb (2-*sec*-butyl-4,6-dinitrophenol) gave selective control only in pea, whereas bromoxynil (3,5-dibromo-4-hydroxybenzonitrile) gave selective control in all crops but pea. Dicamba (3,6-dichloro-*o*-anisic acid) gave selective control in oat and canarygrass. Metribuzin [4-amino-6-*tert*-butyl-3-(methyl-thio)-*as*-triazin-5(4H)one] applied at 0.56 kg/ha during preemergence to 3-cm stage gave selective control only in pea.

Herbicides to control volunteer sunflower were also compared in later planted crops of grain sorghum, soybean, and field bean grown in cultivated rows 76 cm apart. No herbicide applied preemergence gave good sunflower control in sorghum. Metribuzin preemergence at 0.84 kg/ha gave good selective control of sunflower in soybean but injured field bean. Postemergence sprays of atrazine [2-chloro-4-(ethylamino)-6-(isopropylamino)-*s*-triazine] plus oil, cyanazine [2[[4-chloro-6-(ethylamino-*s*-triazin-2-yl] amino]-2-methylpropionitrile], dicamba, bentazon, and 2,4-D sprayed postemergence controlled sunflower in grain sorghum. Bentazon postemergence at 1.12 kg/ha gave excellent selective control of sunflower in soybean and field bean. The lower range of recommended rates for annual weed control was enough for sunflower control by all herbicides except where rates are indicated for bentazon and metribuzin.

Tillage of sunflower crop residues does not control volunteer sunflower. Volunteers will emerge whether or not the soil is inverted by fall or spring plowing or given surface tillage only (R. G. Robinson, unpublished data). If a cultivated row crop is planted in unplowed sunflower stubble with a lister-planter or till-planter, however, the surface soil with sunflower seed is pushed between the newly planted rows. Volunteers that emerge can be killed by cultivation.

Effect of Preceding Crop on Sunflower

In most crop rotation experiments, sunflower produced highest yields following legumes and lowest yields after sunflower (10, 20, 121, 150, 159, 176). In western Canada sunflower gives highest yields after fallow (138), but in Manitoba it usually is grown after small grain to replace or delay a year of fallow. Nearly 80% of the 1971 crop was planted on summerfallow in Saskatchewan. Sunflower yields after fallow and after small grain were equal in Minnesota (150). In contrast with sunflower, corn growth was retarded and P concentrations in the plants were lower after fallow than after small grain. Sunflower yields following sugarbeet were lower than yields after small grain (121, 176). Sunflower following sugarbeet in the Red River Valley of North America often has less seedling vigor than does sunflower after other crops.

Although sunflower yields often are lower after small grains or corn than after legumes, the grain crops are the best preceding crops for sunflower, because they are immune to the major fungal pathogens and weeds that parasitize sunflower. Broomrape (*Orobanche* sp.) is a parasitic weed that lives on sunflower in Europe. It is controlled with resistant cultivars, an insect parasite, and crop rotation (84, 108, 137). An infestation of nearly six broomrape stems/sunflower plant in a sequence of 2 years between sunflower crops was reduced to less than one stem/plant in a 5-year sequence (108).

Sclerotinia sclerotiorum infects sunflower, field bean, soybean, pea, fababean (*Vicia faba* L.), lentil (*Lens culinaris Medikus*), flax, rape (*Brassica napus* L. and *B. campestris* L.), mustard (*B. juncea* (L.) Coss), sugarbeet, potato, alfalfa (*Medicago sativa* L.), sweetclover (*Melilotus alba* Desr.) and many dicot weeds. *Verticillium dahliae* Kleb. infects sunflower, safflower (*Carthamus tinctorius* L.), rape, sugarbeet, potato, alfalfa, red clover (*Trifolium pratense* L.), cotton (*Gossypium* sp.), and many dicot weeds. These fungi, however, include many physiologic races. The races that injury sunflower may not injure all the other crops and vice versa. Soybean, flax, and sugarbeet are grown in rotations with sunflower without noticeable increases in diseases. Low populations of the pathogen rather than host resistance may have prevented epiphytotics in these rotations. Pathologists indicate that the same races of *Verticillium* attack both sunflower and potato, and the same races of *Sclerotinia* attack sunflower, mustard, rape, and field bean.

Comparisons of sunflower in monoculture and in crop rotations up to 7 years long showed that yield generally increased with length of rotation. Downy mildew decreased with increasing length of rotation and its prevalence was negatively correlated with yield (177). On a silt loam soil (Typic Hapludolls, fine-silty over sandy or sandy-skeletal, mixed, mesic) of low fertility, *Verticillium* became the major factor affecting yield of sunflower grown in monoculture for 5 years (147). Alternation with soybean significantly increased sunflower yield and decreased *Verticillium* in the last 3 years of the 5-year trial. A similar trial on a field of high fertility at the same location resulted in *Sclerotinia* and to a lesser extent *Plasmopara* becoming the most prevalent pathogens on sunflower. Disease incidence was less and yields were greater for sunflower grown in alternation with oat and soybean than for sunflower in monoculture.

Sunflower and other crops susceptible to the same pathogen should, ideally, be separated in the crop sequence by resistant crops until the pathogen disappears. Over 10 years are needed for *Verticillium* populations to approach zero in California (75). Although sclerotia of *Sclerotinia* near the soil surface disintegrate rapidly, those buried in the soil may survive at least 5 years (73). Oospores of *Plasmopara halstedii* are reported to survive 6 years in soils in Russia (73). Consequently, crop rotations of practical length can only be expected to reduce the amount of inoculum carryover, not eliminate it. A minimum time between sunflower crops should allow all sunflower stalk, root, and head residues to decompose. On well-drained soil, this takes 1 or 2 years. This minimum time is enough to help control rust (*Puccinia helianthi* Schw.) and other organisms that require sunflower for survival. Three or 4 years are suggested in the USA between sunflower or other crops susceptible to the same races of *Verticillium* or *Sclerotinia*. If a disease becomes serious on a field, a longer time must elapse between susceptible crops. The availability of disease-tolerant hybrids is a new factor

Table 2—Comparative composition of sunflower and corn. From Robinson (153).

Element	Sunflower achenes	Corn caryopses	Sunflower stover	Corn stover + cobs
		%		
N	2.58	1.70**	1.03	0.68**
P	0.39	0.37	0.08	0.06**
K	0.59	0.36**	1.51	0.68**
S	0.17	0.13**	0.20	0.10**
Ca	0.11	<0.01**	1.10	0.32**
Mg	0.23	0.14*	0.58	0.24**
Na	<0.02	<0.02	0.10	<0.02**
		ppm		
Sr	3	<0.5*	59	17*
Al	3	<1.5**	189	131
Fe	33	26**	152	143
Zn	48	26**	34	36
Cu	13	1**	4	3*
Mo	6	4**	15	8**
Mn	14	8**	27	44**
B	14	1**	34	5**

* Crops differ significantly at the 5% level.
** Crops differ significantly at the 1% level.

in planning crop sequences. But resistance to all important soil-borne diseases is needed before making major changes in the crop sequence.

Residues from some herbicides used on the preceding crop may persist in the soil and injure sunflower. Atrazine and simazine (2-chloro-4,6-bis (ethylamino)-s-triazine), commonly used on corn occasionally remain in sufficient amounts to injure the following crop of sunflower. Soils from suspected fields can be tested by comparing sunflower growth in untreated soil with growth in soil mixed with 0.5 g activated carbon/2 kg soil. The carbon inactivates the herbicide. The test must be conducted in sunlight, and the injury symptoms should appear within 3 weeks after emergence. Picloram (4-amino-3,5,6-trichloropicolinic acid), used in wheat and barley, is another herbicide that may remain in the soil and injure sunflower.

Effect of Sunflower on the Following Crop

Sunflower may affect other crops by leaving toxic residues, removal of nutrients, and maintenance of pest populations.

The soil toxin theory was strongly supported by the Chief of the USDA Bureau of Soils in 1906 (211). Although generally discredited for practical application in field crop production, considerable research has shown that plants do secrete substances toxic to other plants. Workers in this science of allelopathy found that sunflower residues in the soil have inhibitory effects on sunflower and many other species (144).

Sunflower removes soil minerals and this affects the following crop. The removal can be beneficial by reducing salinization of irrigated soils where sunflower removed much more Cl and Na than did cotton (19). Elemental depletion of the soil is usually harmful. Sunflower seed and stover have considerably higher concentrations of most elements than do corn

Table 3—Comparative elemental content/ha of sunflower† and corn‡. From Robinson (153).

Element	Sunflower achenes	Corn caryopses	Sunflower stover	Corn stover + cobs
		kg/ha		
N	93	129	57	66
P	14	28	5	5
K	21	27	83	65
S	6	10	11	10
Ca	4	<1	63	31
Mg	8	11	34	23
Trace elements				
Na	0.34–0.65	0.34–0.94	6.00	0.39–2.00
Sr	0.01	0.00	0.27	0.13
Al	0.01	<0.01	1.30	1.58
Fe	0.12	0.20	0.94	1.53
Zn	0.17	0.19	0.22	0.40
Cu	0.04	0.01	0.02	0.02
Mo	0.02	0.02	0.09	0.08
Mn	0.06	0.06	0.16	0.45
B	0.04	0.01	0.19	0.04

† 49,420 plants/ha. Achene yield 3,634 kg/ha. Stover yield 5,693 kg/ha.
‡ 49,420 plants/ha. Caryopses yield 7, 593 kg/ha. Stover + cobs yield 9,844 kg/ha.

grain and stover (Table 2). Both elemental concentration and crop yield determine nutrient removal by crops. Because sunflower yields less than corn, the elemental contents/ha of sunflower seed were generally less than those of corn grain (Table 3). These data indicate that sunflower crops deplete soil fertility less than do corn crops at high yield levels of both crops.

Sunflower is harvested later than small grains; consequently, less time is left for tillage or planting fall crops. Furthermore, sunflower reduced soil moisture reserves more than did most other crops (12, 176, 212). This reduction in soil moisture caused crops following sunflower to yield less than crops following corn in dryland areas of western North America (10, 12). Barnes (12) concluded that sunflower was not a good substitute for fallow in areas where fallow is needed to store soil moisture. Research in areas with more rainfall showed that wheat, oat, flax, rye, corn, and soybean yielded as much after sunflower as after any of the common, nonleguminous field crops (125, 147, 159, 161). Crops following legumes generally yield more than crops following sunflower. But on a soil testing high in NO_3-N (165 kg/ha), preceding crops of soybean, field bean, or sunflower did not significantly affect yields of wheat or corn (R. G. Robinson, unpublished data).

SOIL MANAGEMENT

Tillage and Seedbed Preparation

Improved farm machinery, herbicides, and greater knowledge of seedbed requirements give sunflower growers a choice of many effective tillage systems. Conventional systems involve an initial tillage to dispose of the residue from the previous crop and secondary operations to prepare a seed-bed that provides satisfactory temperature, aeration, and moisture for germination and seedling growth.

Implements used for the initial tillage include the moldboard plow, disk plow, chisel plow (heavy field cultivator), one-way disk (wheatland plow), and rotary tillers. These implements loosen the soil and incorporate crop residues into it. Loosening the soil increases aeration which affects moisture, temperature, and biological activity in the soil. Moldboard and disk plows also invert the soil. Soil inversion results in a bare soil surface. Consequently, other tillage implements are used where erosion is a major problem or where maximum snow retention is desired. Soil inversion generally results in a warmer soil and improved availability of nitrogen and potassium (27, 129). Furthermore, phosphorus is relatively immobile in the soil and tends to accumulate in the surface 3 cm under a continuous mulch system (9, 112). Consequently, sunflower growers using noninversion tillage systems may use a moldboard plow every few years to improve distribution of nutrients in the upper soil profile. Each metric ton/ha of crop residue left on the soil surface reduced average soil temperatures during May and June at the 10 cm depth 0.4 C (27). However, these slightly lower soil temperatures are less harmful to sunflower than to crops like corn or soybeans that require higher temperatures for germination and seedling growth. Furthermore the cooling and moisture conserving effects of mulch are beneficial under dry conditions.

In contrast to the other tillage implements, rotary tillers are powered as well as pulled by the tractor. The rotor chops and incorporates crop residues into the soil. The rotary tiller is much slower and requires more power than the other initial tillage implements. It is the only tillage implement that excels the moldboard plow in controlling perennial weeds. The soil is left in a loose, clod-free condition that is very susceptible to erosion.

Fall plowing of loam and clay soils in humid areas results in better soil tilth and more moisture for seed germination than does spring plowing. However, fall plowing of soybean stubble or other crops that loosen the soil may result in a pulverized soil surface that is vulnerable to wind and water erosion. The danger of wind erosion is particularly great when sandy soils are tilled in the fall. Fields with crop residues on the surface will retain more snow than will bare fields, and moisture from melting snow is beneficial in many dryland areas. Increasing depth of plowing beyond the minimum needed for efficient machine operation, residue incorporation, seedbed preparation, or shattering of a plow-sole generally has not increased crop yields in humid areas where most of the world's sunflower is grown. Plowing depths from 20 to 45 cm did not affect yields of sunflower fertilized with P, but with no P, yields increased when depth increased to 30 or 35 cm (175).

The moldboard plow is the most commonly used of the primary tillage implements in sunflower production, but use of the chisel plow is increasing. Chisel plows leave most of the crop residue at or near the surface. Erosion control, effectiveness in dry soil, and greater speed are their major advantages over moldboard plows. The disk plow requires about the same power as a moldboard plow, but it will function in soils too hard and dry or too sticky for the moldboard plow (202). It differs from the one-way disk in that each disk is mounted on its own axle and set at an angle of 10 to 30° with the vertical. The one-way disk offers relatively trouble-free operation with no plugging or trash coverage problems. It tends to pulverize the soil, and repeated operations leave too little crop residue on the surface to retard erosion.

Implements used for the secondary tillage operations include the common disk harrow, field cultivators, rod weeders, and smoothing harrows of various types. All of these implements except field cultivators pack the seedbed. They break clods into smaller particles resulting in more but smaller air spaces in the soil. These implements leave the soil in good condition for seed germination but very exposed to erosion.

The erosion hazard, soil compaction by heavy machinery, and the paradox of loosening the soil by initial tillage and then packing it with secondary tillage operations has led to the minimum tillage concept. The firm soil needed for seed germination is not the ideal environment for root growth. Consequently many minimum tillage systems develop a firm seedbed for the seed but leave the interrow area loose. This delays germination of weed seeds and provides a better root environment than that provided by a packed seedbed.

Minimum tillage systems are in general use, and the major result has been the elimination of one or more of the secondary tillage operations. Some systems have eliminated all secondary tillage by using tractor and planter wheel tracks to make a firm seedbed in the loose soil left by the initial tillage operation. Strip processing is another minimum tillage system in which only a narrow strip of soil is tilled in front of each planter row. The strips are tilled with narrow blades, narrow rotary hoes, and/or narrow rototillers to kill weeds and firm the soil for the seed.

Minimum tillage systems that eliminate the initial tillage but retain the secondary operations are adapted only where the soil is loose, crop residues are light, and perennial weeds are not prevalent. Such conditions frequently occur in soybean stubble or in sandy soil.

Many minimum tillage systems combine planting with the initial or secondary tillage operations. These include plow-planting, strip processing, loose-ground listing, and the flexible one-way disk tiller with planter attachment which sometimes is called the disker or surflex drill.

Zero tillage systems involve no tillage before planting, and the tillage at planting is just enough to place the seed and agricultural chemicals at the desired depths. These systems leave all the crop residue at the soil surface, minimize soil erosion, and increase storage of water in the soil. Soil temperatures are lower, and some of the systems preclude practical use of some herbicides. Weeds must be controlled with adapted herbicides. Use of general vegetation killers that have no residual effects such as paraquat (1,1′-dimethyl-4,4′-bipyridinium ion) or glyphosate [N-(phosphonomethyl)glycine] before planting or before sunflower emergence is essential.

These general vegetation killers do not replace the need for herbicides normally used in sunflower production. Glyphosate can provide effective control of perennial weeds in zero tillage systems. Zero tillage implements include the lister, the till planter, and the fluted coulter mounted directly in front of each planter row. The lister as a zero tillage implement is effective only following cultivated or other crops where the soil is loose. Seed is planted in relatively cold soil in the lister furrows and heavy rains may cause flooding injury unless the soil drains quickly. Lister furrows on the contour provide excellent control of erosion. The till-planter and fluted coulter leave the field relatively level. Both the lister and till-planter move shattered seed from the previous crop into the interrow area. Consequently, weeds like volunteer sunflower and corn can be controlled by cultivation.

Sunflower planted in fields to be furrow irrigated usually is planted on raised beds constructed with listers. Ridge planting is an adaptation of this practice for continuous row cropping on cold, wet soils. Ridges are made with a lister or with a disk cultivator with disks adjusted to make ridges at the desired row width. Seed planted in the ridges is in warmer soil than that planted in a level field, and the young plants are less likely to be injured by flooding. Sunflower emergence from level plantings was only 83% that of ridged plantings (R. G. Robinson, unpublished data). But lodging averaged 39% in the ridged and 19% in the level plantings. Soil thrown around the sunflower stems by disk cultivation reduced lodging, but ridged plots had insufficient soil between the rows 76 cm apart to permit hilling.

Any of these tillage systems can be effective for sunflower production in specific environments. None has general or world-wide adaptation. If erosion is not a major concern, systems involving the moldboard plow are the least likely to have problems with nutrient availability, stand establishment, herbicide residues, weeds, insects, or diseases. Successful systems result in: (a) placement of seed in firm soil and agricultural chemicals at the proper depth, (b) no green vegetation on the field at either planting or emergence, (c) crop residues that will not interfere with cultivation, herbicide, or fertilizer practices planned for the field, and (d) limited soil erosion.

Elemental Composition and Requirements of Sunflower

Many elements are found in sunflower, but only C, H, O, P, K, N, S, Ca, Fe, Mg, B, Mn, Cu, Zn, Mo, and Cl are essential for growth and seed production.

The sources of carbon, hydrogen, and oxygen are air and water. These elements comprise 95.5% of the dry weight of mature sunflower plants and seed (Table 4). The remaining elements come from the soil matrix or fertilizer and are grouped into primary, secondary, and micronutrient classes. Those most commonly deficient on cropland are the primary elements—N, P, and K. Deficiencies of the secondary elements—Ca, Mg, and S—usually are regional. Deficiencies of micronutrients are sometimes regional but most often occur in parts of fields.

Table 4—Elemental content/ha of sunflower plants containing 10% moisture and yielding 2,000 kg of seed, 3,200 kg of stover, and 800 kg of roots. From Robinson (154).

Composition	Seed	Stover	Root
		kg/ha	
Water	200	320	80
O	918	1,460	375
H	114	181	46
C	689	1,096	282
N	48	31	3
P	7	2	4
K	11	45	4
S	3	6	0.4
Ca	2	32	3
Mg	4	17	1
B, Mo, Zn, Cu, Na, Fe, Mn	0.2	4	3
Other minerals	4	5	2

Table 5—Elemental composition of sunflower plants during five growth periods. From Robinson (149, 153).

	Stem-leaves-receptacle				Roots		Seed	
	Growth periods							
Element	Seedling	Heading	Flowering	Maturing	Heading	Maturing	Maturing	Mature
				%				
N	4.43	3.18	1.69	0.69	1.14	0.35	2.91	2.58
P	0.32	0.36	0.26	0.14	0.35	0.05	0.69	0.39
K	3.22	3.18	2.01	2.37	2.91	0.58	0.82	0.59
S	--	0.36	--	0.18	0.10	0.05	0.24	0.17
Ca	1.62	1.67	1.22	1.34	0.34	0.35	0.18	0.11
Mg	0.91	0.97	0.82	0.71	0.18	0.17	0.31	0.23
Na	0.004	0.04	0.05	0.04	0.29	0.24	<0.02	<0.02
				ppm				
Fe	400	200	70	80	1,793	900	50	33
Zn	51	28	25	12	39	23	46	48
Cu	10	11	11	9	17	8	17	13
Mo	2	2	2	2	22	4	1	6
Mn	79	56	35	31	74	45	18	14
B	38	46	39	39	21	12	18	14
Sr	85	92	66	77	26	32	5	3
Al	--	--	--	--	--	--	--	3

Sunflower seed and stover differ considerably in composition (Table 4). The seed contains much N and relatively little Ca, Mg, and micronutrients, whereas the stover is low in P and relatively high in K, N, Ca, Mg, and micronutrients.

Absorption of elements is rapid in relation to dry matter production during early growth. Consequently, elemental concentrations are high in young plants and decrease to maturity (Table 5). Stem/root ratios about 1 week prior to flowering averaged 2.7 but were much higher at maturity. Gachon (58) reported that 66% of the N, P, and Ca, 75% of the K, and 90% of the Mg were absorbed during the month before and month after the beginning of flowering.

Sunflower is not highly sensitive to soil pH. The crop is grown commercially on soils ranging in pH from 5.7 to over 8. The optimum pH depends on the properties of the soil; no pH is optimal for all soil conditions. The 6 to 7.2 range is an approximate optimum for many soils. Low pH (acid condition) may reduce activity of nonsymbiotic nitrogen-fixing bacteria, reduce availability of P, and increase absorption of Al and Mn to toxic amounts. High pH (alkaline condition) may increase availability of Na to toxic levels, reduce availability of P, Fe, and Mn, and reduce activity of desirable microflora. Thirteen sunflower cultivars grown for 23 days in soil of known Al toxicity and pH 4.1 produced from 11 to 42% as much top growth and 7 to 69% as much root growth as on the same soil adjusted to a pH of 5.5 with $CaCo_3$ (51). $CaCo_3$ decreased Al, Mn, Zn, and Ni in leaves and petioles and increased Ca, P, and Fe.

The seed is an important source of nutrients for the young plant. Sukhareva (185) reported that seeds of high P concentrations (1.64 to 1.87%) also were high in N and K and low in oil. Crops grown from these seeds were high in yield and yields increased with increased P/N ratios. Beleutsev (16) found that sunflower fertilized for two or three generations with 180 kg N and 106 kg P/ha produced seed of high N and P content. Crops grown from this seed yielded 150–290 kg/ha more seed and up to 152 kg/ha more oil than crops from ordinary seed. P alone was nearly as effective as both N and P. Immersion of seed in water for 24 hours before planting increased germination, growth, and seed yield (198). Soaked seeds germinated and seedlings emerged sooner than unsoaked (179). Pre-plant soaking also was successfully used to ensure uniform imbibition of water in a uniform experiment planted the same day at nine locations in North America (157).

Absorption of Nutrients

The sunflower root obtains water and essential elements from the soil. In a comparison of the nine possible root and stem graft combinations among three cultivars, any root adequately supplied nutrients to any stem genotype, but roots did not affect stem or seed phenotype (148).

Elements enter sunflower roots in ionic form and occasionally as complex organic salts. The availability of these elements is influenced by soil management and fertilizer additions. The ions enter the roots by diffusion, by exchange absorption, and by the action of carriers or ion-binding compounds. Although sunflower has extensively branched root systems, it is doubtful that roots directly contact more than a very small percentage of the soil volume. Consequently, direct interception of nutrients by roots represents only part of the total uptake. Barber (11) reported that nutrients move to the roots by mass flow and diffusion. Mass flow is movement of nutrients through the soil in the water flow resulting from transpiration. Nutrient absorption by the root reduces ionic concentration in the water. This results in a concentration gradient along which ions diffuse to the root.

Organic Matter Maintenance

Sunflower roots and stover are raw organic matter. These residues aerate the soil and reduce runoff and erosion. The soil microorganisms feed on the roots and stover which eventually become the humus fraction of the soil organic matter. Humus gives the soil a desirable physical structure and is a nutrient reserve that releases elements for crop growth throughout the season.

Maintenance of soil humus requires additional N beyond that contained in sunflower residues. Consequently, fertilization of the sunflower crop rotation should supply sufficient N for both the desired yield and efficient decomposition of crop residues to humus.

A 2,000 kg/ha yield of sunflower removes 48 kg of N and 689 kg of C in the seed, but 34 kg of N and 1378 kg of C remain in the stover and roots (Table 4). The stover and roots are decomposed by the microorganisms of the soil into humus, CO_2, and water. The 1,378 kg of C are the energy source for the microorganisms. The proportions of the 1,378 kg of C recovered as humus or lost from respiration as CO_2 depend upon availability of N. Without N, nearly all the C will eventually be lost as CO_2 because the C/N ratio of humus is 10/1. With ample N, 65% of the C will be lost as CO_2, because microorganisms retain only about 35% of the C consumed:

$$1378 \text{ kg sunflower C} + O_2 \rightarrow CO_2 \text{ (896 kg C)} + H_2O + 482 \text{ kg microflora C}$$

$$482 \text{ kg microflora C} + 34 \text{ kg sunflower N} \rightarrow 340 \text{ kg humus} + CO_2 \text{ (142 kg C)}$$

The 482 kg C in the microflora required 48.2 kg N to form humus (C/N ratio 10/1). Because only 34 kg N were available in the sunflower residues, 142 kg C were released as CO_2. Consequently, the N budget/ha for a 2,000 kg seed crop amounts to 48 kg for the seed, 31 kg for the stover, 3 kg for the roots and 14 kg for the soil microflora—a grand total of about 100 kg N.

Fertilizer Trials

Numerous research and popular publications show that sunflower responds to fertilizer. Most of the research was with N, but much research involved P and K.

Nitrogen is the most common element limiting sunflower yield (77, 154, 167, 182, 215). Nitrogen fertilizer reduced oil percentage of the seed (34, 77, 93, 131, 154, 174, 178, 182, 184, 191, 194, 199, 215). N increased protein of sunflower grown in solution culture to 38% of dry seed weight (34). The increase in N affected protein quality adversely because the predominant glutamic and arginine amino acids of sunflower protein increased while the more limiting acids—lysine, threonine, histidine, and glycine—decreased. In trials where both N and oil percentages were reported, decreases in oil percentage were accompanied by increases in N percentage. Leaf area was correlated positively with seed yield and was increased most by N-P-K and more by N than by P-K (36). N increased seeds/head (131).

Yields increased with increasing rates of N to 200 kg/ha, but 100 kg N + 44 kg P/ha gave slightly higher yields than 200 kg N (193). More publications in Europe and North America report yield increases from P than from K. Most of the fertilizer trials in North America, however, have been on soils high in K so the lack of response to K was expected. The relatively high K content/ha of sunflower crops compared with corn indicates that an ample supply of K is needed (Table 3). In Illinois, K increased yield from 465 kg/ha to 1,375 kg/ha (210). Oil concentration was not affected, or was increased, by P and K (93, 131, 164, 178, 182, 184, 215). P alone was the most effective fertilizer in increasing oil (182). Saturation of sunflower oil was decreased by P and K (164). The use of N-P fertilizer to hasten maturity was recommended in Canada (138).

Fertilizer Placement

Sunflower yields are highest on fertile soils, so the residual effects of fertilizer applications to other crops in the crop rotation benefit sunflower. Direct application of fertilizer to the sunflower crop may be through soil or foliar treatments.

Soil Application

Fertilizer application is accomplished by broadcasting prior to planting, row placement near or with the seed at the time of planting, or side-dressing between the rows after emergence and usually before the plants are 30 cm high. Sunflower seeds are sensitive to fertilizer salts so not more than 12 kg/ha of highly soluble fertilizer should contact the seed (204). The entire fertilizer application can be placed beside the sunflower row at planting time by using a trailing disk applicator that places the fertilizer 5 cm to the side and 5 cm below the sunflower seeds. Uptake efficiency of P by sunflower was similar to that of wheat and amounted to 24% in band placement at seed level, 20% in band placement below the seed, and 17% when mixed with the soil (206).

Broadcast or row placements of N are equally good because of the solubility of NO_3 and its mass flow in water moving to the sunflower roots as a result of transpiration. Nitrogen near the seed at planting time may slightly stimulate early growth above that caused by broadcast applications. Evidence is lacking, however, that such early stimulus results in higher yields.

Broadcast applications of P and K should be twice that of row placement on soils with low or medium levels of these elements. Most of the P and K is held on the surfaces of soil clay and organic matter. Consequently diffusion, not mass flow, accounts for movement of P and K to the roots. Barber (11) concluded that N, P, and K could be expected to diffuse to roots a distance of 1.00, 0.02, and 0.20 cm, respectively. Some of the P fertilizer should be placed near the row because of the insolubility and low mobility of P. Furthermore, broadcast applications of P should be in advance of plowing or secondary tillage because any downward movement of P is very slow.

The usual practice is to either apply all of the fertilizer in row placement at planting or to use a combination of row and broadcast placement. A common practice is row placement of P, K, and part of the N. The remaining N is broadcast before planting or sidedressed between the rows after emergence.

Foliar Application

Foliar sprays are used to correct deficiencies of micronutrients in many field and tree crops. Essential elements sprayed on the leaves are rapidly absorbed and used. Foliar application of Mn and Mo 10 days after flowering increased sunflower yields by 170 to 420 kg/ha in some years but, B, Cu, and Zn had no effect (171). A 0.005% solution of Zn, Cu, and Mo in the flowering and seed maturation stages increased oil percentage (191).

Foliar sprays of the primary elements have not been a practical alternative to soil application. Gaur et al. (62) obtained higher yields when 25% of the N was applied to the soil and 75% to the foliage than when all of the N was applied to the soil. Galgoczy (59) reported that a 1–3–3 ratio of N–P–K applied at rates of 24 kg/399 liters water/ha or 29 kg/1464 liters water/ha at the beginning of flowering increased seed yield 62%. A second treatment at the time of seed formation resulted in a total yield increase of 97%. Seed oil percentage also increased slightly. These treatments followed soil applications of 27 kg N, 30 kg P, and 30 kg K/ha on infertile sandy soils. Robinson (154) did not obtain significant yield increases from foliar sprays at various growth stages of sunflower on a fertile, silt loam soil using a 15–13–12 ratio of N–P–K at 7 kg/375 liters water/ha. Increasing rates to 14 kg/ha caused some killing of leaf tissue.

Garcia L. and Hanway (60) found that nutrient uptake from the soil did not prevent depletion of N, P, K, and S from soybean leaves during the seed filling period. This deficiency decreased photosynthesis and the leaves turned yellow. Two to four foliage sprays containing the same proportions of N, P, K, and S found in soybean seed kept the leaves healthy and increased number of seeds/plant and yield. Optimum total amounts of all applications amounted to 80N + 8P + 24K + 4S kg/ha. Materials used were urea, K polyphosphate, and K_2SO_4. Single applications should not exceed 20 kg N/ha to avoid leaf killing. This balanced feeding approach combined with the new form of P has not been attempted with sunflower. Results of this new advance in soybean research and the mature sunflower seed composition in Table 5 suggest that an N–P–K–S ratio of 13-2-3-1 is a logical start for foliar fertilization of sunflower.

Fertilizer Recommendations

Data from soil analyses and fertilizer rate experiments are used to formulate recommendations for fields with analyses similar to the experimental areas. Consequently, submission of soil samples to a testing laboratory is the first step to obtain fertilizer recommendations. Common tests are for NO_3-N (NO_3-specific ion activity electrode), extractable P (Bray's No. 1 extractant), and exchangeable K (NH4 acetate extraction). The upper 15 to 20 cm of soil are used for P and K tests and the upper 60 cm for NO_3-N tests.

Recommendations in North Dakota are based on yield goals varying from 1,100 to 3,400 kg/ha (Table 6). The amount of NO_3-N in the soil as indicated by soil test is subtracted from the recommendation (Column 2) for the desired yield goal (Column 1). The fertilizer N can be in urea, ammonia, or NO_3 forms. The simple method of subtracting the amounts of P and K already in the soil from the amount needed does not give a valid recommendation because of fixation and mobility characteristics of P and K. Consequently, the sums of the soil test kg/ha plus the recommended kg/ha differ among the soil test columns in Table 6. The P and K recommendations are for band placement near the seed. If P and K fertilizer is broadcast, the recommendations should be doubled for soils with less than 29 kg/ha of P or 236 kg/ha of K. At least 5 kg/ha of P should be applied with or near the seed on soils where P is needed.

The NO_3-N test has not yet proved to be a valid indicator of sunflower response to N in eastern and southern Minnesota. Consequently, N recommendations are based on soil organic matter levels and the previous crop (50). Recommendations are 20 kg/ha after fallow or legume sod, 65 kg/ha after small grain or soybean, and 110 kg/ha after corn or sugarbeet on land low to medium in organic matter. On land high in organic matter, N recommendations are 10 kg/ha after fallow or legume sod, 35 kg/ha after small grain or soybean, and 65 kg/ha after corn or sugarbeet. P recommendation is 20 kg/ha for soils of less than 22 kg/ha of extractable P. K recommendations are 55 kg/ha for soils testing less than 112 kg/ha of exchangeable K and 37 kg/ha for soils testing from 113 to 224 kg/ha of exchangeable K.

In Manitoba, recommendations for sunflower following other crops are 45 to 67 kg/ha of N and 10 to 15 kg/ha of P. K is not recommended for loam or clay soils, but 23 to 32 kg/ha are recommended for sandy soils (48).

Irrigation

Sunflower is commonly grown as a dryland crop. Crops that are more dependent upon irrigation or of greater economic return/ha are planted on the limited amount of irrigated land. Nonetheless sunflower responds to irrigation and yield increases exceeding 100% are common on droughty soils (151). All nutrient requirements of sunflower must be met for most effective

Table 6. N-P-K recommendations for various yield goals and soil analyses. Adapted from Wagner et al. (204).

Yield goal	NO₃-N in soil + fertilizer N	Extractable P in soil, kg/ha					Exchangeable K in soil, kg/ha				
		0–12	13–19	20–28	29–40	>40	0–56	57–134	135–235	236–336	>336
	kg/ha	recommended P, kg/ha					recommended K, kg/ha				
1,100	55	7	4	0	0	0	18	13	0	0	0
1,300	65	8	5	0	0	0	19	13	0	0	0
1,600	80	8	5	0	0	0	19	15	9	0	0
1,800	90	8	6	0	0	0	19	15	9	0	0
2,000	100	9	7	4	4	0	19	15	9	0	0
2,200	110	9	8	6	5	0	19	15	9	9	0
2,500	125	10	9	6	5	0	19	15	9	9	0
2,700	135	12	10	8	6	0	24	19	13	9	0
2,900	150	12	10	8	6	0	28	24	19	13	0
3,100	165	14	12	10	6	0	37	28	24	19	0
3,400	195	15	12	10	6	0	47	37	28	19	0

use of irrigation water. Robinson (151) found that irrigation, fertilizer, and fertilizer + irrigation increased yields 35, 72, and 474%, respectively. Vitkov and Gruev (201) reported highest yields from irrigation + fertilizer, but no increase from either irrigation or fertilizer alone.

Irrigation prevents premature senescence and allows normal maturation. Consequently seed from irrigated fields is often higher in oil percentage (37, 64, 76, 109, 174, 184, 194, 201). Irrigation reduced P, Mg, Mn, and B concentrations, increased Sr concentrations, and did not affect N, K, Ca, Na, Fe, Zn, Cu, or Mo concentrations in sunflower seed (R. G. Robinson, unpublished data). In other trials, N was decreased by irrigation (109 and R. G. Robinson, unpublished data). Irrigation improved protein quality because methionine, tryptophan, and lysine concentrations were highest at optimum soil moisture (126). Moisture stress at 40% field capacity increased concentrations of arginine and leucine.

Sunflower in northern USA is irrigated with overhead sprinklers. The center pivot system is the most popular. Irrigation is largely on sandy soils and in small amounts to supplement rainfall. In other areas where irrigation is the major source of water, surface irrigation using raised beds between irrigation furrows is practiced. Furrow irrigation had little effect on disease incidence and increased yield, but sprinkler irrigation adversely affected yield and quality by increasing *Septoria helianthi* Ell. and Keller, *Puccinia helianthi* Schw., and *Sclerotium bataticola* Taubenhaus in Europe (203). Irrigation of commercial fields has not caused serious losses from disease in the USA.

Effects of Moisture Stress

Sunflower uses water as a nutrient, as a solvent and carrier of other nutrients, and to keep sunflower cells turgid and growing. The water potential of sunflower foliage must be below that of the water supply for water to move into the leaves. Boyer (21, 22) found that leaf water potential was a good indicator of moisture stress in sunflower. Sunflower leaves with their petioles in pure water had water potentials of -1.5 to -2.5 bars. These may be normal levels for turgid leaves. But leaf growth stopped when leaf water potentials decreased to less than -3.5 bars. Since leaf osmotic potentials were about -10 bars, the minimum turgor required for cell enlargement was 6.5 bars, i.e. 10 minus 3.5. Stomates remain open at turgor pressures below 6.5 bars in sunflower, so low water availability may inhibit leaf enlargement before lack of water affects photosynthesis. In another study, net photosynthesis declined from 30 mg dm^{-2} hour^{-1} at -4 to -8 bars leaf water potential to less than 10 mg dm^{-2} hour^{-1} at -17 bars. But leaf enlargement dropped from 15% at -3 bars to 0 at -4 bars (22). Low leaf water potential affects transpiration and photosynthesis by closure of stomates and by imbibition of oxygen evolution by chloroplasts (25).

Most irrigated sunflower suffers moisture stress between irrigations. Consequently recovery from moisture stress is important for both irrigated and dryland sunflower. Leaf water potential and photosynthesis became normal within 3 to 5 hours after slight desiccation, but there was no recovery from leaf water potentials below -20 bars (24). It took several days for leaf water potentials to recovery from -13 to -19 bars, but photosynthesis recovery was incomplete in high light because stomates failed to open fully. Under low light, photosynthesis became normal within 15 hours after desiccation to -17 bars. Above -12 bars, photosynthesis recovery paralleled that of leaf water potential. Sionit and Kramer (180) found that the lower leaves wilted at a leaf water potential of -6 bars during any stage of growth.

Transpiration rate rapidly adapted to moisture stress and did not completely recover after the period of stress (163). Moisture stress greatly reduced leaf area index.

Auxin activity paralleled the growth rate, and auxin levels in sunflower were highest at 30 to 80% available soil water (69). Auxin level changed within 1 or 2 days when moisture level was reduced to or increased from 10% available water.

Yield losses are generally greatest when moisture stress occurs in the period from 20 days before to 20 days after flowering (132, 145, 154, 162). In other research, sunflower was more sensitive to stress at the slow elongation stage 25 days after planting than at the fast elongation, flowering, or ripening stages of growth (189). Stress at the elongation stages increased the oleic/linoleic acid ratio in the oil.

Soil Characteristics Affecting Irrigation Frequency

Soil consists of solid, liquid, and gaseous phases. The solid phase is a porous matrix. The pore spaces are occupied by the liquid phase (soil water) and by the gaseous phase (soil air). The solid phase of many soils is about 50% of the soil volume. The water and air phases occupy the remaining volume in constantly changing proportions. Water contents of soil often are expressed in terms of field capacities, permanent wilting points, and available water. Field capacity is the water percentage of a well drained soil, initially saturated, after several days of downward drainage. Soil particles hold water at tensions of about 0.1 bar on sandy soils to 0.3 bar on clay soils at field capacity (pF 2.5). At the permanent wilting point (pF 4.2), water is held at a tension of 15 bars and sunflower plants in soil at this moisture level will not recover turgidity when irrigated. Although 15 bars is the generally accepted wilting point for all soils, it is not absolute, and sunflower permanently wilted at −6 to −8 bars in one soil (205). The water available for sunflower growth is that between the field capacity and the permanent wilting point. About 90% of the available water is held at tensions of less than 1 bar in sand. Consequently, the soil moisture percentage of sandy soil can be reduced to near the wilting point percentage without severe injury to sunflower. Sunflower growth on fine textured soils is retarded at soil moisture percentages considerably above the wilting point because tension increases greatly as the soil moisture content decreases.

Water potential in the perirhizal soil around the individual roots may differ from that in the pararhizal soil that is free of roots. Faiz and Weatherley (47) found that the major resistance to water movement into transpiring sunflower plants was in the perirhizal soil. The reductions in water potential across the perirhizal soil and through the plant were about equal and amounted to about 6 bars.

Irrigation and rain are the major sources of soil water, but the two are not equivalent. Irrigation water contains variable amounts of salts in solution. Some salts such as $CaCO_3$, $CaSO_4$, K_2SO_4, or nitrates may be beneficial. For example, sands in central Minnesota are acid but irrigation water contains 18.5 kg $CaCO_3$/ha-cm to help neutralize the acidity. Other irrigation waters may have salts such as NaCl that have harmful effects on crop growth and soil structure.

Unger et al. (194) reported that irrigated sunflower used moisture to a depth of at least 183 cm, and dryland sunflower used moisture to a depth of at least 244 cm. Nearly all of the available water was used to a depth of 120 cm in Saskatchewan, and at normal populations and row spacings 20% of the roots were 0 to 13 cm deep, 52% were 13 to 25 cm deep, 15% were 25 to 38 cm deep, and 13% were below 38 cm (200). Although these data show that sunflower is capable of obtaining deeply stored water, soil conditions determine the desirable depth of irrigation. Gravel or impermeable layers near the surface limit the amount of water that should be applied at each irrigation. Soil texture is a major factor determining frequency of irrigation. On clay soil, highest yields were obtained by irrigation prior to emergence plus three subsequent irrigations of 8 cm each. But, the greatest efficiency of total (irrigation + rainfall + soil) water use was 62 kg seed/ha-cm of water from the irrigation prior to emergence plus an irrigation 2 weeks after the start of flowering (194). On sandy soil (Udorthentic Haploborolls, sandy, mixed), 16 irrigations of 2.6 cm each were needed to maintain available soil moisture above the 50% level (R. G. Robinson, unpublished data). Total water use amounted to 53.8 cm/ha, and the water use efficiency was

Table 7—Feel and appearance of soil with different amounts of water.

Available water	Sand	Loam, silt loam	Clay, silty clay loam
%			
0–25	Dry, loose, 0–2†	Powdery, slightly crusted, easily broken into powder, 0–4†	Hard, baked, cracked, 0–5†
25–50	Dry, will not form a ball when squeezed, 2–4†	Forms a crumbly ball when squeezed, 4–8†	Forms a ball when squeezed, 5–10†
50–75	Forms a crumbly ball when squeezed, 4–5†	Forms a ball, slightly slick, 8–12†	Slick, ribbons out between thumb and forefinger, 10–15†
75	Forms weak ball, breaks easily, 5†	Very pliable, slick, 12†	Slick feeling, easily ribbons out, 15†
100 (field capacity)	No free water appears on soil but wet outline of ball is left on hand		
	7†	16†	20†

† cm of available water/m depth of soil.

50 kg sunflower seed/ha-cm. The most efficient treatment used 33.2 cm/ha with an efficiency of 66 kg seed/ha-cm, but yields were 510 kg/ha below those of the 16-irrigation treatment.

Giminez et al. (64) obtained 3.11 metric tons/ha yields by irrigating to replace all evapotranspiration losses (52 cm/ha), 2.8 metric tons/ha by replacing 44% of evapotranspiration loss, 2.34 metric tons/ha by replacing 22% of evapotranspiration loss, and 1.8 metric tons/ha for no irrigation. Water use efficiencies were 60 kg seed/ha-cm for the 3.11 metric ton yield and 120 kg seed/ha-cm for the 2.8-metric ton yield. Milic (117) found that one or two irrigations of 6 to 7 cm/ha were optimum and gave water use efficiencies of 100 to 140 kg seed/ha-cm.

Irrigation Scheduling

Physiological measurements are direct indicators of irrigation need, but it is easier to use soil moisture or meteorological measurements. Soil moisture percentage is determined by sampling the soil at 30 cm increments to a depth below which few roots penetrate. The samples are weighed, dried at 100 C, weighed again, and soil moisture percentages calculated. Or the "feel" test (Table 7) can be used.

Indirect methods of determining soil moisture require less labor. Tensiometers function to moisture deficits of about −0.8 bar so are primarily useful on sandy soils. Usually they are placed at two depths. The porous clay tensiometer cup is filled with water. Water moves from the cup into the soil as the soil dries. The resulting vacuum in the tensiometer is indicated on a dial at the soil surface in 0.01 bar units. Another indirect method involves measurement of electrical resistance between electrodes embedded in blocks in the soil or at the end of probes placed in the soil. The atomic radiation method uses a probe containing a neutron emitter and a neutron-sensitive tube connected to an electronic meter, so readings can be taken rapidly in the field at various depths.

Table 8—Daily water use by sunflower plants in North Dakota. Adapted from Lundstrom and Stegman (105).

Weeks after emergence	Daily maximum temperatures, °C				
	10–15	16–20	21–26	27–31	32–37
			cm/day		
1	0.10	0.10	0.13	0.15	0.18
2	0.10	0.13	0.18	0.20	0.23
3	0.10	0.15	0.23	0.28	0.30
4	0.13	0.20	0.30	0.36	0.41
5	0.15	0.28	0.38	0.46	0.53
6	0.20	0.33	0.48	0.56	0.66
7	0.23	0.38	0.56	0.66	0.76
8	0.23	0.38	0.56	0.66	0.76
9	0.23	0.38	0.56	0.66	0.76
10	0.20	0.33	0.46	0.56	0.66
11	0.20	0.33	0.46	0.56	0.66
12	0.20	0.33	0.46	0.56	0.66
13	0.20	0.33	0.46	0.56	0.66
14	0.15	0.20	0.28	0.36	0.41
15	0.10	0.13	0.18	0.23	0.28
16	0.10	0.10	0.10	0.10	0.10

Meteorological methods of determining irrigation need usually involve temperature and evaporation factors. Irrigation to replace 50% or less of the water lost by evapotranspiration resulted in lower sunflower yields than did irrigation to replace 75 to 100% of evapotranspiration losses (64, 123). Irrigation to replace 100, 60, and 20% of water lost by evapotranspiration resulted in soil water potentials of 0, -2, and -8 bars, respectively (122). The pan evaporation method suggested for wheat involves a presowing irrigation followed by irrigation whenever the ratio of water applied to pan evaporation = 0.75 regardless of stage of growth (134). Another method involves a continuing daily record of soil moisture deficit based on sunflower age, maximum daily temperature, and rainfall or irrigation (105). The available moisture content of the soil to normal rooting depth is subtracted from the available moisture content at field capacity to give the initial soil moisture deficit. The soil moisture deficit is recalculated each day by subtracting rainfall or irrigation and adding the water used by sunflower (Table 8). Sunflower is irrigated whenever the soil moisture deficit reaches an uneconomic level. Considerable moisture is lost by evaporation from overhead sprinklers and interception by the sunflower foliage. Consequently, cm of water applied should be determined from rain gauges in the sunflower field rather than from output of the pump. This water loss often amounts to 10 to 20%.

Irrigation should maintain soil moisture to at least 50% of field capacity (2, 109, 197). In other research, soil moisture at 80% of field capacity was preferable to lower levels (76, 201). Soil moisture at 80% of field capacity at flowering stages and at 70% of field capacity at other times was optimum (37, 116, 201). A maintenance level of at least 50% available water in the soil often is considered optimum. This is higher than 50% field capacity, and the amount depends on the soil texture.

Irrigation should continue beyond anthesis. Seed yield increased 30% and oil yield 48% from irrigation 16 days after midflower, and seed yield increased 19% from irrigation 22 days after midflower (26). Both numbers and weights of seeds increased. No further improvement resulted from irrigation to physiologic maturity, 31 days after midflower.

WEED CONTROL

Sunflower is both a crop and a weed. As a weed, it is a strong competitor with any field crop. As a crop, it is damaged but rarely destroyed by weeds. Weed research in sunflower has focused on annual weed control. Control of perennial weeds is achieved in fallow periods between crops or in other crops in the crop rotation. Sunflower may be grown successfully in fields with perennial weeds if the weeds are kept under temporary control with row crop cultivation.

Weeds compete with sunflower for moisture, other nutrients in the soil, and occasionally for light and air. The amount of yield reduction varies greatly depending upon the weed species, weed density, time of weed and crop emergence, weather, and soil. Sunflower competes strongly with weeds but does not develop soil cover quickly enough to prevent weeds from establishing. Sunflower given neither cultivation nor herbicide treatment yielded 53% less than weed-free plots in North Dakota (124) and 20 to 50% less in Manitoba (32). Cultivation destroys the weeds between sunflower rows, but the weeds remaining in the row reduce yields. These yield reductions amounted to 12% (124), 20% (154), and 10% (32).

Chubb (32) used height as an index of sunflower growth and found that sunflower grew slowly during the first 2 weeks after emergence and then grew rapidly. Weeds emerging during the early period of slow growth should be more competitive with sunflower than those that emerge later. Yield loss was avoided by weed destruction within 3 weeks after sunflower emergence. If wild oat (*Avena fatua* L.) and sunflower emerged at the same time, however, the critical period for control was reduced to 2 weeks after emergence. Nalewaja (124) found that weeds destroyed within 2 weeks after sunflower emergence did not affect yield, but weeds allowed to grow for 1 month reduced sunflower yield 25%. Maintenance of weed-free rows for the rest of the season was not necessary because cultivation and competition provided by sunflower prevented yield reduction. Johnson (78) reported that one or two cultivations between 2 and 6 weeks after planting were needed to prevent yield loss from weed competition in Georgia.

Weed species exert varying degrees of competition on sunflower depending on their growth habit and density. Sunflower yield losses increased from 8 to 31% as wild mustard [*Brassica kaber* (DC.) L.C. Wheeler var. pinnatifida (Stokes) L. C. Wheeler] numbers/m of row increased from 3 to 52 (124). In the same trial yield losses increased from 6 to 12% as yellow foxtail [*Setaria lutescens* (Weigel) Hubb.] numbers/m of row increased from 20 to 98.

Mechanical Control

Almost all of the North American sunflower plantings are cultivated and/or harrowed for weed control. Probably half of the North American crop also is treated with herbicides. Official surveys show that 51, 40, 59, and 72% of Minnesota's 1973, 1974, 1975, and 1976 crops, respectively, were treated with herbicides. Seed yields in Georgia were as high from one cultivation as from plots given multiple cultivations or herbicide treatment. And weed control from one or two cultivations was as good as from herbicides alone or with cultivation (79). Sunflower yields were low for research plots in these trials. It is doubtful whether the results are applicable to areas where yields are higher or where greater weed problems occur.

Harrowing

Harrowing is of most importance when herbicides are not used, because it is the only practical control for weeds in the sunflower rows. It is also a useful supplement to herbicide treatment especially if the herbicide fails to control weeds. Weeds, that emerge before sunflower, can be killed by harrowing about 1 week after planting but before sunflower emerges. The implements used include the coil spring harrow, spike tooth harrow, rod weeder, and rotary hoe.

The implements used for postemergence harrowing include the coil spring harrow, spike tooth harrow, rotary hoe, and long-tined weeder. Stand losses usually are less than 5% if sunflower has at least two fully expanded leaves beyond the cotyledons and preferably four to six leaves when harrowed. In one trial stand losses were 9% at two-leaf, 7% at four-leaf, and less than 5% at later stages (R. G. Robinson, unpublished data). In other observations, seedlings that had just emerged withstood harrowing as well as those in the four to six leaf stage (138). Sunflower is damaged less by harrowing in dry afternoons when the plants are less turgid than in early morning. The basis of selectivity is differential plant size. Emerging weeds in the "white" stage are uprooted without injury to sunflower plants that have relatively large deep roots. Tractor speed and setting of the harrow or weighting of the rotary hoe to do the most damage to the weeds and the least to sunflower is accomplished on a "try and adjust" basis. Harrowing may be cross, diagonal, or with the rows, but excessive tractor wheel slippage on the rows will cause damage. Several harrowings often are needed, but they should be several days apart to allow sunflower to recover and weeds to sprout.

Postemergence harrowing is most practical on plowed fields. Crop residues left on the surface by some tillage systems are dragged with the harrow and uproot sunflower plants. The coil spring harrow and rotary hoe are the most commonly used implements for selective harrowing. The rotary

Table 9—Cultivation effects on lodging and yield of sunflower. From Robinson (154).

Type of cultivation	Lodging	Yield/ha
	%	kg
None	29	1,590
Scraping less than 3 cm deep	25	1,763
Disk or shovel—not hilled	26	1,723
Disk or shovel—hilled	16	1,885
L.S.D., 5%	5	176

hoe is an excellent weed killer and crust breaker at high speeds of about 16 km/hour. It is less likely to drag crop residues than the other implements. The coil spring harrow and the long tined weeder have vibrating tines which shake out weeds more effectively and do less damage to sunflower than the fixed spikes of a spike tooth harrow. These implements are most effective at 4 to 8 km/hour. The long tined weeder does the least damage to sunflower, but is useless on a crusted soil. The spike tooth harrow is the best of these implements for breaking soil crusts, but the most damaging to sunflower.

Interrow Cultivation

Interrow cultivation is the basic method of weed control in sunflower. It is still the only practical method for control of perennial weeds like quackgrass. Early emerging perennials can be controlled by "blind" cultivation of the planter marks before sunflower emergence. High-speed cultivation without danger of covering sunflower plants can be accomplished at this time.

Sunflower is normally cultivated once or twice and preferably not until it is 15 cm tall so that cultivation can be accomplished at reasonable speed without covering plants. Cultivators use sweeps, shovels, points, disk-hillers, vibra-shanks, and short rotary hoe gangs (rolling cultivator) to destroy weeds and direct the soil. The first cultivation, particularly with disk-hillers, generally directs the soil away from the sunflower row. The last cultivation should push soil into the row to cover weeds and support the stalks.

Weed control is the most common purpose of cultivation, but crust breakage and lodging prevention are very important under some conditions. Soil blown along the surface of crusted fields will destroy sunflower seedlings. Cultivation of the whole field or strips at right angles to the soil movement stops the mass movement of these foliage-cutting particles. Crust breakage by cultivation also affects nitrification, moisture loss, and infiltration of rainfall. Soil NO_3-N was higher in cultivated than in noncultivated corn plots (187). Cultivation of clay soils that shrink and crack deeply on drying reduces evaporative loss of moisture. Evaporation from the sidewalls of shrinkage cracks varied from 33 to 91% of that from surface soil (1). Cultivation increases the maximum infiltration capacity of crusted soils (18). Consequently runoff is reduced and moisture saved during heavy rains. However, cultivation does not have these effects under prolonged wet conditions because it does not affect the percolation rate through the soil profile.

Sunflower may be lodged by heavy rains and wind in midsummer. Such losses are reduced by hilling at the last cultivation. Heads were harvested by hand in the trials reported in Table 9; the relative yield advantage from hilling probably would be even greater from combine harvest.

Flame Cultivation

Flame cultivation, using propane or butane gas to burn small weeds in the rows, did not become a farm practice in sunflower as it did in cotton production. Effective weed control by flaming reduced sunflower yields (154). Flaming is expensive and does not save cultivation expense because the area between the rows is cultivated at the time of flaming, and the number of cultivations often is increased. Furthermore, weeds compete with sunflower before the crop is tall enough to be flamed. The development of equipment to produce water-shielded flame might permit better weed control and less injury (111). The growing scarcity of petroleum products, however, does not encourage adoption of flame cultivation.

Preplanting Tillage

Shallow secondary tillage one or more times during the month prior to seedbed preparation often has been suggested to stimulate weed seed germination. Comparison of tilled vs. untilled strips showed that no such stimulation occurred (158). Soil temperature and moisture determinations before planting failed to show that preplanting tillage made conditions more favorable for weed seed germination than conditions on undisturbed plowed soil. Weed counts, particularly foxtail (Setaria sp.), after crop emergence were higher on tilled than on untilled soil. Soil plowed in the fall usually settles into a compact mass by spring. Early tillage breaks up the mass into clods. Weed seeds trapped in these clods may not have good conditions for germination and therefore remain dormant until rain again settles the soil. Secondary tillage operations prior to final seedbed preparation should normally be confined to the minimum needed to incorporate herbicides, fertilizer, or insecticides.

If sunflower is planted late, early tillage may be needed to prevent dense weed growth that would make preparation of a good seedbed with secondary tillage implements impractical. Preplanting tillage for perennial weed control may be feasible in dry years if sunflower is planted late.

Herbicide Incorporation

Some herbicides must be mixed with the soil for good weed control and to avoid loss from volatilization or photochemical decomposition. Volatilization of herbicide and movement in soil water may further distribute herbicides to give homogeneous concentrations in the soil. Wauchope et al. (209) found that soil mixing during incorporation was not sufficient to change the irregular distribution of herbicide applied by fan sprays. They concluded that individual weeds "sample" much larger volumes of soil than did the researchers' 19 mm soil probe.

The disk harrow is the most commonly used implement for incorporating herbicides. Two passes to the same depth are necessary; the second either at right angles or in the same direction as the first. With two passes, the disk incorporated as deep as it cut but uniform distribution occurred to about 4 cm less than the depth of cut (28). Horizontal distribution of granules was less uniform than vertical and ranged from 6 to 16 kg/ha of an original 10 kg/ha application. High speed and large gang angles improved distribution. A smoothing harrow slightly wider than and pulled behind the disk improves distribution.

Vertical incorporation of granules by the field cultivator was inferior to that of the disk, and horizontal incorporation was much inferior (28). Incorporation of 10 kg/ha gave horizontal distributions ranging from 3.6 to 20.4 kg/ha. Usual recommendations suggest two passes at right angles to each other at a speed of at least 8 km/hour. Sweeps should be used on the shanks, and a spike tooth harrow slightly wider than the field cultivator should be pulled behind. Chisel plows are unsatisfactory for incorporation because their shanks are too far apart to give even distribution of herbicide.

The mulch treader which is a reversed rotary hoe was equal to the field cultivator in horizontal distribution of granules and equal to the disk in vertical distribution when operated 13 cm deep at 8 km/hour and with gang angles at least 16.5° (28). Only one pass of the mulch treader was made. Rolling cultivators also are satisfactory incorporators when operated at 10 to 13 km/hour. Two passes at right angles to each other should be made.

Equipment driven by a power take off (PTO) at speeds of less than 6 km/hour gives good distribution in one pass to nearly the depth of operation so it usually is operated about 6 cm deep.

Preplanting incorporation of in-the-row bands of herbicide can be accomplished with applicator and incorporation units mounted in front of the planter. Preemergence incorporation of narrow bands of herbicide over the planted row is not common, but satisfactory weed control can be achieved in bands as narrow as 26 cm. If granular herbicides are used, the soil must be level behind the planter press wheels to the width of the band, to prevent the granules from rolling into the press wheel tracks. Incorporation is accomplished with narrow treader or weeder sections or PTO-driven incorporators. Untreated soil must not be moved into the treated bands. Consequently early postemergence harrowing is avoided, and the first cultivation should be with disk hillers directing the soil away from the rows.

Herbicides

Selective herbicides used on the sunflower crop kill certain weed species without serious injury to sunflower. The first selective herbicides recommended for the sunflower crop in North America were propham (isopropyl carbanilate) in 1957 and diallate [S-(2,3-dichloroallyl) diisopropylthiocarbamate] for wild oat control in Canada (137). The first herbicide recommended for the sunflower crop in the USA was EPTC (S-ethyl dipropylthiocarbamate) in Minnesota in 1965 (15).

The selective action of herbicides occurs only when specified dosages are applied in specific ways in specific environments. Various public agencies approve these specifications which are printed on the container. The Environmental Protection Agency (EPA) regulates herbicide usage in the USA. Herbicides not registered by the EPA may not be sold in the USA, and those registered must be used only as specified on their container labels. Individual states, however, may occasionally approve herbicide labels on an emergency or temporary basis.

Recommended rates of application are given on the labels in terms of that specific formulation but in general publications application rates for all formulations are converted to a common basis. Consequently recommendations for wettable powders, suspensions, liquids, or granular products of varying concentrations are all expressed as kg/ha of active ingredient. Rates of application are expressed in terms of actual area sprayed. Thus 1 kg of a 1 kg/ha application would cover 3 ha if sunflower rows were 75 cm apart and if only 25-cm bands over the rows were sprayed. Sprayable formulations are applied in enough water for good distribution. The amounts vary from about 20 liters/ha by aircraft to over 400 liters/ha for ground sprayers. The normal range for ground sprayers is 90 to 180 liters/ha. Granular materials for dry application are sold in formulations with inert material to facilitate uniform distribution.

Herbicide applications before planting sunflower are called preplanting treatments. Applications between planting and emergence are called pre-emergence, and those applied after sunflower emergence are called postemergence treatments. At least 1 or 2 cm of water from either rain or irrigation is needed within a few days after planting for good performance of herbicides applied to the soil surface. Water also helps action of incorporated herbicides. Texture and organic matter of the soil influence performance of herbicides in the soil. Rates of application are adjusted to compensate for these soil characteristics. Rates generally increase with increasing amounts of organic matter and from sand to clay textures.

Effectiveness of postemergence herbicides is influenced by the growth stage of the weeds and sunflower and by the aerial environment at the time of spraying. Drought sometimes makes weeds more tolerant to herbicides applied postemergence.

Preplanting Herbicides

EPTC, trifluralin (α,α,α-trifluoro-2,6-dinitro-N,N-dipropyl-*p*-toluidine), dinitramine (N^4,N^4-diethyl-α,α,α-trifluoro-3,5-dinitrotoluene-2,4-diamine), and profluralin [N-(cyclopropylmethyl)-α,α,α-trifluoro-2,6-dinitro-N-propyl-*p*-toluidine] are used as preplanting incorporated herbicides.

These herbicides are most effective on grass weeds and least effective on certain dicot weeds like common ragweed (*Ambrosia artemisiifolia* L.) (Table 10). The herbicides differ in volatility and in recommendations for incorporation. EPTC should be incorporated within minutes after application, profluralin within 4 hours, trifluralin within 8 hours, and dinitramine within 24 hours. However, application of the herbicide and the first incorporation usually are done in one operation. Preparation of the seedbed accomplishes the final incorporation.

Trifluralin is used at rates of 0.56 kg/ha on sandy soil of less than 2% organic matter, 0.84 to 1.12 kg/ha on sandy soil of 2 to 5% organic matter, 0.84 kg/ha on loam and silt loam soils, and 1.12 kg/ha on clay soil or on soil of 5 to 10% organic matter. Applications may be made either in the fall preceding planting or in the spring before planting. Residues from trifluralin applications occasionally remain sufficiently toxic to injure winter wheat following sunflower (130, 195). Residual toxicity has occurred occasionally in spring wheat and sugarbeet following sunflower in drought years in the Red River Valley of North America. Moldboard plowing reduces phytotoxicity by diluting the residual herbicide with a large volume of soil.

EPTC is applied at 3.36 kg/ha in the spring. Fall applications require rates of 4.48 kg/ha on sand to 5.04 kg/ha on fine-textured soils. Dinitramine is applied at the time of seedbed preparation at 0.37 kg/ha on sandy soil, 0.56 kg/ha on loam and silt loam soils, and 0.75 kg/ha on clay soil. Profluralin rates range from 0.56 to 0.84 kg/ha on sandy soil, 0.84 to 1.12 kg/ha on loam and silt loam soils, and 1.12 kg/ha on clay soil. Residues from high rates of profluralin on sunflower in drought years may remain to injure the following sugarbeet crop.

Other preplanting herbicides of current interest but not used commercially on sunflower include: butralin [4-(1,1-dimethylethyl)-N-(1-methylpropyl)-2,6-dinitrobenzenamine], chlorpropham (isopropyl *m*-chlorocarbanilate), diallate, fluchloralin [N-(2-chloroethyl)-2,6-dinitro-N-propyl-4-(trifluoromethyl)aniline], pendimethalin [N-(1-ethylpropyl)-3,4-dimethyl-2,6-dinitrobenzenamine], and triallate [S-(2,3,3-trichloroallyl)diisopropylthiocarbamate].

Preemergence Herbicides

Chloramben is the only herbicide recommended for preemergence use on sunflower in North America. Application rates range from 2.24 kg/ha on sand or loam soils of less than 3% organic matter to 3.36 kg/ha on other soils. Chloramben is more effective on dicot and less effective on grass weeds than are the preplanting herbicides (Table 10). The common method of application is through a spray nozzle or granular applicator mounted behind each planter press wheel. Consequently, a band 25 to 30 cm wide is treated over each row thus reducing costs/ha for herbicide.

Application of preplanting herbicide followed by chloramben preemergence greatly increases the certainty of adequate weed control and controls more species than a single herbicide. The dependence of chloramben activity on moisture has led some farmers in dry areas to apply it in a mixture

Table 10—Weed control by herbicides recommended for sunflower in North America.

Weed species	Preplanting				Pre-emergence	Post-emergence
	Dinitra-mine	EPTC	Pro-fluralin	Tri-fluralin	Chloramben	Barban
Grasses	Weed control†					
Barnyardgrass [*Echinochloa crus-galli* (L.) Beauv.]	E	E	E	E	G	P
Crabgrass (*Digitaria* sp.)	E	E	E	E	G	P
Fall panicum (*Panicum dichotomiflorum* Michx.)	G	E	E	E	F	P
Foxtail (*Setaria* sp.)	E	E	E	E	G	P
Goosegrass [*Eleusine indica* (L.) Gaertn.)	G	E	E	E	F	P
Nutsedge (*Cyperus* sp.)	P	E	P	P	P	P
Proso millet (*Panicum miliaceum* L.)	E	G	F	G	G	P
Shattercane [*Sorghum bicolor* (L.) Moench]	F	E	F	F	P	P
Wildoat (*Avena fatua* L.)	P-F	G	F	F	F	G
Dicots						
Black nightshade (*Solanum nigrum* L.)	F	E	P	P	F	P
Common cocklebur (*Xanthium pensylvanicum* Wallr.)	P	P	P	P	P	P
Common lambsquarters (*Chenopodium album* L.)	G	G	G	G	E	P
Common ragweed (*Ambrosia artemisiifolia* L.)	P	F	P	P	G	P
Jimsonweed (*Datura stramonium* L.)	P	P	P	P	P	P
Kochia [*Kochia scoparia* (L.) Schrad.]	G	F	G	G	F	P
Pigweed (*Amaranthus* sp.)	G	G	G	G	E	P
Prickly sida (*Sida spinosa* L.)	F	P	P	P	F	P
Smartweed (*Polygonum* sp.)	F	P	P	P	G	P
Spurred anoda [*Anoda cristata* (L.) Schlecht.]	P	P	P	P	P-F	P
Velvetleaf (*Abutilon theophrasti* Medic.)	P	P	P	P	F	P
Wild buckwheat (*Polygonum convolvulus* L.)	F	F	F	F	G	P
Wild mustard [*Brassica kaber* (DC.) L.C. Wheeler]	P	P	P	P	F	P

† E = excellent, G = good, F = fair, P = poor.

with their preplanting herbicide. Results have been satisfactory but not as good as preemergence application followed by rain or irrigation.

Other preemergence herbicides that performed well in some research trials but are not recommended include: alachlor [2-chloro-2′,6′-diethyl-N-(methoxymethyl)acetanilide], chlorpropham, diallate, linuron [3-(3,4-dichlorophenyl)-1-methoxy-1-methylurea], metolachlor [2-chloro-N-(2-ethyl-6-methylphenyl)-N-(2-methoxy-1-methylethyl)acetamide], nitrofen (2,4-dichlorophenyl-*p*-nitrophenyl ether), norea [3-(hexahydro-4,7-methanoindan-5-yl)-1,1-dimethylurea], pendimethalin, prometryn [2,4-bis(isopropylamino)-6-(methylthio)-*s*-triazine], terbutryn [2-(*tert*-butylamino-4-(ethylamino)-6-(methylthio-*s*-triazine], and triallate. Diallate and triallate must be incorporated by harrowing to avoid loss from volatilization.

Fluoronitrofen (2,4-dichloro-6-fluorophenyl-*p*-nitrophenyl ether) killed many grass and dicot species including large seed dodder (*Cuscuta indecora* Choisy) at 0.5 kg/ha preemergence, but sunflower was not injured at 16 kg/ha (188). Trifluralin and nitrofen failed to kill dodder at 8 kg/ha. The sunflower hypocotyl is resistant to fluoronitrofen, and the plumule is protected by the seed coat and cotyledons when the seedling emerges through the herbicide layer. This accounts for its high tolerance to this highly phytotoxic herbicide.

Prometryn gave good results in many European trials and, mixed with trifluralin, gave sunflower yields equal to those of weed-free plots and killed weeds like ragweed that are resistant to trifluralin alone (196). But prometryn, terbutryn, and linuron have injured sunflower in other trials (79, 165, 166, and R. G. Robinson, unpublished data).

Postemergence Herbicides

Barban (4-chloro-2-butynyl *m*-chlorocarbanilate) is sprayed when most wild oat is in the 1.5-leaf stage and within 30 days of sunflower emergence at rates of 0.28 to 0.42 kg/ha.

Research is needed to develop good postemergence herbicides for the sunflower crop. Nitrofen controlled several grass and dicot species in research trials from 1966–70. Leaf injury was evident, but yields were not reduced (R. G. Robinson, unpublished data). Herbicides have reduced the number of cultivations, eliminated some harrowing, and improved weed control. But their greatest contribution to sunflower production has been increased assurance that good weed control can be achieved consistently. A postemergence herbicide of wide spectrum weed control and low toxicity to sunflower would be a great advance.

POLLINATION

The sunflower head (inflorescence) usually is composed of about 1,000 to 2,000 individual flowers joined to a common base (receptacle). The flowers around the circumference are ligulate ray flowers with neither stamens nor pistil. The remaining flowers are hermaphroditic and protandrous disk flowers. Anthesis of the disk flowers commences with the appearance of the ray flowers and proceeds from the periphery to the center of the head at a rate of 1 to 4 rows/day. Arrangement of the disk flowers is not in concentric circles as might be implied but in arcs radiating from the center of the head.

Flowering starts with the appearance of a tube partly exserted from the sympetalous corolla. The tube is formed by the five syngenesious anthers, and pollen is released on the inner surface of the tube. The style lengthens rapidly and forces the stigma through the tube. The two lobes of the stigma open outward and are receptive to pollen but out of reach of their own pollen initially. Although this largely prevents self pollination of individual flowers, flowers are exposed to pollen from other flowers on the same head by insects, wind, and gravity. The interval from exsertion of the anthers to fertilization of the flower is 2 days (136).

Cardon (30) found that pericarp development was normal but no ovules developed when heads were enclosed by bags during flowering. When he bagged one head of a multiheaded plant, ovules did not develop on the bagged head but grew normally on the open-pollinated heads that were exposed to pollen from other plants. This low self-compatibility of most of the older sunflower cultivars has long been recognized and used by plant breeders (83, 136). Sunflower cultivars with any degree of self-incompatibility must have interplant movement of pollen for maximum seed set. It is generally accepted that wind is of little importance in interplant transfer of pollen (136) and that bees are the primary pollen movers (113).

Yields of sunflower pollinated by bees usually have been much greater than those of sunflower grown in cages to exclude bees (4, 97, 128). It is important in such studies also to have either cages with bees or cages open to insects, so that the effect of the cages on sunflower yield can be determined.

Most (up to 97%) of the bees on sunflower are nectar collectors, but a small percentage collect only pollen and both types contribute to pollination (17, 53, 97). The nectar gathering bee inserts its tongue and head between the petals and anther tube in order to reach the nectar at the base of the corolla. As a result, the bee is dusted with pollen. About half of the bees pack the pollen into their corbiculae and the rest discard it, usually while hovering in the air (54). Individual disk flowers in the pistillate stage had more nectar and nectar solids than did those in the staminate or post-pistillate stages (52, 57). Highest nectar and nectar solids occurred about 1000 hour (52, 57). Bees visited sunflower most intensively from 0800 to 1000 hour in early plantings (98) and from 0600 to 1100 hour (142). These times coincide with the greatest supply of nectar. For late plantings, peak periods of bee activity did not occur but populations were lowest in the hottest part of the day (98).

Furgala et al. (57) found that hybrids secreted more nectar of higher solid content than did their cytoplasmic male sterile (cms) and restorer parents. Moreover cms lines had less nectar solids than did their pollen-producing maintainer lines. The relatively low nectar content of the parental lines suggests that seed production crossing fields may be less attractive to wild pollinators and in greater need of honey bee colonies than are commercial fields. Oilseed cultivars on the average produced more nectar of higher solid content than did nonoilseed cultivars. Conversely, Frank and Kurnik (52) reported a negative correlation between nectar content and seed oil percentage.

Sunflower honey is a valuable commodity. A. V. Mitchener found that a colony gained 47 kg in 15 days while the bees foraged on sunflower (55). Baculinschi (8) calculated a nectar yield of about 22 kg/ha for the flowering period. Guynn and Jaycox (68) obtained 36 kg honey/colony when 15 colonies were placed alongside a sunflower field of 18 ha.

It is difficult to have precise experimental control in measuring the yield increase due to honey bee colonies placed in sunflower fields. The overwhelming evidence indicates that bee colonies increased yields. Yields were 1,422 kg/ha near the apiary, 1,154 kg/ha at 61 m, 997 kg/ha at 122 m, and 896 kg/ha at 305 m (55). Large-scale experiments in Russia indicated that fields supplied with bees produced 79% more seed than fields without bees (133). With one colony/ha placed by the fields, bees worked well within a radius of 500 m of the apiary.

Seed from heads pollinated by bees was much heavier and of higher germination than seed from self pollinated heads (95). The usual effect of adequate pollination, however, is a decrease in achene weight of the filled achenes (97). The fewer pollinated achenes in partially pollinated heads often grow larger than those from fully pollinated heads.

The recommended number of bee colonies/ha varied from 20 colonies/ha within 3 km of the crop (128) to about 1 colony/ha (33, 181). A field of 50 thousand plants/ha with 1 thousand disk flowers/head will have 50 million disk flowers for pollination. Although pollination of an individual head is completed within 10 days, the total time for a field will be nearly 20 days. Consequently about 3 million disk flowers/day need pollination. If 6 bee visits/disk flower are needed (65), then 18 million flower-visits/day are necessary. If a bee visits 1,080 disk flowers/day (7), about 17,000 bees/ha would be needed to adequately pollinate 1 ha. One colony could supply this number if the bees are not attracted to other crops and do not have to travel too far. Consequently distribution of colonies through the field often is suggested (56, 113). Water supplies scattered through large fields also aid in bee distribution. Water holes about 0.5 × 1 m should be dug about 20 cm deep and lined with plastic sheeting to hold water (R. Bevis, personal communication).

Insecticides are rarely used during the pollination period of sunflower in northern USA and Canada. They are used to a considerable extent in western and southern USA. Insecticides used and their time of application should be treatments that will not reduce bee populations seriously.

There are not enough bee colonies available to pollinate the sunflower crop of North America. Consequently the primary use of honey bees is in the seed production crossing fields where greatest benefits are likely to occur. Highly self-compatible hybrids that will produce a full crop of seed when self pollinated are now available. The need for bees in fields planted to such hybrids is still uncertain. Questions remain as to the adequacy of pollen distribution on the heads without insect activity and the need for bee activity to insure pollination of late emerging stigmas. Research is needed to answer such questions.

The following list of observed pollinators of sunflower was provided by Dr. B. Furgala, Professor, Department of Entomology, University of Minnesota, St. Paul. Many species of Hymenoptera visit the sunflower. In addition to the honey bee, *Apis mellifera* L., the following species have been collected at various localities in Minnesota. (Determinations by F. D.

Parker, Systematic Entomology Laboratory, ARS-USDA, Beltsville, MD 20705.)

HYMENOPTERA

Halictidae

Halictus ligatus Say

Dufourea sp.

Agapostemon virescens (Fabr.)

Agapostemon texanus Cr.

Andrenidae

Andrena helianthi Robertson

Pseudopanurgus illinoiensis Robt.

Megachilidae

Megachile frigida Smith

Anthophoridae

Melissodes agilis Cresson

Melissodes rustica (Say)?

Svastra obliqua (Say)

Triepeolus helianthi (Robertson)

Apidae

Bombus nevadensis Cresson

Bombus ternarius Say

Bombus fervidus Fabr.

Bombus borealis Kirby

Bombus affinis Cresson

Bombus impatiens Cresson

Bombus griseocollis (DeGeer)

Many other species belonging to the five families have been collected, but determinations are still tentative.

LITERATURE CITED

1. Adams, J. E., and R. J. Hanks. 1964. Evaporation from soil shrinkage cracks. Proc. Soil Sci. Soc. Am. 28:281–284.

2. Albinet, E., V. Bulinaru, and M. Vasiliu. 1967-68. Influence of irrigation on sunflower production in the Prut valley. (In Romanian). Rep. Inst. Agron. "Ion Ionescu De La Brad", Iasi 1967–68:47–55.

3. Alessi, J., J. F. Power, and D. C. Zimmerman. 1977. Sunflower yield and water use as influenced by planting date, population, and row spacing. Agron. J. 69:465–469.

4. Alex, A. H. 1957. Pollination of some oilseed crops by honey bees. Texas Agric. Exp. Stn. Prog. Rep. 1960:1–5.

5. Allen, Jr., L. H. 1974. Model of light penetration into a wide-row crop. Agron. J. 66: 41–47.

6. Anderson, W. K. 1975. Maturation of sunflower. Austr. J. Exp. Agric. Anim. Husb. 15: 833–838.

7. Avetisyan, G. A. 1965. Bee pollination of agricultural crops. (In Russian). *His* Pchelovodstvo 209–248. Moskva, Kolos.

8. Baculinschi, H. 1957. Nectar production of sunflowers in the steppe region. (In Romanian). Apicultura 30:9–10.

9. Baeumer, K., and W. A. P. Bakermans. 1973. Zero-Tillage. Adv. Agron. 25:77–123.

10. Baker, G. O., and K. H. W. Klages. 1938. Crop rotation studies. Idaho Agric. Exp. Stn. Bull. 227:1–34.

11. Barber, S. A. 1976. Efficient fertilizer use. p. 13–29. In F. L. Patterson (ed.) Agronomic research for food. Spec. Publ. 26. Am. Soc. of Agron., Madison, Wis.

12. Barnes, S. 1938. Soil moisture and crop production under dryland conditions in western Canada. Can. Dep. Agric. Publ. 595. p. 1–43.

13. Bauer, H. 1970. Sowing dates for sunflower. Results obtained during 3 years of trials carried out at Manfredi (Cordoba). (In Spanish). Bol. Inf. Manisero 20:12–13.

14. Beard, B. H., P. Gyptmantasiri, and K. H. Ingebretsen. 1976. Planting date and spacing effects on yield, oil content, and fatty acid composition of sunflowers. Proc. Sunflower Forum (Fargo, N.D.) 1:29–30.

15. Behrens, R., and G. R. Miller. 1965. Cultural and chemical weed control in field crops 1966. Minnesota Agric. Ext. Folder 212. p. 10–11.

16. Beleutsev, D. N. 1973. The formation of high yielding sunflower seeds. (In Russian). Sel. Semenovod. 1:58–63.

17. Benedek, P., and S. Manninger. 1972. Pollinating insects of sunflower and the activity of honeybees on the crop. (In Hungarian). Novenytermeles 21:145–157.

18. Bertrand, A. R., and J. V. Mannering. 1963. Cut that crust—let in the rain. Crops Soils 15(6):13–14.

19. Bhatt, J. G., and K. N. Indirakutty. 1973. Salt uptake and salt tolerance by sunflower. Plant Soil 39:457–460.

20. Borisov, G. 1969. Effect of preceding crops on yields of sunflower grown on calcareous chernozem soil. (In Bulgarian). Rastenievud. Nauki 6(9):21–28.

21. Boyer, J. S. 1968. Relationship of water potential to growth of leaves. Plant Physiol. 43:1056–1062.

22. ————. 1970. Leaf enlargement and metabolic rates on corn, soybean, and sunflower at various leaf water potentials. Plant Physiol. 46:233–235.

23. ————. 1971. Resistances to water transport in soybean, bean, and sunflower. Crop Sci. 11:403–407.

24. ————. 1971. Recovery of photosynthesis in sunflower after a period of low leaf water potential. Plant Physiol. 47:816–820.

25. ————, and B. L. Bowen. 1970. Inhibition of oxygen evolution in chloroplasts isolated from leaves with low water potentials. Plant Physiol. 45:612–615.

26. Browne, C. L. 1977. Effect of date of final irrigation on yield and yield components of sunflowers in a semi-arid environment. Aust. J. Exp. Agric. Anim. Husb. 17:482–488.

27. Burrows, W. C., and W. E. Larson. 1962. Effect of amount of mulch on soil temperature and early growth of corn. Agron. J. 54:19–23.

28. Butler, B. J., and J. C. Siemens. 1972. Incorporation of surface-applied pesticides. Agrichemical Age 15:14.

29. Canvin, D. T. 1965. The effect of temperature on the oil content and fatty acid composition of the oils from several oilseed crops. Can. J. Bot. 43:63–69.

30. Cardon, P. V. 1922. Sunflower studies. J. Am. Soc. Agron. 14:69–72.

31. Carlson, E. C., P. F. Knowles, and J. E. Dille. 1972. Sunflower varietal resistance to sunflower moth larvae. California Agric. 26(6):11–13.

32. Chubb, W. O. 1975. Weed competition in sunflowers. Manitoba Agronomists Conf. (Winnipeg, Man.) Tech. Papers:129–132.

33. Cirnu, I., and E. Sanduleac. 1965. The economic efficiency of the sunflower (Helianthus annuus) pollination with the aid of the bees. (In Romanian). Lucr. Stiint. Stat. Cent. Seri. Apic. 5:37–51.

34. Coic, Y., C. Tendille, and C. Lesaint. 1972. The nitrogen nutrition of sunflower (Helianthus annuus): effect on yield and grain composition. (In French). Agrochimica 16:254–263.

35. Cotte, A. 1957. Sunflower used as a forage crop. Ann. Amelior. Plant. 7:349–357.

36. Cupina, T., and B. Josic. 1972. Relationship of photosynthetic activity and mineral nutrition in sunflower. (In Serbo-Croat). Savr. Poljopr. 20(7/8):5–12.

37. Delibaltov, I., and I. M. Ivanov. 1973. The effects of irrigation and fertilizer application on yields and quality of sunflower seed. (In Bulgarian). Rastenievud. Nauki 10(7):57–68.

38. Dompert, W. U., and H. Beringer. 1976. Effect of ripening, temperature and oxygen supply on the synthesis of unsaturated fatty acids and tocopherols in sunflower seeds. Z. Pflanzenernaehr. Bodenkd. 1976(2):157–167.

39. Doneen, L. D., and J. H. MacGillivray. 1943. Germination (emergence) of vegetable seed as affected by different soil moisture conditions. Plant Physiol. 18:524–529.
40. Dorrell, D. G. 1976. Chlorogenic acid content of sunflower seed flour as affected by seeding and harvest date. Can. J. Plant Sci. 56:901–905.
41. Downes, R. W. 1975. Physiological and environmental characteristics which affect the adaptation of sunflower cultivars. (In Spanish). Comun. INIA Prod. Veg. 1975(5):7–17.
42. Doyle, A. D. 1975. Influence of temperature and daylength on phenology of sunflower in the field. Aust. J. Exp. Agric. Anim. Husb. 15:88–92.
43. Durksen, D. 1975. Production of sunflowers in western Canada. p. 255–277. *In* J. T. Harapiak (ed.) Oilseed and pulse crops in western Canada—a symposium. Western Cooperative Fertilizers Ltd., Calgary, Alb.
44. Dyakov, A. B., A. Ya. Panchenko, and V. T. Piven. 1974. On the size of sunflower seeds and insect damage to them. (In Russian). Sel. Semenovod. 3:35–38.
45. El-Sharkawy, M. A., and J. D. Hesketh. 1964. Effects of temperature and water deficit on leaf photosynthetic rates of different species. Crop Sci. 4:514–518.
46. Enns, H., and J. E. Giesbrecht. 1971. Spacing of sunflowers, soybeans, and corn. Can. Dep. Agric. Canadex 140:20.
47. Faiz, S. M. A., and P. E. Weatherley. 1977. The location of the resistance to water movement in the soil supplying the roots of transpiring plants. New Phytol. 78:337–347.
48. Fehr, P., G. J. Racz, and R. J. Soper. 1974. Soil fertility and fertilizer practices. p. 108. *In* Principles and practices of commercial farming. Faculty of Agric., Univ. Manitoba, Winnipeg, Canada.
49. Fenelonova, T. M. 1973. Respiration of sunflower seeds during their development and ripening. (In Russian). Fiziol. Rast. 20:372–375.
50. Fenster, W. E., C. J. Overdahl, C. A. Simkins, and J. Grava. 1976. Guide to computer programmed soil test recommendations in Minnesota. Minnesota Agric. Ext. Spec. Rep. 1:1–36.
51. Foy, C. D., R. G. Orellana, J. W. Schwartz, and A. L. Fleming. 1974. Responses of sunflower genotypes to aluminum in acid soil and nutrient solution. Agron. J. 66:293–296.
52. Frank, J., and E. Kurnik. 1970. Investigations on nectar content of domestic and Soviet sunflower varieties. (In Hungarian). Delkeletdunantuli Mezogazd. Kiserl. Intez. Kozl. (Takarmanybazis) 10(2):65–69.
53. Free, J. B. 1964. The behaviour of honeybees on sunflowers (*Helianthus annuus* L.). J. Appl. Ecol. 1:19–27.
54. ————. 1970. Insect pollination of crops. Academic Press, New York. p. 322–331.
55. Furgala, B. 1954a. Honey bees increase seed yields of cultivated sunflowers. Gleanings Bee Cult. 82:532–534.
56. ————. 1954b. The effect of the honey bee, *Apis mellifera* L., on the seed set, yield and hybridization of the cultivated sunflower, *Helianthus annuus* L. Manitoba Entomol. Soc. Proc. 10:28–29.
57. ————, E. C. Mussen, D. M. Noetzel, and R. G. Robinson. 1976. Observations on nectar secretion in sunflowers. Proc. Sunflower Forum (Fargo, N.D.) 1:11–12.
58. Gachon, L. 1972. The kinetics of absorption of the major nutrient elements by sunflower. (In French). Ann. Agron. 23:547–566.
59. Galgoczy, J. 1967. The results of foliar fertilization of sunflower in a 2-year trial on the sandy soils of Szabolcs County. (In Hungarian). Kiserletugyi Kozl. A58(3):71–81.
60. Garcia L., R., and J. J. Hanway. 1976. Foliar fertilization of soybeans during the seed-filling period. Agron. J. 68:653–657.
61. Gaur, B. L., and D. S. Tomar. 1975. Effect of salinity on the germination of sunflower (*Helianthus annuus* L.) varieties. Sci. Cult. 41:429–430.
62. ————, ————, and H. S. Dungarwal. 1975. The efficacy of different methods of N application in sunflower. Indian J. Agron. 20:188–189.
63. Gibbs, M. 1970. The inhibition of photosynthesis by oxygen. Am. Sci. 58:634–640.
64. Giminez, O. R., J. Berengena H., and J. L. Muriel F. 1975. Effect of different rates of water application to a sunflower crop. (In Spanish). An. Inst. Nac. Invest. Agrar. Prod. Veg. 5:197–214.
65. Glukhov, M. M. 1955. Honey plants. (In Russian). Izd. 6, Perer. i Dop. Moskva, Gos. Izd-vo. Selkhoz Lit-ry. 512 p.
66. Goldsworthy, A. 1975. Photorespiration in relation to crop yield. *In* U.S. Gupta (ed.) Physiological aspects of dryland farming. Allanheld, Osmum and Co. Publishers Inc., Montclair, N.J.

67. Graves, C. R., J. R. Overton, T. McCutchen, B. N. Duck, and J. Connell. 1972. Production of sunflowers in Tennessee. Tennessee Agric. Exp. Stn. Bull. 494:1–24.

68. Guynn, G., and E. R. Jaycox. 1973. Observations on sunflower pollination in Illinois. Am. Bee J. 113:168–169.

69. Hartung, W., and J. Witt. 1968. Effect of soil moisture on the growth-substance content of *Anastatica hierochuntica* and *Helianthus annuus*. (In German). Flora Jena 157:603–614.

70. Hesketh, J. 1967. Enhancement of photosynthetic CO_2 assimilation in the absence of oxygen, as dependent upon species and temperature. Planta 76:371–374.

71. Hesketh, J. D., and D. N. Moss. 1963. Variation in the response of photosynthesis to light. Crop Sci. 3:107–110.

72. Hoag, B. K., and G. N. Geiszler. 1971. Sunflower rows to protect fallow from wind erosion. North Dakota Farm Res. 28(5):7–12.

73. Hoes, J. A. 1975. Sunflower diseases in western Canada. p. 425–438. *In* J. T. Harapiak (ed.) Oilseed and pulse crops in western Canada—a symposium. Western Co-Operative Fertilizers Ltd., Calgary, Alb.

74. ———, and H. C. Huang. 1976. Control of *Sclerotinia* basal stem rot of sunflower: a progress report. Proc. Sunflower Forum (Fargo, N.D.) 1:18–20.

75. Huisman, O. C., and L. J. Ashworth, Jr. 1976. Rotation ineffective as *Verticillium* control. California Agric. 30:14–15.

76. Ibragimov, A. 1973. Sunflower under irrigated conditions. (In Russian). Zemledelie 4:54–55.

77. Jocic, B. 1973. Relationships among leaf area, content of some elements in plant tissues, and yield of sunflower at various nutrient levels. (In Serbo-Croat). Savr. Poljopr. 21:57–68.

78. Johnson, B. J. 1971. Effect of weed competition on sunflowers. Weed Sci. 19:378–380.

79. ———. 1972. Weed control systems for sunflowers. Weed Sci. 20:261–264.

80. ———. 1972. Effect of artificial defoliation on sunflower yields and other characteristics. Agron. J. 64:688–689.

81. ———, and M. D. Jellum. 1972. Effect of planting date on sunflower yield, oil, and plant characteristics. Agron. J. 64:747–748.

82. ———, and W. H. Marchant. 1973. Sunflower research in Georgia. Georgia Agric. Exp. Stn. Res. Bull. 126:1–36.

83. Kalton, R. R. 1951. Efficiency of various bagging materials for effecting self fertilization of sunflowers. Agron. J. 43:328–331.

84. Kara, Y. M., and S. M. Lizenko. 1976. Phytomyza in Donetsk province. (In Russian). Zashch. Rast. 7:21.

85. Kawase, M. 1974. Role of ethylene in induction of flooding damage in sunflower. Physiol. Plant. 31:29–38.

86. ———. 1976. Ethylene accumulation in flooded plants. Physiol. Plant. 36:236–241.

87. Keefer, G. D., J. E. McAllister, E. S. Uridge, and B. W. Simpson. 1976. Time of planting effects on development, yield, and oil quality of irrigated sunflower. Aust. J. Exp. Agric. Anim. Husb. 16:417–422.

88. Khan, S. A., and A. Muhammed. 1974. Cultural practices for sunflowers 1. Plant populations. Agric. Pakistan 25:73–76.

89. Kinman, M. L., and F. R. Earle. 1964. Agronomic performance and chemical composition of the seed of sunflower hybrids and introduced varieties. Crop Sci. 4:417–420.

90. Klimov, S. 1968. Three-year results from a trial with different spacings for sunflowers at Ovce Polje. (In Romanian). Godisen Zb. Zemjod. Fak. Univ. Skopje 21:57–65.

91. Klimov, S. V., I. A. Shulgin, and A. A. Nichiporovich. 1975. On the energetic significance of leaf orientation in sunflower. (In Russian). Vestn. Mosk. Univ. 6(3):64–68.

92. Klyuka, V. I., and S. N. Tsurkani. 1975. Effect of temperature on growth and productivity of sunflower in controlled environment. (In Russian). Fiziol. Biokhim. Kult. Rast. 7:493–496.

93. Kordunyanu, P. V., and N. I. Belkin. 1970. Effect of mineral fertilizers on accumulation of fat and fractions of nitrogen in sunflower seed kernels. (In Russian). Agrokhimiya 6:77–83.

94. Krenzer, E. G., D. N. Moss, and R. K. Crookston. 1975. Carbon dioxide compensation points of flowering plants. Plant Physiol. 56:194–206.

95. Kushnir, L. G. 1957. Economic effectiveness of pollination of sunflower by bees. (In Russian). Pchelovodstvo 34(7):23-27.

96. Lammerink, J., and D. A. C. Stewart. 1974. Effects of varying sowing dates on sunflower cultivars. Proc. Agron. Soc. N. Z. 4:9-12.

97. Langridge, D. F., and R. D. Goodman. 1974. A study on pollination of sunflowers (*Helianthus annuus*). Aust. J. Exp. Agric. Anim. Husb. 14:201-204.

98. Lehman, W. F., F. E. Robinson, P. F. Knowles, and R. A. Flock. 1973. Sunflowers in the desert valley areas of southern California. California Agric. 27:12-14.

99. Leocov, M. 1974. Effect of density and fertilizer on the yield and quality of sunflower. (In Romanian). Analele Stiintifice al Universitatii Al. I. Cuza din Iasi, Romania, II. a 20: 155-162.

100. Leon, L. M. 1975. Sowing date of sunflower on irrigated land. (In Spanish). Comun. INIA Prod.Veg. 1975(5):19-26.

101. List, R. J. 1966. Smithsonian meteorological tables. Smithsonian Institution, Washington, D.C. p. 506-510.

102. Lopez, M. de L. 1972. Effect of the date of planting and the row spacing on sunflower crop in Andalucia. Proc. Fifth Int. Sunflower Conf. (Clermont-Ferrand, France):133-136.

103. Luciano, A., and M. Davreux. 1967. Production of sunflowers in Argentina. (In Spanish). Pergamino Agric. Exp. Stn. Tech. Publ. 37. p. 1-53.

104. Ludwig, L. J., and D. T. Canvin. 1971. The rate of photorespiration during photosynthesis and the relationship of the substrate of light respiration to the products of photosynthesis in sunflower leaves. Plant Physiol. 48:712-719.

105. Lundstrom, D. R., and E. C. Stegman. 1976. Irrigation scheduling by the checkbook method. North Dakota Agric. Ext. Service. Mimeo.

106. Lungu, I. 1971. Investigations of soil, seed, plant density and cultural practices, factors which affect the yield of sunflowers. (In Romanian). Probl. Agric. 23:55-63.

107. Majid, F. Z., and M. M. Rahman. 1975. Sunflower as oil seed crop in Bangladesh. 4. Effects of vernalization on a local variety (H-67-6) of sunflower. Bangladesh J. Sci. Res. 10(1/2)142-147.

108. Malykin, I. I., and B. F. Leshcheok. 1973. Control of broomrape in sunflower. (In Romanian). Zashch. Rast. 1973(12):17.

109. Marty, J. R., J. Puech, J. Decau, and C. Maertens. 1972. Effects of irrigation on the yield and quality of sunflower. (In French). Proc. Fifth Int. Sunflower Conf. (Clermont-Ferrand, France):46-53.

110. Massey, J. H. 1971. Effects of nitrogen rates and plant spacing on sunflower seed yields and other characteristics. Agron. J. 63:137-138.

111. Mathews, E. J., and H. Smith, Jr. 1971. Water-shielded, high-speed flame weeding of cotton. Proc. So. Weed Sci. Soc. 24:393-398.

112. McCalla, T. M., and T. J. Army. 1961. Stubble mulch farming. Adv. Agron. 13:125-196.

113. McGregor, S. E. 1976. Insect pollination of cultivated crop plants. USDA Agric. Handb. 496:1-411. U.S. Government Printing Office, Washington, D.C.

114. Mian, A. L., and M. A. Gaffer. 1971. Effect of size of plant population and level of fertilization on the seed yield of sunflower. Sci. Ind. 8:264-268.

115. Mihalyfalvy, I. 1962. Water requirement of sunflower grown as a second crop for green manuring. (In Hungarian). Novenytermeles 11:101-108.

116. Mikhov, I. 1974. On the irrigation regime for sunflower under the conditions of southeastern Bulgaria. (In Bulgarian). Rast. Nauki 11(3):99-109.

117. Milic, M. 1967. Irrigation rate and fertilizer application to sunflower in Metohija. Savr. Poljopr. 15:973-980.

118. Morrison, W. H. 1975. Effects of refining and bleaching on oxidative stability of sunflowerseed oil. J. Am. Oil Chem. Soc. 52:522-525.

119. Moss, D. N. 1966. Respiration of leaves in light and darkness. Crop Sci. 6:351-354.

120. Muhammad, S., and M. I. Makhdum. 1973. Effect of soil salinity on the composition of oil and amino acid and on the oil content of sunflower seed. Pakistan J. Agric. Sci. 10: 71-76.

121. Muresan, T. 1972. The sunflower in Romania. (In Italian). Ital. Agric. 109(1):45-68.

122. Muriel, J. L., and R. W. Downes. 1975. Effect of drought periods during different growth phases of sunflower (*Helianthus annuus* L.) plants grown in a glasshouse. (In Spanish). Comun. INIA Prod. Veg. 1975(5):47-54.

123. Muriel, T. L., R. Jiminez, and J. Berengena. 1975. Response of sunflower (*Helianthus annuus* L.) to different regimes of soil moisture and effect of irrigation during critical development periods. (In Spanish). Comun. INIA Prod. Veg. 1975(5):37–45.

124. Nalewaja, J. D., D. M. Collins, and C. M. Swallers. 1972. Weeds in sunflowers. North Dakota Farm Res. 29(6):3–6.

125. Orlowska, T. 1969. Sunflower as a preceding crop for winter cereals. (In Polish). Roczn. Nauk Roln. 96(1):151–162.

126. Osik, N. S. 1975. Moisture supply and amino acid composition of proteins in sunflower. (In Russian). Doklady Vaskhnil 3:25–27.

127. Pacucci, G., and F. Martignano. 1975. Effect of sowing density on yield and some bio-agronomic characteristics of tall and dwarf sunflower cultivars. (In Italian). Riv. Agron. 9(2/3):180–186.

128. Palmer-Jones, T., and I. W. Forster. 1975. Observations on the pollination of sunflowers. N.Z. J. Exp. Agric. 3:95–97.

129. Parker, D. T., and W. E. Larson. 1965. Effect of tillage on corn nutrition. Crops Soils 17(4):15–17.

130. Petkova, P. 1971. Use of trifluralin for weed control in sunflowers. (In Bulgarian). Rast. Nauki 8(7):157–168.

131. Petrova, M., and I. Kolev. 1976. Fertilization of sunflower on slightly leached chernozem soils in Dobrudzha region 1. Effect of fertilizer rates on yields. (In Bulgarian). Pochvozn. Agrokhim. 11:50–60.

132. Pirjol, L., C. I. Milica, and V. Vranceanu. 1972. A study of drought resistance in sunflower at different growth stages. Proc. Fifth Int. Sunflower Conf., (Clermont-Ferrand, France):36–45.

133. Ponomareva, E. G. 1958. Results of mass experiments on the use of bees as pollinators of entomophilic agricultural plants. (In Russian). Byull. Nauchno-Tekh. Inf. Nauchno-Issled. Inst. Pchelovod 3–4:27–28.

134. Prihar, S. S., K. L. Khera, K. S. Sandhu, and B. S. Sandhu. 1976. Comparison of irrigation schedules based on pan evaporation and growth stages in winter wheat. Agron. J. 68:650–653.

135. Puech, J., P. Lencrerot, and J. Decau. 1975. Effect of a reduction in light intensity on the overall photosynthesis of a crop of sunflower (*Helianthus annuus*). Effect on the oil and protein production in the seed. C. R. Acad. Sci. Ser. D 281:387–390.

136. Putt, E. D. 1940. Observations on morphological characters and flowering processes in the sunflower (*Helianthus annuus* L.). Sci. Agric. 21(4):167–179.

137. ————. 1963. Sunflowers review article. Field Crop Abstr. 16(1):1–6.

138. ————. 1972. Sunflower seed production. Canada Dep. Agric. Publ. 1019. p. 1–30.

139. ————, and J. A. Fehr. 1951. Effect of plant spacings, row spacings, and number of plants per hill on Advance hybrid sunflower. Sci. Agric. 31:480–491.

140. Radenovic, B. 1972. Effect of area per plant on seed yield and oil content of sunflower on smonitza soils in Kosovo. (In Serbo-Croat). Savr. Poljopr. 20(1):77–82.

141. Radford, B. J. 1977. Influence of size of achenes sown and depth of sowing on growth and yield of dryland oilseed sunflower (*Helianthus annuus*) on the Darling Downs. Aust. J. Exp. Agric. Anim. Husb. 17:489–494.

142. Rangarajan, A. V., N. R. Mahadevan, and S. Iyemperumal. 1974. Note on time of visit of pollinating honey bees to sunflower. India J. Agric. Sci. 44:66–67.

143. Renea, S., and F. Olteanu. 1959–60. Some cultivation measures for the growing of sunflower under irrigation. French summary. Anal. Inst. Cercetari Agron. 27B:33–41.

144. Rice, E. L. 1974. Allelopathy. p. 41–48. Academic Press, New York.

145. Robelin, M. 1967. Effects and after-effects of drought on the growth and yield of sunflower. (In French). Ann. Agron. 18:579–599.

146. Robertson, J. A. 1975. Use of sunflower seed in food products. Critical Reviews in Food Sci. and Nutr. 6:201–240. CRC Press Inc., Cleveland.

147. Robinson, R. G. 1966. Sunflower-soybean and grain sorghum-corn rotations versus monoculture. Agron. J. 58:475–477.

148. ————. 1968. Stem or root: which control plant characteristics? Crops Soils 21:20–21.

149. ————. 1970. Sunflower date of planting and chemical composition at various growth stages. Agron. J. 62:665–666.

150. ————. 1970. Management of land diverted from crop production. II. Annual cover crops and fallow. Agron. J. 62:770–772.

151. ————. 1971. New crops for irrigated sandy soils. Minnesota Sci. 27(3):10–11.

152. ————. 1971. Sunflower phenology—year, variety, and date of planting effects on day and growing degree-day summations. Crop Sci. 11:635–638.

153. ————. 1973. Elemental composition and response to nitrogen of sunflower and corn. Agron. J. 65:318–320.

154. ————. 1973. The sunflower crop in Minnesota. Minnesota Agric. Ext. Bull. 299: 1–28.

155. ————. 1974. Sunflower performance relative to size and weight of achenes planted. Crop Sci. 14:616–618.

156. ————. 1975. Effect of row direction on sunflowers. Agron. J. 67:93–94.

157. ————, L. A. Bernat, H. A. Geise, F. K. Johnson, M. L. Kinman, E. L. Mader, R. M. Oswalt, E. D. Putt, C. M. Swallers, and J. H. Williams. 1967. Sunflower development at latitudes ranging from 31 to 49 degrees. Crop Sci. 7:134–136.

158. ————, and R. S. Dunham. 1956. Pre-planting tillage for weed control in soybeans. Agron. J. 48:493–495.

159. ————, F. K. Johnson, and O. C. Soine. 1967. The sunflower crop in Minnesota. Minnesota Agric. Ext. Bull. 299:1–32.

160. ————, D. L. Rabas, L. J. Smith, D. D. Warnes, J. H. Ford, and W. E. Lueschen. 1976. Sunflower population, row width, and row direction. Minnesota Agric. Exp. Sta. Misc. Rep. 141:1–24.

161. ————, and O. C. Soine. 1961. Sunflower production in Minnesota. Minnesota Agric. Ext. Bull. 299:1–12.

162. Rollier, M. 1975. Study of water use in sunflower. (In French). C.E.T.I.O.M. Info. Techn. 44:29–44.

163. ————, and J. G. Pierre. 1975. Study of the water requirements of sunflower. 3. Laboratory studies in 1972 and 1973. (In French). C.E.T.I.O.M. Info. Techn. 46:1–18.

164. ————, S. Trocme, and R. Boniface. 1975. Observations on the application of phosphorus-potassium fertilizer to sunflower. (In French). C.E.T.I.O.M. Info. Techn. 47: 29–37.

165. Sameni, A. M., M. Maftoun, S. M. Hojjati, and B. Sheibany. 1976. Effect of fertilizer -N and herbicides on the growth and N content of sunflower. Agron. J. 68:285–288.

166. Sarpe, N., F. Olteanu, and V. Apostol. 1971. Effectiveness of prometryne applied in bands with SPC-6 machines and some reference to phytotoxic effects in 1965–71 with field-grown sunflower in S. Romania. (In Romanian). Anal. Inst. Cercetari Pentru Cereale Plant. Tehn. Fundulea, B. 39:253–263.

167. Saumell, H. 1972. Sunflower fertilizer application. (In Spanish). Bol. Info. Manisero 27:8–9.

168. Schreiber, K., and L. J. LaCroix. 1967. Manufacture of coated seed with delayed germination. Can. J. Plant Sci. 47:455–457.

169. Schuster, W., and R. Boye. 1971. The influence of temperature and daylength on different sunflower varieties under controlled climate conditions and in the open. Z. Pflanzenzuecht. 65:151–176.

170. ————, and R. Boye. 1971. Productivity of sunflowers with marked physiological differences. 2. Seed yield. Z. Acker Pflanzenbau. 133:321–334.

171. Semikhnenko, P. G., T. E. Guseva, and A. N. Riger. 1973. Effect of trace elements on yield of sunflower grown in the Krasnodar region. (In Russian). Byull. Nauchno-Tekh. Info. Maslichn. Kult. 1973(3):25–27.

172. Shantz, H. L., and L. N. Piemeisel. 1927. The water requirements of plants at Akron, Colo. J. Agric. Res. 34:1093–1190.

173. Shell, G. S. G., and A. R. G. Lang. 1976. Movements of sunflower leaves over a 24-h period. Agric. Meteorol. 16:161–170.

174. Simanskii, N. K. 1961. The effect of fertilizers on yield and oil content of sunflower seeds. (In Russian). Agrobiologija 6(132):849–853.

175. Simeonov, B., and V. Velchev. 1970. Dates and methods of applying superphosphate to sunflower grown on slightly leached chernozem soil in Dobrudzha. Pochvozn. Agrokhim. 5(6):37–44.

176. Sin, G., and C. Pintilie. 1971. Effect of preceding crop on sunflower. (In Romanian). An. Inst. Cercetari Pentru Cereale Plant. Tehn. Fundulea. B.39:245–252.

177. ————, C. Pintilie, and H. Iliescu. 1971. Relationships between rotation, soil moisture dynamics, plant health, and yield of sunflower. (In Romanian). An. Inst. Cercetari Pentru Cereale Plant. Tehn. Fundulea. B.39:265–267.

178. Singh, P. P., Y. K. Sharma, and P. K. Kaushal. 1973. Effect of varying levels of nitrogen and phosphorus on the yield and quality of sunflower. JNKVV Res. J. 7(3):134–136.

179. Singh, R. P., and D. B. Wilson. 1974. A note on sowing soaked and unsoaked sunflower and safflower seeds at various depths and at two root zone temperatures. Ann. Arid. Zone 13:364–369.

180. Sionit, N., and P. J. Kramer. 1976. Water potential and stomatal resistance of sunflower and soybean subjected to water stress during various growth stages. Plant Physiol. 58: 537–540.

181. Smith, H., P. Pankiw, and G. Kreutzer. 1971. Honey bee pollination in Manitoba. Manitoba Dep. Agric. Publ. 525. p. 1–16.

182. Stamboliev, M., and G. Borisov. 1975. Effect and interaction of nitrogen, phosphorus, and potassium on sunflower on calcareous chernozem soil. (In Bulgarian). Rast. Nauki 12(6):102–107.

183. Stoyanova, I. 1969. Determining optimum sowing depth for sunflower. (In Bulgarian). Rast. Nauki 6(2):27–32.

184. ————, M. Petrova, B. Simeonov, P. I. Ivanov, and I. Dimitrov. 1975. Effect of some factors on contents of oil and proteins in sunflower seeds. (In Bulgarian). Rast. Nauki 12(9):25–29.

185. Sukhareva, O. N. 1974. The yielding ability of sunflower seeds in relation to contents of nitrogen, phosphorus, and potassium in them. (In Romanian). Byull. Nauchno-Tekh. Info. Maslichn. Kult. 1974:23–28.

186. Swallers, C. M., and N. Fick. 1973. Performance of large-seeded sunflowers at three plant spacings. North Dakota Farm Res. 31(1):15–16.

187. Swanson, C. L. W., and H. G. M. Jacobson. 1957. Effect of adequate nutrient supply and varying conditions of cultivation, weed control, and moisture supply on soil structure and corn yields. Agron. J. 49:571–577.

188. Takematsu, T., M. Konnai, and Y. Takeuchi. 1975. CFNP a selective herbicide for sunflower. Weed Sci. 23:57–60.

189. Talha, M., and F. Osman. 1975. Effect of soil water stress on water economy and oil composition in sunflower (*Helianthus annuus* L.). J. Agric. Sci. 84:49–56.

190. Tanner, C. B., A. E. Peterson, and J. R. Love. 1960. Radiant energy exchange in a corn field. Agron. J. 52:373–379.

191. Timoshenko, A. G. 1972. Response of sunflower to the application of fertilizers (In Russian). Udobr. Selskokhozyaistvennykh Kult. Moldavii. (Kishinev, Moldavian SSR, "Kartya Moldovyaske") 1972:197–201.

192. Togari, Y., Y. Murata, and T. Saeki. 1969. Photosynthesis and utilization of solar energy. Level I experiments. Report III. JIBP/PP—Photosynthesis. Faculty Agriculture, Tokyo University.

193. Tomov, T. M. 1976. Effect of fertilizers on sunflower yield. (In Bulgarian). Pochvozn. Agrokhim. 11:101–108.

194. Unger, P. W., R. R. Allen, O. R. Jones, A. C. Mathers, and B. A. Stewart. 1976. Sunflower research in the southern high plains—a progress report. Proc. Sunflower Forum (Fargo, N.D.) 1:24–29.

195. Vasilev, D. S., V. A. Degtyarenko, and R. G. Chanukvadze. 1976. Herbicidal activity of Treflan and Nitrofor applied to sunflower. (In Romanian). Zashch. Rast. 2:29.

196. ————, V. A. Degtyarenko, R. G. Chanukvadze, and L. A. Baranova. 1974. The use of treflan and prometryne mixture for sunflower. (In Romanian). Khim. Selsk. Khoz. 12(9):47–49.

197. Vasiliu, M., E. Pascaru, and M. Vasiliu. 1969. Effect of fertilizer and irrigation regime on the yield of sunflower. (In Romanian). An. Inst. Cercetari Pentru Cereale Plant. Tehn. Fundulea B. 37:335–342.

198. Venugopal, N., and N. M. Patil. 1976. Effect of pre-soaking of sunflower seeds in water on the growth and yield of sunflower. Mysore J. Agric. Sci. 10(1):34–37.

199. Vicentini, G., and G. Anelli. 1973. The effect of various rates of N and P_2O_5 on the productivity and oil composition of sunflower (*Helianthus annuus* L.). (In Italian). Agric. Ital. (Pisa) 73(3):175–187.

200. Vijayalakshmi, K., N. K. Sanghi, W. L. Pelton, and C. H. Anderson. 1975. Effect of plant populations and row spacings on sunflower agronomy. Can. J. Plant Sci. 55:491–499.

201. Vitkov, M., and Ts. Gruev. 1973. Relationship between irrigation and fertilizer for sunflower. (In Bulgarian). Rast. Nauki 10:81–92.

202. Vogel, S. L. 1954. Tillage machinery. North Dakota Agric. Ext. Circ AE-49:1–12.
203. Vucic, N., M. Acimovic, J. Vucic, and B. Jocic. 1965. Effect of irrigation method on the intensity of disease occurrence, yield and quality of sunflower in 1964. (In Serbo-Croat). Savr. Poljopr. 13:905–912.
204. Wagner, D. F., W. C. Dahnke, D. E. Zimmer, J. C. Zubriski, and E. H. Vasey. 1975. Fertilizing sunflowers. North Dakota Agric. Ext. Circ. S & F3:1–4.
205. Waisel, Y., and G. Pollak. 1969. Estimation of water stresses in the root zone by the double-root system technique. Israel J. Bot. 18:123–128.
206. Warder, F. G., and K. Vijayalakshmi. 1974. Phosphorus fertilization of sunflower. Can. J. Plant Sci. 54:599–600.
207. Warren Wilson, J. 1967. High net assimilation rates of sunflower plants in an arid climate. Ann. Bot. 30:745–751.
208. ————. 1967. Effect of temperature on net assimilation rate. Ann. Bot. 30:753–761.
209. Wauchope, R. D., J.M. Chandler, and K. E. Savage. 1977. Soil sample variation and herbicide incorporation uniformity. Weed Sci. 25:193–196.
210. Weibel, R. O. 1951. Sunflowers as a seed and oil crop for Illinois. Illinois Agric. Exp. Stn. Circ. 681:1–16.
211. Whitney, M. 1906. Soil fertility. USDA Farm Bull. 257:1–40.
212. Willatt, S. T., and D. T. Coker. 1966. Soil physics. p. 19–21. In Annual Report 1965. Agric. Res. Coun. Centr. Africa:Lusaka 1966.
213. Young, P. A., and H. E. Morris. 1927. Sclerotinia wilt of sunflowers. Montana Agric. Exp. Stn. Bull. 208:1–32.
214. Zelitch, I. 1971. Photosynthesis, photorespiration, and plant productivity. Academic Press, New York and London. 347 p.
215. Zubriski, J. C., and D. C. Zimmerman. 1974. Effects of nitrogen, phosphorus, and plant density on sunflower. Agron. J. 66:798–801.

ROBERT G. ROBINSON: B.S., M.S., Ph.D., Professor, Department of Agronomy and Plant Genetics, University of Minnesota, St. Paul, MN 55108. Research agronomist on sunflower since 1948. Publications on culture, breeding, morphology, physiology, pollination, utilization, and quality of sunflower.

Chapter 5

Harvesting, Handling, and Storage of Seed

R. T. SCHULER, H. J. HIRNING, V. L. HOFMAN,
AND D. R. LUNDSTROM

Because sunflower (*Helianthus annuus* L.) has not been considered a major crop in the USA, equipment designed primarily for other crops has been adapted for sunflower production. Much of this equipment originally was used for corn (*Zea mays* L.) and soybeans (*Glycine max* (L.) Merr.). The drying systems, storage structures, etc., and other equipment for these crops has been very successful. Their adaptation, however, has met with variable success because sunflower is quite different physically and morphologically from corn and soybeans. As a result, some equipment is over-designed for sunflower; for example, storage structures can support larger volumes of sunflower seed than corn or soybeans. Special precautions must be taken by sunflower producers in adapting some equipment, such as crop dryers. Significant modifications also have been necessary in harvesting equipment to facilitate efficient harvesting of sunflower seed.

HARVESTING SUNFLOWER SEED

Sunflower planted in May in the northern production areas of North America usually is ready for harvest in late September or October, a growing season of approximately 120 days. The growing season may vary in length depending upon the summer temperatures during the growing season, relative moisture, and fertility. Sunflower plants are physiologically mature when the back of the head has turned from green to yellow, a change that usually takes place before the heads are dry enough to harvest. In northern regions of North America the sunflower plant frequently is desiccated after a killing frost. If this killing frost occurs before the backs of the heads turn yellow, yield losses will result.

From: Carter, Jack F. (ed.). 1978. *Sunflower Science and Technology.* Agronomy 19. Copyright © 1978 by the American Society of Agronomy, Crop Science Society of America, and Soil Science Society of America, 677 South Segoe Road, Madison, WI 53711 USA.

Fig. 1—A grain combine, with sunflower head attachment, operating in a field.

To reduce seed shattering loss during harvest and loss from birds, many growers in the northern areas prefer to harvest sunflower at moisture contents ranging as high as 20 to 25%. Sunflower seed from the combine then is dried in a grain dryer to 9.5%, which is considered a safe moisture content for bin storage.

Harvesting Attachments

Sunflower is harvested by a combine equipped with a special sunflower header attachment. Harvesting can be done with small grain header, but losses of 46% have resulted (14). All makes and models of combines used to harvest small grains will thresh sunflower. The attachment gathers sunflower heads into the feeder mechanism. Sunflower attachments are available from several equipment manufacturers, but some growers have made their own. If the special attachment is not available commercially, and a grower does not wish to build one, local machine shops usually will build the unit. The attachments are designed to gather only the sunflower heads and reject as much stalk as possible. Major components of this attachment are catch pan, deflector, and small reel. Long pan-like guards, called "catch pans," extend ahead of the cutter bar to catch the seed as it shatters. The deflector mounted above the catch pans pushes the stalk foward until only the heads remain above the cutter bar. As the heads move below the deflector, the stems contact the cutter bar and are cut just below the head. A small reel, mounted directly behind the deflector, moves the heads into the combine feeder. This sunflower attachment operates on the modified stripper

Fig. 2—Sunflower head attachment that utilizes only catch pans and reel.

head principle. Only minor adjustment is necessary after the attachment is mounted on the combine. Usually, only the deflector will need vertical or horizontal adjustment, depending on the size of the sunflower heads and how much stalk is going through the combine. The header of the combine usually runs 0.3 to 0.6 m (1 to 2 ft) above the soil surface depending on how the stalks are standing. Special lifter bars are available to help in raising sunflower stalks that are extremely lodged. A combine with these attachments is shown in Fig. 1. The machine shown in Fig. 2 illustrates a sunflower head without the deflector.

Two pan designs in use are a) narrow pan that adapts to any row spacing, and b) wide pan that adapts to only one specific spacing. The long, narrow pans are usually 0.23 m (9 in) wide and 1.3 m (52 in) long, tapering to a point with the taper starting approximately 0.94 m (37 in) from the back of the pan. The edges of the pan are bent upward to form a tray which is 2.5 cm (1 in) deep. These pans are spaced 0.31 m (12 in) center to center with the slot between them 7.6 cm (3 in) wide. The side pans mounted on the end of the combine head are 0.23 m (9 in) wide but are slightly longer and taper only to one side (9).

The wide pans are built in the same manner as the narrow pans, but should be 0.94 m (37 in) wide for 1.02 m (40 in) or about 8 cm (3 in) less than the row spacing (3). The 8 cm (3 in) spacing between pans allows sunflower plants to flow easily into the cutter bar. The pans are fastened directly over the top of the guards on the cutter bar by installing slightly longer guard bolts. A supporting brace must be installed beneath the pans and fastened to a rigid point beneath the cutter bar. This brace will reinforce the catch pans to make them much more rigid.

Fig. 3—Combine with rotating drum and catch pans.

The deflector consists of a curved piece of sheet metal the full width of the combine head. It is attached to the reel support arms above the catch pans. The reel for the unit is mounted directly behind the deflector and usually consists of only three arms or paddles. These arms, usually 0.31 m (12 in) long, are mounted on the original combine reel bracket. The reel should be mounted approximately 0.10 to 0.13 m (4 to 5 in) above the catch pans so that when the heads come in contact with the reel, they are moved back into the feeder.

In a more recent development, the deflector and reel are replaced by a rotating drum. Projection of triangular-shaped pieces of strap iron are welded to its surface (Fig. 3). As the drum rotates, the projections pass through the slots between the catch pans to dislodge any stalks that may cause clogging (9).

Another commercially available sunflower head for the combine is a stalk walker with the catch pans. A hexagonal reel is mounted above the catch pans. Fingers on the reel feed the stalks between the pans, into the feed conveyor of the combine, and ensure that the stalk is cut just below the head. As a result, a very small quantity of stalks enter the combine.

A grain combine, shown in Fig. 4, equipped with a row crop unit, has been successful harvesting sunflower seed. A pair of gathering belts, one on each side of the row, draw the stalk to a cutting unit and into a cross conveyor. The cutting unit is located below the belts. A large quantity of stalks pass through the machine with this unit.

Fig. 4—Combine equipped with a row crop unit and gathering belts.

Combine Adjustment

Forward Speed

Combine forward speed usually should average between 5 and 8 km/ hour (3 and 5 miles/hour). Optimum forward speed usually will vary depending upon moisture content of the sunflower seed and yield of the crop. Forward speed should be decreased as moisture content of the sunflower seed decreases to reduce the shatter loss as the heads feed into the combine. Faster forward speeds are possible if the moisture of the seed is between 12 and 15%. The higher speeds should not overload the cylinder and the separating area of the combine except in an extremely heavy crop. Seed in the moisture range of 12 to 15% will thresh from the head very well as they pass the cylinder.

Cylinder Speed

After the sunflower heads are separated from the plant, they are threshed in the cylinder and concave area. The normal cylinder operating speed should be about 300 revolutions/min (RPM) depending upon the condition of the crop and the combine being used. This cylinder speed is for a combine with a 0.56 m (22 in) diam cylinder to give a cylinder peripheral speed of 530 m/min (1,725 ft/min). Combines with a 0.47 m (18.5 in) diam cylinder or a 0.60 m (23.5 in) diam cylinder require a different speed adjustment. The smaller diameter cylinder requires an increase in the RPM and a larger diameter requires a decrease in the RPM, to obtain a peripheral speed of about 530 m/min (1,725 ft/min). If a combine cylinder operates at speeds of 400 and 500 RPM giving a peripheral speed of over 760 m/min (2,500 ft/min), very little seed should be cracked or broken if the moisture content of the seed is 11% or higher. Peripheral cylinder speeds of over 910 m/min (3,000 ft/min) should not be used because they cause excessive broken seed and increased dockage. The sieves and the return elevator of the combine may become overloaded with small pieces of broken heads and seeds when excessive cylinder speeds are used (3).

Concave Adjustment

Sunflower threshes relatively easily. When the crop is at 10% or less moisture, the concaves should be set wide open to give a cylinder to a concave spacing of about 2.5 cm (1 in) at the front of the cylinder and about 1.9 cm (0.75 in) at the rear. A smaller concave clearance should be used only if some seed is left in the heads. If the moisture percentage of the crop is between 10 and 12%, rather than increase the cylinder speed, the cylinder to concave clearance should be decreased to improve threshing. If the seed moisture exceeds 20%, a higher cylinder speed and a closer concave setting may be necessary, even though trash in the seed increases. Seed breakage and hulling may be a problem with close concave settings (3).

Fan Adjustment

Oilseed and nonoilseed sunflower often weigh 36.4 to 41.6 kg/hl (28 to 32 lb/bu or 22.4 to 25.6 lb/ft³) and 28.6 to 33.8 kg/hl (22 to 26 lb/bu or 17.6 to 20.8 lb/ft³), respectively. Because the seed is relatively light compared to other crops, excessive wind may blow seed over the chaffer and sieve. Seed forced over the sieve and into the tailings auger will be returned to the cylinder and may be dehulled. Only enough wind should be used to keep the trash floating across the sieve. The chaffer and sieve should be adjusted to minimize the amount of material that passes through the tailings elevator (3).

When the combine is adjusted correctly to thresh sunflower seed, the threshed heads will come through only slightly broken and with only unfilled seed remaining in the head. Cylinder concaves and cleaning sieves usually can be set to obtain less than 5% dockage. Improper settings will crush the seed but leave the hull intact. Proper setting is critical, especially for nonoilseed sunflower that is used for the human food market. The upper sieve should be open enough to allow an average seed to pass through on end or be set at 1.3 to 1.6 cm (0.50 to 0.63 in) opening. The lower sieve would be adjusted to provide a slightly smaller opening or about 0.95 cm (0.38 in) wide. The final adjustments will depend upon the amount of material returning through the tailings elevator and an estimation of the amount of dockage in the grain tank (3).

Field Loss

The harvested yield of sunflower can be increased by making necessary adjustments following a determination of field loss. Three main sources of loss are: (a) loss in the standing crop ahead of the combine, (b) header loss as the crop enters the machine, and (c) threshing and separating loss. The loss found in any of these three areas will give the combine operator a good estimate of sources of seed loss and the adjustments necessary to minimize seed loss.

Loss occurring in any of these areas may be estimated by collecting the seed on the soil surface in a 1 m² (1 yd²) area. If sunflower seed weighs 39 kg/hl (30 lb/bu), a seed mass of 3.9 g/m² (0.17 oz/yd²) will produce a loss of 1.00 hl/ha (1.00 bu/A).

The loss in the standing crop is estimated by collecting and weighing the seed in a 1 m² (yd²) area ahead of the machine at several different places in the field. Header loss can be found by collecting and weighing seed in 1 m² (yd²) behind the head under the combine and subtract the standing crop loss. To find the loss in combine separation, collect and weigh the seed in 1 m² (yd²) area directly behind the rear of the combine and subtract the loss found under the machine. An adjustment must be made to equalize the loss over the entire width of cut. Divide the result by the ratio:

$$\frac{\text{Width of head cut}}{\text{Width of rear end of combine}}$$

This answer is the adjusted separator loss for the width of cut. Divide this result by 3.9 (0.17 if bu/A) to obtain the combine separator loss in hl/ha (bu/A). To calculate the total loss in hl/ha (bu/A), add the seed mass in the standing crop, header loss, and separator loss, and divide this answer by 3.9 (0.17 if bu/A). The percentage loss can be found by dividing the total loss in hl/ha (bu/A) by the yield in hl/ha (bu/A).

Harvest without some seed loss is almost impossible. Usually a permissable loss is about 3%. Loss as high as 15 to 20% has occurred with a well-adjusted combine if the ground speed is too fast, resulting in machine overload (3).

Combine Capacity

Combine capacity often is based on engine power, cylinder width and diameter, straw walker size, sieve area, header size, and grain tank capacity. The true capacity of a combine is its rate of threshing and separating the crop without significant loss of either seed quantity or quality. The capacity is limited by the maximum acceptable seed loss. High capacity in hectares per day can be achieved by keeping the machine operating near full capacity for every possible minute in the field. Operation at full capacity requires uniform feed rate into the combine at all times. Self-propelled combines utilize variable-speed or hydrostatic transmissions to provide speed adjustments that provide uniform feeding into the machine.

A combine operator usually must compromise between speed of operation and seed loss. A seed loss monitor is a recent development that helps to maintain a continuous watch on the amount of seed being lost from the combine. The monitor is attached to the rear of the straw walkers and cleaning shoe to determine the seed being lost. The unit must be calibrated and then will provide the operator with a continuous indication of seed loss and allow him to make adjustments either on the machine or in its forward speed to keep the loss at an acceptable level.

Shaft monitors are available for combines so the operator may maintain a continuous watch on the RPM of critical shafts. If combine cylinder, beater, or fan speeds change outside a given tolerance, an indicator in the combine cab will warn the operator. Some combines have a pressure switch above and at the rear of the straw walkers to warn of buildup. These monitors are very desirable on machines because the operator usually is isolated and not able to observe the operation of machine components such as cylinder, fans, and straw walkers.

Power Requirements

Most self-propelled combines have reserve engine power beyond the usual requirements. A tractor with reserve power is recommended on pull-type combines. This power is needed for hilly fields, soft fields, and for threshing high-moisture sunflower seed. Power required to propel a 6.1 m (20 ft) self-propelled combine is 11 to 15 kW (15 to 20 hp) on level land. In hilly or wet conditions 30 to 37 kW (40 to 50 hp) may be necessary. If a combine has a hydrostatic transmission, more power will be required because this is inefficient. The power requirements to drive the machine for sunflower usually are less than for harvesting grain. At combine capacity, the cylinder requires about 50% of the total power necessary. At times, the power requirement may increase 100 to 200% beyond the average power demand. Sunflower is relatively easy to thresh so all combines or tractor-combine units used for harvesting small grain should have adequate power for the crop.

Desiccants

Chemical desiccants may be applied to the sunflower plants to kill the plants and increase the rate of field drying. A Soviet scientist (2) reports that the desiccant, magnesium chloride, advanced harvesting 8 to 12 days. As a result the need for drying is reduced and losses are reduced by 4 to 5% (2). In the USA, paraquat (1,1'-dimethyl-4,4'-bipyridinium dichloride) is the only preharvest desiccant available for use on oilseed cultivars. No desiccant is cleared for use on nonoilseed sunflower. Some other desiccants are being investigated now for use on sunflower, but are not cleared by the U.S. Environmental Protection Agency (EPA) and are not available commercially.

DRYING AND HANDLING SUNFLOWER SEED

Drying sunflower seed has become an integral part of sunflower production. In some seasons a major portion of the sunflower crop must be dried before the crop can be stored safely and/or transported long distances. When drying is necessary, the sunflower seed must be handled more often. An efficient handling system is imperative to minimize harvest delay.

Some benefits of drying are: harvest before sunflower seed otherwise would be at safe storage moisture percentage; reduced harvesting losses, principally by reduced head shattering; and reduced dependency on favorable weather conditions for harvest. Weather conditions in northern production areas of North America during some harvesting seasons have been unfavorable to field drying so the seed in the mature plant never reached the moisture level for safe storage. The acquisition of drying equipment requires an investment and depreciation cost that becomes an annual production cost regardless of use. Increased operating costs and extra seed handling equipment are required with drying equipment.

Most drying methods use forced air through a mass of seed. The air absorbs moisture from the seed. This drying occurs naturally in the field when drying winds blow through the standing crop on sunny days. Manufactured systems use fan and plenum chamber ducts to force the air through the seed mass.

The principal factors influencing the rate of drying are the air flow, air temperature, relative humidity of the incoming air, and seed moisture percentage. In general terms, the rate of drying may be increased by increasing the air flow, increasing the air temperature, and reducing the relative humidity of the air.

As the temperature of the drying air increases, the moisture carrying capacity of the air increases and the relative humidity decreases. As a general guide, increasing the air temperature 11 C (20 F) doubles the total moisture carrying capacity of the air and reduces the relative humidity to one-half its original value. For example, 280 m³ (10,000 ft³) of air at 16 C (60 F) and 100% relative humidity contains 3.6 kg (8 lb) of water in vapor form, while the same mass of air at 27 C (80 F) and 100% relative humidity contains 7.26 kg (16 lb) of water vapor. Thus, most dryers have heating units which increase the temperature of the ambient air drawn into the drying system.

The rate of drying depends on the difference in the seed moisture per-

Table 1—Equilibrium moisture of oilseed and nonoilseed sunflower seed.

	Equilibrium moisture			
	Oilseed		Nonoilseed	
Relative humidity	25 C	60 C	25 C	60 C
	%			
90	13.3	11.1	20.6	16.1
80	11.1	9.3	16.4	13.0
70	9.6	8.1	13.6	10.9
60	8.6	7.1	11.6	9.2
50	7.4	6.3	9.8	7.8
40	6.5	5.5	8.1	6.5
30	5.5	4.6	6.6	5.3
20	4.4	3.7	5.0	4.1
10	3.1	2.7	3.3	2.7

centage and the equilibrium moisture percentage associated with the drying air. The equilibrium moisture percentage is the seed moisture level reached when the moisture is equalized between the seed and surrounding air. If dry air passes through a mass of seed having a high moisture of 20%, the air will remove moisture from the seed. Conversely, if moist air at 100% relative humidity passes through seed at a low moisture of 9%, the seed will remove moisture from the air. After an extended drying period, the net moisture exchange will be zero. The level of moisture is defined as the equilibrium moisture percentage. This equilibrium moisture percentage also is dependent on temperature and material to be dried. As the temperature increases, the equilibrium moisture percentage decreases (4). The effect of temperature is apparent in Table 1. Oilseed and nonoilseed sunflower have different equilibrium moisture percentages. Therefore, separate sets of values are listed in Table 1 (12). Seed may become overdried if exposed too long to the drying air of low relative humidity and high temperature. An economic loss is realized because more moisture is removed and seed weight reduced more than is necessary. For example, if seed are exposed to air at 50% relative humidity and 60 C (140 F) for several days, the moisture percentage will reach 6.3%.

The drying air is the medium for moisture transport in most drying systems, so the rate of air movement through the seed mass directly influences the drying rate. The factors affecting the air flow are fan design and speed, fan motor size, and resistance of the seed to air flow. The designer of a drying system has control over the fan design and motor selection. The seed depth controls the seed resistance.

When evaluating seed resistance, the static pressure, air flow, and grain depth are interrelated parameters. The static pressure measured in mm of mercury (in of water) with a 'U'-tube manometer, is the air pressure as air enters the seed mass. The air pressure at the exit side of the seed is at atmospheric level. The air flow often is given in terms of $m^3/min/hl$ ($ft^3/min/bu$ or CFM/bu). Table 2 lists the estimated static pressures for various depths and air flow rates used in bin drying systems (12). In Table 3, similar information is listed but at air flow levels used in continuous flow and batch dryers

Table 2—Estimated static pressures for various air flows and seed depths for bin drying in oilseed and nonoilseed sunflower.

Depth of grain	Air flow, m³ sec/hl (CFM/bu)					
	0.08 (1)	0.16 (2)	0.24 (3)	0.40 (5)	0.48 (6)	0.64 (8)
m (ft)	Static pressure, kilopascals (in of water)					
Oilseed						
1.2 (4)	0.01 (0.03)	0.02 (0.10)	0.04 (0.18)	0.10 (0.40)	0.13 (0.52)	0.20 (0.81)
1.8 (6)	0.02 (0.10)	0.07 (0.27)	0.13 (0.51)	0.27 (1.10)	0.36 (1.44)	0.55 (2.23)
2.4 (8)	0.05 (0.20)	0.14 (0.56)	0.26 (1.04)	0.56 (2.26)	0.74 (2.98)	1.15 (4.61)
3.0 (10)	0.09 (0.35)	0.25 (0.99)	0.46 (1.83)	0.98 (3.96)	1.30 (5.22)	2.01 (8.07)
3.6 (12)	0.14 (0.55)	0.39 (1.56)	0.72 (2.89)	1.56 (6.27)	2.05 (8.26)	3.18 (12.77)
Nonoilseed						
1.2 (4)	0.01 (0.04)	0.03 (0.11)	0.05 (0.19)	0.10 (0.40)	0.13 (0.51)	0.19 (0.76)
1.8 (6)	0.03 (0.11)	0.07 (0.29)	0.13 (0.51)	0.26 (1.04)	0.33 (1.34)	0.50 (2.00)
2.4 (8)	0.05 (0.22)	0.14 (0.58)	0.25 (1.02)	0.51 (2.07)	0.66 (2.67)	0.99 (3.99)
3.0 (10)	0.09 (0.38)	0.25 (0.99)	0.43 (1.74)	0.88 (3.53)	1.13 (4.55)	1.69 (6.79)
3.6 (12)	0.14 (0.58)	0.38 (1.53)	0.67 (2.69)	1.36 (5.46)	1.75 (7.04)	2.61 (10.51)

Table 3—Estimated static pressures for various air flows and seed column thicknesses for batch and continuous flow dryers in oilseed and nonoilseed sunflower.

Grain column thickness	Air flow, m³/sec/m³ (CFM/bu)				
	2 (25)	4 (50)	6 (75)	8 (100)	12 (150)
cm (in)	Static pressure, kilopascals (in of water)				
Oilseed					
20 (8)	0.01 (0.05)	0.03 (0.14)	0.06 (0.26)	0.10 (0.41)	0.19 (0.75)
30 (12)	0.03 (0.14)	0.10 (0.40)	0.18 (0.73)	0.28 (1.13)	0.52 (2.09)
40 (16)	0.07 (0.29)	0.20 (0.82)	0.38 (1.51)	0.58 (2.33)	1.07 (4.32)
50 (20)	0.12 (0.50)	0.36 (1.43)	0.66 (2.65)	1.02 (4.09)	1.88 (7.56)
60 (24)	0.20 (0.79)	0.56 (2.26)	1.04 (4.19)	1.61 (6.47)	2.98 (11.97)
Nonoilseed					
20 (8)	0.01 (0.05)	0.03 (0.13)	0.06 (0.24)	0.09 (0.35)	0.15 (0.62)
30 (12)	0.03 (0.13)	0.09 (0.35)	0.15 (0.62)	0.23 (0.93)	0.41 (1.63)
40 (16)	0.07 (0.27)	0.17 (0.70)	0.31 (1.24)	0.46 (1.84)	0.81 (3.24)
50 (20)	0.11 (0.46)	0.30 (1.20)	0.52 (2.11)	0.78 (3.14)	1.37 (5.52)
60 (24)	0.18 (0.71)	0.46 (1.85)	0.81 (3.26)	1.21 (4.86)	2.12 (8.54)

(12). This information is very important when selecting fans for drying systems. The static pressures listed in Tables 2 and 3 are based on clean dry, unpacked seed. Higher static pressures may be expected under actual drying conditions.

The drying process is very complex when individual seed is considered because the drying not only involves moisture removal from its surface but also moisture movement, mostly by diffusion, from the interior of the seed to the seed surface. Often the drying process is evaluated on a larger scale considering the mass of seed in layers. Seed in bin dryers is considered to be in three layers based on moisture percentage during the drying process. The drying air is introduced into the seed at the bottom of most bin dryers. After some time, a drying zone will develop (Fig. 5). The seed below the drying zone will have reached the equilibrium moisture percentage for the condi-

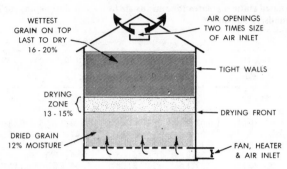

Fig. 5—Typical bin dryer, with several layers of seed based on moisture percentage. After Lundstrom (6).

tions of the air entering the seed mass. This seed is considered to be dry and the moisture level throughout this layer is uniform at equilibrium moisture percentage.

The seed above the drying zone is near the initial moisture percentage and is at equilibrium with air leaving the drying zone. All the drying takes place in the drying zone; the process results in a moisture gradient ranging from the initial moisture percentage at the top of the zone to equilibrium moisture percentage at the bottom of the zone.

The temperature throughout the seed mass has a gradient similar to the moisture. The seed below the drying zone is at the temperature of the incoming air while the seed above the drying zone is slightly warmer than its initial temperature. In the drying zone the temperature ranges from the incoming drying air temperature at the bottom of the drying zone to near the initial seed temperature at the top of the drying zone.

As drying progresses, the drying zone moves through the seed mass to the top layers of the bin. When the drying zone reaches the top layers, drying is complete. The depth of the drying zone and its rate of movement through the seed depend on the air temperature, air flow rate, and initial moisture percentage.

The temperature and humidity of the air entering bin dryers are important because they determine the final moisture percentage of the seeds. If the seeds are overdried, an economic loss results from excessive weight loss due to overdrying.

In bin dryers the location of the drying front often may be determined by pushing a rod 1 to 2 cm (0.4 to 0.8 in) diam, down through the seed from the top. When the force to push the rod eases, the drying front has just been passed. In a batch or continuous flow dryer, the front often may be found by feeling the temperature at the ends of the seed columns. The grain behind the drying zone will feel warm or be near the temperature of the incoming air, while the grain in front of the drying zone will feel cooler, or near the initial temperature of the seed.

Sunflower seed dries very rapidly compared to corn and soybeans because sunflower seed has a much lower volume weight. Most dryers operate on a volume basis so only about half as much moisture must be removed

Table 4—Chart for determining approximate shrinkage of seed during drying. After Lund-strom (6).

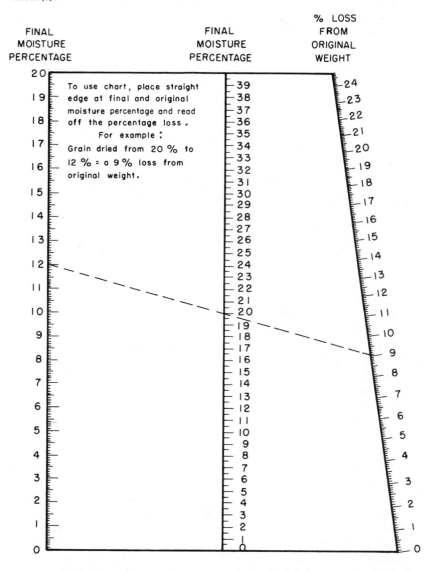

FINAL MOISTURE PERCENTAGE	FINAL MOISTURE PERCENTAGE	% LOSS FROM ORIGINAL WEIGHT

To use chart, place straight edge at final and original moisture percentage and read off the percentage loss .
For example :
Grain dried from 20 % to 12 % = a 9 % loss from original weight.

from a specific volume of sunflower seed as from the same volume of corn. Most crop drying equipment has been designed for corn; therefore, increased seed discharge capacities may be required when drying sunflower seed.

Water evaporated from seed during drying causes a weight loss from the seed mass. The loss depends on the initial and final moisture percentage. Table 4 may be used to estimate the shrinkage of the seed mass during drying. The values in this chart are presented in percent loss and are independent of the units of measurement. Complete sets of shrinkage tables are available commercially or in Extension publications such as AE-94, "Grain Drying Tables" from the Cooperative Extension Service, North Dakota State University (5).

Drying Systems

Four basic drying systems used are in-storage-bin, batch-in-bin, batch, and continuous flow.

In-storage bin drying is one of the simplest methods used. A storage bin fitted with a fan and heater is filled in layers or at one time and dried over a period of days or weeks. When the bin is filled at one time, the temperature of the drying air is raised 5 to 11 C (10 to 20 F) above the outside ambient temperature. When filled in layers, one layer is partially dried before another layer is added. As each additional layer is added it is partially dried before more seed is added. Three or more layers may be added to fill the drying bin. The layers allow for slightly higher temperatures than filling the bin at one time. After the seed has dried, the bin is the storage unit. Thus the drying equipment is used only once per year unless it can be moved from bin to bin. The air flow rate designed for these systems is in the 0.16 to 0.64 m^3/min/hl (2 to 8 CFM/bu) range.

For batch-in-bin drying, a quantity of seed is placed in the bin, dried, cooled, and emptied from the bin so that other batches can follow. Drying temperatures for the batch-in-bin dryers are often in the 49 to 60 C (120 to 140 F) range for commercial sunflower seed, unless seed is used for planting. These systems are commonly designed to dry and cool one batch per day. The air flow rate designed into these systems is in the 0.8 to 2 m^3/min/hl (10 to 25 CFM/bu) range. Often batch-in-bin systems are filled to about 1.2 m (4 ft) deep if no seed mixing equipment is used. But the depth can be increased up to 3.1 m (10 feet) if mixing equipment is available (6).

Two types of mixing are commonly used. Bin stirrers, which also may be used with the in-storage-bin system, consist of one or more vertical augers moving and revolving through the bin to mix the drier seed near the bottom of the bin with the wet seed as shown in Fig. 6. Mixing reduces the possibility of over-drying the bottom layer of seed.

Another method of mixing is the removal of the lower layer of seed with a sweep auger and conveying it to the top of the bin. A vertical auger in the center of the bin carries the seed from the sweep auger to the top of the bin. If the seed moisture content at the bottom is at the desirable percentage, the seed may be conveyed to another bin for cooling. If a small quantity of moisture must be removed, the system may adapt to a continuous flow provided no mixing is required (Fig. 7).

Batch dryers are filled completely at one time. One system commonly used (Fig. 8), is two parallel columns of seed surrounding a plenum chamber. Air is drawn through a fan heater unit and passes through the column of seed. Batch sizes range from 28 to 350 hl (80 to 1000 bu) and column thicknesses range from 0.25 to 0.50 m (10 to 20 in). Drying temperatures range

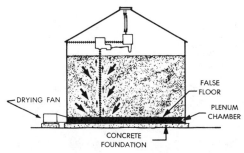

Fig. 6—Bin stirrer. The unit is used to mix dry seed from the bottom of the bin with wet seed near the top. After Lundstrom (6).

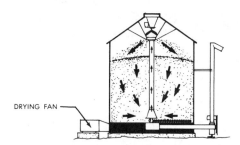

Fig. 7—Sweep auger and a vertical lift auger. These augers are often used to mix seed in recirculating batch dryer. After Lundstrom (6).

Fig. 8—Large batch dryer with one fan and one plenum chamber. These dryers are often installed at farm elevators. After Lundstrom (6).

Fig. 9—Continuous flow dryer with two fans and two plenum chambers, one for drying and one for cooling. After Lundstrom (6).

from 60 to 93 C (140 to 200 F) or somewhat higher than the two previous systems discussed. The common operating sequence is fill, dry, cool, and unload. The operator must manually switch the system through each step on some dryers, but others cycle automatically through the four steps. Some dryers are designed to recirculate the seed during the drying and cooling phases. They have air flows ranging from 4.05 to 8.1 m³/min/hl (50 to 100 CFM/bu) which is higher than the two previous systems (6).

Continuous flow dryers are designed to receive wet seed constantly at the top and discharge cooled dry seed at the bottom. The common configuration of these dryers is quite similar to the batch dryers except that the plenum is divided into two sections and two fans are used (Fig. 9). One fan provides air for a burner and an upper plenum section, while the second fan provides cooling air to the lower plenum section. Most continuous flow dryers are cross-flow type, where the air moves through the seed at right angles to the direction of seed movement. The cooled, dry seed is discharged through an auger at the bottom of the dryer. The seed flow is controlled by adjusting the speed of the feed rolls which discharge the seed into the auger. The speed adjustment is manual in some cases, but many dryers have controls which automatically adjusts the seed flow so that it is discharged at the desired moisture percentage. The designated temperature range is from 60 to 107 C (140 to 225 F), while the column thicknesses range from 0.20 to 0.76 m (8 to 30 in). Air flows are designated from 6.1 to 12.5 m³/min/hl (75 to 150 CFM/bu), which is higher than the previous systems. The conveying system of the continuous flow dryer is important. It must ensure that the wet seed is supplied and the dried seed removed continuously from the dryer. Most dryers have a safety device which turns off the system if an adequate supply of seed is not received (6). Sunflower seed must be dried to 9.5% moisture for long term storage. If seeds are not dried to this safe level, quality reduction will result as discussed in a later section of this chapter.

Drying temperatures up to 104 C (220 F) do not affect the sunflower oil yield or fatty acid composition adversely (13). Farm and commercial, continuous-flow, dryer operation at 71 and 82 C (160 and 180 F) over several seasons has not caused obvious damage to the sunflower seed. If the seed is for planting, drying temperatures should not exceed 43 C (110 F). Drying at temperatures above this level will reduce germination percentage.

Fire is often a hazard while drying sunflower. Handling loosens fine fibers or particles from the seed coat. These particles float in the air around dryers, and they ignite when drawn through the drying fan and burner. If the particles or sparks continue burning until they strike the seed mass, they may cause a fire (11). The probability of a fire increases with increased drying temperature, because the sunflower seed exposed directly to the drying air is drier and is ignited more easily. As a result, recommended drying temperatures are 71 to 86 C (160 to 180 F).

Fire hazards may be reduced by ensuring that clean air enters the dryer via air intakes directed towards the prevailing winds. Most of the particles enter the air in the area where the seed is loaded into the dryer. Another method to ensure that clean air is drawn into the dryer is installation of an intake duct onto the inlet of the fan and burner. Long flexible tubes are available for installation on dryers. Neither of these practices is 100% effective in eliminating dryer fires. If a fire starts, the damage is negligible if it is detected quickly and the equipment turned off and fire extinguished. *Drying equipment always should be attended when drying sunflower seed.* Other procedures to follow while drying sunflower seed are (a) clean around the dryer and in plenum chamber daily, (b) do not overdry, and (c) ensure the seed flow in continuous flow and batch dryers is uniform through all sections. Sunflower stalks and other material sometimes may reduce the seed flow in one section of the dryer producing overdried seed which ignites more readily.

Moisture Determination

The moisture percentage of sunflower seed should be determined as follows: (a) take several random samples with a sampler probe and mix to obtain a representative sample; (b) do not handle with wet or damp hands and place in an airtight container until moisture determination; and (c) follow the directions for the moisture measuring method.

The moisture percentage of sunflower seed may be determined by "direct" or "indirect" methods. Direct methods involve the removal of all the moisture from the seed by heating to drive off the water in the seed. The seed sample is weighed and then heated at 130 C (266 F) usually in a force air oven for 60 min or more until all the moisture is removed (1). The dry weight is determined. The difference of the two weights, which is total moisture, is divided by the initial sample wet weight and multiplied by 100 to obtain percent moisture. Indirect methods use moisture meters to measure electrical conductance or dielectric properties of the sample. The moisture percentage of the seed sample is related directly to the electrical properties. Many available meters indicate moisture percentage directly, but some meters indicate a reading to be converted to a moisture percentage via calibrated charts. The moisture indications of most instruments must be corrected for temperature if the seed temperature is not 21 C (70 F). Not all commercially available moisture meters have a sunflower chart or scale.

Most commercial moisture meters have accuracies of ± 0.5% moisture under normal operating conditions. Seed that has just been dried but not allowed to equilibrate will vary in moisture percentage in the different parts. The seed surface will be drier than the interior and cause the moisture meter to underestimate the moisture percentage. This error can be reduced by equilibrating the seed in a sealed container for 24 hours, which allows the moisture to redistribute evenly in the seed. If the operator cannot equilibrate the sample because the 24 hours might delay the drying, or for other reasons, he must be aware that an error may occur and try to make allowance for it. The estimation of the error is very difficult because it depends on initial moisture percentage, drying air temperature and the humidity, and the length of the drying period. When this error is not considered, the operator drying sunflower seed may assume the seed is at the proper moisture percentage; after equilibrium in storage, the moisture percentage may be too high for long-term storage or sale without penalty for moisture percentage. The risk of this error occurs in other crops, but apparently is greater in sunflower seed because low volume weight and rapid drying allow insufficient time for moisture equilibration.

Seed Handling Systems

Adding seed drying to a farm operation requires additional seed handling equipment. Seed may be handled two or three times, so adequate storage and conveyors are needed to prevent the drying phase impeding the total harvest operation. The only exception is the in-storage-bin system, which uses the same bin for drying and storage.

Three types of handling systems are shown in Fig. 10, 11, and 12. Although these systems include storage, often the seed is not stored on the farm but is transported from the drying location. Therefore, the seed would be loaded into trucks or some other means of transportation instead of into storage bins.

The batch-in-bin dryer in Fig. 10 may be used in a circular arrangement of bins. A short auger conveys the dried seed to the center of the bin circle. Another longer auger then will reach any bin within the circle. A single auger might be used to convey the seed to a holding bin or a truck. Batch and continuous flow dryers also may be part of circular bin systems.

Circular bin arrangements work satisfactorily for use with portable conveyors. If permanently installed handling equipment is planned, how-

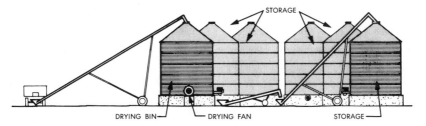

Fig. 10—Circular bin arrangement used with portable conveyors. After Lundstrom (6).

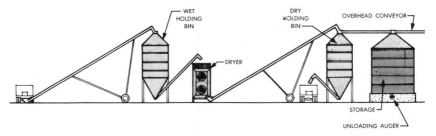

Fig. 11—Continuous flow dryer requiring holding bins and used with portable conveyors. After Lundstrom (6).

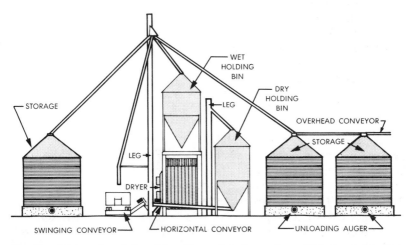

Fig. 12—Completely automated seed handling and drying system utilizing permanently installed equipment. After Lundstrom (6).

ever, a straight line arrangement is more suitable. Straight line arrangements are more adapted to drive-through unloading facilities, ease of expansion, and efficient use of permanently installed loading and unloading equipment.

When using a portable batch or continuous flow dryer, storage for both wet and dry seed should be provided. Figure 11 shows a continuous

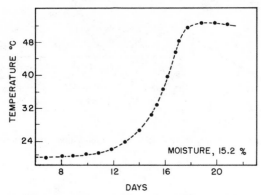

Fig. 13—Increased temperature with time of whole sunflower seed stored under adiabatic conditions. After Sallans et al. (10).

flow dryer used in this system. Dried seed may be conveyed to a hopper-bottom, holding bin or conveyed directly to storage with an overhead conveyor. The seed from the holding bin may be loaded into trucks. The holding bin for wet seed provides a continuous source of seed to the dryer and should have four to six hours of drying capacity.

A completely mechanized seed handling and drying system is shown in Fig. 12. A permanently installed, high-speed, vertical elevator conveys incoming wet seed to a hopper bin which feeds the dryer by gravity. Dried seed is conveyed to a shorter elevator to convey it to the dry holding bin. Dried seed from the dry holding bin feeds back to the main leg directly into storage or a waiting truck.

All conveyors must be of adequate size for efficient dryer operation and to ensure the drying system does not impede the harvesting operation.

STORAGE OF SUNFLOWER SEED

Sunflower seed is relatively easy to store when compared with other farm crops because it has a lower density. The large bulky seed weigh only 36.4 to 41.6 kg/hl (28 to 32 lb/ft³) and 28.6 to 33.8 kg/hl (22 to 26 lb/bu) for oilseed and nonoilseed, respectively, compared with corn at 77 kg/hl (48 lb/ft³). Seed of the heavier oilseed cultivars weighs about the same as oats (*Avena sativa* L.). The low volume weight of sunflower facilitates the choice of storage structures. Any structure suitable for storing small grain will hold sunflower seed.

The maximum moisture percentage considered safe for storage of sunflower is 9.5% which results in a relative humidity of the air between the seed of approximately 75%. The growth of molds and other microflora is possible at relative humidities higher than 75%. The growth of microflora causes a loss of weight and an increase in moisture percentage and temperature of the stored seed. Sallans et al. (10) in 1941 found that certain fungi began to grow in sunflower seeds stored at 11% moisture. Under adiabatic conditions with a starting seed moisture at 15.2% and temperature of 20 C (68 F) after about 6 days, the temperature increased rapidly and reached about 50 C (122 F) after 10 days as shown in Fig. 13 (10). Then the rate of temperature increase slowed, corresponding with the apparent thermal death point for several types of fungi and bacteria. Experiments with cereal grains have shown that the temperature continues to increase beyond 50 C due to thermophilic bacteria and possible spontaneous combustion of the grain (10).

Clean sunflower seed in storage have been used in all of the experiments on spontaneous heating. Sunflower seed from the combine usually contain green plant material and weed seeds. These foreign materials create drying problems even if the seed moisture is below 9.5%. In bins filled without grain spreading devices, the foreign material tends to accumulate directly under the filling spout, therefore, the moisture percentage in the center of the bin is higher than the average for the bin. Mold growth may start in the center of the bin producing heat, carbon dioxide, and water. If this growth continues uncontrolled, the total bin of seed may overheat. Hence, sunflower seed should be cleaned before storage to remove as much of the trash as possible. The bin also should be equipped with a spreading device to distribute seed and remaining foreign material uniformly throughout the bin.

Although sunflower seed is stored at less than 9.5% moisture, spontaneous heating may occur if the temperature is above 5 C (40 F). As the air temperature outside the bin drops during the fall and winter months, air currents occur within seed mass in the bin. The seed near the center of the bin remains at about the same temperature as when placed in the bin. As the seed near the sidewalls cools, the air surrounding this seed also cools. This cooler air near the sidewalls sinks to the floor and forces the warmer air near the center to rise through the sunflower seed. As the air moves through the sunflower seed, it picks up available moisture. When the warm moisture-laden air reaches the cool zone at the surface, moisture condenses on the seed. When the air temperature rises above freezing, molds begin to grow in this rewetted area at the top center of the bin and cause damage comparable to that occurring in seed initially stored at high moisture percentage.

To prevent moisture migration through the sunflower seed, aerating equipment should be installed in any bin containing more than 550 hl (1,500 bu) of seed. A small fan inserted into the center of the bin will aerate through ducts in the floor or through a perforated drying floor in the bin. Air flow of 1 m³/min (1 CFM) for each 180 hl (15 bu) of sunflower seed usually will prevent the moisture migration. The aerating equipment should be operated in the fall until the seed temperature reaches 2 C (35 F). The sunflower seed may be cooled below 2 C (35 F) without damage to seed. If the sunflower seed is to be stored into the next season and has been cooled below freezing temperatures, problems with condensation can be expected as the cold seed is removed from storage in above freezing temperatures.

Water may drip from conveyors and spouting. If moisture samples are taken as the seed is removed, the moisture meter will give false readings due to surface moisture on the seed. The aerating equipment should be operated during spring months to warm the seed gradually and prevent condensation problems.

Insects have not been a serious problem in stored sunflower seed. Bins should be checked regularly for insect infestation. If insects are found, the seed should be treated with insecticides approved for sunflower seed or cooled below 0 C (32 F), if possible.

SUNFLOWER FOR SILAGE

Sunflower has been ensiled for many years. Ensiling is done when 50 to 65% of the plants are in bloom. Later harvesting results in a less palatable feed. Some early studies indicate that sunflower silage has two-thirds of the total digestible nutrient of corn silage (8).

The Northwest Experiment Station at Crookston, Minn., found excellent acceptability of sunflower silage by cattle (7). Average daily gain was 0.95 kg (2.12 lb) per day for animals fed sunflower silage versus 1.05 kg (2.31 lb) per day for animals fed alfalfa silage.

The best sunflower silage is produced when the whole plant is harvested. Length of cut is similar to corn silage ideally about 2 to 4 cm (0.75 to 1.5 in). In the Minnesota study (7), moisture percentage was in the range of 65 to 70% and storage was in an oxygen-limiting structure.

Sheaffer et al. (15) suggest double cropping sunflower following barley in Maryland. As second crop in a double cropping system in east-central U.S., sunflower can produce as much dry matter as corn. However, Sheaffer et al. (15) reported that as a second crop in a double-cropping system, corn produced more than twice as much total digestible nutrients as sunflowers harvested at full bloom.

LITERATURE CITED

1. American Society of Agricultural Engineers Yearbook. 1977. Moisture measurement-grains and seeds. Standard 353, p. 417.
2. Dvoryadkin, N. I. 1972. Sunflower production in the USSR. Proc. 5th Int. Sunflower Conf. (Clermont-Ferrand, France). p. 468–473.
3. Friesen, O. H. 1971. Combines—operation and adjustment. Extension Service Branch., Manitoba Dep. Agric., Publ. 470.
4. Larmour, R. K., H. R. Sallans, and B. M. Craig. 1944. Hygroscopic equilibrium of sunflower seed, flaxseed and soybeans. Can. J. Res. 22F:9–18.
5. Lundstrom, D. R. 1973. Grain drying tables. Cooperative Extension Service, North Dakota State Univ., Circular AE94.
6. ————. 1974. Grain drying on the farm. Cooperative Extension Service, North Dakota State Univ., Circular AE98.
7. Marx, G. D. 1974. Sunflowers for forage. Holstein Echoes 10:3.
8. Morrison, F. G. 1948. Feeds and feeding. The Morrison Publishing Co., Ithaca, N.Y. p. 456–457.

9. Putt, E. D. 1972. Sunflower seed production. Can. Dep. of Agric., Publ. 1019.
10. Sallans, H. R., G. D. Sinclair, and R. K. Larmour. 1944. The spontaneous heating of flaxseed and sunflower seed stored under adiabatic conditions. Can. J. Res. 22F:181-190.
11. Schuler, R. T. 1972. Investigations in sunflower drying. North Dakota Farm Res. 29(6): 10-14.
12. —————. 1974. Drying related properties of sunflower seeds. ASAE Paper No. 74-3534. St. Joseph, Michigan.
13. —————, and D. C. Zimmerman. 1973. Effect of drying on sunflower seed oil quantity and fatty acid composition. Trans. ASAE 16(3):520-521.
14. Shadden, R. C., J. A. Mullins, and T. McCutchen. 1970. Mechanical harvesting of sunflowers in Tennessee. Proc. 4th Int. Sunflower Conf. (Memphis, Tennessee). p. 265-270.
15. Sheaffer, C. C., J. H. McNemar, and N. A. Clark. 1976. The sunflower as a silage crop. Agric. Exp. Stn. Misc. Publ. 893. Univ. of Maryland.

RONALD T. SCHULER: B.S., M.S., Ph.D.; Assistant and Associate Professor (Agricultural Engineer—Power and Machinery), Agricultural Engineering Department, North Dakota State University, Fargo, 1970-1976; Associate Professor (Agricultural Engineer—Power and Machinery), Department of Agricultural Engineering, University of Minnesota, St. Paul, 1976 to date.

HARVEY J. HIRNING: B.S., M.S., Ph.D.; Extension Agricultural Engineer (Rural Electrification), Agricultural Engineering Department, North Dakota State University, Fargo, 1975 to date.

VERNON L. HOFMAN: B.S., M.S.; Extension Agricultural Engineer (Power and Machinery), Agricultural Engineering Department, North Dakota State University, Fargo, 1975 to date.

DARNELL R. LUNDSTROM: B.S., M.S. Extension Agricultural Engineer (Rural Electrification 1967-1974, Soil and Water 1974 to date), Agricultural Engineering Department, North Dakota State University, Fargo.

Chapter 6

Insect Pests

J. T. SCHULZ

While nearly everyone has observed annual wild sunflower (*Helianthus annuus* L.), many persons are unaware that it is native to North America. Heiser (51) reported that a few artifacts of early wild sunflower were recovered from archaeological sites in the southwestern U.S. and Colorado. Sunflower grows in many different habitats in North America. Hence a moderate to large diversity of insect species have become associated with this host. Walker (98) recorded 66 species of insects attacking wild and cultivated sunflower in Kansas. Phillips et al. (67) recorded 48 species of insects in five orders attacking sunflower in Texas.

The paucity of literature on insect associations with sunflower before economic cultivars suggests cursory attention. Worldwide demand for high quality vegetable oils, coupled with the development of cultivated sunflower suitable for cropping in the USA, has resulted in large plantings. Increased production has been accompanied by a corresponding increase in insect infestations. Insect depredation is now an important deterrent to production in the Upper Midwestern States of North Dakota, South Dakota, and Minnesota. Significant losses to insects also have been reported from the other two major production areas in the USA, the High and Rolling Plains of Texas and California.

Since most of the insect-sunflower interactions have been documented in North America, this chapter deals primarily with these associations. Head infesting, foliage and stem feeding, and root-infesting species, as well as other insect pests, are discussed. Life stages and cycle, seasonal activity, damage characteristics, host preference, and control of each species is considered. Existing literature dictated to a great extent the inclusion of the species reviewed. Documented references to sunflower insect pests from other areas of the world are scarce, but the pests are discussed in this chapter where appropriate. Color photos of insects and damage follow page 210.

From: Carter, Jack F. (ed.). 1978. *Sunflower Science and Technology.* Agronomy 19. Copyright © 1978 by the American Society of Agronomy, Crop Science Society of America, and Soil Science Society of America, 677 South Segoe Road, Madison, WI 53711 USA.

HEAD-INFESTING SPECIES

Sunflower Moth, *Homoeosoma electellum* (Hulst)
[Lepidoptera: Pyralidae]

The sunflower moth, *Homoeosoma electellum* (Hulst), is distributed widely throughout the USA, Cuba, Canada, and Mexico and is possibly the most devastating species to sunflower. This insect has been the subject of more research than any other species of the complex.

Life Stages

Egg. Drake and Harris (29) and Bird and Allen (9) provided the initially described life stages of *H. electellum* (then referred to as sunflower webworm). Eggs were reported in both studies as small, elongated, and yellowish. Satterthwait and Swain (82) reported the egg of the sunflower moth as pearly white, elliptical, finely reticulated, from 0.63 to 0.80 mm long and 0.23 to 0.27 mm in diam, and placed singly or in small groups among or within the sunflower florets. These observations (82) concur with those of Drake and Harris (29) who reported females laid 30 or more eggs either singly or in groups of four or five within or among the corolla tubes of individual florets.

Satterthwait and Swain (82) obtained up to 91 eggs from a single female, most of them laid in a single day. Development of the egg was completed in 40 to 72 hours. Randolph et al. (71) found egg deposition to be quite variable from recent in vitro studies. They reported a record 221 eggs from one female deposited in 4 days. Eggs usually were laid by individuals ranging from 26 to 173 with a mean of 97.8. Most larvae emerged within 48 hours, the remainder within 72 hours.

Larva. Satterthwait and Swain described larvae of *H. electellum* (82) as purplish or reddish brown with four longitudinal, light bluegreen stripes on the dorsum of its abdomen. Larvae were 19 mm long. Adams and Gaines (1) in Texas observed sunflower moth larvae that were approximately 25 mm long with a greenish-yellow body and five brown dorsal longitudinal stripes. Carlson (18) noted additional variation in the color pattern and indicated larvae were yellowish green and had five brown to black dorsal longitudinal abdominal stripes. He reported larvae measured from 16 to 19 mm long. The reviewer observed larvae in North Dakota similar to both those described by Satterthwait and Swain (82) and Adams and Gaines (1).

Pupa. Satterthwait and Swain (82) observed pupae measured about 10 mm long and were reddish-yellow. They further reported pupae were slender, without spines but with a few fairly long setae at the caudal end.

Adult. Bird and Allen (9) reported adults they collected in Manitoba as "delicate grey moths." Adams and Gaines (1) described the adult as a gray moth with a wing span of approximately 20 mm (1). Satterthwait and Swain (82) reported *H. electellum* had a wingspread of 21 mm and body length of about 11 mm.

Life Cycle

Drake and Harris (29) initially described the life cycle of the sunflower moth and reported it averaged 25 days in Iowa. Adults lived only a few days. Larvae fed for 12 to 14 days on sunflower heads and molted four or five times.

In Manitoba, Bird and Allen (9) observed only four larval instars in their rearing studies. Duration of the third instar ranged from 3 to 6 days, the fourth instar, 7 to 16 days.

Satterthwait and Swain (82) observed five instars with stadia of each as follows: first, less than 4 days; second, 3 to 5 days; third, 5 days; fourth, 1 to 3 days; and fifth, 10 to 12 days. Observations of larval development reported by both Bird and Allen (9) and Satterthwait and Swain (82) are in close agreement.

A pupal stage lasting 6 to 7 days was reported by Drake and Harris (29). A prepupal and pupal period of 3 to 6 and 16 to 24 days, respectively, was observed by Bird and Allen (9). Satterthwait and Swain's (82) report in 1946 that the mature larvae construct a tough, open-mesh, close-fitting cocoon in a silken tunnel among the achenes or within a single achene is at variance with the more recent publications of Carlson (18) in California and Teetes and Randolph (93) in Texas. Carlson found pupae of *H. electellum* in the soil in Northern California, not in sunflower seeds. Teetes and Randolph reported mature larvae migrated from the sunflower heads to the stem and leaves and lowered themselves to the soil by means of a strand of silk. Larvae then constructed tough, silken cocoons in the soil which became covered with soil particles. They also reported a tunnel led from the cocoon toward the soil surface through which the adult emerged. These researchers indicated that, while pupae were found as close to the soil surface as 50 mm and as deep as 200 mm, over 50% of them were situated 75 to 100 mm below the soil surface.

Teetes and Randolph (93) recorded the duration of the leaf stages of the sunflower moth reared in laboratory and field. Mean number of days for development were 42.5 and 43.1 days, respectively.

Seasonal Activity

Satterthwait and Swain (82) reported in 1933 on the seasonal activity in Missouri of *H. electellum*. Adults were observed from 28 June to 4 September, although moths were collected from cages until 5 November. They reported that, with a duration of life cycle ranging from 24 days in midsummer on plants growing outdoors to 34 to 38 days in midsummer and autumn on caged individuals, at least four broods of *H. electellum* could be expected in the St. Louis, Missouri, vicinity. First generation moth flight occurred between 15 to 21 May; second generation, 28 June; third, about 2 August; and fourth, about 4 September. They also reported observations in Cuba from September 1932 to February 1933. Duration of the life cycle was 24 to 26 days. Despite mild winter conditions in Cuba, activity of the moth nearly ceased during December. Further observations revealed that while heavy infestations occurred in September, infestation levels were light in June 1933. The increased duration of the larval stages recorded in December was presented as evidence that the species normally overwintered as a larva.

In 1968 Teetes and Randolph (89) observed seasonal trends of *H. electellum* on cultivated and wild sunflower and Indian blanket, *Gaillardia pulchella* (Foug.), at McGregor, Texas. Weekly collections made from April to November, contained the first adults, 24 April, in areas where *G. pulchella* was flowering. Larval infestation levels on this host were 10.4% on this date and increased to a maximum of nearly 60% by 2 May.

Second generation activity was confined largely to wild and cultivated sunflower flowering in early June. Heads of cultivated sunflower were 100% infested with larvae from 20 June to 11 July. Maximum infestation of wild sunflower heads reached 83%.

After cultivated sunflower was harvested, Teetes and Randolph (89) reported larval populations were maintained in wild sunflower, although at extremely low infestation levels. They indicated there were at least two overlapping generations per year. They speculated that a third generation occurred at low population levels.

Beckham and Tippins (5) reported from Georgia that larval infestations of cultivated sunflower increased from 36% 30 July to 80% by 1 Sept. 1970. No indication of time and duration of moth activity was reported.

Johnson (53) reported that E. C. Carlson in California had observed a small number of moths in early June. A second flight of moderate number was detected in the second and third weeks of July. A heavy flight was observed during the initial 2 weeks of August, followed by a small flight in early September.

In the first major outbreak of *H. electellum* in North Dakota in 1975, moth activity was observed during the first week of July and continued in some locales until early September. (Oseto, C. Y. 1975. Annual report, Department of Entomology, North Dakota State Univ., Fargo). In 1976, infestations of *H. electellum* were nearly nonexistent in North Dakota and the neighboring province of Saskatchewan. Studies to assess overwintering capabilities of this insect in both these areas have revealed the sunflower moth apparently is incapable of overwintering in both North Dakota and Saskatchewan. (Oseto C. Y., and A. P. Arthur. 1976. Personal communication, 16 November. Canada Research Station, Saskatoon, Saskatchewan.)

Principal activity of the sunflower moth was found from November to April in the vicinity of Monterrey, Mexico. (Graziano, J. V. Personal communication. Chapingo, Mexico.)

Damage Characteristics

Frass from florets and larvae are progressively webbed together by a material over the face of the developing head, and gives the infested head a diagnostic characteristic. Satterthwait and Swain (82) reported larvae laid a delicate silk over the surface of composite flowers which bound the dying parts of the florets with larval frass and gave the flower a "trashy appearance." They reported that early instar larvae fed predominantly on the florets rather than the achenes. The two researchers also indicated that when the larva is observed easily, it forms and lives in a tube of silk that is green-tinged. Larvae fed on the developing achenes after cutting a small (0.8 mm diam) hole through the hull. All the achenes in an entire head may be destroyed. Larvae also were found to tunnel in the receptacle and occupied the hollow stalk beneath it.

Carlson (18) determined single larva damaged an average of nine achenes in a 3-week period. Serious seed loss resulted from moderate to severe infestations of 12 to 24 larvae per head.

Carlson (18) also developed a method for assessing the severity of damage to the heads. This rating was developed originally as a means of evaluating the efficacy of insecticides in control of *H. electellum* larvae, but also provides a means of assessing differences in susceptibility of cultivars to this species. Damage was rated (5 class values) on the day of the last insecticidal application and at 14, 21, and 35 days after the last application. "Spots" of damage on heads consisted of fairly discrete clumps of webbing and frass. Carlson's scale of class values is as follows:

0—no evidence of worm damage
I—1 to 2 damaged spots
II—3 to 5 damaged spots
III—6 to 8 damaged spots
IV—9 to 11 damaged spots
V—12 or more damaged spots

Carlson calculated an index of severity by multiplying class × frequency, summing these products for each of the total number of inspection dates and then dividing this sum by total number of heads observed.

Host Preference

The sunflower moth has been collected from many plant species. One of the earliest records of infestation by sunflower moth of hosts other than *Helianthus* sp. was provided by Forbes (37). He noted infestation of the buds of gumweed (*Grindelia* sp.). Records of host-sunflower associations by Drake and Harris (29) revealed the following host plants infested in Iowa: African marigold, *Tagetes erecta* L.; French marigold, *T. patula* L.; golden wave, *Coreopsis tinctoria* Nutt. tickseed, *C. lanceolata* L.; orange coneflower, *Rudbeckia fulgida* Alt.; yellow chamomile, *Anthemis tinctoria* L. and *Helianthus* sp. Bottimer (10) in Eastern Texas found *Rudbeckia maxima* (Nutt.) to be infested.

Satterthwait and Swain (82) found the earliest seasonal infestations of *H. electellum* in Missouri in *Coreopsis grandiflora* Hogg. Infestations were recorded subsequently from ornamental sunflower, *Heliopsis helianthoides* var. *pitcheriana* Fletcher, *H. annuus* L., *H. tuberosus* L., Jerusalem artichoke, a hybrid *Helianthus* whose flowers resembled dahlias and *Verbesina encelioides* Cav. These researchers also reported the most important weed host in Cuba was *Bidens pilosa* L., romerillo blanco, but it was not heavily infested.

Wene (100) reported *H. electellum* larvae were observed on young citrus at Edinburg, Texas. Larvae infested new growth and usually bored into the lower half of the twig and tunnelled downward for a short distance. Death of the new growth resulted, and was a serious problem.

In California Knowles and Lange (60) recorded damage to safflower (*Carthamus tinctorius* L.) and other wild and cultivated composites.

Teetes and Randolph (89), in early spring, 1967, at College Station and McGregor, Texas, systematically recorded host infestations of sunflower moth. They estimated the magnitude of infestation and assigned a rating of 1 to 5 (5-most highly infested). The most important early season hosts were rosering gaillardia (*Gaillardia pulchella* Foug.), lanceleaf gaillardia (*G. aestivalis* (Walt)], and golden crownbeard (*Verbesina enceloides* Cav.). These three hosts had an infestation rating of 4 and probably served as primary hosts for first generation larvae. Additional hosts were infested as they reached the bloom stage and included the following: *Helianthus debilis* T. and G., cucumber leaf sunflower; *H. annuus* L., common sunflower; *H. petiolaris,* prairie sunflower; *H. maximiliani* Schrad., Maximilian sunflower; *Coreopsis basalis* (Otto and Dietr.), Goldenman; *Engelmannia pinnatifida* Gray, Engelmann daisy; *C. grandiflora* Hogg., vig flower; *Cassia roemeriana* Scheete, two leaf senna (single larva only collected). Abundance ratings for these hosts ranged from 1 to 4. The magnitude of infestation levels generally reflected the earliness of the onset of bloom for each plant species.

Teetes and Randolph also found *H. electellum* in small cotton bolls in Glassock County, Texas, in 1968. They cited records by the USDA Plant Pest Control Division which indicated the sunflower moth had been found feeding on sweetclover (*Melilotus alba* Desr.); corn (*Zea mays* L.); *Citrus* cotton (*Gossypium* sp.); globemallow (*Sphaeralcea* sp.); and safflower (*C. tinctorius* L.). The sunflower moth was found infesting musk thistle (*Carduus nutans* L.) in North Dakota. (Scholl, C. G. 1976. Personal communication, July. North Dakota State Department of Agriculture, Fargo.) Identification of the larvae was confirmed by D. M. Weisman, U.S. National Museum, Washington, D.C.

Although the sunflower moth is capable of infesting and feeding on hosts in various plant families, its main hosts are species of the family Compositae.

Control

Significant literature deals with control of the sunflower moth. Early literature dealt with the association of parasites and predators of the sunflower moth but more recent literature reflects increased cultivation of sunflower and management and control techniques. Published accounts record control efforts by use of biological, cultural, and chemical methods.

Parasites and Predators. Initial observations of parasites of the sunflower moth resulted from a survey conducted by Satterthwait and Swain (82) during the period, April to mid-September, 1933. Their observations were confined to species of Compositae in the Missouri Botanical Garden and other flower gardens in the St. Louis, Missouri, vicinity.

The following parasites were identified from their studies: *Lixophaga variabilis* Coq., *Erynnia* (=*Anachaetopsis*) *tortricis* (Coq.) *Leskiomima tenera* Wd., *Clausicella* (=*Siphophyto*) *floridensis* Towns., Bracon (=*Microbracon*) *mellitor* (Say), *B. nuperus* (Cress.), *Chelonus altitudinus* Vier., *Apanteles homoeosomae* Mues., *Agathis buttricki* Vier., *Macrocentrus ancylivorus* Roh., *"Angitia"* n. sp., *Creamastus epagoges* Cush. and *Perilampus epagoges* Cush. Highest levels of parasitism were observed from two of the tachinids, *E. tortrices* and *C. floridensis,* and two braconids, *A. homoeosomae* (USA and Cuba) and *A. buttricki.* A predaceous clerid, *Hydnocera pubescens* Lec., was the only predator observed in this survey. *Metarrhizum anisopliae* (Metsch.), an entomogenous fungus, was found infecting larvae at one collection site.

Teetes and Randolph (90) studied the seasonal abundance and parasitism of *H. electellum.* They obtained 12 species of parasites from sunflower moth larvae collected from infested *Gaillardia pulchella*; six species were collected from larvae which infested cultivated sunflower. They noted a relationship between the species of parasite collected, the host plant from which larvae were obtained, and the time of year.

Principal parasites collected from larvae infesting *G. pulchella* were the braconids, *C. altitudinis* and *A. epinotiae. C. altitudinis* was the primary species reared from larvae collected in the early season; it did not parasitize larvae after 14 May. *A. epinotiae* was the predominant larval parasite collected 23 May to 11 July. Maximum level of parasitism induced by the parasite complex of sunflower moth larvae on *G. pulchella* was about 42%. The predominant parasites obtained from larvae collected from cultivated sunflower head were the tachinid, *C. neomexicana,* and the braconid, *C. altitudinis.* The highest level of parasitism induced by the parasite complex of those larvae that infest cultivated sunflower was 52.7%.

Johnson (53) cited an unpublished report of E. C. Carlson in which four families of insects that parasitized sunflower moth larvae in Northern California were noted: Ichneumonidae, *Mesotimes gracilis* (Cress.);

Pristomerus pacificus (Cress.); Braconidae, *Apanteles homoeosoma* (Mues.) Otididae, *Euxesta anna* (Harris); and an unidentified tachinid. Only three or four of the five species of parasites were found in any one area of California. The impact of these parasites on sunflower moth larvae was not determined.

Recently two *tachinids, Lixophaga plumbea* Aldrich and *Erymnia tortricis* (Coq.), were identified by researchers at the U.S. National Museum as parasites of *H. electellum* in North Dakota (Oseto, C. Y. 1976. Personal communication, November).

Planting Dates. Muma et al. (65) made the initial report on the effects of planting date on infestation density of the sunflower moth. Sunflower planted in Nebraska on 2 May averaged less than one larva per head whereas untreated sunflower planted on 8 June averaged nearly three larvae per head.

Teetes and Randolph (91) reported studies conducted in 1968 sunflower plots planted after 15 April at College Station, Texas, sustained significantly lower sunflower moth infestations than those planted prior to that date. The number of larvae per head ranged from 63 to 178 for the four planting dates preceding 15 April and from 0 to 8 for the eight dates between 15 April and 30 July. Availability of moisture was cited as a major problem encountered with late planted sunflower since both yield reductions and undeveloped seed were noted from these plots. A similar study was conducted in 1969 by Teetes and Randolph (94) and an assessment was made of the role of planting dates on infestation leels of *H. electellum.* Plots were established at both College Station and McGregor, Texas. Results from this study closely paralleled those of the previous study. Plots planted after 11 April at College Station had significantly fewer larvae per head than those planted earlier. Plots planted after 25 April at McGregor and significantly fewer larvae than those planted earlier.

Beckham and Tippins (5) reported an early planting of cultivated sunflower in Spalding County, south Georgia, sustained only a 4% infestation in heads whereas 16% of the heads were infested in a late planting.

Insecticides. Chemical definition of insecticides cited in this section are given in an appendix table at the end of the chapter. Use of chemicals to control *H. electellum* was reported initially by Muma et al. (65), Wene (100), and Adams and Gaines (1).

Although larval control obtained by these three groups of researchers was less than satisfactory, Muma and coworkers indicated use of benzene hexachloride in their 1949 studies resulted in significant reductions in larval populations. Wene reported use of 5% DDT dust in an aerial application eliminated the sunflower moth from the citrus grove infested by this species. Adams and Gaines reduced injury caused by larval *H. electellum* with application of 10% DDT-40% sulfur or 3% lindane-5% DDT-40% sulfur when each were applied at the rate of 16.2 kg per ha. They also reported efficacy of the chemicals was greatest when the first application was made at the onset of bloom and followed by two or three applications at weekly intervals. Knowles and Lange (60) reported a mixture of 5% DDT and 2% parathion dust was effective when applied at the 60% bloom stage. When two applications were to be made, they suggested use of 10% DDT dust at 44.8 kg/ha for both applications. Initial application was made at the 20% bloom stage and the second application 10 days later. Carlson (18) evaluated a series of both synthetic organic insecticides and bactericides from 1964 to 1966. Three applications of endosulfan, 1.12 kg ai/ha (ai = active ingredient), resulted in excellent moth control even though it was not significantly better than two applications. Two applications of either endosulfan or diazinon, 1.68 kg ai/ha, commencing at the onset of bloom also resulted in satisfactory control. All treatments significantly increased the size and number of undamaged seeds and yields. In the 1965 and 1966 trials, Carlson also obtained excellent control with the experimental chemical, GS13005 (methidathion) after three applications, 1.12 kg ai/ha. Maximum control from use of the three most satisfactory insecticides evaluated was obtained when three applications were made at 5 to 7-day intervals. Initial application was recommended at the onset of bloom. Two products containing the bacterium, *Bacillus thuringiensis* Berliner, were not effective in controlling the sunflower moth.

Carlson (19) reported three applications of endosulfan, 1.12 kg ai/ha, reduced larval infestation in sunflower heads and damage to seeds. Initial application was made at the onset of bloom and later applications made at intervals of 1 to 2 weeks. Three applications of Gardona® , methyl parathion, and carbofuran also satisfactorily controlled *H. electellum* larvae when the spraying regimen outlined for endosulfan was followed.

Teetes and Randolph (88) reported that sunflower at College Station, Texas, sprayed twice with methyl parathion, 1.12 kg ai/ha, had significantly fewer larvae than those treated with either endosulfan, malathion, or diazinon. All four insecticides reduced the number of larvae and increased yield over that of an untreated crop. In similar tests at McGregor, Texas, endosulfan and carbaryl reduced number of larvae after one application; yield, however, was not increased. With two applications, methidathion (GS13005), azinphosmethyl, and methyl parathion were among several chemicals which increased yield. Larval infestation in plots treated three times were lower than that of untreated plots in all but the malathion ultra-low volume plot. Yields were increased in all but the trichlorfon plot.

They also found that with multiple applications of methyl parathion, endosulfan, and carbaryl at 5-day intervals, yield was significantly increased by spraying. Insecticides applied at the 50% and 100% bloom stages were most effective in reducing larval infestations. Results from additional studies evaluating the effectiveness of 14 insecticides in control of *H. electellum* were reported by Teetes and Randolph (91) at McGregor, Texas. Two or three applications of each of six insecticides were made at weekly intervals to Peredovik; all reduced larval infestations. Three applications of the insecticides, endosulfan, diazinon, methidathion (GS 13005), and methyl parathion applied at 0.56 kg ai/ha reduced larvae and significantly increased yields.

Two conventional, low-volume applications of monocrotophos, carbaryl, malathion, azinphosmethyl, mevinphos, chlorpyrifos, toxaphene, and parathion were applied to plots at College Station; they reduced numbers of larvae and increased yields.

Teetes and Randolph (93) again reported results of their studies to determine efficacy of 16 insecticides in control of sunflower moth larvae. Their results with methyl parathion, endosulfan, methidathion, and carbaryl were similar to those reported from earlier studies as either two or three applications resulted in reduced larval numbers. Initial application was made at the McGregor and College Station at the 20 and 30% bloom stage, respectively. Ultra-low volume applications of methyl parathion (1.12 kg ai/ha) and azinphosmethyl (0.56 kg ai/ha) reduced numbers of larvae and increased yields in the College Station test.

Carlson (20) recently reported the use of carbofuran, 1.12 and 2.24 kg ai/ha, in small plots resulted in satisfactory reductions of the damage caused by the sunflower moth. Seed weight per head increased significantly only with the 2.24 kg rate. Initial application was made when bloom stage was 25 to 50% with the subsequent two or three applications made at 5 to 7-day intervals during August and September.

Tests with both the miscible and encapsulated formulations of methyl parathion (1.12 kg ai/ha) also were made. Satisfactory reduction of damage and increased seed weights were recorded using the miscible formulations. The encapsulated formulation neither increased seed weights nor provided the level of control obtained with the miscible formulation.

Carlson (20) further reported that Dipel® , HD-1 strain of *Bacillus thuringiensis* Berliner var. *kurstaki* de Borjac and Lemille, did not control *H. electellum* larvae.

Fucikovsky (38) reported they were following Carlson's recommendation for use of endosulfan to control *H. electellum* in Mexico.

Three insecticides now are approved by the Environmental Protection Agency in the USA for control of the sunflower moth. They are endosulfan, 1.12 kg ai/ha; methyl parathion, 1.12 kg ai/ha, and methidathion, 0.56 kg ai/ha.

Genetic Resistance. Kinman (59) reported that in his 1965 plantings at McGregor, Texas, he found inbred lines descending from a single plant selected from the Morden line 953-102-1-1-22-12 sustained only 10 to 15% damage by *H. electellum* larvae. Other lines which flowered on the same date were destroyed completely. These lines were subsequently used interchangeably as the male parent of the hybrid T56002.

Teetes and coworkers (87) reported on the susceptibility of T56002, and other lines from the USDA sunflower breeding program, to the sunflower moth. This research, conducted from 1967 to 1969, was designed to evaluate differences in susceptibility among cultivars and hybrids. Hybrids T66001 and T56002 and the Russian cultivars, Armavirec and Kubanec, were damaged less severely by sunflower moth larvae than other lines or cultivars tested. Damage in Armavirec, however, was attributed to its early flowering after early planting, permitting it to escape heavy infestations. Later plantings of this cultivar were heavily infested; hybrids that sustained the lowest infestation in early plantings were heavily infested. The hybrids that sustained the lowest infestations were fast-flowering hybrids.

Carlson et al. (21) identified resistance or tolerance in some selections of *H. annuus* cultivars in their 1970 studies at Davis, California. This tolerance was associated only with those selections which possessed a phytomelanin layer of the pericarp. This layer, first identified about 1900, was referred to as the "carboniferous layer" by Hanausek (45). Molisch (64) identified this layer as a "black amorphous mass which is embedded in the pericarp." Putt (70) characterized this layer as an inner, pigmented substance which fills the intercellular spaces between the outer layer and the adjoining sclerenchyma tissue. Kienwick (57) indicated the Russians used the term "armored layer" to define the hardened layer which forms 3 to 4 days after fertilization.

In 1971 Carlson evaluated progeny from several selections and confirmed that a reduction in damage was associated with selections which possessed the black armored layer. He reported progeny of selection 526-1 were consistently the most resistant, and they were rated from 0.63 to 1.0 on the damage index (5 = high damage rating). Confirmation of the presence of the armored layer for some of these selections was made by a chemical soaking test in which seeds were allowed to remain for 30 min in a mixture of 3 parts saturated solution of potassium dichromate and 1 part sulfuric acid by volume.

In a more recent report, Carlson and Witt (22) delineated their methods of rearing *H. electellum* and their procedures for using these larvae to evaluate sunflower heads for resistance.

In a laboratory evaluation of selfed heads in 1972 tests, heads with the phytomelanin layer had a reduction in number of severely damaged seeds by over 97% when compared to heads which did not possess this layer. Selections H2131 and H2135, which possessed the phytomelanin layer, had no damage. With open-pollinated heads, severe damage was reduced by 72% in those heads with the armored layer when compared from those with no armored layer. Carlson and Witt (22) indicated seed damage was reduced by 83% when the best lines having this layer were compared to only the susceptible check line. Lines H2052 and H2059 were found in their 1973 tests to have the greatest reduction in seed damage when they evaluated heads from lines having the armored layer.

The current research endeavors of Carlson and coworkers would suggest that we will develop agronomically acceptable cultivars of *H. annuus* with tolerance or resistance to the sunflower moth. If resistant cultivars are developed, the use of chemicals for control of this species would be minimized or eliminated and the potential adverse effects of chemicals on other biota reduced.

Laboratory Rearing and Physiology

Teetes et al. (86) successfully reared larvae of *H. electellum* on the wheat germ diet originally developed by Adkisson et al. (2) and modified by Vanderzant et al. (97). They maintained their parental stock on this diet with a photophase of 14 hours at 27 C.

Adult *H. electellum* were maintained in wide-mouth gallon jars to which 25 mm of moist sand covered with a paper towel had been placed. A honey solution, in a small vial plugged with cotton, was food. Eggs were laid on small balls of absorbent cotton placed in the jar. The jar was covered with a screen wire lid.

When larvae hatched, they were transferred with a fine camel's hair brush from the gallon jar to 4-dram clear plastic vials containing artificial diet. Racks of the smaller vials were then placed in bioclimatic cabinets where temperature and photoperiod was maintained. They reported that induction of a larval diapause was dependent on an interaction of both temperature and photoperiod. In their studies the Texas researchers evaluated two temperatures, 21 C and 27 C, and three photoperiods, L10:D14; L12:D12; L14:D10 (L = light period, D = dark period, in hours). Diapause was induced more easily in larvae maintained at 21 C than at 27 C, but only if the light period was less than 11 hours/day. These researchers reported the greatest percentages of diapausing larvae were obtained when they were subjected to 10 hours or less of light/day at 21 C regardless of the photoperiodic exposure given eggs or adults. Termination of diapause was dependent on both photoperiod and temperature, and required less time at 27 C than at 21 C. Larvae on a regimen of 11 hours or more of light/day terminated diapause sooner than those maintained at shorter light periods.

Teetes and Randolph (92) reported adult sunflower moths tested in traps exhibited no preference among the colors, red, white, blue, yellow, and green.

They obtained an average of 117 male sunflower moths in each trap baited with 5 virgin females within a sunflower field; a strong sex aggregation response was demonstrated. A sex pheromone was suggested as the probable cause of the attraction.

European Sunflower Moth, *Homoeosoma nebulella* Hbn.
[Lepidoptera: Pyralidae]

The common species of sunflower moth in Europe and Russia, *Homoeosoma nebulella* Hbn., was of little economic importance prior to the culture of sunflowers in that area of the world (57). Reh (72) of Rumania, Gabor (39) of Hungary, and Kienwick (57) of Germany, have reported on the life cycle of *H. nebulella* in their countries and the damage to sunflower caused by this species.

H. nebulella and *H. electellum* have some similar characteristics. Identifying characters and biology of *H. nebullela* provided by Kienwick (57) are included.

Life Stages

Larva. Larvae are a "dirty" green-yellow, with "dirty" purple stripes on the back and two stripes on the sides. Mature larvae are about 1.5 cm long and fusiform in shape. Head is brown, mandibles darker brown, and legs are black.

Adult. Forewings of moths are a light yellow, 11 to 12 mm long with leading edges whitish underneath and two brown spots in the middle. Hind wings are transparent and whitish with gray leading edges. Overall body color is a pale yellow.

Life Cycle

Kienwick (57) reported *H. nebulella* overwinters as larvae in a cocoon. Pupation occurs the following spring. Females lay approximately 200 eggs on the inner wall of the ring of coalesced stamens of early bloom sunflower. The first two larval stages feed only on pollen and apparently the remaining larval stages feed only on the seeds, constructing a webbing around the damaged inflorescence similar to *H. electellum*. In about 3 weeks after hatching, larvae lower themseles on a silken thread and pupate in a silk-like cocoon about 3 cm below the soil surface. Duration of the life cycle under optimum conditions ranged from 35 to 45 days. Up to three generations per year have been reported (57, 72).

Banded Sunflower Moth, *Phalonia hospes* Wlshm.
[Lepidoptera: Phaloniidae]

Westdal (101) originally reported the banded sunflower moth, *Phalonia hospes* Wlshm., a head-infested species, caused seed destruction in 1948 ad 1949 of 1.7 and 3.5%, respectively, in Manitoba sunflower fields. Schulz and Lipp (84) predicted this species would become of major importance with increased production of sunflower in North Dakota.

While early reports of activity of this species reflected potential concern that *P. hospes* would have a significant impact on sunflower production in Manitoba and North Dakota, increased production of sunflower in both areas has not been accompanied by increased activity of this species. In North Dakota annual banded sunflower moth activity has ranged from nearly nonexistent to moderate. Damage associated with the highest levels of infestation has been low. (Schulz, J. T. 1970–1976. Annual reports. Department of Entomology, North Dakota State Univ., Fargo.)

Life Stages

Egg. The only description of the life stages of *P. hospes* was provided by Westdal (101). He reported eggs were oval, somewhat flattened, with reticular markings, about 0.45 mm in length, 0.29 mm in width. Freshly laid eggs are white but become light brown at maturity. Eggs are deposited singly on the bracts of the sunflower head; largest numbers are laid on the outer whorl of the involucre of the bracts. Heads just prior to flowering are preferred for oviposition.

Larva. Newly hatched larvae are white and about 1.0 mm long. Westdal reported the head capsule is dark brown. After the first molt larval coloration changes from light pink or yellow to reddish or purplish, and finally to green at maturity. Larvae have five instars; a fifth instar larva is about 10 mm long.

Newly emerged larvae move from the bracts to the florets, feed on the pollen until the third instar is reached, and then tunnel into the achene, where they may consume part or all of the contents. Westdal (101) noted that while the seed coat remained soft, larvae tunnelled through it to reach adjacent seeds. As the seed coat hardened, larvae entered and left via a hole in the top of each seed. Each larva destroyed an average of 2.5 seeds during its larval development.

Pupa. Pupation of *P. hospes* occurred in late June in Manitoba after overwintering in the cocoon stage in the soil. The pupa is dark brown and about 6.0 mm long. Prior to adult emergence, pupae work themselves to the soil surface.

Adult. Westdal and Barrett (103) reported this small moth is about 6 mm long with a wing span of about 13.5 mm. The straw-colored forewing possesses a dark brown, roughly triangular area in the median portion of the wing. Hind wing is light gray brown and without any distinctive markings.

Moths are nocturnal and strongly attracted to light. During daylight, they are quiescent and remain on the undersides of lower leaves of sunflower plants or other objects.

Life and Seasonal History

Westdal (101) and Westdal and Barrett (103) reported that fifth instar larvae leave sunflower heads and enter the soil near the base of the plant from late August to the end of September. After larvae burrow to about 50 mm deep, they spin a cocoon and overwinter in this state. Larvae remain inactive within the cocoon until pupation commences in late June. Westdal (96) determined the pupal period was about 12 days under laboratory conditions. Initial emergence of adults occurred in early July; emergence continued until the latter part of August. Egg laying began in Manitoba about 15 July and continued for about 6 weeks. No eggs were found in the field after 25 August. Eggs hatched in 5 to 8 days. About 3 weeks are required for larvae to attain the fifth instar stage.

Westdal (101) noted that despite oviposition by *P. hospes* over a long period, larvae in any individual head are usually within one or two instars of the same stage of development. This finding was attributed to the short period during which the head is susceptible to infestation. Larvae in all stages of development, however, could be found in most fields because development of the open-pollinated cultivars grown at that time did not bloom uniformly, and susceptible plants were available for prolonged periods.

Control. Westdal (102) reported at least two species of parasites, *Chelonus phaloniae* Mason and *Glypta* sp., attacked the banded sunflower moth in Manitoba and restricted populations of this species to a subeconomic level. Earlier Westdal and Barrett (103) found 80% of overwintering larvae were parasitized. They also reported that deep tillage of stubble fields in the fall reduced spring emergence of *P. hospes* by about 80%.

With noneconomic levels of *P. hospes* also experienced in North Dakota (Schulz, J. T. 1970–1973. Annual reports. Department of Entomology, North Dakota State Univ., Fargo), no management or control practices have been needed to control this insect. Since the parasite complex identified with other moth species of the sunflower pest complex is not host-specific, it is probable that some of these species help maintain *P. hospes* at noneconomic levels. Normal disking or chisel-plowing to a depth of 10 to 15 cm by North Dakota farmers also contributes to increased larval mortality.

No insecticides are registered now for use in control of *P. hospes* in the USA.

Sunflower "Budworm", *Suleima helianthana* (Riley) [Lepidoptera: Olethreutidae]

While no approved common name for this species has been established by the Entomological Society of America's Committee on Common Names of Insects, *Suleima helianthana* (Riley) has been referred to as the sunflower budworm (88). Riley (73) originally described adults of *S. helianthana* as a Tortricidae, *Semasia helianthana*. Dyar (30) listed the species as *Thiodia helianthana* Fernald (Tortricidae). Heinrich (49) reported this species as *Eucosma helianthana* and noted that larvae had an arrangement of setae on the ninth abdominal segment which was characteristic of the Olethreutidae. Heinrich (50) and Forbes (37) defined this species as *Suleima helianthana* (Riley).

Distribution of *S. helianthana* was reported from Maryland to Texas and west to California (30, 37, 50, 73). Walker (98) reported large numbers of *S. helianthana* in Kansas. More recently Schulz (Schulz, J. T. 1971. Annual report. Department of Entomology, North Dakota State Univ., Fargo) and Robinson (75) reported this species occurred in North Dakota and Minnesota.

Life Stages

Egg. Satterthwait (81) reported eggs of *S. helianthana* are about 0.3 mm wide, 0.5 mm long, and about 0.2 mm thick and strongly attached to the plant surface by a secretion. Eggs are marked distinctly with rounded hexagonal impresses, concave in end areas, convex elsewhere, and arranged in alternating rows. Chorion was delicate and similar to waterproof transparent paper. Satterthwait observed a single female in midsummer laid 130 eggs. He reported the egg stage required about 4 days.

Satterthwait's early observations were corroborated by Ehart (31), who found freshly laid eggs translucent white and slightly ovoid, with a mean size of 0.58 ± 0.06 mm long and 0.41 ± 0.05 mm in diam. The chorion was sculptured irregularly, giving the egg a distinctly wrinkled appearance.

Ehart (31) observed eggs usually were laid singly and found on the growing point and upper leaves of immature sunflower and at the base of heads of mature sunflower. Brassard (13) also found that oviposition on sunflower seedlings was concentrated near the terminal region. Of 428 eggs counted on plants placed in oviposition cages in the greenhouse, 34.8% were located on upper leaf surfaces, 59.8% on lower leaf surfaces, and 5.4% on the stem. Mean number of eggs oviposited by 12 females reared in vitro was 237 (range 59 to 439).

Larva. Satterthwait (81) found fully grown larvae were 8 to 11 mm in length, with head and body widths 1.27 mm and 2 mm, respectively. Cuticle was white and contained small red-brown setae flared at the base. Head and cervical shield were dark brown, anal plate, light brown.

MacKay (62), in her monograph of larval characteristics of North American Olethreutidae, reported length of the largest *S. helianthana* larvae was 14 or 15 mm, somewhat larger than previously reported by Satterthwait (81). Head was yellow brown usually with some darker pigmentation in the ocellar area and at the postgenal juncture. Thorax and thoracic appendags possessed varying degrees of brownish pigmentation. Body color was pale. Ehart found that larvae collected in the Red River Valley of North Dakota fit the larval descriptions provided by MacKay. Fifth instar larvae were approximately 15 mm long with yellow-brown heads, spine-like setae, and pale yellow-brown abdomen. *S. helianthana* possesses a spinneret separating this species from other presently known *Suleima* sp. Located between the labial palps, the spinneret was about five times as long as wide with a broad distal end.

Pupa. Satterthwait (81) reported pupae were about 1.32 mm wide and 5 mm long and often larger. Sheaths of the upper and hind wings and the third pair of legs reached to the middle of the fourth abdominal segment. The head and dorsum of the first abdominal segments were unarmed, the dorsum of the second abdominal segment smooth; other segments had many spines. Duration of the pupal stage was 6 days. Ehart found pupae of *S. helianthana* averaged 1.8 ± 0.07 mm wide and 9.5 ± 0.03 mm long.

Color of pupae varied from light tan just after pupation to dark brown prior to adult emergence.

Adult. Riley (73) reported adult male *S. helianthana* had a wing expanse of 15 to 20 mm. Wings were whitish gray with forewings marked with two graduated blackish patches on the inner margin and a black dash and white triangle at the apical angle. He indicated that the head and thorax were pale gray, and the abdomen gray. Satterthwait's (81) and Riley's description of adult *S. helianthana* was precisely the same.

Life and Seasonal Cycle

Satterthwait (81) provided the first account of the life cycle of *S. helianthana.* He reported this species overwintered as mature larvae and had two or more generations per year. Overwintering larvae pupated in early spring. First generation adults were active by mid-June and continued to 4 July. Satterthwait observed adults until 4 October in Missouri. He concluded that since the insect attacked terminal and axillary buds, eggs were placed in crevices near the buds and bracts as well as leaves. Females placed in cages with sunflower plants laid eggs on upper and lower leaf surfaces and along and away from the ribs and veins.

Newly hatched larvae were observed feeding as leaf miners, webbing and entering a leaf rib. In one instance Satterthwait observed a larva webbing and feeding on parenchyma of a leaf surface. Older larvae entered and fed on buds, leaf axils, and bracts or burrowed in leaf axils. Mature larvae pupated near the entrance opening and protected it by enclosing the pupa with silkbound masses of frass. Duration of the life cycle was 50 days.

Phillips et al. (67) noted fairly large larval numbers of this species during most of the growing season in Texas. Three peaks of adult *S. helianthana* activity were observed, mid-June, early July, and mid to late August.

Ehart (31) recorded two generations of *S. helianthana* occurred in North Dakota and Minnesota in studies conducted in 1972 and 1973. Adults from overwintering larvae appeared in the last week of May and were observed until mid-June. While second generation adults were observed as early as the second week of July, peak emergence occurred the third week. Adult activity persisted until mid-August. No adults were collected after this period.

Larval activity for the first generation extended from about 12 June to mid-late July. Larval development was completed in five instars, each stadium requiring 4 to 6 days. First generation larvae usually were collected initially on volunteer immature sunflower and later on infestations from current season plantings. Second generation larvae were found in late July to mid-August and on late planted sunflower and immature plants. Some larvae completed their development by mid-August, others not until mid-September. Regardless of when the fifth instar development stage was reached, all individuals remained as fully formed larvae which overwintered. Before larvae tunnelled into undisturbed pith to overwinter, entrance holes were plugged with frass, plant material, and silken strands, an observation which corroborates a similar finding of Satterthwait (81). No *S. helianthana* larvae or pupae were found in 122 soil samples examined by Ehart.

Damage Characteristics

Satterthwait (81) reported that damage symptoms caused by feeding of larvae *S. helianthana* included misshapen gall-like heads of *Helianthus*. He reported a black frass, bound together with silk, marked the entrance of each larva into the plant.

Satterthwait (81) indicated damage to the crop is greatest when the terminal buds of small young plants are available. Ehart (31) noted that all initial feeding by first instar, first generation larvae was found on terminal growth or new leaf tissue surrounding terminals. First instar feeding was not noticeable by cursory examination of the plant. When oviposition occurred on unfolded leaves and conditions favored plant growth, larval infestations were recorded in leaf axils or stems. When oviposition occurred on terminal buds or when eggs hatched near terminal buds, infestations occurred in buds or leaf axils of upper leaves. Damage symptoms associated with second instar larvae were observable. If infestations were in leaf axils or stems, they were characterized by neatly cut holes surrounded by a prominent amount of black frass similar to that reported by Satterthwait (81).

Larvae infesting terminal regions fed between the unfolded leaves or directly on leaf petioles or terminal buds. When larvae fed on older plant tissue, they remained on the surface of stalks until they were capable of penetrating the cuticle and epidermal tissue of stems. Penetration by larvae usually occurred at leaf axils or in internodes, and infrequently in leaf petioles.

Frequency of sites infested in cultivated sunflower were reported by Ehart (31). Of 4,122 infestations observed from 1971 to 1973, an average of 53.1% occurred at the juncture of the leaf axil and stem, 24.6% in the stem between leaf axils, 16.4% in the head or terminal, and 5.9% in the leaf petiole. Each larva that established in stem tissue damaged an average of 3.18 cm^2 of stem tissue (range 1.98 to 7.23 cm^2). Severe deformity associated with stem infestations resulted in stalk breakage and reduced yields when heads did not produce seeds.

When infestations of the terminals occurred prior to bud development, the entire head often was lost. When infestation occurred after bud development, damage varied from partially deformed heads to total abortion of the terminal regions.

Second generation infestations were confined largely to stem and head tissue of late-developing *H. annuus* cultivars. Larvae occasionally entered receptacles through disk flowers and developing seeds, and dislodged them. Both Heinrich (49, 50) and Phillips et al. (67) reported *S. helianthana* larvae fed on seeds.

Host Preference

Satterthwait (80) observed *S. helianthana* infesting common sunflower (*Helianthus annuus* L.), Jerusalem artichoke (*H. tuberosus* L.), *Coreopsis grandiflora* Hogg, and "bitter root."

Ehart (31) reported that in his survey of 136 plant species (36 families) in North Dakota and Minnesota in 1972 and 1973, *S. helianthana* was found to infest only five species: common ragweed (*Ambrosia artemesiifolia* L.); Canada thistle [*Cirsium arvense* (L.) Scop.]; common sunflower (*H. annuus* L.); Maximilian sunflower (*H. maximiliani* Schrad.); and prairie sunflower (*H. petiolaris* Nutt.).

While *Helianthus* sp. appear to be the predominant hosts for *S. helianthana,* Brassard (13) recorded an 11.5% survival of 174 larvae bioassayed on *A. artemisiifolia* in the greenhouse. In a similar study with giant ragweed (*A. trifida*), none of the 45 larvae bioassayed survived on this host.

Laboratory Rearing

Brassard (13) evaluated 32 artificial diets before he found one that was suitable for rearing *S. helianthana.* He achieved an 80% survival of larvae of this species when they were reared on a shredded diet developed to rear the tobacco hornworm, *Manduca sexta* (Johannson). The diet was developed originally by Yamamoto (109) and modified by R. A. Bell (Reinicke, J. P. 1975. Personal communication. USDA, Metabolism and Radiation Research Laboratory, Fargo, N.D.). Survival rates of all stages, larval and pupal weights, and development times were similar to those of larvae reared on sunflower.

Larvae were reared in 28 g jelly cups to which the diet had been added previously. Diets were acclimatized to incubator temperatures prior to capping with either cardboard or plastic caps. Other conditions consisted of temperatures at 25 ± 2 C and a 16 hour:8 hour light-dark regimen. No more than two larvae/cup could be reared successfully. Pupae were allowed to remain in the diets until adults emerged to reduce mortality associated with handling.

Adults were transferred to 9 × 14 cm clear, plastic, cylindrical containers which had been lined with green blotter paper. They were maintained at 24 ± 8 C and a 16/8 light-dark regimen. Seedlings or excised seedlings were provided as resting sites and as an ovipositional stimulus in a 100 ml beaker filled with moist soil. A mixture of brown sugar, yeast hydrolyzate, and distilled water was provided through openings in the top of the cage by inverted 4-dram vials. One to five pairs of male and female moths were placed in these oviposition cages. Eggs were removed daily.

Control

Although *Suleima helianthana* has been listed as one of the species of the sunflower insect pest complex of importance in Texas, North Dakota, and Minnesota, the magnitude of infestations have not made control measures necessary. No insecticides are registered now to control this species in the USA.

Differences in infestation levels were observed among cultivars and breeding lines evaluated from 1971 to 1973 (Schulz, J. T. 1971 to 1973. Annual reports. Department of Entomology, North Dakota State Univ., Fargo). Infestation levels ranged from 0 to 42.5% in 43 cultivars in 1971; 2 to 54% in 60 cultivars in 1972; and 3 to 56% in 73 cultivars in 1973. Since many of the parental lines and cultivars evaluated from 1971 to 1973 possessed undesirable agronomic characteristics, few have reached commercial status.

Schulz also observed that late planted sunflower in 1974, (late May to early June) had very low infestations of *S. helianthana*. Low infestation levels continued during the 1975 and 1976 growing seasons. (Schulz, J. T. 1974 to 1976. Annual reports. Department of Entomology, North Dakota State Univ., Fargo.)

Parasites and predators have been identified from parasitized *S. helianthana* larvae. Satterthwait and Swain (82) and Satterthwait (81) reared and identified the following species: *Hydnocera pubescens* LeC.; *Lixophaga variabilis* Coq.; *Clausicella* (= *Siphophyto*) *floridensis* Towns.; *Helicobia helicis* Towns.; *Muscina stabulans* (Fall.); Bracon (= *Microbracon*) *caulicola* Gahan; *Bracon* sp.; *Apantales epinotiae* Vier.; *Pristomerus euryptchiae* Ashm.; and *Macrocentrus ancylivorus* Roh. Impact of these parasites and predators on *S. helianthana* infestation levels was not determined.

Brassard (13) identified three species of parasitic wasps, *Macrocentrus* sp. (probably *ancylivorus*), Hymenoptera: Braconidae; *Glypta* sp. and *Temelucha* sp. both Hymenoptera: Ichneumonidae from parasitized *S. helianthana* larvae in North Dakota and Minnesota. The period of activity of each of these three species was delineated from larval collections and subsequent rearing of parasites. *Macrocentrus* sp. parasitized first generation larvae of *S. helianthana* in early July and reached the peak of its activity in mid to late July. *Glypta* sp. parasitized both first and second generation larvae from mid-July to early September. *Temelucha* sp. parasitized first generation larvae in mid-July. Levels of parasitism of *S. helianthana* ranged from 1.1 to 20.5% and depends on location of the infestations and the time of year. Highest level of parasitism recorded during 1973 and 1974 was in mid-July when all three parasites were active.

Seed Weevils, *Smicronyx fulvus* LeC. and *S. sordidus* LeC.
[Coleoptera: Curculionidae]

Infestations of sunflower seeds by the weevils, *Smicronyx fulvus* LeC. and/or *S. sordidus* LeC. have been reported at irregular intervals for more than 50 years. Forbes (36) provided the earliest account of the magnitude of damage caused by *S. fulvus,* "sunflower weevil is a new insect to agriculture which has led to the virtual abandonment of the growing of sunflower seed for oil." Cockerell (26) reported *S. fulvus* common on sunflower in the Boulder, Colorado, area, but indicated it was not a serious pest. Phillips et al. (67) reported the "sunflower weevil," *S. sordidus,* fed on sunflower seeds; they considered this species of potential economic importance to sunflower growers in Texas.

Recent outbreaks of this seed weevil complex in North Dakota have tended to corroborate the potential damage that can be caused by these species. Up to 80% of seeds in some commercial sunflower fields in southeastern North Dakota were infested by larvae of *S. fulvus* and *S. sordidus.* (Oseto, C. Y. 1975. Annual report. Department of Entomology, North Dakota State University, Fargo.)

Taxonomy

In early publications dealing with these two insect species, both were placed in the genus *Desmoris* and referred to as *D. fulvus* and *D. constrictus.* Anderson (4) reevaluated the morphology and taxonomy of all species in the genus *Smicronyx* and reclassified all forms in the genus *Desmoris* to *Smicronyx.* Recent literature dealing with these seed weevils reflects this change in classification.

Life Stages

Egg. Satterthwait (80) provided the only description of this life stage of *S. fulvus.* He reported eggs of this species were elliptical, white, and measured from 0.45 to 0.66 mm long, 0.38 to 0.56 mm in thickness. They are deposited singly in the feeding scars near the kernel. Oseto also found eggs of *S. fulvus* oviposited between the achene and seed coat. Since these feeding punctures close tightly, external visible evidence of oviposition is generally lacking.

Satterthwait did not observe eggs of *S. sordidus.* No other description is available.

Larva. S. fulvus larvae are white and undergo five developmental instars. Mature larvae were 0.58 mm long. Developmental period for the
initial four larval stadia range from 7 to 14 days. Preliminary results from
laboratory rearing studies indicated the fifth instar undergoes diapause
(Oseto, C. Y. 1976. Personal communication).

Pupa. No published descriptions of the characteristics of the pupa of
either species have been located.

Adult. Satterthwait (80) reported *S. fulvus* (= *D. fulvus*) was about 2.5
mm long and a bright rufous or iron rust in color. Anderson (4), in his comprehensive description of this developmental stage, reported a mean length
of 2.85 mm for *S. fulvus* males, 2.94 mm for females. Elytra were black to
ferruginous, with small, ovate scales. Underside of the thorax was covered
with grayish white scales; scales of the underside of the abdomen, light
ochreous.

Satterthwait reported *S. sordidus* (= *D. constrictus*) was slightly larger
than *S. fulvus* and gray. Anderson (4) reported a mean length of 3.52 mm
for *S. sordidus* males, 3.59 for females. Elytra were piceous to black with
small elliptical white to yellowish-white scales. Underside of thorax and abdomen were covered with rounded, white scales.

Life and Seasonal History

While most larvae of both species overwintered in the soil (80), some
remained within the hull until the following spring and summer. Satterthwait's observations from his studies in eastern Missouri and Central
Illinois were corroborated by Oseto, who also observed the majority of
larvae of *S. fulvus* and *S. sordidus* overwintered in the soil in North
Dakota. Oseto also documented larval overwintering in sunflower seed.
Satterthwait found a few larvae of *S. fulvus* remained in this stage for nearly 11 months. A 10-month larval period was considered normal.

Pupation took place in the soil and was completed in about 8 days.
Oseto's observations in North Dakota were similar; he reported a pupal
period of 7 to 9 days. Satterthwait (80) observed adult *S. fulvus* on
Helianthus sp. from 17 July to 29 Sept. 1934. *S. sordidus,* from 28 June to
15 Aug. 1934. In his adult collections Satterthwait obtained from 20 to 30 *S.
fulvus* adults to every *S. sordidus* adult. Oseto recorded the presence of
adult *S. sordidus* in sunflower fields earlier in the season than adult *S. fulvus.* However, the predominant species of cultivated sunflower during the
season was *S. fulvus.* First adults were collected on 18 July 1975, with peak
emergence of *S. fulvus* the week of 20 July 1975. Adults initially fed on the
involucral bracts. As the plant matured, feeding activity became concentrated on the pollen.

Both Satterthwait (80) and Oseto reported these two species had one
generation per year.

Host Preference

Satterthwait (80) reported infestations of both species only on *Helianthus* sp.

Control

Attempts by Satterthwait (80) to control the *Smicronyx* species by applications of chemicals in 1934 and 1935 were not successful. Muma et al. (65) reduced the weevil population for 5 days following applications of benzene hexachloride at 1.12 kg ai/ha. Corresponding decrease in percentage of infested seed was not obtained. Applications of DDT did not result in satisfactory control. Oseto reported no significant reduction in larval damage to seeds following ground and aerial application of methadithion at 0.56 kg ai/ha in 1975. He attributed the poor control to insufficient coverage by the insecticide on the plant or ineffective timing of the application.

Satterthwait (80) also reported his attempts to develop lines of sunflower resistant to infestation by the weevil complex. These studies, conducted from 1936 to 1940, resulted in isolation of seven lines, developed by sibbing and selfing, which are "related resistant." Infestations in the most resistant selections ranged as low as 5% of the plants and less than 1% of the seed.

The *Smicronyx* species have several parasites. Bigger (7) observed parasitism of *S. fulvus* by *Callimone albetarse* Huber, a chalcid wasp, in Illinois. Bigger (8) again reported additional parasites were observed in association with or parasitizing *S. fulvus* and included another chalcid, *Zatropis incertus* (Ashm.) and *Bracon mellitor* (Say) [Hymenoptera: Braconidae]. The magnitude of parasitism by the latter species led Bigger (8) to believe *B. mellitor* was an important natural control agent of *S. fulvus* in the field. Satterthwait (82) confirmed the earlier observation of Bigger regarding the potential importance of *B. mellitor*. He found this parasite in 65% of the heads infested by *S. fulvus* at one Illinois site.

More recently Oseto found an unknown species of Therevidae (Diptera) in the soil with larvae of *S. fulvus* and *S. sordidus*. The impact of this group of insects as a mortality factor in control of the *Smicronyx* larvae is not known. Currently we have no truly satisfactory control of either species of the seed weevil complex on cultivated sunflower.

A Sunflower Curculio, *Haplorhynchites aeneus* (Boh.) [Coleoptera: Curculionidae]

Haplorhynchites aeneus (Boh.) is the most common and widely dispersed species of this genus of weevils. It occurs throughout the eastern U.S. and westward to the Continental Divide, although it is most numerous in the Midwest. The species also has been recorded from Manitoba and Saskatchewan, Canada.

Damage associated with infestations of *H. aeneus* generally has been minimal in cultivated sunflower although losses to 10% have been recorded in localized areas of North Dakota, especially in fields adjacent to those planted to sunflower the previous year. (Schulz, J. T. 1970 to 1972. Annual reports. Department of Entomology, North Dakota State Univ., Fargo).

Taxonomy

Early literature dealing with this species referred to *Rhynchites aeneus* (Boh.). Hamilton (44) revised the species included in *Rhynchites* and placed six species, including *aeneus,* in the genus *Haplorchynchites.* Recent published literature reflects the acceptance of this revision.

Life Stages

No published descriptions of the egg, larval, or pupal stages of *H. aeneus* were found. This reviewer found eggs in the developing head and excised them from the base of the disk flowers. Eggs ranged from a pearly white to cream color and were ovoid to round. Hamilton (43) also noted placement of the eggs at the base of disk flowers. He studied the biology of this species on *Helianthus divaricatus* L., *H. annuus* L., and *H. grosseserratus* Martens in Illinois.

Larvae also were obtained from the head and appeared white or grayish white. They were 'C'-shaped and ranged from about 4 to 6 mm in length.

Adults. Adult males and females are black; males, 3.9 to 6.6 mm long, 2.0 to 3.4 mm wide; females 4.0 to 6.5 mm long, 2.1 to 3.4 mm wide. Head, pronotum and elytra are punctate.

Life and Seasonal History

Hamilton (43) published the only account of the life cycle of *H. aeneus.* He observed male and female adults on sunflower heads in mid to late July. Copulation occurred on the developing heads. After or during copulation, females moved from the head to a point on the stem 12 to 25 mm below the head and cut the stem with their mandibles. The cut is not complete, and the flower head remained attached for varying periods of time. Eggs, which are found at the base of the disk flowers, hatch in about 1 week. First instar larvae feed on pollen from a single disk flower. As the larvae matured, adjacent disk flowers were consumed. By the time larvae reached maturity, 20 to 30 days after hatching, heads were broken from the plant and had fallen to the ground. Larvae left the cut heads and entered the soil to a depth of about 30 cm and overwintered there. Highest concentration of larvae were in the soil near the host plants.

Pupation occurred the first week of July of the following year. Duration of this period was about 10 days. Hamilton (43) observed a single generation per year.

The life cycle documented by Hamilton in Illinois appears to closely reflect the cycle of this species in North Dakota (Schulz, J. T. 1970 to 1972. Annual reports. Department of Entomology, North Dakota State Univ., Fargo). Adults were observed initially in cultivated sunflower on 20 July 1970, 16 July 1971, and 21 July 1972. No adults were observed in the field after the second week of August.

Infestation density of *H. aeneus* fluctuated markedly in North Dakota. Moderate infestations were observed from 1970 to 1972 with "head clipping" from 0 to 22% recorded in parental lines and cultivars of *H. annuus*. From 1973 to 1976, *H. aeneus* infestation levels have been low to nonexistent.

Damage

Primary damage that is associated with adult feeding on *Helianthus* sp. is the result of stem-feeding by the female. Heads either hang limply attached to the stem at the point of fracture or ultimately break from the stem and drop to the ground. Similar damage caused by *Rhynchites mexicanus* Gyll. has been recorded by Fuchikovsky (38) from sunflower in Mexico.

Host Preference

Haplorhynchites aeneus has been collected from the following hosts: *Helianthus annuus* L., *H. grosseserratus* Marteus, *H. divaricatus* L., *H. microcephalus* T. & G., *Silphium perfoliatum* L., *S. lacinatum* L., *S. terebinthinaceum* Jacq., *S. integrifolium, Heliopsis helianthoides* (L.), and *Psoralea pedunculata* (Mill.).

Control

No management or control measures have been required to reduce damage caused by *H. aeneus,* nor are there any insecticides registered for control of this species on cultivated sunflower in the USA. Since the heaviest infestations to date have occurred in fields planted adjacent to those of the preceding year, placement of current plantings at some distance from previous plantings may reduce the potential for damage by this species.

This reviewer noted marked variation in head clipping among 132 sunflower accessions evaluated for tolerance or resistance to this species from 1970 to 1972. Heads clipped in the hybrid (P-21 VRI × P-21 VRS) × HA60, averaged less than 1% during those 3 years. Many of the lines evaluated during this period that had HA60 as a parent sustained head clipping of 2% or less. Parental lines HA64, HA65 and HA113 were promising sources of resistance or tolerance. Low population levels of *H. aeneus* since 1972 have prevented further studies with this species.

Sunflower Seed Midges
[Diptera: Cecidomyiidae]

Three species of midges [Diptera: Cecidomyiidae] have been recorded as pests of sunflower in the USA. Breland (14) noted infestations of sunflower seed by *Lasioptera murtfeldtiana* Felt and *Asphondylia globulus* Osten Sacken in Texas. Satterthwait (81) reported *L. murtfeldtiana* infested sunflower in Missouri and Illinois. More recently Schulz (74) recorded severe damage to sunflower grown in North Dakota and northwestern Minnesota by *Contarinia schulzi* Gagné.

Of the three species *L. murtfeldtiana* and *C. schulzi* appear to be the most important pests of sunflower.

L. murtfeldtiana has been recorded from Texas, Illinois, Kansas, and Tennessee; *C. schulzi* from North Dakota, Minnesota, and Texas.

1. *Lasioptera murtfeldtiana* Felt

Satterthwait (81) provided the principal description of the life stages of this species.

Life Stages

Egg. Eggs of this species were reported as similar to those of the Hessian fly, *Phytophaga destructor* (Say), in size, form, and color. They are placed in crevices in the head and in the florets. Satterthwait (81) observed hatching in less than 46 hours.

Larva. Mature larvae are about 2 mm long and 2.5 times long as wide. They are white, with a distinctly segmented body. Duration of the larval stage was not reported.

Pupa. Pupation occurs within the achene, sometimes within a pocket in the lining of the achene. Pupa is about 2.5 mm long and ranges from white to black. A prominent, bifid, chitinous horn is present on the front of the pronotum. Duration of the pupal stage was 5 to 9 days.

Adult. Flies are about 2 to 2.5 mm long, brown and covered with silvery scales at margins of the abdominal segments. Newly emerged adults possess wings with dark scales. These scales are soon lost, and the wings are then hyaline. The abdomen of the gravid female appears red, due to the color of the eggs.

Life and Seasonal History

Satterthwait (81) reported the only stage of *L. murtfeldtiana* that overwintered in the St. Louis area was the larva in atrophied seed in soil, trash, or on the dead stalk. There are no reports of the number of generations per year for this species. Females appear to have a high reproductive potential since 157 eggs were dissected from one gravid female.

Damage

Satterthwait (81) observed newly hatched larvae migrated to undeveloped sunflower ovaries and burrowed through all parts of the developing flower. A single achene may be invaded by several larvae. He indicated that larvae were capable only of infesting immature achenes. Breland (14) noted achenes showed no external evidence of infestation, and they ultimately dried and were blown away as chaff during the threshing operation. No estimates of damage caused by this species have ben reported.

Host Preference

The only hosts reported to be attacked by *L. murtfeldtiana* were *Helianthus* sp., including *H. annuus* L., *H. grosseserratus* Martens, *H. mittali* Torrey & Gary, *H. uniflorus* Nutt., and *H. chrysanthemiflorus* Hort. ex Bailey (81).

Control

Measures were not undertaken to control this species. Satterthwait (81) suggested fall plowing of trash as one principal means of reducing infestation levels. One predator, *Orius insidiosus* (Say) [Insidious plant bug], and three parasites are found attacking this species. Parasites included *Platygaster* sp., a wasp, and two chalcids, *Torymus obscura* (Breland) and *Rileya n. sp.* (81).

2. *Contarinia schulzi* (Gagne)

Infestations of a new Cecidomyiid species were discovered in cultivated sunflower in northwestern Minnesota and eastern North Dakota in 1971 by Schulz and coworkers. Gagné (40) described adults and mature larvae and reported this was probably a native species unknown until its discovery in 1971. Additional descriptions of eggs, early larval instars, and the biology of this species are provided by Samuelson (79).

Life Stages

Egg. Individual eggs of *C. schulzi* are extremely small. The actual size was not determined. Samuelson observed they were laid singly or in masses averaging about 47 eggs per mass. Numbers of masses ranged from 1 to 11 per plant. Eggs or masses of eggs were laid on the upper and lower epidermis of the inner bracts of the developing sunflower head. Eggs initially laid were white, opaque and with a sculptured chorion. Eggs became orange as they reached maturity. Duration of this cycle was 3 days.

Larva. C. schulzi has three larval instars (79). First and second instars are translucent to opaque white and ranged from 0.13 to 0.7 mm long. Last instar larvae are opaque to pinkish white, 1.0 to 1.5 mm long, and possess the typical cecidomycid clove-shaped "breastbone" or sternal spatula.

Newly emerged larvae concentrated in clumps at the bases of the bracts on the margin of the head. As the bud matured, larvae migrated from the periphery to the center of the head. Duration of the larval stage was 10 days. When mature larvae dropped from the head to the soil, they formed cocoons in 4 to 5 days and a "nymph" 10 to 12 days later.

Adult. Adult males and females are small flies whose wing length ranges from 1.38 to 1.55 mm. Wing and body color are gray. The reader is referred to Gagne's description for additional details.

Life and Seasonal History

C. schulzi overwintered in soil only as a larval cocoon. Formation of this life stage is governed by the percentage soil water present with larger numbers of larvae becoming cocooned in soils at 40 and 50% water content. Adult emergence also was dependent on the percent soil water content with maximum emergence at 50 and 60% soil water. In soils with less than 40% soil water, adults did not emerge. Adults emerged in early July in North Dakota and northwestern Minnesota; larval infestations were noted generally from 10 to 17 July in volunteer and cultivated *H. annuus*. Adults live from 3 to 5 days.

Females oviposited on developing buds, with maximum damage recorded to plants whose bud diameter at the time of oviposition was about 2.0 mm. Oviposition in leaf axils also was noted in plants whose bud development was not advanced enough for oviposition. Duration of the life cycle was 30 to 35 days. Number of generations per year has not been determined precisely.

Damage

Schulz (83) reported initial damage by larval feeding was manifested by necrotic spots that appeared on the flower bracts. If infestations preceded opening of the inflorescence, distorted ray petals or their total absence indicated possible presence of midge larvae. Infestations within developing buds that were 25 to 50 mm in diam frequently resulted in heads that did not open or develop normally. Such heads appear gnarled. Since larvae tended to concentrate initially in an area between the bract and developing inflorescence, seed on the periphery of the head was damaged first. As infestations increased, the entire head became damaged. Center areas of 25 to

100 mm on many heads were totally devoid of seed. Yield losses in commercial fields have ranged to 20%.

Host Preference

Infestations by *C. schulzi* have been restricted only to *Helianthus annuus* L., *H. maximiliani* Schrad., and *H. petiolaris* Nutt. No other hosts have been identified as reservoirs for this species.

Control

Properly timed applications of insecticides have controlled larval infestations of *C. schulzi*. Applications just prior to emergence of adults in early to mid-July, protected sunflower. Methadithion is the only insecticide registered for use to control midge on sunflower. When larvae are established in developing heads, economic control cannot be obtained from applications of insecticides (Schulz, J. T. 1972 to 1974. Annual reports. Department of Entomology, North Dakota State Univ., Fargo).

No resistance has yet been identified in parental lines or hybrids developed from either public or private breeding programs, although differences in susceptibility have been observed (Schulz, J. T. 1972 to 1976. Annual reports. Department of Entomology, North Dakota State Univ., Fargo).

A Tephritid Complex
[Diptera: Tephritidae]

Surveys to determine the characteristics and magnitude of damage caused by insects to cultivated sunflower were conducted from 1967 to 1970 in North Dakota. These surveys resulted in the identification of a tephritid complex of six species associated with sunflower, three of which appeared to have the potential for causing host injury. Two of these species, *Gymnocarena diffusa* (Snow) and *Neotephritis finalis* (Loew), infest the inflorescence of sunflower (61).

Gymnocarena diffusa (Snow). This species is predominantly restricted to the Great Plains. The eastern limit of its range is Missouri, the northernmost record is Montana; this species ranges into the Southwest as far as Arizona (33).

Life Stages

Kamali and Schulz (55, 56) provided the initial description of the immature stages of *G. diffusa* and defined the biology of this species.

Egg. G. diffusa eggs are light brown, tapered at both ends, with smooth chorion, 1.40 to 1.65 mm long, 0.37 mm diam. They are laid singly and inserted between the involucral bracts of sunflower heads.

Larva. Mature larvae are yellow to light brown, 7.50 to 7.80 mm long and 2.60 to 3.25 mm wide. They are sharply tapered anteriorly and truncated caudally. Mature larvae in the prepupal stage are heavily pigmented, the greatest concentrations between the body segments. Larvae of *G. diffusa* inhabit and feed on the spongy tissue of the sunflower receptacle.

Puparium. Pupation of *G. diffusa* takes place on the sunflower receptacle, on the disk among the seeds, or in the soil. The pupa remains within modified last larval skin. Puparium is egg-shaped, black and heavily sclerotized, 6.2 to 7.5 mm long and 3.25 to 3.60 mm wide.

Adult. Adult is a fairly robust fly, 7.6 to 7.8 mm long. Body color ranges from a yellow brown to light brown. The wing is distinctive in having an almost continuous, transverse, hyaline band from the costa to the posterior margin. There are other hyaline markings.

Life Cycle

G. diffusa overwintered in the soil at a depth of about 3 cm as a pupa from late August or early September until mid-June. Adults emerged in late June and fed on glandular secretions of the disk flowers of open buds and involucral bracts of volunteer and wild *Helianthus* species. Oviposition on sunflower cultivars commenced in mid-July on unopened buds. Most eggs were laid between the second and fourth layer of bracts during the afternoon. Oviposition period ranged to 10 days with up to 94 eggs per female laid. Larvae emerged 6 to 8 days after eggs were laid and moved to the spongy tissue of the receptacle to complete entire larval development in about 30 to 32 days. Larvae complete their development in three instars. Mature larvae normally formed a hole of about 1.5 mm diam and dropped to the soil to pupate in mid to late August or early September. Occasionally mature larvae tunneled toward the disk and pupated among the seeds. These pupae later dropped to the soil.

Damage

Moderate to extensive tunnelling by larvae in receptacle tissue caused the predominant host damage. Larvae do not feed on developing seeds; no adverse effect on head development, seed set or on yields was noted.

Host Preference

Gymnocarena diffusa (Snow) has been found infesting only *H. annuus* L. and *H. maximiliani* (Schrad.) in North Dakota.

Control

Damage caused by this species has been insignificant. Control measures have not been required. A parasitic wasp, *Perilampus* sp. (Hymenoptera: Perilampidae) was reared by Kamali from pupae of *G. diffusa* in North Dakota. An *Aspergillus* sp. also was found to infect pupa of *G. diffusa* in the field. Impact of these two organisms on the populations of *G. diffusa* was not assessed.

Neotephritis finalis (Loew). Of all the species of the tephrited complex, *Neotephritis finalis* (Loew) has been the most widely cited for its association with and damage to sunflower. Walker (98) observed this species in moderate members feeding on sunflower seeds in Kansas. Breland (14) reported the nature of damage caused by *N. finalis* in Texas. Schulz and Lipp (84) reported only localized infestations present at the time of their report, but indicated this species might be of potential economic importance in the future. Beckham and Tippins (5) reported their observations in 1969 and 1970 indicated this species to be potentially the most destructive pest in reducing seed yields in North Georgia. More recently Fuchikovsky (38) reported *N finalis* a major pest in many parts of central Mexico.

Foote and Blanc (35) reported the geographic range of this species to include the continental U.S., except Alaska and New England, southern Canada, and northern Mexico.

Taxonomy

There are numerous synonyms for *N. finalis*. Early literature referred to this species as *Tephritis afinis, T. finalis,* and *Trypeta finalis*.

Life Stages

Egg. Kamali (54) reported eggs of *N. finalis* were white to light yellow, 1.13 mm long and 0.2 mm diam. Eggs normally were attached to the corolla of incompletely opened sunflower inflorescences; occasionally they were laid between the involucral bracts of unopened buds. Incubation of eggs was 4 days under greenhouse conditions of 25 to 28 C.

Larva. Mature larvae are dark yellow, 4.5 mm long and 1.5 mm wide and feed at the base of the disk flowers and newly-formed seeds. Duration of larval period was 14 days.

Pupa. Puparium is light brown, 3.5 mm long and 1.5 mm diameter. Kamali (54) reported pupation occurred within the seeds on disk. Duration of this period was 8 to 9 days.

Adult. Foote and Blanc (35) reported *N. finalis* is a small to medium sized fly. Wings are dark brown color with hyaline spots of varying size.

Life and Seasonal History

Kamali (54) provided the only report on the biology of this species. He reported *N. finalis* overwintered as a pupa in the soil of North Dakota. Observation from 1970 to 1973 indicated adults emerged as early as the first week of July. Initial larval infestations were recorded in late July and early August. Newly hatched larvae on young or immature blooms crawled to the base of the corolla and tunnelled through the ovarian wall.

A second generation of adults was collected from mid to late August in North Dakota. In 1973, after an abnormally warm fall, a partial third generation of adults was observed in early October. Flies which emerged at that time were not able to find suitable hosts and perished with the onset of cold weather. No *N. finalis* infestations of consequence were noted during the 1974 and 1975 growing seasons.

Damage

Kamali (54) reported a single *N. finalis* larva feeding on flower parts at an early developmental stage tunnelled through 12 ovaries. Mature larvae feeding on older sunflower destroyed only one to two seeds.

Host Preference

Unlike *Strauzia longipennis* (Wied.) and *Gymnocarena diffusa* (Snow), *Neotephritis finalis* (Loew) is a polyphagous insect. Phillips (68) reported *N. finalis* larvae were found in seeds of *Actinomeris* sp. and flower heads of *Encelia californica* Nutt. and *Dahlia* sp. Foote (34) and Foote and Blanc (35) added a wide range of hosts from which *N. finalis* was collected. Among those listed were *Eriophyllum lanatum, Helianthella uniflora, Wyethia mollis, Balsamorhiza* sp., *Chrysothamnus* sp., and various species of *Helianthus.*

Control

N. finalis has not required active control measures to be undertaken in the major production areas of the USA, and no insecticides are registered. Fuchikovsky (38) recommended use of malathion 1000E, 51 liters/ha, or endosulfan 35%, 2.01 liters/ha, for control of *N. finalis* in Mexico. Sporadic occurrence of this species in the northern sunflower production areas of the USA to date suggests this species probably will not be a major problem in that area. In the southern, southwestern, and western U.S. sunflower production areas, however, *N. finalis* infestations occasionally may reach levels requiring control measures to prevent crop damage.

FOLIAGE AND STEM FEEDING SPECIES

Sunflower "Budworm"—*Suleima helianthana* (Riley)
[Lepidoptera: Olethreutidae]

Refer to discussion of this species in the section on Head-Infesting Species.

Sunflower Beetle—*Zygogramma exclamationis* (Fab.)
[Coleoptera: Chrysomelidae]

The predominant leaf feeding species of the sunflower pest complex in the northern U.S. sunflower-growing region is the sunflower beetle, *Zygogramma exclamationis* (Fab.). Criddle (28) reported *Z. exclamationis* widespread in Manitoba feeding on leaves of wild sunflower. He reported this species had spread to cultivated sunflower where it bred as readily as on the wild sunflower. Westdal and Barrett (103) included this species as one of the five most important pests of sunflower in Manitoba. Schulz and Lipp (84) reported the sunflower beetle as one of the six most important species of the sunflower pest complex in North Dakota. Distribution of the sunflower beetle is not restricted to northern U.S. Brisley (16) noted feeding by this species on *H. petiolaris* Nutt. in Arizona.

Life Stages

Description of the life stages of this species was provided initially by Criddle (28). More recently Westdal and Barrett (103) and Westdal (102) have defined characteristics of the stages more precisely.

Egg. Eggs are cigar-shaped, yellow to orange, and 1.5 to 2 mm long.

Larva. Larvae are dull yellow-green and humpbacked. At maturity they are about 10 mm long.

Pupa. Pupae are yellowish and about the size of the adult.

Adult. The sunflower beetle is similar in appearance to the Colorado potato beetle, *Leptinotarsus decimlineata* (L.), but is only about two-thirds as large, 6 to 8 mm long and 4 to 5 mm wide. Head and thorax are dark brown. Elytra are characterized by alternate light and dark and light brown stripes, the lateral dark stripe resembling an exclamation mark from which the specific name is derived.

Life and Seasonal History

Westdal and Barrett (103) and Westdal (102) provided the most comprehensive information on the biology of this species.

Adults overwinter in the soil and emerge at about the time sunflower seedlings appear and commence feeding, mate, and lay eggs. Eggs are laid singly on stems and undersides of the leaf. Duration of the egg stage is about 7 days. Larvae feed on leaves only at night. During the day they congregate under the bracts of the flower bud. Duration of the larval stage is about 6 weeks. Mature larvae drop from the plant into the soil and pupate in earthen cells. The pupal stage lasts from 10 days to 2 weeks. Adults of the succeeding generation may emerge briefly near the latter stages of the growing season and feed on uppermost leaves of the plant before returning to the soil to overwinter. There is one generation per year in Manitoba and the U.S. Northern Plains.

Damage

Principal damage associated with adults occurs soon after they emerge from hibernation. Their feeding activity on the leaf edges of seedlings has caused severe damage. Westdal (102) noted adults seldom feed on cotyledons. Fields may be defoliated severely if beetles are numerous. If beetles are controlled with insecticides at this time, plants that have been damaged severely apparently recover from this injury (106), but maturity is delayed and seed yield was reduced significantly (102).

Larval damage is evident as holes in leaves. Westdal (102) reported that plants with 25 or more larvae may be defoliated completely resulting in yield reductions of nearly 30%.

Schulz recorded a 52.6% reduction of seed in a commercial field of Peredovik sunflower which was under drought stress at the time the plants averaged 50% defoliation at the 70 to 80% bloom stage. (Schulz, J. T. 1973. Annual report. Department of Entomology, North Dakota State Univ., Fargo.)

Host Preference

Criddle (28) and Westdal (102) reported Z. exclamationis was found infesting only Helianthus sp.

Control

Westdal (102) indicated natural control factors usually kept down populations of the sunflower beetle. He reported a complex of Coccinelidae, Chrysopidae, and a beetle, Libia atriventris Say (Coleoptera: Carabidae), fed on eggs and larvae in Manitoba. Despite the effectiveness of predators and parasites, chemical control was necessary on three occasions in Manitoba in 1952, and during 1957 to 1959 and 1971 to 1974. Westdal (102) indicated adult control should be initiated when the average population exceeds one beetle per two plants. A single application of an approved insecticide usually was found adequate. Westdal recommended larval control to be initiated after all eggs had hatched. A single application of a currently recommended insecticide should be adequate. Westdal did not provide the current Manitoba insecticides because continued review required periodic changes in recommendations.

The pattern of sunflower beetle activity in North Dakota and Minnesota has been similar to that in Manitoba. The most recent major outbreak of sunflower beetle occurred in 1973 in conjunction with an outbreak of the painted lady (*Cynthis cardui* L.). Recommendations for use of insecticides are similar to those outlined by Westdal.

In the USA, toxaphene is the only chemical currently registered for sunflower beetle control. The future status of toxaphene is in doubt, because this insecticide is under review by the Environmental Protection Agency.

Painted Lady, *Cynthia cardui* L.
[Lepidoptera: Nymphalidae]

The painted lady (thistle caterpillar), *Cynthia cardui* L., is widespread throughout much of the USA and Canada. While numbers of this species are usually low and generally are no problem to successful culture of sunflower, occasional outbreaks result in serious infestation that may require control. The most recent outbreak of *C. cardui* occurred in 1973 in the western half of the USA, when the largest numbers of the painted lady in 25 years were recorded.

Life Stages

Westdal (102) provided the following description of the life stages of *C. cardui.*

Egg. Eggs of the painted lady are laid on the food plants and are small, spherical, and white.

Larvae. Caterpillars are brown to black and spiny. A pale yellow stripe occurs on each side of the body. Mature larvae are 30 to 35 mm long.

Chrysalid. The pupa of *C. cardui* is a molten gold in color, about 25 mm long.

Adult. The painted lady butterfly is about 25 mm long with a wingspread of about 50 mm. Upper wing surfaces are brown with red and orange mottling and white and black spots. Undersides of wings are marble-gray, buff, and white. Each hind wing possesses a row of four distinct and one indistinct eyespots.

Life and Seasonal History

C. cardui is indigenous to the southern regions of the USA and migrates annually to the northern regions and Canada. Westdal (102) reported the painted lady breeds in Canada, migrates south for the winter, and returns to Manitoba in early June.

Eggs hatch in about 1 week in Manitoba. Larvae fed on leaves until they reached maturity in early July. Chrysalids that are formed hang from the leaves of the plant. Duration of the chrysalid stage was about 10 days in Manitoba. Second generation was reported by Westdal as usually small in numbers, with little damage to sunflower.

Damage

Larvae of the painted lady feed chiefly on Canada thistle, *Cirsium arvense* (L.) Scop. When large migrations occur, such as those observed in 1973, larvae defoliated cultivated crops, including sunflower. Larvae produce a loose, silk webbing that covers them during their feeding activity. Effect of defoliation by *C. cardui* on seed yields is comparable to that described for defoliation by larvae of the sunflower beetle. (Westdal, 102; Schulz, J. T. 1973. Annual report. Department of Entomology, North Dakota State Univ., Fargo.)

Control

Since *C. cardui* is seldom a problem on sunflower, control measures are not normally necessary. Infestation levels of this species were extremely high in 1973 in North Dakota, South Dakota, and portions of northwestern Minnesota. Emergency use clearance of toxaphene was then obtained for use in control of *C. cardui*.

Westdal (102) reported the painted lady usually is heavily parasitized in Manitoba and also is infected by bacterial pathogens.

Sunflower Maggot, *Strauzia longipennis* (Wied.) [Diptera: Tephritidae]

The sunflower maggot, *Strauzia longipennis* (Wied.), has been the most widely studied species of the complex of tephritid flies known to be associated with sunflower. This genus of flies is strictly North American in distribution; geographic range of *S. longipennis* included nearly every contiguous state in the USA (35). Beirne (6) reported *S. longipennis* occurred wherever sunflower was grown in Canada, New Brunswick, Prince Edward Island, Quebec, Ontario, Alberta, and particularly, Manitoba.

Many variants of *S. longipennis* exist. Among the characters which have been used to distinguish these variants is the variability of the wing patterns. Kamali (54) recorded 12 different wing patterns in *S. longipennis* specimens collected in North Dakota.

Walker (98) reported the sunflower maggot as a root feeder in sunflower in Kansas. Allen et al. (3) reported *S. longipennis* was first observed in Manitoba in 1948 and by 1951 had infested 96.4% of sunflower examined. They reported pith tissue of stems was nearly destroyed. Schulz and Lipp (84) reported larvae infestating 100% of the plants in cultivated sunflower in some locales in North Dakota.

Life Stages

Egg. Westdal and Barrett (104) reported eggs of *S. longipennis* were white, smooth, elongate oval about 1.0 mm long and 0.35 mm wide. Egg stage was 7 to 8 days in duration.

Larva. Mature larvae of the sunflower maggot are typical of most tephritid larvae. They are elongate, about 9 mm long and 2.5 mm wide, pale yellow and live in a habitat surrounded by a firm, juicy food medium (67). The sunflower maggot is differentiated from other tephritids by the unique structure of the anterior respiratory organs which occur on the first thoracic segment; large organs comprised 49 to 60 tubules scattered over a wide flat distal surface (104). There are three larval instars. Duration of the larval period was 6 weeks in the field (104).

Pupa. The pupa is enclosed in a pale yellow puparium comprised of 11 segments. It is approximately 6 mm long and 2.25 mm wide.

Adult. The adult, originally described by Wiedemann (107) is a showy, yellow fly with picture wings. Westdal and Barrett (104) reported females were about 8 mm long; males, 7 mm. A row of strong black bristles on each side of the vertex is a distinguishing feature in both sexes. While wing patterns are variable, that portion of the pattern terminating at the wing tips is a design similar to the letter F. Wing markings in the male are frequently darker and coalesce; the distal portion appears almost black.

Life and Seasonal History

The most exhaustive study of the life history of *S. longipennis* was conducted by Westdal and Barrett (104) in Manitoba. They reported *S. longipennis* overwintered as a pupa in the soil after maggots leave the sunflower in mid-August. Greatest concentration of puparia were near the roots of the host plant. In North Dakota, Kamali (54) did not find puparia until late August or early September.

Adults emerged as early as the first week of June in Manitoba, although more normally the second week. Kamali recorded adult sunflower maggot in field margins and shelterbelts in the latter part of May during his studies, 1970 to 1973.

Westdal and Barrett (104) reported adults were very active during the day. This observation is at variance with the observations of Steyskal who indicated adult *S. longipennis* were crepuscular in habit and oviposited only during the early evening hours (Steyskal, G. C. 1970. Personal communication, September. Systematic Entomology Laboratory, U.S. National Museum, Washington, D.C.). Kamali (54) reported major adult activity and oviposition during the early evening period, 1800 to 2000 hours CDT, in North Dakota. He also recorded significant copualtion and oviposition during the period 0900 to 1100 CDT.

Oviposition occurs on sunflower when plants are about 30 cm high. Eggs are laid singly, usually near the apical meristem, above the puncture, just beneath the epidermis, and among and parallel to the vascular tissues.

The maggot emerges from the upper end of the egg and rasps a tunnel into the pith. Sunflower maggot larvae may tunnel up or down. They may be one or two long tunnels, but normally are several short tunnels. Maggots seldom feed in roots. When fully developed, they emerge from the stalk and leave a fairly characteristic exit hole. Westdal and Barrett (104) noted that there is generally no outward manifestation of the presence of maggots within the stalk. A minute oviposition scar may be visible.

Researchers in the Northern U.S. report only one generation per year.

Damage

Damage caused by feeding of *S. longipennis* larvae ranged from slight tunnelling in the pith tissue to complete destruction of the pith. Westdal and Barrett (104) reported no apparent detrimental effect on the plant. Brink (15), however, had indicated a high percentage of seeds failed to develop because of maggot injury.

In a series of studies over the period 1948 and 1950 to 1957, Westdal and Barrett (105) developed a 6-point rating scale for assessment of damage caused by maggot feeding. The scale was as follows: 0—no visible maggot injury; 1—one to three tunnels less than half the length of the plant; 2—two to three tunnels more than half the length of the plant; 3—25 to 50% of pith destroyed; 4—51 to 75% of pith destroyed; and 5—76 to 100% of pith destroyed. Their major criteria of maggot damage were head diameter, seed weight, number of unformed seeds, and percentage seed set.

In these studies, the percentage plants infested ranged from about 55 to 97%. They reported the coefficient of correlation between the percentage of plants infested and the damage index (sum of the ratings per sampled plants per field) was + 0.99, significant at the 1% level.

They concluded that injury to the pith of the sunflower stalk had no apparent effect on head diameter, seed yield, number of unformed seeds, or seed of the plant. They also concluded that since severe infestations of the pith resulted in this tissue being completely consumed, the primary function of the pith is that of a supporting tissue. The plant remained healthy and vigorous despite complete pith destruction. They also noted that oilseed sunflower rarely broke except in strong winds, and then only if the plant had been attacked by stalk-rotting organisms.

Kamali (54) reported the life cycle of *S. longipennis* in North Dakota closely paralleled the cycle reported by Westdal and Barrett (104). Kamali noted that the male to female ratio in North Dakota during the 3 years of his study was 70:30; Westdal and Barrett had reported a 51:49 ratio. Damage ratings to sunflower accessions evaluated by Kamali ranged from 0.3 to 2.7 on the Westdal-Barrett scale, and no relationship was detected between stem diameter and damage ratings.

Control

Chemical control of larval *S. longipennis* was attempted initially by Allen et al. (3). They reported 100% protection to sunflower following application of 0.1% demeton emulsion. Application was made as a total plant spray or as a soil treatment after sunflower maggot females had oviposited in the plants. Larvae were killed in the early instars. About 98% control also was achieved when leaves were soaked in a 0.1% solution of demeton for 22 hours.

Kamali (54) reduced infestation levels from 30 to 88% after one and two applications of methadithion, 1 kg/ha. Applications were made while adult females were still active. Reductions of 50%, 46%, and 38% in larval infestations were obtained after a single application of phorate (2 kg ai/ha), aldicarb (2 kg ai/ha), and carbofuran (3 kg ai/ha), respectively.

Westdal and Barrett (104) reported a parasite, *Psilus* sp. probably *atriconis* (Ashm.) [Hymenoptera: Braconidae], parasitized 4.7% of 362 puparia examined in 1952 and 9.9% of 1825 specimens in 1953.

Research conducted to date on the sunflower maggot, *Strauzia longipennis* (Wied.), suggests that the need to control or manage of this species in cultivated sunflower will be minimal.

Stem Weevil
Cylindrocopterus adspersus LeC.

Infestations of sunflower by the weevil, *Cylindrocopterus adspersus* LeC. were first reported by Newton (66) when he indicated larvae were feeding in stalks of wild and commercial sunflower in Colorado. More recently Phillips et al. (67) reported heavy infestations of *C. adspersus* in commercial and wild sunflower in Texas. Heaviest infestations were noted in mid-August.

Infestations of this weevil were first observed in two fields of commercial sunflower in North Dakota in 1973. An 80% yield loss occurred in one field (Schulz, J.T. 1973. Annual report. Department of Entomology, North Dakota State Univ., Fargo).

Studies have been conducted since 1973 at North Dakota State University to define the biology, ecology, and control of this species. Information reported in the remainder of this section (unless otherwise indicated) was provided by Casals (24).

Life Stages

Egg. C. adspersus eggs are laid singly beneath the epidermis of the plant. Newly laid eggs are glossy white, variable in shape, avoid to elongated, and rounded at both ends. Eggs range in length from 0.52 to 1.0 mm (avg 0.6 mm). At maturity eggs are pale brown, with a prominent dark spot on one end. Duration of the egg stage is about 1 to 2 weeks.

Larva. Mature fourth instar larvae are creamy white, 5 to 6 mm long with an inverted 'Y'-shaped epicranial suture. Head is deeply pigmented. Duration of the larval stage is about 11 months.

Pupa. Pupae of *C. adspersus* are white and glabrous when newly formed with the extremities and eyes gradually turning brown with age. Pupae range from 5.2 to 7.0 mm long. Prior to adult emergence, head snout and wing buds become light brown. Duration of the pupal stage is about 21 days.

Adult. Both male and female adults are brown, with dark brown and white spots on the elytra. Females averaged 5 mm long, ranging from 3.5 to 6 mm, and males 4 mm, ranging from 2.8 to 4.5 mm. Beak and antennae are black. Duration of the adult stage is about 3 weeks.

Life and Seasonal History

C. adspersus overwinters as diapausing fourth instar larvae in basal areas of sunflower stalks or in soil and plant trash. Fallen stalks do not provide a suitable overwintering site. Pupation by overwintering larvae occurred in early June in North Dakota. Adults emerged 3 or 4 weeks after pupation. Initial appearance of adult *C. adspersus* was 23 June 1974, 5 July 1975, 2 June 1976, and 9 June 1977. Above-normal temperatures in late May and June 1976 reduced the length of the pupal period by 10 days. Emerging adults feed on epidermal and palisade tissues of foliage and upper stem. Mating occurred during a 2-week feeding period which preceded oviposition. Just prior to oviposition females descended to the lower portion of the plant where eggs are laid in epidermal tissues. *C. adspersus* females oviposited up to 20 eggs/plant over a 17-day period.

Initial larval emergence was observed 9 July 1974, 14 July 1975, 9 June 1976, and 20 June 1977. Newly emerged larvae fed on subepidermal tissues and tunnelled into fibrovascular bundles. As larvae matured, feeding was concentrated in the pith, with larvae tunnelling upward in about one-half of the stem. As the fourth instar stage was reached, larvae commenced downward in the stalk feeding on pith and other contents of the inner stem. Tunnels were filled with frass, and the pith was destroyed. By the last week of August in North Dakota, larvae fed to about 0.2 to 0.8 mm above the soil surface. A rudimentary chamber was constructed in the pith; larvae over-wintered in this chamber. *C. adspersus* is univoltine in North Dakota.

Damage

Larval feeding in the stem is the predominant cause of damage. Feeding by the mobile first and second instars resulted in small tunnels in pith tissue. Damage by these stages usually is not severe, although xylem and phloem vessel systems are affected. Feeding by the last two instars completely destroyed the pith. Subsequent invasion by fungi further enhance breakdown of stem tissues. Stems are weakened and break over about 10 cm above the ground prior to harvest.

Host Preference

In addition to wild and cultivated sunflower, *H. annuus* L., *C. adspersus* has been collected from *Helianthus multiflorus, Ambrosia trifida,* and *Xanthium* sp. (69). In southern California, *C. adspersus* was collected as adults from *Ambrosia acanthicarpa* Hooker, *A. chenopodifolia* (Bentham) Payne, *A. chamissonis* (Lessing) Green (Goeden, R. D. 1975. Personal communication to P. Casals Bustos) and *A. psilostachya* DC. (42). It was reared from larvae and pupa in mines in pith of stems and branches of *A. acanthicarpa.*

C. adspersus was occasionally collected by Casals Bustos from *Amaranthus albus* and *A. retroflexus* L. He also collected this species at least once from *A. graecizans* L., *Ambrosia artemesifolis* L., *A. coronopigifolia* T. and G., *Centaurea repens* L., *Helianthus rigidus* Deaf., *Senecio aureus* L., *Sonchus arvensis* L., *Chenopodium album* L., *Trifolium pratense* L., and *Beta vulgaris* L.

Control

No specific measures have been defined for control of *C. adspersus.* Fall tillage to a depth of 15 cm reduced overwintering larval populations by about 50% in North Dakota (24). Adult control with properly timed applications of insecticides appears feasible, although no insecticides are registered now for *C. adspersus* control in the USA.

Cerambycid Complex
[Coleoptera: Cerambycidae]

A complex of long-horned beetles has been identified by Phillips et al. (67) as infesting sunflower in Texas. This complex includes *Mecas saturnina* LeC., *Dectes texanus* LeC., and *Ataxia* sp.

The Texas researchers (67) confirmed prior observations of Stride and Warwick (85) that female *M. saturnina* made two girdles 25 to 35 mm apart near the apex of the sunflower stem and deposited the egg in a cavity between the girdles. Because the plant is weakened, the top portion is lodged, and seed failed to develop. Damage associated with *M. saturnina* is comparable to that caused by the weevil, *Haplorhynchites aeneus*.

In 1970 Phillips et al. (67) reported *M. saturnina* destroyed 62% of the sunflower plants during the early growing season in their study area. They indicated that a single female could destroy more than 100 sunflower plants.

Further studies to define the biology of this insect on sunflower in Texas are in progress.

In initial observations reported by Phillips et al. (67), *Dectes texanus* was found to make more than two girdles on the plant stem several centimeters below the head. Damage was similar to that caused by *M. saturnina*. Larvae of *D. texanus* tunnelled from the girdled area to the roots and extensively damaged the vascular tissue. One insect was found to girdle more than one sunflower plant. Although this species has been studied for its impact on soybean, *Glycine max* L. Merr. (46, 47), it has been studied in sunflower only recently.

While *Ataxia* sp. has been recorded previously as a pest of wild sunflower, thistles, and ragweed (32) and commercial sunflower in Nebraska (65), no information is known about the bionomics of this species.

Extensive study of this beetle complex will be needed before the full impact of these three species on sunflower can be assessed. Initial reports by Phillips et al. (67) suggested that one or more of these species may be a serious problem to the successful cultivation of sunflower in Texas.

ROOT INFESTING INSECTS

Carrot Beetle, *Bothynus gibbosus* (DeGeer)
[Coleoptera: Scarabeidae]

A widely distributed insect in the USA, northern Mexico, and southern Canada (74), the carrot beetle, *Bothynus gibbosus* (DeGeer), was characterized by Bottrell et al. (11) as a serious pest of oilseed sunflower grown in the Texas High Plains. Early literature citations which delineate crops attacked by this species are numerous and include, among others, the following: Comstock (27), dahlia and sunflower in New York; Webster (99), carrot in Indiana; Howard (52), corn in Louisiana and carrots and dahlia in Wisconsin, and Chittenden (25), carrot and parsnip in New Jersey and Long Island.

More recently Burkhardt (17) reported extremely severe infestations of carrot beetle (90 adults/plant) on "domesticated" *Helianthus annus* L. in Kansas. Beckham and Tippins (5) reported adult *B. gibbosus* attacked roots of sunflower in Georgia.

The number of different crops attacked by this species indicated a large number of hosts are acceptable to the carrot beetle.

Fig. 1—Sunflower moth, *Homoeosoma electellum.*

Fig. 2—Sunflower moth larva and damage to head caused by larval feeding.

Fig. 3—Sunflower budworm, *Suleima helianthana.*

Fig. 4—External damage symptom indicating presence of budworm larval feeding.

Fig. 5—Sunflower seed weevil, *Smicronyx fulvux.*

Fig. 6—Sunflower seed weevil, *Smicronyx sordidus.*

Fig. 7—Damage to sunflower achene caused by *Smicronyx* larvae.

Fig. 8—Emergence hole produced by seed weevil larvae dropping to soil.

Fig. 9—The "head clipper" weevil, *Haplorhynchites aeneus.*

Fig. 10—Feeding scar caused by female *H. aeneus* which becomes the fracture point when sunflower head is severed.

Fig. 11—Abnormal ray petal development caused by feeding of sunflower midge larvae, *Contarinia schulzi.*

Fig. 12—Damage to receptacle caused by sunflower midge.

Fig. 14—Sunflower maggots feeding in pith tissue of sunflower stem.

Fig. 13—Adult sunflower maggots, *Strauzia longipennis.*

Fig. 15—Sunflower beetle, *Zygogramma exclamationis.*

Fig. 16—Sunflower beetle larva.

Fig. 17—Sunflower stem weevil, *Cylindrocopturus adspersus.*

Fig. 18—Sunflower root weevil, *Baris strenua.*

Fig. 19—Overwintering site of *C. adspersus* larva.

Fig. 20—Stalk breakage induced by *C. adspersus* larvae.

Fig. 21—Root damage induced by *B. strenua* larvae.

Fig. 22—Long horned beetle, *Mecas inornata* and typical feeding scars on stem. (Picture courtesy of C. R. Rogers).

Fig. 23—Carrot beetles, *Bothynus gibbosus* (Picture courtesy of C. R. Rogers).

Fig. 24—Sunflower damage caused by carrot beetle. (Picture courtesy of C. R. Rogers).

The wide distribution of *B. gibbosus,* coupled with its broad host diversity, suggests this species could be a major pest of sunflower in many regions of this country, although it is classified now as a major pest only in the High and Rolling Plains of Texas and adjoining areas of Oklahoma.

Life Stages

Hays (48) described the life stages of the carrot beetle.

Egg. Eggs are globular in shape, white and smooth when newly laid. At maturity they are about 2.5 mm long and dull white. Eggs are laid at the bases of plants, particularly in soil with a large amount of decaying organic matter. Mean duration of the egg stage was 11 days with a range of 7 to 22 days.

Larva. Newly hatched larvae (grubs) are white, with the head and mandibles darkening after a few hours to a characteristic brown color. The body color has a bluish tinge throughout the larval cycle. Mature *B. gibbosus* larvae are about 31 mm long and 9 mm wide. The last abdominal segment bears a ventral patch of short, straight hairs arranged triangularly, but without a double row of spines, which occur in some of the other genera of Scarabaeidae.

Pupa. Pupae of *B. gibbosus* are about 15 mm long and 5 mm wide. Overall color is light brown. Pupae generally are found lying on their backs in the soil. Hayes (48) found the average length of the pupal stage to be 19.1 days (range 11 to 29 days).

Adult. Descriptions of the adults have been provided by Hayes (48) and Cartright (23). Adult *B. gibbosus* are fairly large, cumbersome, oblong-oval, reddish-brown beetles. They often are mistaken for one of the June bugs or May beetles although the presence of the anterior pronatal depression with small tubercle and strongly punctuate elytra distinguishes *B. gibbosus* from them. Adults range from 10 to 17 mm long to 6 to 11 mm wide. Beetles burrow into the soil or hide beneath objects that will shelter them from light during the day. They are attracted to lights at night.

Life and Seasonal History

Bionomics and behavior of the carrot beetle have been extensively studied by Bottrell et al. (12) and Rogers (76) in the High and Rolling Plains of Texas.

Bottrell et al. (12) found hibernating carrot beetles in soil to a depth of less than 30 cm, although they reported most adults were entangled within roots of sunflower plants at depths of 25 to 75 mm. Rogers (76) reported overwintered adults were most commonly found around the roots of prairie sunflower (*Helianthus petiolaris* Nutt), "carelessweed" (*Amaranthus palmeri* Wats.), horseweed (*Conyza canadensis* L.), and sowleaf daisy (*Prionopsis ciliata* Nutt.) in the Texas Rolling Plains. Spring emergence of overwintering adults was reported from a 2-year study by Rogers (76) to have occurred on 6 April 1972 and 13 April 1973. Spring emergence apparently occurred suddenly, because Rogers reported light trap catches ranged from 20/trap per night to 1,000/trap per night in a 1-week period. King (58) reported rainfall of 10.91 cm during the 14-day period preceding spring emergence of adult *B. gibbosus* resulted in the most simultaneous emergence in South Carolina, but Rogers (76) indicated precipitation was not a critical factor in Texas. King reported temperature was not a controlling factor in spring emergence in South Carolina; Rogers suggested a certain mean temperature threshold must be reached before emergence occurred.

Hayes (48) reported mating occurs underground and in the darkness of the seclusive site of the adult habitat. Although Rogers was unable to define the precise date of the onset of spring oviposition, it appeared beetles commenced egg-laying in late April in the Rolling Plains. In 1973 Rogers recovered eggs and first instar larvae from a soil sample near light traps on 12 June. Eggs, all larval stages, and pupae were recovered on 18 July. Summer emergence of adults was recorded 20 July. Both Bottrell et al. (12) and Rogers (76) reported two distinct peaks of adult emergence in Texas, an observation that reflected the spring and summer emergence patterns of *B. gibbosus*. Rogers (76) found these population peaks occurred 7 to 10 days earlier in the Rolling Plains than those in the High Plains, with up to 45,000 beetles trapped in a single night. Adults continued feeding on suitable hosts until the onset of winter at which time they overwintered in the soil at the site at which they were located.

Damage Characteristics

Bottrell et al. (11) reported the adult caused the principal damage to sunflower and other hosts. Beetles burrowed to the root system and fed on lateral roots. They found the mean depth of the feeding zone was 3.55 cm. One adult can destroy all lateral roots of a small plant, resulting in wilting and ultimate death of the plant. When adults emerged from feeding they made a circular exit hole within 2.5 to 7.5 cm of the plant. In a study conducted in observation cages, no adult that emerged was found to reenter the soil to attack additional plants; these researchers (11), however had not determined whether beetles had attacked other plants prior to their observations. Rogers and Howell (78) suggested that while the observations recorded by Bottrell et al. (12) were the common mode of injury in larger plants, the most significant injury incurred by small plants resulted from adults chewing through the cortex or completely severing the taproot, 2.5 to 2.7 cm below the soil surface. Both Bottrell et al. (12) and Rogers (76) reported larvae did not feed on roots of sunflower. Rogers found more larvae around the roots of *A. palmeri* than any other host, although they usually were associated with a detritus layer 15 to 20 cm below the surface rather than the plant roots.

Rogers and Howell (78) also reported the indirect loss of sunflower plants attacked by *B. gibbosus* might be as important as the attack itself. Large numbers of carrot beetles provide food for skunks, oppossum, raccoons, and coyotes. Excavation for beetles by these mammals destroyed many large plants in Knox County, Texas.

Host Preference

Rogers (76) found the carrot beetle has a large number of natural hosts in the Rolling Plains of Texas. In addition to common and prairie sunflower, "carelessweed", horseweed, and sawleaf daisy, which serve as overwintering and food hosts for spring emerging adults, silver nightshade (*Solanum elaeagnifolium* Cav.), white rosinweed (*Silphium albiforum* Gray), prairie pepperweed (*Lepidium densiflorum* Schrad.), wild carrot (*Daucus carota* L.), and guar (*Cyamopsis tetragonolba* Taub) also were attacked by the adults emerging in spring.

Rogers (76) also found annual sunflower more susceptible to carrot beetle infestations than perennial *Helianthus* sp. *H. hirsutus* Raf., *H. tuberosus* L., and *H. mollis* Lam. exhibited no evidence of feeding by the carrot beetle.

Control

Bottrell et al. (11) evaluated efficacy of nine insecticides in a series of studies initiated in 1969. Granular formulations of selected insecticides were applied to the seed furrow at planting in one field test. In two additional field tests, surface sprays or subsoil granular applications were made. Insecticides included for evaluation included aldrin, Landrin® , carbofuran, carbaryl, heptachlor, chlordane, aldicarb, disulfoton, and diazinon. None of the insecticides evaluated prevented severe plant damage. In a more recent attempt to evaluate control of *B. gibbosus* with insecticides, Rogers and Howell (78) obtained results similar to those obtained by Bottrell et al. (11). Applications of three experimental organophosphate and carbamate insecticides, Nemacur® and fensulfothion, were applied at planting. Rogers concluded that although the percentage of sunflower plants destroyed was smaller than recorded by Bottrell et al. (11), none of the treatments protected plant roots.

Bottrell (12) evaluated response of selected sunflower cultivars to *B. gibbosus* and found all sustained severe damage as a result of *B. gibbosus* infestations. Cultivars evaluated included Peredovik, VNIIMK 8931, Arrowhead, Greystripe, NKHO 1, GOR 101, Krasnodarets, and P-21VR1 × MENN. RR18-1.

Rogers (76) reported about 1.32% of overwintered adults were parasitized by the sarcophagids, *Gymnoprosopa argentifrons* Towns. and *Sarcophaga sarracenioides* Aldrich. He also reported overwintered adults were infested by unidentified bacteria. Two or more species of Carabidae, not named in this report, attacked carrot beetle larvae.

Fungi of undetermined species induced heavy mortality among eggs and larvae of the carrot beetle in both the field and laboratory.

No satisfactory means appears to have evolved that will reduce economically limiting infestations of *B. gibbosus* below the economic threshold. This insect continues to pose a threat to the sunflower industry in the High and Rolling Plains of Texas.

A Root-Infesting Weevil, *Baris strenua* LeC.
[Coleoptera: Curculionidae]

A third species of a weevil complex which infests cultivated sunflower in the USA is *Baris strenua* LeC. Gilbert (41) reported this species distributed from Montana to Guatemala, westward to California, and eastward to Illinois. He also reported this species was of no economic importance.

Prior to 1973, Gilbert's observations apparently were justified because only one feeding of *B. strenua* on commercial sunflower was reported (93). Since 1973, *B. strenua* has been one species of a weevil complex causing increased damage to cultivated sunflower in North Dakota. One commercial field in 1973 sustained nearly 50% seed loss due to infestations by this species. (Schulz, J. T. 1973. Annual report. Department of Entomology, North Dakota State Univ., Fargo.) Increased activity by this species has been observed since 1973.

Studies commenced in late 1973 to obtain this information, since there was no published literature on the biology and ecology of *B. strenua*.

Data provided in the ensuing paragraphs were provided from studies conducted by Casals (24).

Life Stages

Egg. Eggs of *B. strenua* were creamy white, ovoid, and average 1 mm long. They were inserted singly into the epidermis of the root of sunflower. Mature eggs were dark brown, and the head capsule of the future larva is well insinuated. Duration of the egg stage is about 7 days.

Larva. Newly hatched larvae are white with a pale yellow sclerotized head and a dark eye spot. First instar larvae had a mean length of 1.5 mm; mature fourth instar larvae, 0.7 mm. Fourth instar larvae exhibit the typical 'C'-shaped body characteristic of the curculionids, three segmented thorax with enlarged swelling, no legs, and inconspicuous spiracles. Duration of the larval stage is about 11 months.

Pupa. B. strenua pupae are elongated, obovate, white to yellow white in color, and averaged 5 mm long. Duration of the pupal stage is about 16 days.

Adult. Adult *B. strenua* are black, oval, robust weevils with punctate head and rostrum from one-half to three-fourths as long as the pronotum. Males ranged from 4 to 6 mm long, females 0.2 to 0.4 mm larger. Duration of the adult stage is about 4 weeks.

Life and Seasonal History

Similar to findings with *Cylindrocopturus adspersus,* Casals (24) determined that *B. strenua* overwintered as fourth instar diapausing larvae. Unlike *C. adspersus, B. strenua* overwintered as a larval cocoon near roots of infested plants. The larval cocoon is constructed from small soil particles adhered by a larval secretion. Cocoons are small, round to ovoid, 10 mm long, 4 mm wide, and are similar to other soil particles. Larval cocoons have been recovered to a depth of 45 cm in the soil. Larvae pupated during the first week of June in North Dakota. Adult *B. strenua* were active in sunflower fields from late June to late July. Newly emerged adults feed on the epidermal tissue of leaves during early morning and late afternoon. Adults were difficult to detect on plants during midday. About a week after emergence, adults congregated in the root zone near the soil surface, continued feeding, and copulated. Oviposition commenced about 2 weeks after emergence in callous tissue formed as a response to adult feeding. *B. strenua* females laid from 5 to 7 eggs/plant during about 32 days. Emergence of larvae was recorded on 11 July 1973, 9 July 1975, and 7 July 1976. *B. strenua* larvae are less mobile than *C. adspersus* larvae and confined their feeding areas to the same area they hatched. Larvae fed on cortical cells of the root and the epidermis.

Damage

Primary damage to sunflower is caused by feeding of larvae of *B. strenua* on roots. The root stele and other root tissues near the soil surface are destroyed. Stems of severely infested sunflower plants break over near the soil surface.

Host Preference

Tuttle (96) reported adult *B. strenua* fed on terminal growth of *Helianthus grosseserratus* Martens. Casals obtained *B. strenua* from cultivars and wild *Helianthus annuus* L., *H. maximiliani* Schrad.), and *H. petiolaris* Nutt. He periodically collected *B. strenua* from *Amaranthus albus* L., *Ambrosia artemesiifolia* L., and *Sonchus arvensis* L. Less frequent collections were made from *Amaranthus graecizans* L., *A. retroflexus* L., *Ambrosia coronopifolia* T. and G., *Centaurea repens* L., and *Salsola kali* L.

Control

No control measures have been defined for control of *B. strenua*. Because of the overwintering behavior of this species, fall tillage did not significantly increase overwintering mortality of the larvae. Less than 2% of larvae were found parasitized by the wasp, *Bracon baridii* Marsh, in North Dakota.

OTHER INSECT PESTS

Cutworms, Grasshoppers, Stored Product Complex

Cutworms [Lepidoptera: Noctuidae] are among the most important pests of field crops in many areas of the USA and Canada. Outbreaks of the dark-sided, *Euxoa messoria* (Harris) and red-backed, *E. ochrogaster* (Guenee) cutworms in 1974 and 1975 in North Dakota and Minnesota caused severe seedling loss and necessitated emergency clearance for use of the insecticide, toxaphene. While these are the two principal species infesting sunflower in these two states, cutworm species which attack sunflower vary among regions.

Damage caused by cutworms usually consists of the cutting of the stem of the plant at or just below the soil surface. Many species come to the soil surface at night to feed, especially in years of high soil moisture. During the day, larvae burrow just below the soil surface near the plants attacked. Wilted or dying plants frequently indicate presence of cutworms. Since severed plants may dry and blow away, bare patches in the field remain as evidence of cutworm infestations.

Cutworm activity is greatest from May through the end of June in the major sunflower production areas of North Dakota, South Dakota, Minnesota, and the prairie provinces of Canada. Species predominant in this region overwinter as eggs in the soil which are laid in late July and early August and remain dormant until the onset of warm weather the following spring. There is no generation per year.

Toxaphene is the only insecticide currently approved for use in control of cutworms on sunflower in the USA at this writing. Cutworms are subject to attack by parasites, predators, and disease pathogens which may reduce the magnitude of infestations.

Grasshoppers [Orthoptera: Locustidae] occasionally reach numbers sufficient to cause damage to sunflower especially in the Dakotas and Minnesota. Defoliation of the plant and destruction of developing seed in the head are the predominant damage sustained by sunflower.

Toxaphene is the only insecticide currently approved for use in control of grasshoppers on sunflower in the USA. Additionally, use of the three currently approved insecticides, endosulfan, methyl parathion, and methidathion, for control of other sunflower insects will reduce grasshopper infestations. Late season use of insecticides must be restricted to field margins.

Stored sunflower seed may be infested by several species of the stored-product insect complex. Species of the complex are those that attack other commodities in addition to sunflower.

Conditions which favor such infestations are common to all commodities. They include, among others, poorly maintained and inadequately cleaned storage facilities and storing seed at high moisture levels.

While reports of insect attack on stored sunflower seed have not been numerous to date in the USA, the predominant insect reported has been the Indian meal moth, *Plodia interpunctella* (Hbn) [Lepidoptera: Pyralidae]. This species feeds on nearly all dry foodstuffs. In severe infestations, the characteristic webbing produced by larvae of *P. interpunctella* is observed easily.

Adult Indian meal moths are recognized easily. Forewings are reddish brown with the front or base a grayish-white. Hind wings are grayish-white. The adult is about 1.5 cm from tip to tip of wings and is active at night or in dark places. Females deposit from 40 to 350 or more eggs singly or in clusters of 12 to 30 on the sunflower seed or on adjacent objects. Larvae emerge from 2 days to 2 weeks. Larvae generally are white, often with a greenish or pinkish tinge and light-brown head. Mature larvae range from 1 to 1.5 cm long. Pupation is in a silken cocoon on the surface of the food material. Depending on the temperature maintained in the storage facility, a life cycle is completed in 4 to 6 weeks (63).

Other insects reported to infest stored sunflower seed includes the sawtoothed grain beetle (*Oryzaephilus surinamensis* L.), confused flour beetle (*Tribolium confusum* Jay du Val), and meal snout moth (*Pyralis farinalis* L.)

Sunflower seed stored in facilities that are constructed and maintaned properly are less likely to become infested. Carryover stocks of sunflower seed should be monitored continuously for presence of species of the stored insect complex. If fumigation becomes necessary, the only fumigant registered in the USA for use on sunflower seed is aluminum phosphide (Phostoxin®).

POTENTIAL PEST SPECIES

Many insect species have been collected from sunflower and are reported in the literature, and all are potential pests. None of these species now appear to be of major economic importance in sunflower production areas of the USA, Canada, or Mexico.

In a review of this magnitude, however, it probably is desirable to list some of these species in the event they should become of greater significance as the culture of sunflower continues to expand. A selected listing is provided in Table 1.

ACKNOWLEDGMENT

The reviewer extends his sincere thanks to colleague Dr. C. Y. Oseto, for his critical review of the manuscript and for assistance with the selection and photography of insects appended to this chapter.

Table 1—Insects collected from sunflower for which information on their status as a pest remains to be determined.

Order and species	Citation
COLEOPTERA	
Rhodobaenus tredecimpunctatus Ill.	Walker (98)
	Beckham and Tippins (5)
	Phillips et al. (67)
Mordellistena pustulata Melsh.	Walker (98)
	Bekcham and Tippins (5)
Anacentrinus deplanatus Casey	Phillips et al. (67)
Lixus fimbriolatus Boh.	Williams (108)
HEMIPTERA	
Leptoglossus phyllopus (L.)	Adams and Gaines (1)
	Beckham and Tippins (5)
	Phillips et al. (67)
Hercias sexmaculatus (Barber)	Phillips et al. (67)
Carythucha morrilli Osborn and Drake	Rogers (77)
Hymenarcys nervosa (Say)	Adams and Gaines (1)
	Phillips et al. (67)
Xynonysius californicus (Stal)	Beckham and Tippins (5)
HOMOPTERA	
Aphis helianthi Mon'l	Walker (98)
Macrosteles fascifrons Stål	Annu. Rep. Dep. Entomology,
	North Dakota State Univ., Fargo(1969;1971)
LEPIDOPTERA	
Heliothis zeae (Boddie)	Teetes et al. (95)
	Beckham and Tippins (5)
H. virescens (F.)	Teetes et al. (95)
Estigmene acrea (Drury)	Phillips et al. (67)
Spodoptera exigua (Hbn)	Phillips et al. (67)

APPENDIX
Chemical Identity of Cited Insecticides†

Insecticide	Chemical definition
aldicarb	2-methyl-2 methylthio propeonaldehyde O-(methylcarbamayl) oxime
aldrin	not less than 95% of 1,2,3,4,10,10 hexachloro-1,4,4a,5a,8a-hexahydro-1,4-endo-exo-5,8-dimethanonaphthalene
azinphosmethyl	0,0-diethylphosphorodithioate S-ester with 3-(meracaptomethyl)-1,2,3-benzotriazin-4 (3H)-one
benzene hexachloride	1,2,3,4,5,6-hexachlorocyclohexane, consisting of several isomers and containing a specified percentage of gamma
carbaryl	1-naphthyl methyl carbamate
carbofuran	2,3-dihydro-2,2 dimethyl-7-benzyl-furanyl methylcarbamate
chlorpyrifos	0,0-diethyl O-(3,5,6-trichlor-2-pyridyl) phosphorothioate
DDT	1,1,1-trichloro-2 2 bis (p-chlorophenyl) ethane
diazinon	0,0-diethyl O-(isopropyl-4 methyl-6 pyrimidinyl) phosphorothioate
disulfoton	0,0-diethyl S[2-(ethylthio) ethyl] phosphorodithoate
endosulfan	6,7,8,9,10,10-hexachloro-1,5,5a,6,9,9a-hexahydro-6,9-methano-2,4,3 benzodioxathiepin 3-oxide
fensulfothion	0,0-diethyl 0 p-[(methylsulfinyl) phenyl] phosphorothioate
heptachlor	1,4,5,6,7,8,8-heptachloro-3a,4,7,7a tetrahydro-4,7-methanoindene
Landrin® ‡	3,4,5-trimethylphenyl methylcarbamate, 75%; 2,3,5-trimethylphenyl methylcarbamate, 18%
malathion	S-[1,2-bis (ethoxycarbomyl) ethyl] 0,0-dimethyl phosphorodithioate
methidathion	S-[(2-methoxy-5oxoΔ²-1,3,4-thiadiazolin-4-yl) methyl] 0,0-dimethyl phosphorodithioate
methyl parathion	0,0-dimethyl O-(-p-nitrophenyl) phosphorothioate
mevinphos	methyl 3-hydroxy-alpha-crotonate dimethyl phosphate
monocrotophos	dimethyl phosphate ester with (E)-3-hydroxy-N-methylcrotonamide
Nemacur‡	ethyl 4-(methylthio)-m-tolyl isopropylphosphoramidate
parathion	0,0-diethyl 0-p-nitrophenyl phosphorothioate
toxaphene	chlorinated camphene containing 67–69% of chlorine
trichlorfon	dimethyl (2,2,2-trichloro-1-hydroxyethyl) phosphonate

† Chemical definition established by Committee on Insecticide Terminology, Entomological Society of America.

‡ No common names approved; citation by proprietary name only

LITERATURE CITED

1. Adams, A. L., and J. C. Gaines. 1950. Sunflower insect control. J. Econ. Entomol. 43: 181–184.

2. Adkisson, P. L., E. S. Vanderzant, D. L. Bull, and W. E. Allison. 1960. A wheat germ medium for rearing the pink bollworm. J. Econ. Entomol. 53:759–762.

3. Allen, W. R., P. H. Westdal, C. F. Barrett, and W. L. Askew. 1954. Control of the sunflower maggot, *Strauzia longipennis* (Wied.) (*Diptera: Trypetidae*) with demeton. Ann. Rep. Entomol. Soc. Ont. 85:53–55.

4. Anderson, D. A. 1962. The weevil genus *Smicronyx* in America North of Mexico (Coleoptera: Curculionidae). Proc. U.S. Natl. Mus. 113:185–372.

5. Beckham, C. M., and H. H. Tippins. 1972. Observations of sunflower insects. J. Econ. Entomol. 65:865–866.

6. Beirne, B. P. 1971. Pest insects of annual crop plants in Canada. Mem. Entomol. Soc. Can. 78:66.

7. Bigger, J. H. 1930. A parasite of the sunflower weevil. J. Econ. Entomol. 23:287.

8. –––––. 1931. Another parasite of the sunflower weevil, *Desmoris fulvus* (LeC). J. Econ. Entomol. 24:558.

9. Bird, R. D., and W. R. Allen. 1936. An insect injuring sunflower in Manitoba. Can. Entomol. 68:93–94.

10. Bottimer, L. J. 1926. Notes on some Lepidoptera from eastern Texas. J. Agric. Res. 33:803.

11. Bottrell, D. G., R. D. Brigham, B. C. Clymer, and J. R. Cate, Jr. 1970. Evaluation of insecticides for controlling the carrot beetle in sunflower. Tex. Agric. Exp. Stn. Prog. Rep. 2831, 1–17.

12. –––––, R. D. Brigham, and L. B. Jordan. 1973. Carrot beetle: Pest status and bionomics on cultivated sunflower. J. Econ. Entomol. 66:86–90.

13. Brassard, D. W. 1976. Bionomics and rearing of *Suleima helianthana* (Riley) (Lepidoptera: Olethreutidae). M.S. thesis. Department of Entomology, North Dakota State Univ., Fargo.

14. Breland, O. P. 1939. Additional notes on sunflower insects. Ann. Entomol. Soc. Am. 32: 719–726.

15. Brink, J. E. 1922. The sunflower maggot *Strauzia longipennis* (Wied.). Ann. Rep. Entomol. Soc. Ont. 53:72–74.

16. Brisley, H. R. 1925. Notes on the Chrysomelidae (Coleoptera) of Arizona. Trans. Am. Entomol. Soc. 874:167–182.

17. Burkhardt, C. C. 1957. Carrot beetle control in domesticated sunflowers. J. Econ. Entomol. 50:369.

18. Carlson, E. C. 1967. Control of sunflower moth larvae and their damage to sunflower seeds. J. Econ. Entomol. 60:1068–1071.

19. –––––. 1971. New insecticides to control sunflower moth. J. Econ. Entomol. 64: 208–210.

20. –––––. 1975. Pesticides for controlling sunflower moth larvae. Calif. Agric. 29:12–13.

21. –––––, P. F. Knowles, and J. E. Delle. 1972. Sunflower varietal resistance to sunflower moth larvae. Calif. Agric. 26:12–13.

22. –––––, and R. Witt. 1974. Moth resistance of armored-layer sunflower seeds. Calif. Agr. 28:12–14.

23. Cartright, O. L. 1959. Scarab beetles of the genus *Bothynus* in the United States (Coleoptera: Scarabaeidae). Proc. U.S. Natl. Mus. 108:515–541.

24. Casals Bustos, P. 1976. Bionomics of *Cylindrocopturus adspersus* LeC. and *Baris strenua* LeC. (Coleoptera: Curculionidae). Ph.D. thesis. Department of Entomology, North Dakota State Univ., Fargo. (Diss. Abstr. Int. 38(4):1542B.)

25. Chittendon, F. H. 1902. Some insects injurious to vegetables. USDA Bur. Entomol. Bull. 33:32–37.

26. Cockerell, T. D. A. 1915. Sunflower insects. Can. Entomol. 47:280–282.

27. Comstock, J. H. 1881. Report of the entomologist. Part I. Miscellaneous insects. Rep. N. Y. Comm. Agric. for 1880. N. Y. State Dep. Agric. p. 235–275.

28. Criddle, N. 1922. Popular and practical entomology. Beetles injurious to sunflowers in Manitoba. Can. Entomol. 54:97–98.

29. Drake, C. J., and H. M. Harris. 1926. The flower webworm, a new flower pest. Iowa State Hort. Soc. Trans. 61:223-26.

30. Dyar, H. G. 1902. A list of North American Lepidoptera and key to the literature of this order of insects. U.S. Natl. Mus. Bull. 52. 1-723.

31. Ehart, O. R. 1974. Biology and economic importance of *Suleima helianthana* (Riley) on sunflower cultivars, *Helianthus annuus* L. in the Red River Valley. M.S. thesis. Department of Entomology, North Dakota State University, Fargo.

32. Fisher, W. S. 1924. A new species of *Ataxia* from the United States. (Coleoptera: Cerambycidae). Can. Entomol. 95:253-254.

33. Foote, R. H. 1960. Notes on some North American descriptions of two new genera and two new species. (Diptera). Proc. Biol. Soc. Wash. 73:107-118.

34. ————. 1960. The species of the genus *Neotephritis* Hendel in America north of Mexico. (Diptera: Tephritidae). J.N.Y. Entomol. Soc. 68:145-151.

35. ————, and F. L. Blanc. 1963. The fruit flies or Tephritidae of California. Bull. Calif. Insect. Surv. 7:1-117.

36. Forbes, S. A. 115. Report of the Illinois State Entomologist for 1913 and 1914. p. 4.

37. Forbes, W. T. M. 1923. The Lepidoptera of New York and neighboring states. Cornell Univ. Agric. Exp. Stn. Mem. 68.

38. Fuchikovsky Zak, L. 1976. Enfermedades y plagas del gerosol en Mexico. Colegio de Postgraduados, Escuela Nacional de Agricultura, Chapingo, Mexico.

39. Gabor, Reichart. 1959. A napraforgomology (*Homoeosoma nebullebum* Hb.) nevelesevel Kapcsolatos meg figyelesek. Rovar Kozl. 12(34):497-510.

40. Gagné, R. J. 1972. A new species of *Contarinia* (Diptera: Cecidomyiidae) from *Helianthus annuus* L. (Compositae) in North America. Entomol. News 83:279-281.

41. Gilbert, E. E. 1964. The genus *Baris* Germar in Germany. Univ. Calif. Publ. Entomol. 34:1-153.

42. Goeden, R. D., and D. W. Richer. 1968. The phytophagous insect fauna of the ragweed, *Ambrosia confertifolora,* in Southern California. Environ. Entomol. 4(2):301-306.

43. Hamilton, R. W. 1973. Observations on the biology of *Haplorhynchites aeneus* (Boheman) [Coleoptera: Rhynchitidae]. Coleopt. Bull. 27:83-85.

44. ————. 1974. The genus *Haplorhynchites* (Coleoptera: Rhynchitidae) in America North of Mexico. Ann. Entomol. Soc. Am. 67:787-794.

45. Hanausek, T. F. 1902. Zur Entwicklungsgeschichte des Perikarps von *Helianthus annuus.* Ber. Dtsch. Bot. Ges. 20:449-454.

46. Hatchett, J. H., D. M. Daugherty, J. C. Robbins, R. M. Barry, and E. C. Houser. 1975. Biology in Missouri of *Dectes texanus,* a new pest of soybean. Ann. Entomol. Soc. Am. 68:209-213.

47. ————, R. D. Jackson, and R. M. Barry. 1973. Rearing a weed cerambycid, *Dectes texanus,* on an artificial medium, with notes on biology. Ann. Entomol. Soc. Am. 66:519-522.

48. Hayes, W. P. 1917. Studies on the life-history of *Ligyrus gibbosus* DeG. (Coleoptera). J. Econ. Entomol. 10:253-261.

49. Heinrich, C. 1921. Some Lepidoptera likely to be confused with the pink bollworm. J. Agric. Res. 20:807-836.

50. ————. 1923. Revision of the North American moths of the subfamily Eucosminae of the family Olethreutidae. U.S. Natl. Mus. Bull. 123:1-298.

51. Heiser, Charles B. Jr. 1976. The sunflower. Univ. of Oklahoma Press, Norman. 198 p.

52. Howard, L. O. 1898. Recent injury by the sugar cane beetle and related species. USDA Bur. Entomol. Bull. 18.

53. Johnson, A. L. 1975. The inheritance of resistance to the sunflower moth, *Homoeosoma electellum* (Hulst.), in cultivated sunflower, *Helianthus annuus.* Ph.D. Thesis. Univ. of California, Davis. (Diss. Abstr. Int. 36(11):5430B).

54. Kamali, K. 1973. Biology and ecology of the Tephritid complex on sunflower (Diptera: Tephritidae). Ph.D. Thesis. Department of Entomology, North Dakota State University, Fargo. (Diss. Abstr. Int. 35(02):872-B).

55. ————, and J. T. Schulz. 1973. Characteristics of immature stages of *Gymnocarena diffusa* (Diptera: Tephritidae). Ann. Entomol. Soc. Am. 66:288-291.

56. ————, and J. T. Schulz. 1974. Biology and ecology of *Gymnocarena diffusa* (Diptera: Tephritidae) on sunflower in North Dakota. Ann. Entomol. Soc. Am. 67:695-699.

57. Kienwick, L. 1964. Die phytomelanschicht in perikarp von *Helianthus annuus* L. als

resistenzschicht gegen *Homoeosoma nebulella* Hb. Z. Pflanzenkr. Pflanzenschutz 71: 294–301.

58. King, E. W. 1963. The effect of temperature and rainfall on the spring emergence of the carrot beetle, *Ligyrus gibbosus*. Ann. Entomol. Soc. Am. 56:826–829.

59. Kinman, M. L. 1966. Tentative resistance to the larvae of *Homoeosoma electellum in Helianthus*. Proc. 2nd Int. Sunflower Conf. (Morden, Manitoba) p. 72–74.

60. Knowles, P. F., and W. H. Lange. 1954. The sunflower moth. Calif. Agric. 8:11–12.

61. Lipp, W. V. 1972. The magnitude and characteristics of damage caused by insects injurious to sunflowers. M.S. thesis. Department of Entomology, North Dakota State Univ., Fargo.

62. MacKay, M. R. 1959. Larvae of the North American Olethreutidae (Lepidoptera). Can. Entomol. Suppl. 10.

63. Metcalf, C. L., W. P. Flint, and R. L. Metcalf. 1962. Destructive and Useful Insects. 4th Ed. McGraw Hill, Inc., N.Y. 1–1087.

64. Mollisch, H. 1923. Michrochemie der Pflanze. Abhandlungen des staatwissen schaftliches seminar. Jena Universitat 11:358–360.

65. Muma, M. R., R. N. Lyness, C. E. Claasen, and A. Hoffman. 1950. Control tests on sunflower insects in nebraska. J. Econ. Entomol. 43:477–480.

66. Newton, J. H. 1921. A sunflower pest. 12th Annu. Rep. State Entomol. Colorado State Dep. Agric. 35:37–38.

67. Phillips, R. L., N. M. Randolph, and G. L. Teetes. 1973. Seasonal abundance and nature of damage of insects attacking cultivated sunflowers. Texas Agric. Exp. Stn. Publ., MP-1116, 1–7.

68. Phillips, V. T. 1946. The biology and identification of Trypetid larvae. Mem. Am. Entomol. Soc. 12:1–161.

69. Pierce, W. D. 1916. Notes on the habits of weevils. Proc. Entomol. Soc. Wash. 18:6–10.

70. Putt, E. D. 1944. Histological observations on the location of pigments in the achene wall of sunflower (*Helianthus annuus* L.). Sci. Agric. 25:185–190.

71. Randolph, N. M., George L. Teetes, and M. C. Baxter. 1972. Life cycle of the sunflower moth under laboratory and field conditions. Ann. Entomol. Soc. Am. 65:1161–1164.

72. Rhe, L. 1919. *Homoeosoma nebulella* Hb. als sonnenblumenschadling in Rumanien. Z. Angew. Entomol. 5:267–277.

73. Riley, C. V. 1881. Descriptions of some new Tortricidae. Trans. St. Louis Acad. Sci. 4: 316–324.

74. Ritcher, P. O. 1966. White grubs and their allies. Oregon State Univ. Press, Corvallis.

75. Robinson, R. G. 1973. The sunflower crop in Minnesota. Minnesota Agric. Ext. Serv. Bull. 299. Rev.

76. Rogers, C. E. 1974. Bionomics of the carrot beetle in the Texas Rolling Plains. Environ. Entomol. 3:969–974.

77. ————. 1977. Laboratory biology of a lace bug on sunflower. Ann. Entomol. Soc. Am. 70:144–145.

78. ————, and G. R. Howell. 1973. Behavior of the carrot beetle on native and treated commercial sunflower. Texas Agric. Exp. Stn. Prog. Rep. 3249.

79. Samuelson, C. R. 1976. Biology and economic importance of *Contarinia schulzi* Gagne on sunflower cultivars, *Helianthus annuus* L. in the Red River Valley. Ph.D. thesis. Department of Entomology, North Dakota State Univ., Fargo. (Diss. Abstr. Int. 37(10): 4869B).

80. Satterthwait, A. F. 1946. Sunflower seed weevils and their control. J. Econ. Entomol. 39:787–792.

81. ————. 1948. Important sunflower insects and their insect enemies. J. Econ. Entomol. 41:725–731.

82. ————, and R. B. Swain. 1946. The sunflower moth and some of its natural enemies. J. Econ. Entomol. 30:575–580.

83. Schulz, J. T. 1973. Damage to cultivated sunflower by *Contarinia schulzi* Gagné. J. Econ. Entomol. 66:282.

84. ————, and W. V. Lipp. 1969. The status of the sunflower insect complex in the Red River Valley of North Dakota. Proc. North Cent. Branch Entomol. Soc. Am. 24:99–100.

85. Stride, G. O., and E. P. Warwick. 1962. Ovipositional girdling in a North American cerambycid beetle, *Mecas saturnina*. Anim. Behav. 10:112–117.

86. ————, P. L. Adkisson, and N. M. Randolph. 1969. Photoperiod and temperature as factors controlling the diapause of the sunflower moth, *Homoeosoma electellum*. J. Insect Physiol. 15:755–761.

87. ————, M. L. Kinman, and N. M. Randolph. 1971. Differences in susceptibility of certain sunflower varieties and hybrids to the sunflower moth. J. Econ. Entomol. 64:1285–1287.

88. Teetes, G. L., and N. M. Randolph. 1968. Chemical control of the sunflower moth on sunflowers. J. Econ. Entomol. 61:1344–1347.

89. ————, and N. M. Randolph. 1969. Some new host plants of the sunflower moth in Texas. J. Econ. Entomol. 62:264–265.

90. ————, and N. M. Randolph. 1969. Seasonal abundance and parasitism of the sunflower moth, *Homoeosoma electellum*, in Texas. Ann. Entomol. Soc. Am. 62:1461–1464.

91. ————, and N. M. Randolph. 1969. Chemical and cultural control of the sunflower moth in Texas. J. Econ. Entomol. 62:1444–1447.

92. ————, and N. M. Randolph. 1970. Color preference and sex attraction among sunflower moths. J. Econ. Entomol. 63:1358–1359.

93. ————, and N. M. Randolph. 1970. Hibernation spring emergence, and pupation habits of the sunflower moth, *Homoeosoma electellum*. Ann. Entomol. Soc. Am. 63:1473–1475.

94. ————, and N. M. Randolph. 1971. Effects of pesticides and dates of planting sunflowers on the sunflower moth. J. Econ. Entomol. 64:124–126.

95. ————, N. M. Randolph, and M. L. Kinman. 1970. Notes on noctuid larvae attacking cultivated sunflower. J. Econ. Entomol. 63:1031–1032.

96. Tuttle, D. M. 1952. Studies of the bionomics of Curculionidae. Ph.D. thesis. University of Illinois, Urbana.

97. Vanderzant, E. S., C. D. Richardson, and S. W. Fort, Jr. 1962. Rearing the bollworm on artificial diet. J. Econ. Entomol. 55:140.

98. Walker, F. H., Jr. 1936. Observations on sunflower insects in Kansas. J. Kans. Entomol. Soc. 9:16–25.

99. Webster, F. M. 1889. *Ligyrus gibbosus* injuring carrots in Indiana. Insect Life 1:382–383.

100. Wene, G. P. 1950. Sunflower moth larva injuring to young citrus. J. Econ. Entomol. 43:948.

101. Westdal, P. H. 1949. A preliminary report on the biology of *Phalonia hospes* Wlshm. (Lepidoptera: Phaloniidae), a new pest of sunflowers in Manitoba. 80th Annu. Rep. Entomol. Soc. Ont.

102. ————. 1975. Insect Pest of sunflower. p. 475–495. *In* J. T. Harapiak (ed.) Oilseeds and pulse crops in Western Canada—A symposium.

103. ————, and C. F. Barrett. 1955. Insect pests of sunflowers in Manitoba. Can. Dep. Agric. Publ. 944, 1–8.

104. ————, and C. F. Barrett. 1960. Life-history of the sunflower maggot, *Strauzia longipennis* (Wied.) (Diptera: Trypetidae) in Manitoba. Can. Entomol. 92:481–488.

105. ————, and C. F. Barrett. 1962. Injury by the sunflower maggot, *Strauzia longipennis* (Wied.) (Diptera: Trypetidae) to sunflowers in Manitoba. Can. J. Plant Sci. 42:11–14.

106. ————, W. Romanow, and W. L. Askew. 1973. Some responses of sunflowers to defoliation. Proc. Manitoba Agron. Annu. Conf. 1973:98–99.

107. Wiedemann, C. R. W. 1830. Aussereuropäisehe zweiflügelige insektan. Vol. 2, 644 p. Hamm.

108. Williams, R. W. 1942. Notes on the bionomics of *Lixus fimbriolatus* Bot. Ann. Entomol. Soc. Am. 35:366–372.

109. Yamamato, R. T. 1969. Mass rearing of the tobacco hornworm. J. Econ. Entomol. 62:1427–1431.

J. T. SCHULZ: B.S., M.S., Ph.D.; Professor and Chairman, Department of Entomology, North Dakota State University, Fargo; Project Leader, Biology, ecology and management of sunflower insects, 1967–present; Board of Directors(ex officio), Sunflower Association of America, 1975–present.

Chapter 7

Diseases

D. E. ZIMMER AND J. A. HOES

Diseases of sunflower, *Helianthus annuus* L., are infectious or noninfectious. Infectious diseases are incited by fungi, bacteria, nematodes, mycoplasms, and viruses. Noninfectious diseases are caused by physical agents such as extreme temperatures, deficiencies or excesses of water or mineral nutrients in the soil, injurious gases in the atmosphere, and lightning.

Sunflower is the known host of more than 35 infectious microorganisms, mostly fungi, which may, under certain climatic conditions, impair the normal physiology of the plant so that yield and quality are reduced significantly. No exact data are available, but it is estimated that infectious diseases cause an average annual loss of 12% in yield from the nearly 12 million ha of sunflower in the world.

Sunflower diseases have been classified historically as major or minor based upon regional, national, and global losses inflicted. Downy mildew, *Plasmopara halstedi* (Farl.) Berl & de Toni, rust, *Puccinia helianthi* Schw., Sclerotinia white mold, *Sclerotinia sclerotiorum* (Lib.) de Bary, and Verticillium wilt, *Verticillium dahliae* Klebahn, are considered major diseases of sunflower in North America. These diseases often have inflicted serious yield losses over the past three decades and threatened sustained sunflower production in some areas of the world, especially in the USA and Canada where wild sunflower abounds as a native disease host and where climatic factors often are favorable for disease development. Other diseases, although observed in North America, rarely have inflicted significant losses.

The relative importance of sunflower diseases varies annually with biological and climatic factors and management practices. A disease such as rust is worldwide and occurs with variable intensity in all principal sunflower producing areas. Other diseases such as charcoal rot, *Macrophomina phaseoli* (Maubl.) Ashby, and downy mildew, *Plasmopara halstedi* (Farl.) Berl. & de Toni, are restricted in distribution by climatic factors. Charcoal rot generally is confined to the warmer regions of the world where high temperatures and restricted soil moisture predispose sunflower

From: Carter, Jack F. (ed.). 1978. *Sunflower Science and Technology.* Agronomy 19. Copyright © 1978 by the American Society of Agronomy, Crop Science Society of America, and Soil Science Society of America, 677 South Segoe Road, Madison, WI 53711 USA.

to attack. Downy mildew generally is found in the more temperate regions where emerging seedlings are exposed to low temperatures and abundant precipitation.

Sunflower plants are susceptible to pathogenic organisms from seed germination to harvest. Biological and climatological factors, however, restrict the activities of some pathogens to specific growth stages of the sunflower plant. Other pathogens are not as environmentally and biologically sensitive, and diseases they incite may occur throughout the growing season. Infection may be systemic, such as in Verticillium wilt and aster yellows, or localized on leaves, stems, and heads, such as in rust and Phoma black stem. Some diseases, such as downy mildew, exhibit both localized and systemic phases. A pathogen such as *Sclerotinia sclerotiorum* is nonspecific and causes disease of roots, stems, heads, and seeds.

Color photos of disease effects follow page 258.

DOWNY MILDEW

Causal Organism

Downy mildew is incited by *Plasmopara halstedii* (Farl.) Berl. & de Toni, a binomial name conventionally used for a closely related group of fungi attacking not only *Helianthus annuus* L. but also several other species of the subfamilies Asteroidae and Cichorioideae of the family Compositae (92, 112). The genus *Plasmopara* belongs to the family Peronosporaceae (Order Peronosporales, Class Phycomycetes) and is composed entirely of obligate parasites that produce intercellular mycelia with globular haustoria, and conidiophores which emerge and become aerial through the stomata (108). The conidiophores are slender and monopodially branched, at nearly right angles, bearing conidia (zoosporangia) singly at the tips of the branches. Conidia germinate by the formation of biflagellate zoospores or by germ tubes. Like sunflower, its major host, *P. halstedii* is considered to be of North American origin (92). Its worldwide distribution as a seedborne organism in more temperate regions and its geographic isolation have caused this complex species to split into microspecies, forms, or races which some taxonomists believe are sufficiently distinct to subdivide the species. Based on rather obscure morphological characteristics, Novotelnova (109) classified the downy mildew pathogen of sunflower to specific rank as *Plasmopara helianthi* Novot., a name widely used in Eastern Europe but not in the Western World, where *Plasmopara halstedii* is preferred. Based upon biological specialization, Novotelnova (109) distinguished formae *helianthi, perennis,* and *patens.* These subspecific names have not been used widely in the pathological literature.

Life Cycle

The life cycle of *P. halstedii* begins with oospores, thick-walled, sexually produced resting structures, which are essential for its perpetuation. Oospores occur on infested residue from the preceding sunflower crop as well as within the pericarp and testa of seed harvested from systemically infected plants (29, 109). Overwintering oospores germinate mostly under wet conditions the following spring. Some oospores, however, remain dormant, and germination may be delayed for up to 14 years (109). Oospores germinate by producing thin-walled zoosporangia which, in turn, produce biflagellate zoospores. The zoospores may be attracted to rootlets, primary roots, and hypocotyls of emerging sunflower seedlings. Upon contacting host tissue, the zoospore encysts and penetrates into a cell by means of an infection peg giving rise to an intracellular network of hyphae and intercellular haustoria (107, 109). Infection of subterranean parts may cause seedlings to damp-off when the disease proceeds rapidly (51, 172), or produce a basal gall (170) when infections remain localized. However, infection usually causes systemic invasion in which symptoms appear on aboveground plant parts (29, 35, 172). The pathogen sporulates on the surface of invaded tissues, producing further zoosporangia responsible for secondary infections of subterranean tissues and of foilage. As the season advances, sexual organs—oogonia (female and antheridia (male)—are formed in the intercellular spaces of roots, stems, and often seeds (109). A fertilization tube from the antheridium passes into the oosphere contained in the oogonium. The oosphere develops into the thick-walled oospore. Ultimately oospores return to the soil, thus completing the life cycle.

Symptomology

Age of tissue, cultivar reaction, moisture and temperature, and level of inoculum are factors affecting infections by *P. halstedii* and the kind of severity of symptom expression.

Damping-Off

Subterranean infection of susceptible plants in the early seedling stage under cool and moist conditions results in damping-off and seedling blight (172). Seedlings may be killed before or soon after emergence, and plant stands are reduced. Seedling blight due to downy mildew often is not recognized because infection strikes early in the season, and dead, dry plants are windblown.

Systemic Downy Mildew

Systemic infection is characterized by stunted plants, leaves that are chlorotic and abnormally thick, brittle stems, and erect, usually sterile, flower heads (Fig. 1; photos follow page 258). Systemic disease is caused primarily by infection of young seedlings by zoospores liberated from overwintering oospores, or from wind-borne zoosporangia (107). First symptoms are yellowing of the first pair of true leaves, often at the base or along the midrib. As the plant grows, the fungus spreads to younger tissue, the chlorotic area expands, and chlorosis appears on leaves successively up the stem. At flowering time when healthy plants are 1.5 to 1.8 m tall, the height of severely, systemically infected plants may vary from 0.1 to 1.0 m, and the upper leaves will be entirely chlorotic. Under conditions of high humidity and cool temperatures, the lower surfaces of chlorotic leaves develop a white downy growth composed of conidiophores and conidia protruding through the stomata (Fig. 4).

The appearance of visible symptoms due to systemic infection depends on the reaction of the cultivar, the age of the seedling when infection occurred (172), and on the environmental conditions during the incubation period (109). The older the seedling at the time of infection, the more delayed the expression of the symptom. Symptom expression may be delayed under some conditions until flowering, and the distinctive color differentiation is lacking (30, 149).

Localized Downy Mildew—Angular Leafspot and Basal Gall

The localized, secondary phase is caused by zoospores liberated from windborne conidia (zoosporangia). Plants are susceptible to localized infection for a longer period than to systemic infections. However, leaves become increasingly resistant with age. Infection appears to occur through stomata and, since conidia usually germinate by producing zoospores, water congestion may play an important role in establishing an avenue for infection. Secondary infection is observed first as random, small, greenish-yellow, angular spots on young leaves (Fig. 2). These spots may enlarge and coalesce to infect a large part of the leaf. Conidia and conidiophores appear at the lower surface of the discolored area and persist for some time at high relative humidity but soon disappear under dry conditions.

Secondary infection rarely culminates in systemic disease in the field (172), and therefore is of minor economic importance. Cohen and Sackston (29), however, have shown that under controlled conditions secondary infection can produce systemic infection descending to the roots, and ascending to the subsequently formed foliage.

Sunflower is susceptible for an extended period of time to root infection causing basal gall. The primary roots of infected plants are discolored, scurfy, and hypertrophied, and the number of fibrous secondary roots is reduced, thereby increasing susceptibility to drought (170).

Epidemiology

The nature of inoculum, the age of the plant, moisture and temperature, and host-parasite interaction are factors that direct the course of invasion, determine the incidence of disease, and the kind and severity of symptoms.

The low percentage of plants with systemic symptoms which results from planting seed from systemically infected plants (30, 51, and 172) makes it highly unlikely that dramatic outbreaks of systemic downy mildew can be attributed to seedborne inoculum. Soilborne inoculum from a previous sunflower crop or windborne inoculum from systemically infected seedlings in surrounding fields probably are more important in such outbreaks (169).

Seedlings become increasingly resistant to systemic invasion with age; thus any environmental condition which favors rapid seedling development shortens the interval of maximum susceptibility (172). Although seedling development is directly proportional to soil temperature, the range of soil temperature that normally prevails during the spring planting season is not a factor limiting infection by downy mildew. In North Dakota, Zimmer (172) found that early, midseason, or late planting had little effect on the incidence of systemically infected plants. Rainfall and its intensity during the period of maximum susceptibility to systemic invasion was the major factor. If the interval of maximum susceptibility was relatively free of rainfall, little or no systemic infection occurs. If, however, rainfall was moderate or excessive during this period, the incidence of systemic infection was high.

Root infection that can result in basal gall (Fig. 3) may be influenced less by seedling age and environmental conditions than systemic infection.

Distribution and Economic Importance

Although of North American origin, downy mildew has been distributed, presumably through seed, to all major sunflower producing areas of the temperate zones, except Australia, where strict quarantine regulations have precluded its introduction. The biology of the pathogen, however, suggests that spread to new production centers likely will be more restricted by inhospitable climates than by human endeavors (138).

Yield losses from downy mildew vary widely according to region, year, and date of planting. The incidence of systemically infected plants in fields may range from 1% to near 50% or higher, depending upon the occurrence and intensity of rainfall in relation to seeding time (172). In the major sunflower producing areas of Eastern Europe and North America, losses to 80% have occurred when long periods of precipitation and cool weather followed planting (109, 168, 172). Yield losses may not be proportional to seedling losses because of yield compensation by neighboring healthy plants. Yield losses associated with root and basal stem infection (170) are more difficult to assess. Root infections, however, reduce yield in fields receiving heavy rainfall after the plants have advanced beyond the stage of susceptibility to systemic infection, and before they have become resistant to root infection.

Races and Host Resistance

Two races of the *P. halstedii* of sunflower are known. They are the North American race, apparently confined to the North American continent, and the European race, occurring throughout the major sunflower producing area of Europe (171). Fortunately, quarantine programs until recently had prevented the introduction and establishment of the more virulent North American race in Europe. Observations made by D. E. Zimmer in 1976, however, while he was visiting Eastern Europe, and unpublished research results of scientists in Romania suggest that the North American race or a race with similar virulence has been introduced or has developed in Romania. This new race is a serious threat to many of the European hybrid cultivars.

Resistance to downy mildew is conditioned by four genes, Pl_1, Pl_2, Pl_3, and Pl_4 (156, 158, 159, 171, 175). Each of the four genes conditions resistance to the European race, but only the Pl_2 and the Pl_4 genes condition resistance to the more virulent North American race (91, 156, 157, 171). Zimmer (171) reported no segregation in F_2 populations of crosses between lines homozygous for the Pl_2 gene and the Pl_4 gene, thus suggesting that they are identical. The "group immunity" cultivars developed at the Pustovoit All-Union Oilseed Institute at Krasnodar, USSR, from interspecific crosses of *H. annuus* and *H. tuberosus* L. are resistant to downy mildew in Europe, but the identity of the resistance genes is unknown. The absence of the North American race as a test organism in the Soviet Union, coupled with the repeated backcrossing to the downy mildew susceptible cultivar VNIIMK 8931, strongly suggests that genes conditioning resistance of the group immunity lines developed by G. V. Pustovoit may be susceptible to the North American race. Preliminary results of D. E. Zimmer show that several breeding lines allegedly tracing to such interspecific crosses, and reportedly resistant in Europe, are susceptible to the North American race, thus suggesting that such cultivars are not useful as sources of downy mildew resistance in North America.

P. halstedii completes the sexual cycle annually, therefore, maximum opportunity exists for the recombination of virulence genes and the development of new races. The horizontal or general type of resistance reported by Vear (156) may be valuable to restrict the development of new races which in other host-parasite systems have caused difficulty in developing cultivars with long-term resistance.

The nature of resistance to downy mildew is not known. Some observations show limited disease development culminating in sporulation on hypocotyls and occasionally cotyledons of otherwise resistant cultivars (105). This behavior suggests that some physiological or morphological barrier restricts the invading fungus to the cotyledonary sheath, thus excluding it from the meristematic tissue of the epicotyl. Unpublished work of G. Wehjte and D. E. Zimmer suggests that the resistance of hybrid cultivars with the Pl$_2$ gene may be manifested by the lack of effective zoospore encystment or cessation of the infection processes before penetration is achieved.

Control

The only effective control for downy mildew is to use disease resistant cultivars. Rigid enforcement of strict quarantine regulations may limit the spread of downy mildew into areas of the world where it does not occur now. All presently available open-pollinated cultivars are highly susceptible to downy mildew. The open-pollinated cultivars, 'Novinika' and 'Progress,' with downy mildew resistance derived from *Helianthus tuberosus* are being evaluated in the USSR and may be released (120). Many of the presently grown F_1 hybrids of USA, Romanian, and French origin, however, possess genes for resistance to one or both known races of *P. halstedii* and can be grown in their respective countries without loss from downy mildew. The resistance of most of these hybrids traces to material released cooperatively by USDA and the Texas and North Dakota Agricultural Experiment Stations (42).

The seed-borne, soil-borne, and wind-borne nature of *P. halstedii,* combined with the long-term survival of its oospores in crop residue and the presence each spring of volunteer seedlings heavily infected with downy mildew precludes effective control by cultural means in most parts of the world. In much of Eastern Europe and the Soviet Union using rotations of 8 to 10 years, sunflower does not volunteer excessively in the subsequent crop and diseased plants are removed routinely from the fields, minimizing losses from downy mildew (109). Under conditions highly favorable to downy mildew in the Soviet Union, even these cultural practices have little effect and losses of 25% or more have been reported.

RUST OF SUNFLOWER

Causal Organism

Rust of sunflower is caused by the basidiomycete, *Puccinia helianthi* Schw. *P. helianthi* is composed of races of varying pathogenic specialization. A race may attack only certain cultivars of the same species, or only certain annual and/or perennial species (15, 23, 56, 137, 176). *P. helianthi* is a macrocyclic, autoecious species, i.e., telial, pycnial, and uredial stages are produced on one host. The uredial, summer or repeating stage and the telial, overwintering stage, are conspicuous. The uredia are chestnut brown, round, and distributed on leaves, petioles, stems, and involucral bracts of the capitulum. Urediospores are brown, and vary from subglobose to obovate. They are echinulate, with four median germ pores, and average 20.8 to 23.7 μ (15). Telia are oval and brownish black. The teliospores are two-celled, smooth, brown, oblong elliptical, and slightly constricted at the septum; their average size is 23.9 by 39.1 μ (15).

The pycnia and aecia are inconspicuous, usually occur on young seedlings early in the spring, and rarely are found on older plants. The pycnia occur in small clusters, frequently on the cotyledons, or first or second true leaves. They are flask-shaped, amber-colored at first, later becoming orange. The pycniospores are small, oval, hyaline, and appear shiny and viscous in mass. The aecia appear 5 to 7 days after the pycnia and always in close proximity to them, frequently on the lower leaf surface, and opposite, or nearly opposite a cluster of pycnia. The aecia are orange-red with white laciniate margins. Aeciospores are orange, typically ellipsoidal, and finely echinulate with four median germ pores. They vary in size from 21 to 28 μ by 18 to 20 μ (15).

Symptomology

Rust of sunflower is characterized by small, circular, powdery, orange to black pustules scattered over the entire vegetative surface, but is most common on the leaves (Fig. 5). The uredial or repeating stage is the most damaging and conspicuous. The first uredial pustules usually appear on the lower leaves and, as the season advances, the disease spreads and pustules are produced on progressively younger leaves. Chlorotic areas usually surround the pustules even on the susceptible cultivar. In highly resistant varieties no uredia are produced and only small chlorotic or necrotic flecks develop at the point of infection. With severe infestations the stems, petioles, floral bracts, and floral parts may become rusted and uredia coalesce to occupy most of the leaf surface. The leaves senesce prematurely under such conditions, and yield and seed quality are reduced. Rust severity varies with environment, age of host, and inherent resistance of the host species or cultivar.

As the host plant approaches maturity or is subjected to physiological stress, teliospores appear in old uredia, develop into telia, and the "black rust" stage appears.

Distribution and Economic Importance

Sunflower rust occurs wherever sunflower is cultivated extensively. Serious losses in yield and quality have been reported from Argentina (94), Canada (124), USSR (150), USA (177), and Australia (25). The extensive cultivation of the high oil Soviet cultivars 'Peredovik' and 'VNIIMK 8931,' which have some "field resistance," has minimized losses from rust in some areas (59, 174). The general decline of losses from rust in the Soviet Union (150) was attributed to the "general resistance" of the Russian oilseed cultivars. Long-term crop rotation and early season control of volunteer sunflower in the Soviet Union also may have been involved in the decline of rust.

Nonoilseed cultivars in the USA were highly susceptible to rust until 1974 when the cultivar 'Sundak' was introduced (173). Most of the present hybrids are resistant to the predominant races of rust of North America. Prior to the introduction of Sundak, losses from rust were occasionally so severe that the nonoil sunflower industry was threatened. Sunflower rust reduced yield (44), and seed quality by reducing test weight and oil percentage, and increasing hull-to-kernel ratio (177). Moderate to heavy rust infection had little influence, however, on the fatty acid values of the seed oil (177).

Life Cycle

The sunflower rust fungus has the typical life cycle of other macrocyclic, autoecious, heterothallic species of the genus (*Puccinia* (34). The teliospore is the only truly diploid stage, becoming so as the two (+ and −) nuclei in each cell fuse during the maturation or pregermination process. Each cell of the teliospore germinates by producing a promycelium which produces four haploid sporidia via meiosis. The sporidia are thinwalled and fragile, and incapable of surviving long range dissemination. When a sporidium contacts the surface of a sunflower cotyledon, leaf petiole, or hypocotyl, under favorable conditions it produces a fine germ tube which penetrates directly, i.e., through leaf surfaces other than through stomata. The mycelia from such infections produce the flask-shaped pycnia.

During meiosis in the promycelia, the haploid lines segregate into two intercompatible and nonintercompatible (+) and (−) mating groups. Mating between pycnia of opposite mating groups is necessary for the life cycle to proceed from the pycnial stage to the dicaryotic aecial stage. Mating usually occurs after insects or rain transfer pycniospores from one haploid thallus to another. Mating does not lead immediately to nuclear fusion, but to dicaryotization of the thallus and production of aecia with binucleate aeciospores. Aeciospores become airborne and infect sunflower foliage usually near where they are produced. They are relatively insensitive to the environment, however, and can be disseminated long distances without losing viability. The thallus resulting from aeciospore infection is dicaryotic and produces uredia and dicaryotic urediospores. The urediospores are the repeating stage and the spore form primarily responsible for the explosive nature of rust outbreaks. Both aeciospores and urediospores penetrate the host through the stomata.

Disease Cycle

Teliospores are usually the only overwintering structure in the northern region; however, urediospores could survive in regions where winters are mild. The fungus also may overwinter as mycelia in winter annual or perennial sunflowers. Teliospores become detached from the dried sunflower foliage during threshing and adhere to the seeds. These spores are long lived and may germinate in subsequent years when the seed is planted.

Teliospores normally exhibit a dormancy period of variable length depending on storage, the condition of the plant on which they are produced, and harvest time. Miah and Sackston (102) reported erratic results in attempting to germinate teliospores. Some teliospores germinated immediately but others were dormant and germinated sporadically for 2 or more years.

Sporidia, pycnia, and aecia appear in early spring on volunteer seedlings among plant debris of the previous year's crop (124) as well as on emerging plants of wild species (176). During the severe rust years of 1971, 1972, and 1973 in North Dakota, pycnial infections were observed by D. E. Zimmer as early as 15 May. Aeciospores were observed 10 to 14 days later and rapidly spread the infection to surrounding plants and to emerging seedlings in planted fields.

Aeciospores, like urediospores, usually germinate within 4 hours after inoculation if free moisture is present. Germination is by a single germ tube from one of the germ pores (147). Germ tubes form irregularly shaped appressoria over stomata 6 to 8 hours after inoculation. A peg, formed from the lower surface of the appressorium penetrates the substomatal cavity and enlarges into a substomatal vesicle. Twenty-four hours after inoculation, two or more infection hyphae arise from the vesicle. When the infection hyphae contacts a cell, a septum is formed and a haustorial mother cell is produced. A tube enters the host cell and produces a knob-like haustorium. After establishing a nutritional link with the host, the mycelia grow rapidly in susceptible cultivars and culminates in the massing of hyphae under the epidermis and the development of uredial pustules. Teliospores are produced within the uredial pustules as the host plant ages, thus completing the disease cycle.

Host Range

P. helianthi is confined to the genus *Helianthus* where it occurs on more than 35 annual and perennial species (13, 14, 56, 176). Although restrictive pathogenicity exists among collections of rust from wild annual and perennial sunflower, such specificity probably is not sufficiently restrictive to support the establishment of biological forms (varieties) (14, 24, 176). Arthur and Cummins (14) concluded that no culture of *P. helianthi* is distinctive enough to be called a variety. Wild populations of annual sunflower contain some rust resistant plants (56, 176). Completely resistant or susceptible populations rarely are found.

Physiological Races and Host Resistance

The occurrence of distinct physiologic races of *P. helianthi* on cultivated sunflower was established first by Sackston (137). Some of the pathogenic differences observed by earlier workers on a variety of *Helianthus* species probably were due to physiologic specialization (11, 12, 82). Inbred lines of *H. annuus* used to differentiate pathogenic races were not available until the 1960's when Sackston (137) used two resistant accessions to identify four races. Later work established that each accession contained a single dominant gene for resistance. Accession 953-102 is the source of the R_1 gene; accession 953-88 is the source of the R_2 gene (126). Miah and Sackston (101) suggested that the rust resistance of these two inbreds is considerably more complex.

Collections of wild *Helianthus*, especially those of *H. annuus* and *H. petiolaris*, were shown by Hennessy and Sackston (56) and by Zimmer and Rehder (176) to represent a vast reservoir of rust resistance genes that plant breeders can use to broaden the rust protection of domestic cultivars. Wild annual sunflower provides a breeding sanctuary for the rust fungus in the absence of susceptible cultivars. Studies on wild annual sunflower suggest that many races occur, their number limited only by the number of differentials used. Probably rust of sunflower can be controlled effectively for long periods by using specific genes for resistance, especially when a free breeding rust population exists on wild sunflower. The "slow rusting" of certain Soviet cultivars with field resistance (150) may be of value to reduce losses to an acceptable level if specific genes for resistance should lose their effectiveness.

Rust resistance derived from interspecific hybrids of *H. annuus* and *H. tuberosus* and incorporated into some experimental Soviet cultivars may provide stability against new races (120, 121). The discovery of rust in the USA on both *H. annuus* and *H. tuberosus*, however, suggests that recombination of virulence genes could produce a race virulent on both *H. annuus* and *H. tuberosus* (172).

Control

Several sources of rust resistance are known and the R_1 and R_2 genes have been used widely to develop resistant cultivars (126, 174). All oilseed and nonoilseed hybrid cultivars produced by using the parental lines developed and released by USDA are moderately resistant or fully resistant to race 1, the predominant race in the principal sunflower producing area of North America (174). These hybrids are resistant in other areas of the world, suggesting that race 1 is a predominant worldwide race.

The sunflower rust fungus completes the sexual cycle annually; therefore, maximum opportunity exists for combining new mutant virulence genes and developing new races (176). Races already are known that attack cultivars with the R_1 and R_2 genes (137). As cultivars with the R_1 and R_2 genes are grown more extensively, races attacking them are expected to have a selective pathogenic advantage and become predominant.

While breeding for resistance is the best method to control rust, crop rotation and early elimination of volunteer and wild sunflower near the production field will reduce primary inoculum and decrease the risk of severe yield losses from rust. The relative low unit value of sunflower, difficulty in obtaining complete foliage coverage by aerial application of pesticides, and the lack of registration of fungicides for the control of foliage diseases of sunflower in North America now prohibit the use of fungicides to control rust.

SCLEROTINIA WILT AND HEAD ROT

Causal Organism

Taxonomy

Sclerotinia sclerotiorum (Lib.) de Bary is a major pathogen of sunflower that infects roots, stems, and flower heads to cause root rot, wilt, and stalk and head rot. The same organism often is called *Sclerotinia libertiana* Fuckel in eastern European literature, but Whetzel (160) placed it in synonymy with *S. sclerotiorum.* Korf and Dumont (86) assigned the species to *Whetzelinia,* a new genus, but objections to this disposition have been raised (36). *Sclerotinia minor* Jagger (133), and *S. trifoliorum* Fuckel (32), are also pathogenic to sunflower. Purdy (119) found a continuous intergradation of sclerotium size in these species and proposed that the name *S. sclerotiorum* be applied to all. More recently, however, after ontogenetic studies of sclerotia (161), as well as mycelial interaction and electrophoretic investigations, Wong and Willetts (165) concluded that *S. sclerotiorum, S. trifoliorum,* and *S. minor* are three distinct species.

Description of the Pathogen

S. sclerotiorum is an Ascomycete and produces tiny, mushroom-like structures or apothecia as the perfect, sexual stage. Spermatia or microconidia, also produced in senescent cultures in the laboratory, may be functional in sexual reproduction but are not known to infect plants.

Apothecia of *S. sclerotiorum* produced in a diseased sunflower field in Manitoba in 1973 were light brown, 2 to 8 mm in diam, sessile or stalked, and arose from some sclerotia only 3 × 5 mm. A single sclerotium produced as many as eight apothecia (Fig. 6). Asci, carried in tightly packed masses at the upper surface of the apothecium, measured 90 to 120 μ in length and 6.2 to 10 μ in width. The hyaline, one-celled ascospores, numbering eight per ascus, were 10 to 11.3 μ wide (J. A. Hoes, unpublished data). These measurements agree well with those for other isolates of this sunflower pathogen studied by Jones (78) and Bisby (23).

Apothecia developed from sclerotia which are black, hard propagules that overwinter in the soil. The mature sclerotium consists of an outer pigmented rind about three cells wide, a layer of smaller, thin-walled, hyaline pseudoparenchymatous cells two to four cells across, and a medulla of pro-senchymatous tissue partly embedded in a gelatinous matrix (161). The sclerotium provides nutrients at germination. Two modes of germination occur, i.e., myceliogenic, forming only hyphae, and carpogenic, producing apothecia.

Host Range

S. sclerotiorum has a vast host range. It attacks field, forage, vegetable, and ornamental crops, trees and shrubs, and numerous herbaceous weeds. The pathogen causes drop in lettuce (*Lactuca sativa* L.), white mold and crown rot in beans (*Phaseolus vulgaris* L.) and forage legumes, stem blight in peanuts (*Arachis hypogaea* L.), stem rot in rapeseed (*Brassica* spp.), and storage rot in beans and other vegetables. Hosts belong to 190 species, placed in 130 genera of 45 plant families (4). The pathogen caused root rot and wilt of wild *H. annuus* (67), and crown rot of *H. californicus* DC (8) and *H. tuberosus* L. (23). Cruciferae, Leguminosae, and Solanaceae are important host families.

Little or no evidence is available of physiological specialization in *S. sclerotiorum* (118). Isolates, however, varied widely in pathogenicity.

H. annuus also is a host of other species of *Sclerotinia* but records from North America are rare. Sackston (133) observed serious root rot and wilt of sunflower incited by *S. minor* Jagger in Chile. Cormack (32) noted an isolate of *S. minor* from lettuce in New York, one of *S. trifoliorum* Eriks from red clover (*Trifolium pratense* L.) in Kentucky, and one of *S. sclerotiorum* from sweetclover (*Melilotus* sp.) in British Columbia all caused serious disease in 'Mennonite' sunflower. In studies by Keay (81), *H. tuberosus* was killed by isolates of *S. sclerotiorum* from hop (*Humulus lupulus* L.) and swede (*Brassica napus* L.) as well as by two isolates of *S. minor,* while one of 10 isolates of *S. trifoliorum* was only slightly pathogenic.

Life Cycle

Sclerotia commence and conclude the life cycle of *S. sclerotiorum.* Myceliogenic germination of sclerotia that can occur during the entire growing season causes infection of underground plant tissue, and produces root rot, basal stem rot, and wilting of plants. In carpogenic germination of sclerotia, apothecia carried on stalks arise from sclerotia in the upper soil layers, emerge at the soil surface, and discharge ascospores that become airborne. Ascospores germinate under favorable conditions, infect the host tissue, causing stem rot and mainly head rot. Sclerotia produced within and on the surface of affected parts return to the soil with the sunflower residue to repeat the cycle in the next season.

Symptomology

Root Rot and Wilt

Sclerotinia wilt usually appears near flowering time when seasonal temperatures are reaching a maximum. However, plants may be killed much earlier. Diseased plants occur scattered in the rows, often singly at first, soon in groups of two or more plants, until near harvest when continuous portions of rows are diseased. In warm weather when the disease develops rapidly, a plant appearing healthy one day may be wilted completely the next day. A brown, wet, soft lesion develops, which at first partially surrounds the stem base but later completely girdles the stem and extends upward. Lesions are covered partially or completely by white mycelia (Fig. 7). The lesion may vary from a few to 50 cm in length depending on the stage of plant development at infection, and on temperature and moisture conditions. Plants infected late in the season may not wilt and the only exterior symptom may be a small brown lesion at the stem base. Stems of severely diseased plants shred into vascular strands, turning straw-colored as they dry. Such stems are weak and the plants lodge easily. Sclerotia of variable size and shape, white at first then later black, develop on the surfaces and in the pith cavities of roots and basal stems.

Head Rot and Stalk Rot

These diseases, incited by airborne ascospores, develop in the adult plant. Heads are infected during the long period from flowering to maturity. Infection may begin in any part of the fleshy receptacle. The pathogen produces a conspicuous, white mycelial mat and the rot spreads throughout the flower head (Fig. 8), often extending downward from the receptacle to the upper stem. Ultimately, the entire head may be destroyed and most of the host tissue converted to a continuous mat of sclerotial tissue. Severely diseased heads disintegrate until only a skeleton of shreded, straw-colored to white fibrous tissue remains (Fig. 9).

Stalk rot infections can develop at any part of the stem, often in the upper half. Infection usually occurs at a node and the lesion may extend up and/or down. Stalk rot is most visible when the stems mature and the infected tissue appears lighter than the brown color of normal maturity. Exterior mycelium seldom is conspicuous and sclerotia may or may not develop inside the affected stem parts.

Disease Cycle

Development of Root Rot and Wilt

Wilt develops when mycelia from germinating sclerotia infect underground parts. Hypocotyls, especially when succulent, are invaded readily (54) but the pathogen penetrates the host mainly through roots (167). This course is evident from field observations that root systems of diseased plants usually are destroyed almost completely by the time that plants wilt and lesions appear at the stem base. Injury is greatest in the fibrous roots which are concentrated in the upper 20 cm of soil. Spread of disease through root contacts (66, 167) is further evidence that roots are important sites of infection in wilt development. Infection is favored by wounds caused when branch roots emerging from the pericycle rupture the cortex (167). Disease incidence is greatest in dense stands (22, 78, 167).

The first step in pathogenesis is mobilization of inoculum, i.e., germination of sclerotia. Mycelia of *S. sclerotiorum* grow slightly at 5 C, and best at 20 to 25 C (1). Conditions of temperature, oxygen, and moisture in the upper soil layers are adequate for both root development of the host and germination of the sclerotia. Proximity of the sclerotium to host tissue and a food base are vital to penetration and establishment of the fungus inside the host. Mycelial growth of *S. sclerotiorum* in soil is sparse. In sterilized soil mycelial extension never exceeded 30 mm from the parent sclerotium, and in unsterilized soil growth was restricted, the hyphae never being longer than 5 mm (162). Newton and Sequeira (106) found that the fungus was unable to extend more than 2 cm from a cornmeal food base in muck soil and was unable to infect lettuce unless the inoculum was within 2 mm of the host. Abawi and Grogan (1) reported that infection occurred only when a food base was placed in direct contact with the sclerotium and the bean seed or stem. With sunflower, contact with inoculum occurs at random when actively growing branch and fibrous roots grow near a sclerotium. Extracts from seed and other tissue of sunflower supported growth of *S. sclerotiorum* (1). Root excretions and cortical material dislodged from roots in soil presumably provide adequate nutrients to bridge the distance between germinating sclerotia and host tissue.

According to Bisby (22), plants are "usually attacked at or near the surface of the ground". This may happen, especially in young seedlings (54), with abundant moisture and if a nearby food base with mycelium is present. We have found, however, that attack at or near the stem base is rare and that infection through roots is most common. The pathogen probably is indiscriminate and can invade roots at any site. Ultimately, it reaches the tap root and continues growth downward and upward, decomposing parenchymal tissue and consuming the cortex. The rot spreads simultaneously to other roots of the same plant, either by contact, or presumably via points of root attachment along the tap root and the lower hypocotyl. As the disease progresses, the pathogen also reaches the pith of the tap root and expands further upward and into the pith of the stem. Sclerotia develop wherever mycelium occurs.

Development of Head Rot

Abundant moisture and a food base are essential to infection of aboveground plant organs by ascospores (1). In 1968 in Manitoba, prominent head rot of sunflower was associated with abnormally high rainfall of over

25 cm during July and August when sunflower blooms and senescent flower parts are present (59). Birds carrying mycelium caused wounds and transmitted the disease to a sunflower crop that developed head rot in a dry summer (167).

Abawi and Grogan (1) studied temperature as an epidemiological factor. The range for apothecial production was 10 to 25 C, with an optimum of about 10 C, and ascospores germinated equally well from 10 to 30 C, whereas germ tubes grew best at 20 to 25 C (1). Ascospore discharge occurred from 4 to 32 C but was greatest at 22 C (106). Generally, moderate temperatures of 20 to 25 C appear optimum for infection.

Mechanism of Pathogenesis

Host tissue is macerated and cell walls are dissolved in development of root rot and head rot. The middle lamella of sunflower cells consists mainly of pectic substances, cellulose and hemicellulose, and cell-wall-degrading enzymes attacking these substrates occur in increased concentration in Sclerotinia-diseased plants. High pectin-degrading polygalacturonase activities were associated with pathogenesis by *S. sclerotiorum* (52), and extracts from diseased plants degraded hemicelluloses and pectic substances, but extracts from healthy sunflowers were inactive (53). Cellulase probably contributes to pathogenesis (95). Phosphatidase, possibly responsible for changes in host cell permeability, occurred early in pathogenesis and was correlated with symptom expression in bean (96). Oxalic acid also is involved. Growth rate of *S. sclerotiorum* in culture and during disease development in bean was correlated positively with oxalate production, while polygalacturonase activity also increased (97). Oxalic acid alone bleached bean tissue and caused marked injury, but oxalic acid with a polygalacturonase preparation was extremely toxic and hypocotyls soon collapsed (17).

Distribution and Economic Importance

S. sclerotiorum is distributed worldwide and is prevalent in temperate as well as in tropical and subtropical regions. The pathogen is important outside of North America in all major sunflower production areas, including Argentina (94), France (90), Mexico (48), and eastern European countries.

Sunflower has been grown as an oilseed crop in North America on a large scale for only a decade, and to date diseases caused by *S. sclerotiorum* have been of relatively minor importance. Most of the crop has been grown in fields which never or rarely have been cropped to sunflower, and the vast areas available allow for long rotations. Low precipitation in some production centers does not favor development of Sclerotinia diseases. The importance of Sclerotinia diseases in the USA may increase as sunflower culture expands and intervals between sunflower crops become shorter.

Nonoilseed and oilseed sunflower production in Canada has been concentrated in the Red River Valley in Manitoba where crop rotations, including sunflower and other host crops, are short. Under these conditions in Manibota, the most important diseases of sunflower now are caused by *S. sclerotiorum* (67). Hoes (60), in a 1971 survey in Manitoba covering 25% of the fields, found the average infection to be 14% and some fields were practically destroyed. Again, in 1975, Sclerotinia wilt occurred in 29 of 57 Manitoba fields with an incidence of 10 to over 80% in 16 fields (67). Yield reductions occur, and seed quality may be affected seriously in fields with severe wilt. Yield reduction and other damage depends on when infection occurred and whether the pathogen incited wilt or head rot. Generally, high fertility and moisture supplies favoring vegetative growth of the plant also favor root infection and development of wilt. Long periods of precipitation with short, rainless intervals during which heads dry increase the incidence of head rot. A head affected by Sclerotinia rot usually will be a complete loss.

Sclerotia may be economically important because of their toxicity. Ruddick and Harwig (129) studied rats fed with sclerotia from sunflower with head rot. Delay of ossification, decrease in maternal weight, and lowered food consumption were ill effects initiated with 2 to 8% sclerotia in the rations. Properly cleaned sunflower seed, however, is unlikely to contain concentrations of sclerotia high enough to create a health hazard.

Survival of Sclerotia

The saprophytic ability of the pathogen, and moisture and temperature conditions are principal factors affecting the survival of *S. sclerotiorum.* Williams and Western (163) found that sclerotia in soil increased in weight and that sclerotia regenerated and produced daughter or secondary sclerotia. Secondary sclerotia developed in soil at depths of 5 to 30 cm but not at the soil surface (31). Mycelium of the pathogen grew saprophytically (22) and over-wintered on sunflower stalks (167). Krüger (87) found that during September 1968 in Germany sclerotia developed in mycelium-infested stems of rapeseed buried in soil. Residue of sunflower, rapeseed, peas (*Pisum sativum* L.) or other host crops, in swath or soil-borne, may provide host tissue for development of mycelia and sclerotia when moisture and temperature are favorable.

Sclerotia in soil are exposed to variable moisture and temperature conditions. Conditions near the soil surface are the most variable. Soil type also is important because the texture affects waterholding capacity and the organic content affects microflora including antagonists. Desiccation of sclerotia caused loss of nutrients and predisposed them to rapid microbial colonization and rot (145). Deterioration is affected by temperature. At 27 C after 1 and 3 months in wet soil, 51 and 73% of the sclerotia, respectively, had deteriorated, but at 5 C only 7 and 15%, respectively, had rotted. Little deterioration occurred at 27 C in dry soil (31). High moisture combined with high temperatures are thus most detrimental to sclerotial survival.

Variable longevity of sclerotia has been recorded (31, 100, 115). McLean (100) found that a few sclerotia were still viable and produced apothecia after 5 years burial in moist soil, although destruction was nearly complete. In New York, less than 6% of the sclerotia survived two mild winters at 5 cm depth but over 40% survived at 17.5 cm depth (115). Disintegration was greatest nearest the soil surface (115), an observation also made by McLean (100). In Nebraska under different climatic conditions, 78% of sclerotia buried at soil depths of 5 to 20 cm were recovered after 3 years of burial, but none was recovered from the soil surface (31). Survival of sclerotia below the soil surface is much greater during severe than during mild winters.

Apothecia arise primarily from sclerotia at 1 to 2 cm soil depth, whereas few apothecia are produced from sclerotia at 5 cm depth (162). Sclerotia below the soil surface for 17 months formed more apothecia than those remaining on the soil surface (31).

Control

Effective control of wilt and head rot of sunflower can be achieved only by a multipronged approach using cultural, biological as well as chemical, and possibly genetic methods.

Cultural Methods

The upper layers of the soil and the soil surface appear to be the least favorable sites for sclerotia survival. Cultural control in fields by shallow harrowing retains infested residue on or near the soil surface and should accelerate the reduction of inoculum potential of the pathogen. Plowing buries sclerotia (1) and tends to conserve them. Plowing may cause increased root infection and wilt but decreased aerial infection and head rot. Care should be taken to prevent introduction of diseased residue into other fields. Also a good practice is to isolate sunflower from stubble of host crops to reduce the danger of windblown ascospores.

Sunflower and other host crops should not be used in the same rotation. In the severe Sclerotinia epidemic on sunflower in 1971 in southern Manitoba (60), the crop histories of 19 fields with varying amounts of disease showed that the most severe disease occurred when host crops had been grown in two or three seasons in the 5-year period during 1966–1970 (P. Bergen, personal communication). Disease occurrence was associated negatively with the frequency of nonhost crops such as barley (*Hordeum vulgare* L.), and wheat (*Triticum aestivum* L.), and with fallowing. Disease incidence also was associated negatively with the frequencies of beets (*Beta vulgaris* L.) and flax (*Linum usitatissimum* L.) crops not attacked by Sclerotinia disease in southern Manitoba. Disease incidence was associated positively with the frequency of highly susceptible crops such as rapeseed, yellow mustard (*Brassica* sp.), and sunflower, as well as with field peas which the pathogen attacks without producing serious disease. The data suggested that 5 years should separate sunflower from a preceding host crop. Host crops in short rotation also were associated with severe Sclerotinia wilt of sunflower in 1975 (67).

Wider plant spacing controls Sclerotinia wilt in two ways. It hastens drying of the soil and apothecia, thus reducing development of head rot. Also, wider plant spacing decreases the chances of contact between roots of a diseased plant and of adjacent plants, thus reducing incidence of wilt disease (66, 167). Sunflower plants spaced 20 cm apart in the row had 28 to 45% wilt, whereas only 3 to 5% disease occurred in plants spaced at 30 cm (167). Yield reductions due to wilt were lower at densities of 27,500 to 55,000 plants/ha (11,000 to 22,000 plants/acre) than at densities of 82,000 plants/ha (32,500 plants/acre) and higher (66). Yield reductions due to wilt would be minimized if farmers did not overplant and if plants were spaced uniformly (66).

Biological Control

A large epidemiological difference exists between sclerotia produced on plants with head rot and those produced as a result of wilt disease. The former are nearly always healthy and viable, and therefore have maximum disease potential. However, sclerotia recovered from soil (65) or produced in roots and stem bases (72) often are decomposed by soil-borne hyperparasites, especially *Coniothyrium minitans* Campbell (73). Huang (72) showed that during development of sunflower wilt by *S. sclerotiorum,* the pathogen, in turn, is invaded and sclerotia destroyed by *C. minitans.* Inoculation of soil naturally infested with *S. sclerotiorum* reduced wilt incidence and increased yield of sunflower in the field (71). Biological control of Sclerotinia diseases of sunflower using this hyperparasite appears promising, especially since it is not pathogenic to sunflower (72).

Genetic Control

Research workers have made only slight progress in breeding for resistance to Sclerotinia wilt and head rot. Putt (122) first noted differences in susceptibility to Sclerotinia wilt in sunflower. Recurrent selection for wilt resistance within the cultivar Peredovik has resulted in populations with reduced susceptibility (66). Resistance to Sclerotinia wilt observed in the field in Canada is not expressed in greenhouse tests (H. C. Huang, personal communication). Field resistance appears to be heritable, however, and thus may prove useful in reducing losses from Sclerotinia wilt. Variation among cultivars for the incidence of head rot has been observed, but differences appear to be due to plant height (90). Presumably, plants escaped head rot because of their growth habit where heads dried faster and ascospores had a greater distance to travel (90).

Chemical Control and Sanitation

Benomyl [methyl 1-(butylcarbamoyl)-2-benzimidazolecarbamate] and other fungicides have been used with success to control air-borne infections in beans and other crops (21, 49). Fungicidal control of sunflower head rot, however, has not been studied.

Seedstocks without sclerotia contamination aid in reducing wilt epidemics (67). Apparently healthy seed may carry mycelium of the pathogen (136).

VERTICILLIUM WILT

Causal Organism

Verticillium dahliae Klebahn is the incitant of this important wilt disease. The fungus is classified in the Fungi Imperfecti, and a sexual stage is not known. The organism produces new variants asexually, however, by processes of heterokaryosis and parasexuality (116). The mycelium of the pathogen is septate and hyaline. Elliptical to ovoid, hyaline, mostly one-celled conidia measuring about $3 \times 5 \mu$ (37) are produced at the tips of phialides, specialized structures occurring in whorls along the main conidiophore which is completely hyaline (146). The fungus is typified morphologically by variably shaped microsclerotia (Fig. 10), which are overwintering structures. Electron microscopic studies (26) reveal that the microsclerotium consists of tightly compressed branching chains of nearly globose hyphal cells, 8 to 15 μ in diam. The cells embedded in a mucilaginous matrix, are of two types: thin-walled, degenerated hyaline cells that contain cytoplasm but no nuclei, and thick-walled pigmented cells that contain nuclei, cytoplasm, food reserves, and other materials. Only the pigmented cells can germinate.

The sunflower pathogen was identified originally as "*Verticillium albo-atrum* Reinke & Berth., microsclerotial plus mycelial strain" (140). However, the pathogen was recognized later as *V. dahliae* (37). Morphological and other distinctions between the two species are described clearly (146).

Symptomology

Initial symptoms usually appear near or at flowering time. Symptoms appear first on the lower leaves and then develop successively on the leaves higher on the stem. Prominent yellow, interveinal patches appear, usually first in the center or near the periphery of the leaf (Fig. 11). These chlorotic patches, probably caused by a toxin (128), gradually enlarge and coalesce while their centers turn brown and necrotic. Sackston et al. (140), who first described the symptoms on sunflower, suggested the name "leaf mottle" because of the mottled appearance effected by the contrast between the yellow and the persisting green areas. Ultimately, the entire leaf may turn brown and wither but "halos" of yellow tissue persist around the necrotic areas. Black, streaky patches occur at the base of the stem. Cross sections of the lower stem show a brown discoloration of the vascular system. The pathogen becomes distributed throughout the plant, and can be isolated from roots, all levels of the stem, petioles, midribs of the upper leaves, and from the receptacle of the inflorescence (140).

Severely diseased plants show stunting, small flower heads, and destruction of root systems by secondary invaders. At the end of the season, centers of tap roots of infected plants, even of resistant cultivars, usually are discolored black from the production of masses of microsclerotia. Actual wilting of plants with typical leaf mottle symptoms rarely occur even in highly susceptible cultivars. However, yellowing and wilting of leaves, and flaccid heads typify plants of the more tolerant 'Krasnodarets' and 'Peredovik' that become affected by Verticillium wilt (67).

Distribution and Economic Importance

V. dahliae is reported from most major sunflower production areas including Argentina (94), Mexico (48), the Soviet Union (139), and more recently from France (5) where sunflower has been grown extensively for only the last decade. The nature of the seedborne pathogen (136) may have contributed to its worldwide distribution. Verticillium wilt threatened sunflower production in Canada in the 1950's (140) and the disease has remained important (67). In the USA the increase in sunflower plantings in North Dakota and Minnesota has been accompanied by an increase in Verticillium wilt (43). The disease has been reported also from California (85).

Losses from Verticillium wilt can be significant. Hoes and Putt (68) surveyed 36 fields and estimated that in 10 fields with 50 to 100% wilt, nearly complete loss occurred in two fields, 50% loss in five fields, and 10 to 15% loss in three fields. Hoes (61) obtained more detailed information on losses due to Verticillium wilt in 1971. Plants with severe, moderate or mild symptoms showed average yield reductions of 80, 55, and 18%, respectively, while respective head diameter was reduced by 42, 28, and 16%. Reduction in head size was the most important factor in yield reduction. Oil concentration of the kernels was reduced from 51.4 to 46.2%, hectoliter weight was reduced about 10%, and seeds of diseased plants were smaller.

Damage due to Verticillium wilt is a function of the level of inoculum, relative resistance of the cultivar, weather conditions, and possibly the specific pathogenicity of local strains of the pathogen.

Host Range and Specialization

Verticillium dahliae (Verticillium albo-atrum, microsclerotial type) has a wide host range. It attacks many different plants including field and forage crops, ornamentals, trees, vegetables, and weeds. Until 1957, over 350 species, all dicots, belonging to over 160 genera had been recorded as hosts of either *V. dahliae* or *V. albo-atrum* (139). Most of the records probably relate to *V. dahliae* because of its vast host range (116, 146, 166).

Lacy and Horner (89) classified hosts as "immune", "resistant", and "susceptible". Immune plants are symptomless, i.e., immune to the wilt syndrome and with only superficial penetration by *V. dahliae.* In resistant and susceptible hosts, the infection becomes systemic, but symptoms are absent or mild in resistant hosts and severe in susceptible hosts. Immune hosts include dicotyledonous weeds as well as agriculturally important monocotyledonous crops such as barley (*Hordeum vulgare* L.), corn (*Zea mays* L.), oats (*Avena sativa* L.), onion (*Allium cepa* L.), and wheat (*Triticum aestivum* L.) (20, 39).

Strains of *V. dahliae* differing in host range and in pathogenicity occur. Of two isolates of *V. dahliae* from sunflower in Manitoba, one caused little or no disease in 'Norgold' potato (*Solanum tuberosum* L.), whereas the other was very pathogenic. Neither isolate was pathogenic to tomato (*Lycopersicon esculentum* L.) while both caused severe disease in sunflower (58). An isolate of *V. dahliae* from *Chrysanthemum morifolium* Ramab. (128) and 13 isolates from potato, tomato, cotton (*Gossypium* sp.), and strawflower (*Helichrysum bracteatum* Ndr.) (110) caused disease in sunflower inbred CM 162. Sackston et al. (140) reported that strains of *V. dahliae* from cotton in Africa were pathogenic to sunflower.

Sunflower production in the USA is expanding into areas where Verticillium susceptible crops such as cotton and safflower (*Carthamus tinctorius* L.) (84) are commonly grown. This expansion may be followed by an increase in the importance of Verticillium wilt in sunflower because pathogenic strains already may be present.

Disease Cycle

V. dahliae is a soil-borne organism and survives as microsclerotia, occurring in plant residue of the previous crops or weeds. Mechanical damage by tillage and further breakdown by soil organisms release the microsclerotia into the upper soil layers. In nature microsclerotia usually are dormant due to fungistasis. Root exudates break dormancy, however, and microsclerotia germinate to produce hyphae and conidia. Infection is initiated with direct penetration of host tissue by hyphae and germinating conidia (142). Penetration of immune hosts is superficial and the pathogen soon dies without producing microsclerotia (39).

Garber and Houston (50) describe in detail the infection of susceptible cotton by the microsclerotial type of *V. albo-atrum*. The pathogen penetrated the root cap or the epidermis in the region of elongation and maturation, either between or directly through the epidermal cells. The lower hypocotyl below the soil surface also was invaded. Hyphal advance was inter- and intracellular, involving either a few hyphae or mycelial strands when inoculum potential was high. Progress of hyphae was impeded in the cortex and at the endodermis. Few hyphae overcame the endodermal barrier, but once overcome, vessel members were penetrated, invariably through the pits. An appressorium sometimes was formed. Upon invasion of the xylem vessels of roots or hypocotyl, conidia were produced within 3 days after inoculation. The fungus then become systemic, either by continued mycelial growth or by upward movement of conidia which germinated in leaf traces and at sites higher along the stem. No essential differences occurred in fungal distribution throughout plants of susceptible or highly tolerant cultivars, except for more extensive fungal development in the susceptible cultivar.

Control

Control measures include breeding for resistance, crop rotation, and sanitation.

Breeding for Resistance

This is the most effective means of control. Hoes and Putt (69) noted apparent genetic control of wilt in commercial cultivars in which the incidence and severity of wilt was much less in fields of 'Peredovik' than in fields of 'Mennonite.' Putt (123) studied the inheritance of resistance. Inbred lines used as parents varied widely in wilt reaction. Qualitative or complete dominance of resistance was evident in one cross. Lack of dominance of resistance, heterosis for resistance, and dominance of susceptibility appeared from other crosses.

Dominant resistance, based on a single gene, also has been identified by Fick and Zimmer (43) in several inbreds including HA 89, a line used in successful production of wilt resistant hybrids using the cytoplasmic male sterility-fertility restorer system. The relationship of that gene to gene V_1 identified by Putt (123) is not known, however. The use of dominant resistance in producing wilt resistant hybrids is the most convenient. Heterosis for resistance using suitable parents offers another effective system to produce wilt resistant hybrid cultivars.

Resistance to Verticillium wilt is widespread in wild sunflower (70). Collections of *H. annuus* from Manitoba, Saskatchewan, and North Dakota generally were less resistant than those from more southern latitudes, e.g., Colorado and Kansas. *Helianthus petiolaris* Nutt. generally was more resistant than *H. annuus*. A center of resistance was postulated to exist in the central and southern Great Plains, coinciding with the hypothetical center of origin of *H. annuus* and *H. petiolaris*. Dominance of resistance, and also lack of dominance of resistance were suggested by the frequency distributions in hybrid and segregating progenies (70).

Crop Rotation

The objectives of crop rotation are to reduce pathogenic inoculum thus reducing disease incidence and yield losses. Crop rotation may not successfully control Verticillium wilt (40). Microsclerotia of *V. dahliae* are long-lived and the persistence of these propagules is a prime factor. Other factors are the years between susceptible crops, the number of microsclerotia in the soil, the nature (wilt-immune or not) and/or cultivar of the intervening crops, and undefined characteristics inherent to soils. Hoes (63) observed severe wilt in Mennonite sunflower in Manitoba fields that had been in cereal crops for at least 12 years since the preceding, apparently healthy, sunflower crop. The high wilt susceptibility of cultivars and the shortness or lack of crop rotations were factors which rapidly increased levels of inoculum so that serious outbreaks of wilt in Manitoba occurred in a few years (140).

The role of susceptible or resistant hosts is recognized in the maintenance and increase of inoculum potential (40, 89, 142), but the role of resistant or immune crops is not well defined. Benson and Ashworth (20) found that parasitic colonization of barley by *V. dahliae* in California was not a significant survival mechanism. Microsclerotia were not found in root residues of immune crops such as alfalfa (*Medicago sativa* L.), barley, corn, and grain sorghum (*Sorghum vulgare* Pers.) that had been grown in infested fields. The organism did colonize the roots of these crops but colonies were superficial and short-lived only. Huisman and Ashworth (74) concluded that immune-host culture had little influence on survival of *V. albo-atrum* (microsclerotial) in soils and that the pathogen's rate of attrition is too low to make short-term rotations useful for wilt control. Study of the importance of nonhost crops such as Gramineae to sunflower Verticillium wilt seems justified.

Seed Sanitation

Seed lots of susceptible cultivars occasionally contained up to 7% seeds contaminated with microsclerotia (139). Using seeds of healthy plants prevents introduction of the pathogen.

CHARCOAL ROT

Charcoal rot, so-called because of the gray to black discolored stems, is a disease affecting at least 284 plant species in warm, semiarid climates. The causal agent produces very small, black sclerotia, and is best known as *Sclerotium bataticola* Taub., the imperfect stage. *Macrophomina phaseoli* (Maubl.) Ashby is the perithecial, sexual stage. The disease is observed only occasionally in North Central U.S. where the climate does not favor its development. In the Southwestern U.S. and abroad in other areas with arid climates, charcoal rot has inflicted significant losses to sunflower (2, 6, 111, 135, 151). The pathogen can cause a destructive seedling blight, and root or basal stem rot of sunflower (2, 151). The seed-borne nature of the pathogen suggests that it can spread to new sunflower production areas (41).

The most common symptom is shredding of tissue at the basal stem and the top of the rap root. Affected stems are largely hollow and rotted, and the stalk crushes easily under pressure. Masses of sclerotia discolor the basal stem (Fig. 12). Obvious symptoms usually are not apparent until onset of flowering, but plants may have become infected very early in the season. High soil temperatures favor the growth of the pathogen (28). Moisture and heat stress predispose plants to attack (38, 93), induce premature ripening, and reduce yield and seed quality. Charcoal rot in the Krasnodar region of the USSR reduced head diameter by 30%, seed yield by 18 to 64%, test weight by 13 to 36%, and oil content by 5 to 8%. Diseased plants are also more subject to lodging (151).

Since charcoal rot is associated with moisture stress and high temperature, any practice that reduces exposure of plants to those conditions will reduce loss. Early season planting or choosing early maturing cultivars to escape midsummer heat and drought are practices which could reduce losses from charcoal rot. Irrigation also could be employed to minimize losses. Although little is known about the relative resistance of most sunflower cultivars and parental lines, investigations of Orellana (111) and more recently by Alabouvette and Bremeersch (5) demonstrated that cultivars respond differently to artificial inoculation as well as to natural infections in the field thus suggesting the control of charcoal rot by breeding for resistance.

WHITE BLISTER RUST

White blister rust of sunflower caused by the strictly obligate fungus *Albugo tragopogi* (Pers.) Schroet. has been reported in Argentina (141), Australia (79), Uruguay (134), USSR (108), and in several other countries.

White blister rust is considered one of the most important diseases in several South American countries (M. L. Kinman, personal communication). Although the disease has been reported on both wild annual and wild perennial sunflower in the USA (9) extensive surveys have not detected it in the present area of sunflower production. As North American sunflower production intensifies and expands, blister rust may become a significant disease where environmental conditions favor disease development.

Early symptoms consist of raised, yellow, isolated spots 1 to 2 mm in diam on the lower surface of leaves and sometimes on the petioles (Fig. 13). The pustules rupture and appear as blisters that release masses of conidia that become airborne. These conidia spread the disease to surrounding plants. The conidia behave primarily as zoosporangia producing 7 to 11 biflagellate zoospores that penetrate the host through the stomata, encyst therein, and produce intracellular hyphae (79). Severely infected leaves often die and turn brown giving a blighted appearance to the plant.

Infection of sunflower by *A. tragopogi* appears to be dependent upon an adequate supply of free water from rainfall or heavy dews (79). Sporangia germinate over a wide range of temperatures from 4 to 35 C but zoospores remain viable only at temperatures of 4 to 20 C (79). These rather restrictive moisture and temperature requirements may be major factors limiting the occurrence of white blister rust.

Control practices for white blister rust have not been developed because of its sporadic occurrence and limited economic importance in most countries. Resistance to white blister rust is known in some plant species but has not been documented in sunflower.

ALTERNARIA DISEASES

Leaf and stem spots, seedling blight, and head rot are diseases caused by several species of *Alternaria* (Fungi Imperfecti) that occur on sunflower in various parts of the world. *Alternaria zinniae* Pape (99) occasionally is important on sunflower in North America. The pathogen attacks the leaves causing dark colored, target-like lesions which are roughly circular (Fig. 14). Brown flecks and streaks appear on stems and petioles, and, when numerous, form large necrotic areas (Fig. 15). The fungus is seed-borne, and can cause damping-off and seedling blight. The beaked conidia are borne singly on the conidiophores. Conidial dimensions including the beak are 36.6 to 236.4 × 8 to 22 μ. The filamentous beak alone may be as long as 146 μ (99), and this feature distinguishes *A. zinniae* from other species of *Alternaria* pathogenic to sunflower.

Alternaria helianthi (Hansf.) Tubaki & Nishihara (*Helminthosporium helianthi* Hansf.) produces dark brown lesions on leaves, stems, petioles, and flower parts (155). This pathogen occurs in Argentina, India, Japan, African countries (155), and Brazil (127). It causes a serious disease of heads and is a threat to sunflower in Yugoslavia (76). *A. helianthi* has not been reported in North America, but seedborne inoculum from other continents is a potential danger to sunflower on this continent. Symptoms appear to be similar to those described for *A. zinniae.* The conidia are non-beaked and are borne singly on the conidiophores. Their average dimensions are 74 × 19 μ. The conidia are cylindrical to long-ellipsoid and are lightly colored. They average five transverse septa and longitudinal septa are less common on conidia from the host than from culture (155). Islam et al. (76) found that some sunflower lines and hybrids were moderately resistant, although no genotype was immune. Chemical control by benomyl appears to be effective (76). *A. helianthi* also is pathogenic to chrysanthemum (155). *Drechslera helianthi* Iliescu et al. isolated from a sunflower stalk recently has been described from Romania (75), and might be confused morphologically with *A. helianthi.* However *D. helianthi* was not pathogenic to sunflower, and Iliescu et al. (76) considered it a saprophyte.

An unidentified species which closely resembles *A. leucanthemi* Nelen causes serious sunflower disease in Yugoslavia (3). It may be seed-borne similar to other *Alternaria* pathogens and is therefore a potential hazard to sunflower in North America.

Alternaria alternata (Fr.) Keissler, synonymous with *A. tenuis* Nees (144), is reported as a pathogen attacking leaves, bracts, and receptacles of sunflower in Iran (55). The conidia are pyriform and yellowish brown to brown in color. Two or three conidia occur together in a chain. They have three to seven transverse septa and one to three longitudinal septa. Spores produced in culture measure 15 to 60 × 9 to 13 μ (55). The pathogen also is reported from the Soviet Union (148).

PHIALOPHORA YELLOWS

Yellows, a systemic disease, has thus far been found only in Canada. Hoes (62) described the disease, isolated the causal fungus, and assigned it to the genus *Phialophora* of the Fungi Imperfecti. The parasite recently was identified as *P. asteris* (Dowson) Burge & Isaac f. sp. *helianthi* (153). Symptoms appear near flowering time when leaves turn dull light green. Large areas of the leaf soon turn a dull yellow, usually starting at the apex and leaf margins and extending inwards (Fig. 16). Severely diseased plants are stunted and flower heads may be sterile. The vascular tissue turns brown. As in Verticillium wilt, symptoms develop first on the lower leaves and then on leaves higher on the stem. Symptoms of yellows may be confused with those of nitrogen deficiency and also resemble those of mild infections of Verticillium wilt, but there are characteristic differences (62). Yellows is a rather inconspicuous disease which may explain the lack of recorded occurrences. The organism is soil-borne and overwinters in Manitoba. Pathogenicity studies indicate that cultivar Commander is highly susceptible and the disease could become economically important. A source of resistance is known, and transfer of resistance to new cultivars is not difficult (64).

PHOMA BLACK STEM

McDonald (98) described black stem in Canada and identified the causal organism as *Phoma oleracea* var. *helianthi-tuberosi* Sacc. The pathogen produces black lesions on heads, leaves, and stems (Fig. 17, 18). Head infections are superficial and blackened areas are produced on the receptacle and the bracts. Unlike head diseases caused by several other pathogens, the head does not disintegrate and is not flaccid. Lesions on leaves are black, vary in shape, and are not distinctive. However, the characteristic lesions on the stems are shiny, jet-black. Infections start at the base of the petioles and spread along the stems. Severe infections may kill young plants, and weaken, stunt, and reduce head size of older plants. The light to dark brown pycnidia may be inconspicuous in the field, but form abundantly under moist conditions. Pycnidia are 155 to 308 μ in diam. The conidia are hyaline, reniform to oblong, and measure 4.3 to 7.2 $\times$ 1.4 to 2.9 μ (98). Frezzi (46) described the perithecial or perfect stage, *Leptosphaeria lindquistii* Frezzi in Argentina.

SEPTORIA LEAF SPOT

Septoria helianthi Ell & Kell. is distributed widely on sunflower and can reduce yield. Infection may be severe under moist conditions, especially at high plant populations and with few years between sunflower crops in the rotation. The pathogen belongs to the Fungi Imperfecti and produces abundant conidia in pycnidia. Conidia are filiform, hyaline with three to five septa, and measure 40 $\times$ 70 μ $\times$ 2 to 4 μ (45). The disease is characterized by necrotic lesions up to 15 mm in diam. They are mostly angular to diamond-shaped and brown on the upper leaf surface, and lighter grey-brown on the lower leaf surface (57), (Fig. 19). Small, black pycnidia are produced on both surfaces of the lesion. In early stages, the lesions often are surrounded by a narrow yellow halo which merges gradually with the surrounding green tissue. The lesions may coalesce in later stages and the leaf may wither and dry. The disease develops first on the lower leaves and gradually spreads to the upper leaves. Septoria leaf spot was damaging in 1962 in Manitoba (57) but has not been of economic importance since. Crop rotation and clean seed are suggested control measures (45). The pathogen is physiologically specialized (19).

BOTRYTIS HEAD ROT OR GREY MOLD

Head rot due to *Botrytis cinerea* Pers. ex Fr. (Fungi Imperfecti) in North America is not serious and yield reductions may occur only when wet weather delays harvest. Initially, a brown lesion appears at the back of the head, often on bracts and at the edge of the fleshy receptacle but commonly also near the center (Fig. 20). Under moist conditions the soft rot spreads over the entire back of the head and lesions become gray with the abundant production of conidiophores and conidia. Seedling blight due to internal infection or external contamination may result from seeds harvested from plants with Botrytis head rot (136). Water promotes the disease. Moisture collects at the seed surface when heads are incompletely fertile, and on the back surface when heads arch during maturation. Old pollen and senescent floral parts and bracts facilitate infection (33). Dichlofluanide or thiophanate-methyl + maneb applied at the beginning of flowering controlled the head rot and also reduced the amount of seed-borne contamination (33). Heads that inclined slightly and with a flat rather than arched flowering surface were least predisposed to Botrytis attack (33). Tircomcinu (152) has reported that inbreds differed in reaction to the pathogen.

RHIZOPUS HEAD ROT

Rhizopus head rot in North America is a sporadic disease which may be important during wet weather (177), especially when heads are predisposed to infection from mechanical injury. Birds (10), insects (103), or hail may have caused the wounds. The specific identity of *Rhizopus* sp. (Mucorales, Phycomycetes) in North America is yet to be determined (177). However, *R. nigricans* Ehrenb., synonymous with *R. stolonifer* (Ehrenb. ex Fr.) Vuill., and *R. arrhizus* Fischer, synonymous with *R. nodosus* Namyl. (10) are storage rot pathogens well known in North America (9), and these two species cause serious sunflower head rot in the Soviet Union (77). Irregular, water-soaked spots appear at the back of the head, enlarge and turn brown, while the receptacle becomes soft and pulpy (103) (Fig. 21). Woolly masses of whitish mycelium interspersed with numerous black-colored, spore-producing sporangia may be externally evident and usually are especially conspicuous inside the receptacle at the stem attachment. Mishra et al. (103) found that susceptibility of the heads increases with age, and maximum infection occurs at the soft dough stage. Infection results in seeds of lighter weight and lower oil content with discolored hulls and kernels as well as some dropping of heads (177). The fungus invades seeds and reduces germination (77). Application of copper-8-quinolinolate as flowering ended effectively controlled head rot due to *R. arrhizus* in Israel (10).

DAMPING-OFF, ROOT ROTS, AND STEM ROTS

Several soil-borne fungi acting individually or in complexes cause damping-off, root rots, and stem rots of sunflower. Such diseases usually are of minor economic importance but can cause economic damage under stress conditions (16, 113). Fungi usually associated with such diseases include *Phytophthora drechsleri* Tucker, *Pythium aphanidermatum* (Edson)

Fitz., *Phymatotrichum omnivorum* (Shear) Duggar, *Fusarium moniliforme* Sheldon, *Rhizoctonia solani* Kuehn, *Macrophomina phaseoli, Sclerotinia sclerotiorum,* and *Plasmopara halstedii.* Diseases caused by the last three organisms have been discussed earlier.

Damping-off is a disease of young seedlings. Most damping-off organisms are weak parasites and cause upon invasion a rapid killing and collapse of tissue. Sunflower seedlings often may be destroyed before emergence, or the disease continues to develop after emergence, causing a rather sudden collapse and death of seedlings. Plants become more resistant with age. Although damping-off is caused by a number of readily identified fungal species, the symptomology of most damping-off diseases is such that positive identification of the causal agent depends on culturing.

Fungi causing damping-off also can cause root rots and stem rots of adult sunflower. Plants grown under stress may be more susceptible to such diseases than plants of high vigor. The fibrous root system is attacked and rootlets are destroyed. Above-ground symptoms are stunting, premature ripening, and wilting. In severe cases plants may be killed before seed is set.

POWDERY MILDEW

Powdery mildew of sunflower is incited by the obligate ascomycete, *Erysiphe cichoracearum* DC (9). The disease appears as white to gray mildewy areas on aerial parts of the plant, usually on the leaves but occasionally on stems and bracts. These areas may enlarge and coalesce until most of the plant surface is infected (Fig. 24). As the season progresses microscopically visible black dots scattered over the white mildewy areas may occur. These black dots are cleistothecia, overwintering structures of the pathogen, that contain asci and ascospores.

Powdery mildew of sunflower is worldwide in distribution, but occurs in greatest intensity in the more tropical areas where it occasionally has caused senescence of the plant at the flowering stage or soon afterwards. In the more temperate areas of the world, powdery mildew usually is not observed until flowering and its intensity seldom is of any economic importance.

Little emphasis has been placed upon breeding for resistance although cultivars of sunflower appear to differ widely in reaction to the pathogen.

BACTERIAL DISEASES

Leaf spots, vascular wilt, and head rot are diseases caused by bacteria reported from various sunflower growing areas in the world. Bacterial diseases in North America are of minor importance and records are scarce. Leaf spot caused by bacteria belonging to the "*Pseudomonas syringae* group" potentially may be more important. Piening (117) identified the pathogen, established its pathogenicity, and studied the disease in Alberta.

Incipient lesions on young leaves are 1 mm in diam., pale green, angular water soaked spots, which turn brown to black in 3 to 4 days (Fig. 22). Lesions may enlarge on older leaves and coalesce to form large, irregular patches of necrotic tissue. Lesions also occur on petioles and stems, and their vascular tissue grows more slowly than laminar tissue, resulting in leaves that are wrinkled. All 11 sunflower cultivars tested were susceptible. Bacterial leaf spot also occurred on indigenous hosts such as *Helianthus nuttali* and *Aster* sp., and isolates from these plants and from *H. annuus* were reciprocally pathogenic. The rod-shaped cells of the pathogen measured 3.0 to 7.0 $\times$ 0.6 to 1.5 μ. The bacteria were motile and had one to many polar flagella. Colonies on nutrient agar were circular, translucent or pale cream in color, with fimbriated edges. Sackston (131) implicated unidentified bacteria in soft rot of sunflower stems in Manitoba but did not determine the etiology of the diseases. Stem rot is not important in oilseed cultivars. *Pseudomonas solanacearum* (E. F. Sm) Smith caused wilt in Puerto Rico (9). This pathogen of importance on crops such as tobacco and tomato, is restricted to the southeastern United States where the climate is warm and moist, and where sunflower is not grown commercially.

Bacterial pathogens are commonly seed-borne and could be introduced from elsewhere into North America on contaminated seed. In Japan, *Bacterium helianthi* Kawam. caused leaf spots on *H. annuus* and *H. debilis* Nutt., and was described by Kawamura (80). In Mexico, *Pseudomonas* sp. introduced into vascular tissue by a dipterous larva caused leaf spot (104). Another *Pseudomonas* sp., also in Mexico, is transmitted by larvae of a fly and caused sunflower head rot (47). Species of *Erwinia* caused soft rot of stems are reported from Tanzania (7), the Soviet Union (27), and Yugoslavia (11).

MYCOPLASMS AND VIRUSES

Aster yellows originally believed to be incited by a virus and recently established as incited by a mycoplasm is the only common "virus like" disease of sunflower in North America. Aster yellows occurred frequently in sunflower fields in Manitoba in the mid-1950's (132) and reached sufficient intensity to enable Putt an Sackston (125) to evaluate sunflower breeding lines for resistance in the field. Since that time the disease incidence has declined and sunflower with aster yellows is observed rarely.

Aster yellows is transmitted primarily by the six-spotted leaf hopper, *Macrosteles divisus* Uhl. (88), and occurs on a wide variety of plant species. The incidence of aster yellows, however, is not always correlated with the vector population. The relative abundance of reservoirs of the causal agent in infected perennial weed species and climatic and biologic factors which attract the vectors to the reservoirs also are important. It is common for the disease to be present on one host but nearly absent from another known host in the same area. This condition results in part from the host preference of the vector and not necessarily because of the relative resistance of the host.

Sunflower leaves infected with aster yellows become increasingly chlorotic. The chlorosis may affect leaves unilaterally and sectors of the entire plant including the capitulum may show symptoms while the remainder appears healthy. The normal color of the ray petals is altered and a greenish tinge appears. The infected capitulum will frequently show a wedge-shaped portion or sector with hypertrophy which remains green and frequently shows phyllody (Fig. 23).

The limited incidence of asteryellows on sunflower has not justified any control effort. If control becomes necessary, the vector population could be suppressed by insecticides. Putt and Sackston (125) concluded from differences in response of breeding lines in naturally infected plots that aster yellows could be controlled through breeding. The fact that resistance is dominant would facilitate the production of resistant F_1 hybrid cultivars.

Other sunflower diseases caused by mycoplasms have been reported from France and Israel. Sunflower phyllody caused stunt, leaf deformation, and sterility (143). Safflower phyllody caused witches broom, yellows, and sterility, and was transmitted by the leafhopper *Neoalitirus fenestratus* (Herrich-Schaffer) (83).

Except for a mechanically transmitted mosaic reported in Argentina (154) believed to be introduced from the USSR, and isolated cases of sunflower mosaic, ringspot, and yellowing viruses in Africa (164), England (130), India (18), Uruguay (134), and the USA (114), no viruses have been reported to occur extensively on cultivated sunflower.

LITERATURE CITED

1. Abawi, G. S., and R. G. Grogan. 1975. Source of primary inoculum and effects of temperature and moisture on infection of beans by *Whetzelinia sclerotiorum*. Phytopathology 65:300–309.

2. Acimovic, M. 1962. *Sclerotium bataticola* as an agent of sunflower wilt in Vojvodina. Zast. Bilja No. 69 to 70, p. 125–138.

3. ———. 1969. *Alternaria* sp. novi parzit suncokreta u Jugoslaviji. Zast, Bilja 106: 305–309. (1975, 3243, Abstr.).

4. Adams, P. B., R. D. Lumsden, and C. F. Tate. 1974. *Galinsoga parviflora*: a new host for *Whetzelinia sclerotiorum*. Plant Dis. Rep. 58:700–701.

5. Alabouvette, C., and P. Bremeersch. 1975. Two diseases new to France in sunflower crops. C. R. Seances Acad. Agric. Fr. 61:626–636 (In French; Abstr. in Rev. Plant. Pathol. 1975, No. 3243).

6. ———, and ———. 1976. *Macrophomina phaseolina* parasite du tournesol: etat actuel de nos connaissances et orientation des recherches. Cent Tech. Interprob. Oleagineaux Metrop. Inform. Tech. 52:13–17.

7. Allen, D. J. 1974. Diseases of sunflower (*Helianthus annuus*) in Tanga Region, Tanzania. Plant Dis. Rep. 58:896–899.

8. Anon. 1939. Canadian Plant Disease Survey, 18th Annu. Rep., Research Branch, Canada Dep. Agric., Ottawa. p. 103.

9. ─────. 1960. Index of plant diseases in the United States. USDA-ARS Handb. 165, Washington, D.C. p. 531.

10. Arnan, M., M. J. Pinthus, and R. G. Kenneth. 1970. Epidemiology and control of sunflower head rot in Israel caused by *Rhizopus arrhizus*. Can. J. Plant Sci. 50:283–288.

11. Arsenijevic, M. 1970. A bacterial soft rot of sunflower (*Helianthus annuus* L.). Rev. Plant Pathol. 51:No. 568. 1972. (Abstr.).

12. Arthur, J. C. 1903. Cultures of Uredineae in 1902. Bot. Gaz. 35:10–23.

13. ─────. 1905. Cultures of Uredineae. J. Mycol. 11:53–54.

14. ─────, and G. B. Cummins. 1962. Manual of the rusts in United States and Canada. Hafner, New York. p. 438.

15. Bailey, D. L. 1923. Sunflower rust. Minnesota Agric. Exp. Stn. Tech. Bull. 16. p. 35.

16. Banihashemi, Z. 1975. Phytophthora black stem of sunflower. Plant Dis. Rep. 59:721–724.

17. Bateman, D. F., and S. V. Beer. 1965. Simultaneous production and synergistic action of oxalic acid and polygalacturonase during pathogenesis by *Sclerotium rolfsii*. Phytopathology 55:204–211.

18. Battu, A. N., and H. C. Phatak. 1965. Observations on a mosaic disease of sunflower. Ind. Phytopathol. 18:317.

19. Beach, W. S. 1919. Biologic specialization in the genus Septoria. Am. J. Bot. 6:1–33.

20. Benson, D. M., and L. J. Ashworth, Jr. 1976. Survival of *Verticillium albo-atrum* on nonsuscept roots and residues in field soil. Phytopathology 66:883–887.

21. Beute, M. K., D. M. Porter, and B. A. Hadley. 1975. Sclerotinia blight of peanuts in North Carolina and Virginia and its chemical control. Plant Dis. Rep. 59:697–701.

22. Bisby, G. R. 1921. Stem-rot of sunflowers in Manitoba. Sci. Agric. 2:58–61.

23. ─────. 1924. The Sclerotinia disease of sunflowers and other plants. Sci. Agric. 4:381–384.

24. Brown, A. M. 1936. Studies on the infertility of four strains of *Puccinia helianthi* Schw. Can. J. Res. Sec. C. 14:361–367.

25. Brown, J. F., P. Kajornchaiyakul, M. Q. Siddiqui, and S. J. Allen. 1974. Effects of rust on growth and yield of sunflower in Australia. p. 639–646. *In* Proc. 6th Int. Sunflower Conf. (Bucharest, Romania).

26. Brown, M. F., and T. D. Wyllie. 1970. Ultrastructure of microsclerotia of *Verticillium albo-atrum*. Phytopathology 60:538–542.

27. Bushkova, A. N., and S. P. Alekseeva. 1968. A new bacterial disease of sunflower. Rev. Plant Pathol. 49: No. 2964, 1970. (Abstr.).

28. Chan, Y. H., and W. E. Sackston. 1966. Factors affecting pathogenicity of *Sclerotium bataticola*. Proc. Can. Phytopathol. Soc. 33:12–13.

29. Cohen, Y., and W. E. Sackston. 1973. Factors affecting infection of sunflowers by *Plasmopara halstedii*. Can. J. Bot. 51:15–22.

30. ─────, and ─────. 1974. Seed infection and latent infection of sunflower by *Plasmopara halstedii*. Can. J. Bot. 52:231–236.

31. Cook, G. E., J. R. Steadman, and M. G. Boosalis. 1975. Survival of *Whetzelinia sclerotiorum* and initial infection of dry edible beans in western Nebraska. Phytopathology 65:250–255.

32. Cormack, M. W. 1946. *Sclerotinia sativa*, and related species as root parasites of alfalfa and sweet clover in Alberta. Sci. Agric. 26:448–459.

33. Courtillot, M., C. Lamarque, M. Juffin, and F. Rapilly. 1973. Recherche de moyens de lutte contre le *Botrytis* du tournesol (*B. cinerea* Pers.): choix des methodes et des dates d'intervention en fonction des aspects biologiques et epidemiologiques de la maladie. Phytiatr. Phytopharm. 22:189–199.

34. Craigie, J. H. 1927. Discovery of the function of the pycnia of the rust fungi. Nature 120:765–767.

35. Delanoe, D. 1972. Biologie et epidemiologie du mildiou du tournesol (*Plasmopara helianthi* Novot.). CETIOM Inform. Tech. 29:1–49.

36. Dennis, R. W. G. 1974. *Whetzelinia* Korf & Dumont, a superfluous name. Kew Bull. 29:89–91.

37. Devaux, A. L., and W. E. Sackston. 1966. Taxonomy of *Verticillium* species causing wilt of horticultural crops in Quebec. Can. J. Bot. 44:803–811.

38. Edmunds, L. K. 1964. Combined relation of plant maturity, temperature and soil moisture to charcoal stalk rot development in grain sorghum. Phytopathology 54:514–517.

39. Evans, G., and A. C. Gleeson. 1973. Observations on the origin and nature of *Verticillium dahliae* colonizing plant roots. Aust. J. Biol. Sci. 26:151–161.

40. ————, and C. D. McKeen. 1975. Influence of crops on numbers of microsclerotia of *Verticillium dahliae* in soils and the development of wilt in southwestern Ontario. Can. J. Plant Sci. 55:827–834.

41. Fakir, G. A., M. H. Rao, and M. J. Thirumalachar. 1976. Seed transmission of *Macrophomina phaseolina* in sunflower. Plant Dis. Rep. 60:736–737.

42. Fick, G. N., M. L. Kinman, and D. E. Zimmer. 1975. Registration of RHA 273 and RHA 274 sunflower parental lines. Crop Sci. 15:106.

43. ————, and D. E. Zimmer. 1974. Monogenic resistance to Verticillium wilt in sunflowers. Crop Sci. 14:895–896.

44. ————, and ————. 1975. Influence of rust on performance of near-isogenic sunflower hybrids. Plant Dis. Rep. 59:737–739.

45. Frandsen, N. O. C. 1949. *Septoria helianthi* Ell. et Kell. als Erreger einer Blattfleckenkrankheit auf Sonnenblumen. Phytopathol. Z. 15:88–91.

46. Frezzi, M. J. 1968. *Leptosphaeria lindquistii* n. sp. forma sexual de *Phoma oleracea* var. *helianthi-tuberosi* Sacc., hongo causal de la "mancha negra del tallo" del girasol (*Helianthus annuus* L.) en Argentina. Patol. Veg. 5:73–80.

47. Fucikovsky, L. 1972. Pudricion bacteriana del capitulo del girasol (*Helianthus annuus*). Phytopathology 62:758 (Abstr.).

48. ————. 1976. Enfermedades y plagas del girasol en Mexico. Colegio Post-Grad. Escuela Nacional de Agricola, Chapingo, Mexico. p. 77.

49. Gabrielson, R. L., W. C. Anderson, and R. F. Nyvall. 1973. Control of *Sclerotinia sclerotiorum* in cabbage seed fields with aerial application of benomyl and ground application of cyanamide. Plant Dis. Rep. 57:164–166.

50. Garber, R. H., and B. R. Houston. 1966. Penetration and development of *Verticillium albo-atrum* in the cotton plant. Phytopathology 56:1121–1126.

51. Goossen, P. G., and W. E. Sackston. 1968. Transmission and biology of sunflower downy mildew. Can. J. Bot. 46:5–10.

52. Hancock, J. G. 1966. Degradation of pectic substances associated with pathogenesis by *Sclerotinia sclerotiorum* in sunflower and tomato stems. Phytopathology 56:975–979.

53. ————. 1967. Hemicellulose degradation in sunflower hypocotyls infected with *Sclerotinia sclerotiorum*. Phytopathology 57:203–206.

54. ————. 1972. Changes in cell membrane permeability in sunflower hypocotyls infected with *Sclerotinia sclerotiorum*. Plant Physiol. 49:358–364.

55. Hedjaroude, G. A. 1973. La maladie des taches brunes du tournesol (*Helianthus annuus* L.) en Iran. Phytopathol. Z. 78:274–277.

56. Hennessy, C. M. R., and W. E. Sackston. 1972. Studies on sunflower rust. X. Specialization of *Puccinia helianthi* on wild sunflowers in Texas. Can. J. Bot. 50:1871–1877.

57. Hoes, J. A. 1962. Diseases of sunflowers in 1962. p. 16. *In* Proc. Manitoba Agronomy Conf. (12–13 Dec. 1962, Winnipeg, Canada).

58. ————. 1968. Verticillium wilt of sunflower and potato. p. 46–51. *In* R. A. Gallop (ed.) Old Dutch potato. 6th Report, Chip Potato Seminar. Univ. of North Dakota, Grand Forks, Old Dutch Co., Minneapolis, Minn.

59. ————. 1969. Sunflower diseases in Manitoba in 1968. Can. Plant Dis. Surv. 49:27.

60. ————. 1971. Sclerotinia root rot of sunflowers. p. 66. *In* Proc. Manitoba Agronomy Conf. (14–15 Dec. 1971, Winnipeg, Canada).

61. ————. 1972. Losses due to Verticillium wilt in sunflower. Phytopathology 62:764 (Abstr.).

62. ————. 1972. Sunflower yellows, a new disease caused by *Phialophora* sp. Phytopathology 62:1088–1092.

63. ————. 1975. Sunflower diseases in western Canada. p. 425–438. *In* Oilseed and pulse crops in western Canada. A symposium. Western Co-op Fertilizers, Calgary.

64. ————, and H. Enns. 1974. Inheritance of resistance to Phialophora yellows of sunflower. p. 309–310. *In* Proc. 6th Int. Sunflower Conf. (Bucharest, Romania).

65. ————, and H. C. Huang. 1975. *Sclerotinia sclerotiorum*: viability and separation of sclerotia from soil. Phytopathology 65:1431–1432.

66. ————, and ————. 1976. Control of Sclerotinia basal stem rot of sunflowers: a progress report. p. 18–20. *In* Proc. Sunflower Forum (Fargo, N.D.).

67. ————, and ————. 1976. Importance of disease to sunflower in Manitoba in 1975. Can. Plant Dis. Surv. 56:75–76.

Fig. 1—Discoloration and stunting of sunflower, typical symptoms of plants systemically infected with the downy mildew fungus, *Plasmopara halstedii*.

Fig. 2—Isolated angular lesions caused from localized infection of sunflower by windborne spores of the downy mildew fungus.

Fig. 3—Basal stem gall of sunflower caused from infection of subterranean portion of the plant by zoospores.

Fig. 4—White downy growth (conidiophores and conidia) occur on the lower surface of leaves of plants with systemic symptoms of downy mildew when exposed to high relative humidity.

Fig. 5—Rust of sunflower incited by *Puccinia helianthi* is characterized by scattered chestnut brown pustules at both leaf surfaces, most prominent at the lower surface.

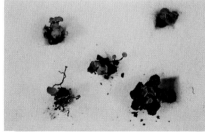

Fig. 6—Sclerotia of the white mold fungus, *Sclerotinis sclerotiorum*, overwinters in the soil and may germinate by producing from 1 to 8 tiny apothecia.

Fig. 7—Stalk rot of sunflower incited by the white mold fungus is characterized by brown, wet, soft lesions at the stem base. The lesions frequently girdle the stem and are frequently covered by white mycelia which later mass to form overwintering black sclerotia.

Fig. 8—Sunflower heads infected with white mold fungus exhibit a white mycelial mat and black sclerotia.

Fig. 9—Sunflower heads severely infected with white mold fungus often disintegrate leaving only a skeleton of shredded fibrous tissue.

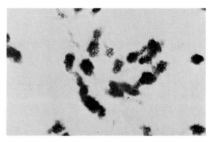

Fig. 10—*Verticillium dahliae*, the incitant of wilt of sunflower, is typified morphologically by microsclerotia composed of tightly compressed globose cells.

Fig. 11—Prominent yellow, intraveinal patches frequently located near the leaf margins are typical symptoms of Verticillium wilt of sunflower.

Fig. 12—A gray-black discolored area at the base of the stem (plant on left) is characteristic of sunflower plants infected with the charcoal rot fungus, *Macrophomina phaseoli*.

Fig. 13—White blister rust of sunflower incited by *Albugo tragopogi* is characterized by raised isolated yellow pustules at the upper leaf surface.

Fig. 14—Dark colored leaf lesions, frequently target-like, roughly circular and often associated with the leaf veins are characteristic of Alternaria leaf spot.

Fig. 15—*Alternaria zinnae* frequently produces brown flecks and streaks on stems and petioles, sometimes caused premature drying of leaves.

Fig. 16—Yellows, incited by *Phialophora asteris* f. sp. *helianthi*, is characterized by progressive discoloration of the leaves from the margins inward, stunting, and frequently sterility.

Fig. 17—Large dark to jet black lesions with relatively regular margins normally beginning at the junction of the petiole and stem frequently encircling the complete stem and causing premature ripening characterize Phoma black stem.

Fig. 18—Superficial blackened lesions on the head and bracts also are characteristic of infection by *Phoma oleracea* var. *helianthi tuberosi*.

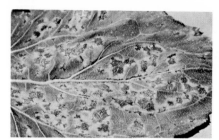

Fig. 19—Brown to straw-colored, diamond-shaped lesions containing small black pycnidia are characteristic of leaf spot, incited by *Septoria helianthi*.

Fig. 20—A brown lesion originating frequently at the edge of the fleshy receptacle, rapidly spreading over the entire head, and becoming grey with the production of conidiophores and conidia characterizes Botrytis head rot or grey mold.

Fig. 21—Rhizopus head rot incited by *Rhizopus arrhizus* is typified by large, rapidly spreading, brown lesions causing the head to become soft, pulpy, and deteriorated.

Fig. 22—Powdery mildew of sunflower incited by *Erysiphe cichoracearum* is characterized by whitish gray mildewy areas principally on the leaves. These areas may enlarge and coalesce until the whole leaf is covered.

Fig. 23—Bacterial leaf spot of sunflower incited by *Pseudomonas syringae* is characterized by pale-green, angular, water soaked spots which turn brown with age and coalesce to form large irregular lesions.

Fig. 24—A sunflower head with a wedge-shaped sector which frequently grows larger than the surrounding area, remains green, and shows phyllody is likely infected with the aster yellow mycoplasm.

68. ————, and E. D. Putt. 1962. Diseases of sunflowers in Manitoba in 1962. Can. Plant Dis. Survey 42:256-257.

69. ————, and ————. 1964. Diseases of sunflower in western Canada in 1964. Can. Plant Dis. Surv. 44:236-237.

70. ————, ————, and H. Enns. 1973. Resistance to Verticillium wilt in collections of wild *Helianthus* in North America. Phytopathology 63:1517-1520.

71. Huang, H. C. 1976. Biological control of Sclerotinia wilt in sunflowers. p. 69-72. *In* Proc. Manitoba Agronomy Conf. (Winnipeg, Canada).

72. ————. 1977. Importance of *Coniothyrium minitans* in survival of sclerotia of *Sclerotinia sclerotiorum* in wilted sunflower. Can. J. Bot. 55:289-295.

73. ————, and J. A. Hoes. 1976. Penetration and infection of *Sclerotinia sclerotiorum* by *Coniothyrium minitans*. Can. J. Bot. 54:406-410.

74. Huisman, C. O., and I. J. Ashworth. 1976. Influence of crop rotation on survival of *Verticillium albo-atrum* in soils. Phytopathology 66:978-981.

75. Iliescu, H., A. Hulea, and S. Bunescu. 1974. *Drechslera helianthi* nov. sp. isolated from a sunflower stalk. p. 665-672. *In* Proc. 6th Int. Sunflower Conf. (Bucharest, Romania).

76. Islam, U., A. Maric, I. Cuk, and D. Skoric. 1976. Studies on leaf, stem and head spotting of sunflower caused by *Alternaria helianthi* in Yugoslavia. p. 140-141. *In* Proc. 7th Int. Sunflower Conf. (Krasnodar, USSR).

77. Ivanchenko, M. Y. 1976. Sunflower dry rot and methods to reduce its damage. p. 137-138. *In* Proc. 7th Int. Sunflower Conf. (Krasnodar, USSR).

78. Jones, E. S. 1923. Taxonomy of the *Sclerotinia* on *Helianthus annuus* L. Phytopathology 13:496-500.

79. Kajornchaiyakul, P., and J. F. Brown. 1976. The infection process and factors affecting infection of sunflower by *Albugo tragopogi*. Trans. Br. Mycol. Soc. 66:91-95.

80. Kawamura, E. 1934. Bacterial leaf spot of sunflower. Ann. Phytopathol. Soc. Japan 4:25-28.

81. Keay, M. A. 1939. A study of certain species of the genus *Sclerotinia*. Ann. Appl. Biol. 26:227-246.

82. Kellerman, W. A. 1905. Uredineous infection experiments. J. Mycol. 11:30-32.

83. Klein, M. 1970. Safflower phyllody—a mycoplasma disease of *Carthamus tinctorius* in Israel. Plant Dis. Rep. 54:735-738.

84. Klisiewicz, J. M. 1975. Survival and dissemination of *Verticillium* in infected safflower seed. Phytopathology 65:696-698.

85. ————, and B. H. Beard. 1976. Diseases of sunflower in California. Plant Dis. Rep. 60:298-301.

86. Korf, R. P., and K. P. Dumont. 1972. *Whetzelinia*, a new generic name of *Sclerotinia sclerotiorum* and *S. tuberosa*. Mycologia 64:248-251.

87. Kruger, W. 1975. Uber die Bildung von Sklerotien des Rapskrebserregers (*Sclerotinia sclerotiorum*) (Lib. de Bary) im Boden. Mitt. Biol. Bundesanst. Land und Forstwirtsch. Berlin Dahlem. 163:32-40.

88. Kunkel, L. O. 1926. Studies on aster yellows. Am. J. Bot. 13:646-705.

89. Lacy, M. L., and C. E. Horner. 1966. Behaviour of *Verticillium dehliae* in the rhizosphere and on roots of plants susceptible, resistant and immune to wilt. Phytopathology 56:427-430.

90. Leclercq, P. 1973. Influence de facteurs hereditaires sur la resistance apparente du tournesol a *Sclerotinia sclerotiorum*. Ann. Amelior. Plant. 23:279-286.

91. ————, Y. Cauderon, and M. Dauge. 1970. Selection pour la resistance au mildiou du tournesol a partis d'hybrides topinambour X tournesol. Ann. Amelior. Plant. 20:363-373.

92. Leppik, E. E. 1966. Origin and specialization of *Plasmopara halstedii* complex on the Compositae. FAO Plant Prot. Bull. 14:72-76.

93. Livingston, J. E. 1945. Charcoal rot of corn and sorghum. Univ. of Nebraska Agric. Exp. Stn. Res. Bull. 136. p. 32.

94. Luciano, A., and M. Davreux. 1967. Produccion de girasol en Argentina. Inst. Nac. Technol. Agropecu. (I.N.T.A.). Publ. Tech. No. 37, Pergamino. p. 53.

95. Lumsden, R. D. 1969. *Sclerotinia sclerotiorum* infection of bean and the production of cellulase. Phytopathology 59:653-657.

96. ————. 1970. Phosphatidase of *Sclerotinia sclerotiorum* produced in culture and in infected bean. Phytopathology 60:1106-1110.

97. Maxwell, D. P., and R. D. lumdsden. 1970. Oxalic acid production by *Sclerotinia sclerotiorum* in infected bean and in culture. Phytopathology 60:1395–1398.

98. McDonald, W. C. 1964. Phoma black stem of sunflowers. Phytopathology 54:492–493.

99. ————, and J. W. Martens. 1963. Leaf and stem spot of sunflowers caused by *Alternaria zinniae*. Phytopathology 53:93–96.

100. McLean, D. M. 1958. Some experiments concerned with the formation and inhibition of apothecia of *Sclerotinia sclerotiorum* (Lib.) de By. Plant Dis. Rep. 42:409–412.

101. Miah, M. A. J., and W. E. Sackston. 1970. Genetics of rust resistance in sunflower. Phytoprotection 51:1–16.

102. ————, and ————. 1970. Genetics of pathogenicity in sunflower rust. Phytoprotection 51:17–35.

103. Mishra, R. P., U. S. Kushwaha, M. N. Khare, and J. N. Chand. 1972. Rhizopus rot of sunflower (*Helianthus annuus*) in India. Ind. Phytopathol. 25:236–239.

104. Montano, J., E. Olivas, and L. Fucikovsky. 1972. Bacteriosis foliar en girasol (*Helianthus annuus*). Phytopathology 779 (Abstr.).

105. Montes, F., and W. E. Sackston. 1974. Growth of *Plasmopara* within susceptible and resistant sunflower plants. p. 623–628. *In* Proc. 6th Int. Sunflower Conf. (Bucharest, Romania).

106. Newton, H. C., and L. Sequeira. 1972. Ascospores as the primary infective propagule of *Sclerotinia sclerotiorum* in Wisconsin. Plant Dis. Rep. 56:798–802.

107. Nishimura, M. 1922. Studies on *Plasmopara halstedii*. J. Coll. Agric. Hokkaido Imp. Univ. 11:185–210.

108. Novotelnova, N. S. 1962. White rust on sunflower. Zashchita Rastenii, Moskva. Rev. Appl. Mycol. 42:266 (Abstr.).

109. ————. 1966. Downy mildew of sunflower (In Russian). Izgatelsivo 'Nauka' Moscow-Leningrad. p. 149 (English transl., 1976, USDA-ARS and NSF, by Indian Nat. Sci. Document. Cent., New Delhi).

110. Orellana, R. G. 1969. Relative virulence and selective pathogenesis among isolates of *Verticillium* from sunflower and other hosts in relation to sunflower wilt. Phytopathol. Z. 65:183–188.

111. ————. 1970. The response of sunflower genotypes to natural infection by *Macrophomina phaseoli*. Plant Dis. Rep. 54:891–893.

112. ————. 1970. Resistance and susceptibility of sunflowers to downy mildew and variability in *Plasmopara halstedii*. Bull. Torrey Bot. Club 97:91–97.

113. ————. 1973. Sources of resistance to a soil-borne fungal disease complex of sunflower. Plant Dis. Rep. 57:318–320.

114. ————, and A. Quacquarelli. 1968. Sunflower mosaic caused by a strain of cucumber mosaic virus. Phytopathology 58:1439–1440.

115. Partyka, R. E., and W. F. Mai. 1962. Effects of environment and some chemicals on *Sclerotinia sclerotiorum* in laboratory and potato fields. Phytopathology 52:766–770.

116. Pegg, G. F. 1974. Verticillium diseases. Rev. Plant Pathol. 53:157–182.

117. Piening, L. J. 1976. A new bacterial leaf spot of sunflowers in Canada. Can. J. Plant Sci. 56:419–422.

118. Price, K., and J. Colhoun. 1975. Pathogenicity of isolates of *Sclerotinia sclerotiorum* (Lib.) de Bary to several hosts. Phytopathol. Z. 83:232–238.

119. Purdy, L. H. 1955. A broader concept of the species *Sclerotinia sclerotiorum* based on variability. Phytopathology 45:421–427.

120. Pustovoit, V. S. 1960. Interspecific rust resistant hybrids of sunflowers (In Russian). Otdalennaya Hibridization Rastenii. Moscow. p. 376–378.

121. Pustovoit, G. V., V. P. Ilatonsky, and E. L. Slyusar. 1976. Results and prospects of sunflower breeding for group immunity by interspecific hybridization. p. 26–28. *In* Proc. 7th Int. Sunflower Conf. (Krasnodar, USSR).

122. Putt, E. D. 1958. Note on differences in susceptibility to Sclerotinia wilt in sunflowers. Can. J. Plant Sci. 38:380–381.

123. ————. 1964. Breeding behaviour of resistance to leaf mottle or Verticillium in sunflowers. Crop Sci. 4:177–179.

124. ————, and W. E. Sackston. 1957. Studies on sunflower rust. I. Some sources of rust resistance. Can. J. Plant Sci. 37:43–54.

125. ————, and ————. 1960. Resistance of inbred lines and hybrids of sunflower (*Helianthus annuus* L.) in a natural epidemic of aster yellows. Can. J. Plant Sci. 40:375–382.

126. ————, and ————. 1963. Studies on sunflower rust. IV. Two genes, R_1 nd R_2, for resistance in the host. Can. J. Plant Sci. 43:490–496.

127. Ribeiro, I. J. O., O. Paradela Filho, J. Soave, and G. da Silva Corvellini. 1974. Occorencia de *Alternaria helianthi* (Hansf.) Tubaki e Nishihara sobre girassol (*Helianthus annuus* L.). Bragantia 33:81–85.

128. Robb, J., L. Busch, J. D. Brisson, and B. C. Lu. 1975. Ultrastructure of wilt syndrome caused by *Verticillium dahliae*. II. In sunflower leaves. Can. J. Bot. 53:2725–2739.

129. Ruddick, J. A., and J. Harwig. 1975. Prenatal effects caused by feeding sclerotia of *Sclerotinia sclerotiorum* to pregnant rats. Bull. Environ. Contam. Toxicol. 13:524–526.

130. Russell, G. E., P. H. L. Cook, and E. S. Bunting. 1975. New or uncommon plant diseases and pests. Plant Pathol. 24:58–59.

131. Sackston, W. E. 1952. Sunflower diseases in Manitoba in 1951. Canadian Plant Disease Survey. 31st Annu. Rep. Research Branch, Canada Dep. Agric., Ottawa. p. 36–39.

132. ———. 1954. Sunflower diseases in Manitoba in 1953. Canadian Plant Disease Survey. 33rd Annu. Rep. Research Branch, Canada Dep. Agric., Ottawa. p. 45–48.

133. ———. 1956. Observations and speculations on rust (*Puccinia helianthi* Schw.) and some other diseases of sunflower in Chile. Plant Dis. Rep. 40:744–747.

134. ———. 1957. Diseases of sunflowers in Uruguay. Plant Dis. Rep. 41:885–889.

135. ———. 1959. Black root rot of sunflowers in Uruguay caused by *Sclerotium bataticola*. Proc. Can. Phytopathol. Soc. 20:14 (Abstr.).

136. ———. 1960. *Botrytis cinerea* and *Sclerotinia sclerotiorum* in seed of safflower and sunflower. Plant Dis. Reptr. 44:664–668.

137. ———. 1962. Studies on sunflower rust. III. Occurrence, distribution, and significance of races of *Puccinia helianthi* Schw. Can. J. Bot. 40:1449–1458.

138. ———. 1974. Downy mildew (*Plasmopara*) of sunflowers: a policy for seed importation and plant breeding. p. 619–622. *In* Proc. 6th Int. Sunflower Conf. (Bucharest, Romania).

139. ———, and J. W. Martens. 1959. Dissemination of *Verticillium albo-atrum* on seed of sunflower (*Helianthus annuus*). Can. J. Bot. 37:759–768.

140. ———, W. C. McDonald, and J. Martens. 1957. Leaf mottle or Vericillium wilt of sunflower. Plant Dis. Rep. 41:337–343.

141. Sarasola, A. A. 1942. Sunflower disease. Publ. Direction Agric. Buenos Aires. 14 p. Rev. Appl. Mycol. 26:376. Abstr.

142. Schreiber, L. R., and R. J. Green, Jr. 1963. Effect of root exudates on germination of conidia and microsclerotia of *Berticillium albo-atrum* by the soil fungistatic principle. Phytopathology 53:260–264.

143. Signoret, P. A., C. Louis, and B. Alliot. 1976. Mycoplasma-like organisms associated with sunflower phyllody in France. Phytopathol. Z. 86:186–189.

144. Simmons, E. G. 1967. Typification of *Alternaria, Stemphylium* and *Ulocladium*. Mycologia 59:67–92.

145. Smith, A. M. 1972. Biological control of fungal sclerotia in soil. Soil Biol. Biochem. 4:131–134.

146. Smith, H. C. 1965. The morphology of *Verticillium albo-atrum, V. dahliae*, and *V. tricorpus*. N.Z.J. Agric. Res. 8:450–478.

147. Snood, P. N., and W. E. Sackston. 1970. Studies on sunflower rust. VI. Penetration and infection of sunflowers susceptible and resistant to *Puccinia helianthi* race 1. Can. J. Bot. 48:2179–2181.

148. Svetov, V. G. 1975. Alternaria blight of sunflower along the Kuban River. Rev. Plant Pathol. 55:No. 1397, 1976. Abstr.

149. Tikhonov, O. I. 1972. Peculiarities of infection and forms of sunflower downy mildew. p. 156–158. *In* Proc. 5th Int. Sunflower Conf. (Clermont-Ferrand, France).

150. ———. 1975. Diseases and pests of sunflower and means of controlling them. p. 391–456. (In Russian). *In* V. S. Pustovoit (ed.) Sunflowers. Kolos Moscow.

151. ———, V. K. Nedelko, and T. A. Perestova. 1976. Infection period and spreading pattern of *Sclerotium bataticola* in sunflower tissue p. 134. *In* Proc. 7th Int. Sunflower Conf. (Krasnodar, USSR).

152. Tircomnicu, M. 1976. New aspects of parasitic and saprophytic activity of *Botrytis cinerea* and *Alternaria* spp. in sunflower crops in Romania. p. 143. *In* Proc. 7th Int. Sunflower Conf. (Krasnodar, USSR).

153. Tirilly, Y., and C. Moreau. 1976. Etude comparative de quelques *Phialophora* parasites vasculaires de plantes. Bull. Soc. Mycol. France 92:349–358.

154. Traversi, B. 1949. Estudio inicial sobre una enfermedad del girasol (*Helianthus annuus* L.) in Argentina. Rev. Inst. Agric. 3:345–351.

155. Tubaki, K., and N. Nishihara. 1969. *Alternaria helianthi* (Hansf.) comb. nov. Trans. Brit. Mycol. Soc. 53:147–149.

156. Vear, F. 1974. Studies on resistance to downy mildew in sunflowers (*Helianthus annuus* L.). p. 297–302. *In* Proc. 6th Int. Sunflower Conf. (Bucharest, Romania).

157. ————, and P. Leclercq. 1971. Deux nouveaux genes de resistance au mildiou du tournesol. Ann. Amelior. Plant. 21:251–255.

158. Velkov, V., and T. Shopov. 1975. Development of sunflower lines resistant to *Plasmopara helianthi* Nov. I. Inheritance of resistance. (In Bulgarian). Plant Sci. 9:103–107.

159. Vranceanu, V., and F. Stoenescu. 1970. Immunity to sunflower downy mildew due to a single dominant gene. (In Romanian). Probl. Agric. 2:34–40.

160. Whetzel, H. H. 1945. A synopsis of the genera and species of the Sclerotiniaceae, a family of stromatic inoperculate discomycetes. Mycologia 37:648–714.

161. Willetts, H. J., and A. L. Wong. 1971. Ontogenetic diversity of sclerotia of *Sclerotinia sclerotiorum* and related species. Trans. Br. Mycol. Soc. 57:515–524.

162. Williams, G. H., and J. H. Western. 1965. The biology of *Sclerotinia trifoliorum* Erikss. and other species of sclerotium-forming fungi. I. Apothecium formation from sclerotia. Ann. Appl. Biol. 56:253–260.

163. ————, and ————. 1965. The biology of *Sclerotinia trifoliorum* Erikss, and other species of sclerotium-forming fungi. II. The survival of sclerotia in soil. Ann. Appl. Biol. 56:261–268.

164. Wiltshire, S. P. 1955. Plant diseases in the British Colonial Dependencies. FAO Plant Prot. Bull. 3:140.

165. Wong, A. L., and H. J. Willetts. 1973. Electrophoretic studies of soluble proteins and enzymes of *Sclerotinia* species. Trans. Br. Mycol. Soc. 61:167–178.

166. Woolliams, G. E. 1966. Host range and symptomatology of *Verticillium dahliae* in economic, weed, and native plants in interior British Columbia. Can. J. Plant Sci. 46:661–669.

167. Young, P., and H. E. Morris. 1927. Sclerotinia wilt of sunflowers. Univ. Montana, Agric. Exp. Stn. Bull. No. 208. p. 32.

168. Zimmer, D. E. 1971. A serious outbreak of downy mildew in the principal sunflower production area of the United States. Plant Dis. Rep. 55:11–12.

169. ————. 1972. Field evaluation of sunflower for downy mildew resistance. Plant Dis. Rep. 56:428–431.

170. ————. 1973. Basal gall of sunflower initiated by *Plasmopara halstedii*. Plant Dis. Rep. 57:647–649.

171. ————. 1974. Physiological specialization between races of *Plasmopara halstedii* in America and Europe. Phytopathology 64:1465–1467.

172. ————. 1975. Some biotic and climatic factors influencing sporadic occurrence of sunflower downy mildew. Phytopathology 65:751–754.

173. ————, and G. N. Fick. 1973. Registration of Sundak sunflower. Crop Sci. 13:584.

174. ————, and ————. 1974. Some diseases of sunflowers in the United States—their occurrence, biology, and control. p. 673–679. *In* Proc. 6th Int. Sunflower Conf. (Bucharest, Romania).

175. ————, and M. L. Kinman. 1972. Downy mildew resistance in cultivated sunflower and its inheritance. Crop Sci. 12:749–751.

176. ————, and D. A. Rehder. 1976. Rust resistance of wild *Helianthus* species of the North Central United States. Phytopathology 66:208–211.

177. ————, and D. C. Zimmerman. 1972. Influence of some diseases on achene and oil quality of sunflower. Crop Sci. 12:859–861.

DAVID E. ZIMMER: B.S., Eastern Illinois University, M.S. and Ph.D., Purdue University; Supervisory Plant Pathologist and Oilseed Research Leader, SEA-USDA, 1961–1977; Area Research Director, Georgia, South Carolina Area, SEA-USDA, Tifton, Georgia, 1977.

JOHN A. HOES: B.S., M.S., Ph.D., Research Scientist, Plant Pathology, Research Station, Agriculture Canada, Morden, Manitoba; sunflower pathologist since 1961; author or co-author of more than 50 scientific and general publications on plant diseases.

Chapter 8

Birds and Sunflower

JEROME F. BESSER

Many species of birds eat sunflower seeds. Most persons who have operated a feeding station for birds, especially in midwinter, are aware of the fondness of many birds for sunflower seed. Producers of birdfeed for these stations include sunflower seed in all of their high quality bird feeds. One reason for reintroduction of commercial sunflower plantings into North America during the 1950's and 1960's was to supply an increasing demand for birdfeed. Schaffner and Taylor (20) reported that about 65% of the sunflower seed produced in 1965 was used in bird feeding stations and indicated that demand exceeded 1,360 metric tons (30 million 1b) annually for this use. Farmers should not be surprised that many species of birds consider sunflower fields as incredibly large feeding stations. That this may be an accurate appraisal is confirmed by many vehement complaints from farmers planting sunflower adjacent to favored bird habitats, such as marsh and tree cover.

Sunflower seed is a preferred bird food because the seed contains many proteins and fats essential to their growth, molt, fat storage, and weight maintenance processes. Unfortunately for North American producers, sunflower ripens in the postbreeding season of the birds, when the food demands of bird populations are near their annual peak. Bird numbers at this time are at their annual high, because the fledged young have been added recently to the adult population. The energy reserves of adult birds are minimal, and they have a high food demand at this time, having spent long periods during the nesting season in defending breeding territories, laying eggs, and raising their young. They badly need high energy foods to complete replacement of their feathers, which are molted annually in August and September, and to store fat that will be needed for the annual fall migration of hundreds or thousands of kilometers through regions where foods may not be as plentiful nor weather conditions as suitable for feeding.

The Great Plains of northern North America has long been a banquet table for birds. Sunflower growers recently have added a large and pre-

From: Carter, Jack F. (ed.). 1978. *Sunflower Science and Technology.* Agronomy 19. Copyright © 1978 by the American Society of Agronomy, Crop Science Society of America, and Soil Science Society of America, 677 South Segoe Road, Madison, WI 53711 USA.

Fig. 1—Blackbirds, such as this flightline of Red-winged Blackbirds arriving to feed in a field of ripening sunflowers in Steele County, North Dakota, are the principal family of birds that cause damage in North America. (Photo by the author.)

ferred dish to this table. A similar dish, the wild sunflower (*Helianthus annuus* L.) has been present for millenia, but not in such quantity, or at least in such concentrated quantity as cultivated sunflower. Wild sunflower has been found in the diets of many species of birds in numerous food habit studies conducted since the Great Plains were first plowed. Since the advent of both effective herbicides and improved cultural practices, patches and scattered plants of wild sunflower are seen less commonly. Thus, it is conceivable that cultivated sunflower is replacing wild sunflower in the diet of many species.

Birds prefer certain cultivars of cultivated sunflower. Surprisingly, the oilseed cultivars are much preferred by most birds over the confectionery cultivars or even those cultivars specifically planted for the birdfeed market. Apparently persons buying sunflower seed for feeding stations buy the large or striped seeds more for appearance than from knowledge of bird preference or nutritive needs.

BIRD SPECIES THAT DAMAGE SUNFLOWER

Blackbirds are the principal family of birds that damage ripening sunflower in North America (Fig. 1). The Red-winged Blackbird (*Agelaius phoeniceus* L.) probably damages sunflower more than all other species of blackbirds combined. Other important species of blackbirds causing damage to sunflower are the Common Grackle (*Quiscalus quiscula* L.) and the Yellow-headed Blackbird (*Xanthocephalus xanthocephalus* Bonaparte). Both the Rusty Blackbird (*Euphagus carolinus* Muller) and the Brewer's Blackbird (*E. cyanocephalus* Wagler) also feed in sunflower fields, but appear to feed mostly on sunflower seed on the soil surface. Few Brown-headed Cowbirds (*Molothrus ater* Boddaert) have been observed in sunflower fields.

Both Red-winged and Yellow-headed Blackbirds nest extensively in and around the marshes of the Northern Great Plains. Red-winged Blackbirds also utilize roadside and drainage ditches, pastures, and other upland sites extensively. Common Grackles use shelterbelts, tree groves, shrub thickets, and other planted and natural woody vegetation for nesting. The young of these species produced in these areas, together with the adults, form large roosting congregations each August and September. Late summer roosts of 10,000 to 100,000 blackbirds are common in the Northern Great Plains, and some may contain a million or more individuals. On several occasions, flightlines of approximately one quarter million birds leaving or entering roosts have been observed to pass over a single sunflower field, different flocks entering and leaving the field simultaneously. A population estimated at 62,000 blackbirds was once observed in a single 16 ha (40 A) field near Hope, North Dakota.

The Red-winged Blackbird comprises most of the individuals in flocks feeding on sunflower in September in the Northern Great Plains of North America. The most damage to sunflower occurs in September in this region. Many of the Yellow-headed Blackbirds already have begun their migration to central Mexico at this time as have Common Grackles to their wintering areas in the lower Mississippi Valley and the U.S. Gulf Coast.

The Red-winged Blackbird is the principal species in these flocks and it is the most abundant breeding blackbird species in the Dakotas and in Canadian provinces. Stewart and Kantrud (22) sampled breeding habitats throughout North Dakota in 1971 and estimated that 2.1 million males bred in that state. Biologists of the U.S. Fish and Wildlife Service conducted four censuses of breeding male Red-winged Blackbirds in a 78,000 km² (30,000 mi²) area of North and South Dakota, centered on the James and Souris Rivers, between 1964 and 1971 and found that 1.6 to 2.1 million males bred in the sample area each year. Despite the opinion of some farmers that blackbird populations are increasing each year, they found the breeding population was quite stable. In years with below-normal precipitation, however, a higher percentage of the population nested along roadsides and lower percentages in potholes and fields. The reverse was true in years of above-normal precipitation.

Many other passerines also feed on sunflower heads, but none, except the House Sparrow (*Passer domesticus* L.), causes major damage to commercial sunflower plantings. Biologists of the Fish and Wildlife Service have witnessed the passerines listed in Table 1 taking sunflower seed from standing heads. Many of the fringillids that are known to damage heads of

266   BESSER

Table 1—A list of bird species observed feeding on sunflower seeds in commercial fields in Minnesota, North Dakota, and South Dakota.

Common name	Scientific name
Feeding on standing heads	
Red-winged Blackbird	*Agelaius phoeniceus* (Linnaeus)
Yellow-headed Blackbird	*Xanthocephalus xanthocephalus* (Bonaparte)
Common Grackle	*Quiscalus quiscula* (Linnaeus)
Brewer's Blackbird	*Euphagus cyanocephalus* (Wagler)
Rusty Blackbird	*Euphagus carolinus* (Muller)
House Sparrow	*Passer domesticus* (Linnaeus)
Pine Siskin	*Spinus pinus* (Wilson)
Common Goldfinch	*Spinus tristis* (Linnaeus)
Blue Jay	*Cyanocitta cristata* (Linnaeus)
Brown Thrasher	*Toxostoma rufum* (Linnaeus)
Northern Oriole	*Icterus galbula* (Linnaeus)
Common Crow	*Corvus brachyrhynchos* (Brehm)
Feeding on the soil surface on unharvestable seed	
Mourning Dove	*Zenaida macroura* (Linnaeus)
Gray Partridge	*Perdix perdix* (Linnaeus)

cereal grains in other regions of North America, e.g., the Dickcissel (*Spiza americana* Gmelin) and the Bobolink (*Dolichonyx oryzivorus* L.) probably could be added to this list. Some species of birds feed heavily on sunflower seed in sunflower fields, but take all or nearly all their seed from the soil surface or from lodged or fallen heads. These species also are listed in Table 1.

The parakeet and dove families of birds seriously damage sunflower in South America. Population reduction programs were initiated in Argentina in 1965 and Uruguay in 1971 (14) specifically to relieve damage to sunflower heads by the Monk Parakeet (*Myopsitta monachus* Boddaert) and the Eared Dove (*Zenaida auriculata* Des Murs) (Fig. 2). Doves of the genus *Streptopelia* are also a problem in sunflower in Kenya, Africa (De Grazio, personal communication). Both the Spanish Sparrow (*Passer hispaniolensis* L.) and House Sparrow create problems in ripening sunflower in Morocco (De Grazio, 1974, personal communication). In the USSR, Dementev and Gladkov (12) state that the Spanish Sparrow or Turkestan Blackbreasted Sparrow (*Passer hispaniolensis transcaspicus* Tschusi) causes a major bird problem in sunflower. The extensive poisoning campaign directed at this species and at the House Sparrow in the USSR may be attributable in part to damage to sunflower heads. Sparrows, probably the House Sparrow and the Spanish Sparrow, also damage sunflower in Turkey, along with crows (Ilter, 1971, personal communication). Camprag (7) found that the House Sparrow and the European Tree Sparrow (*Passer montanus* L.) heavily damaged sunflower in Yugoslavia and reported that these sparrows caused similar problems in France, Hungary, Poland, and Romania. The Rose-ringed Parakeet (*Psittacula krameri* Scopoli), the Alexandrine Parrot (*Psittacula eupatria* L.), the House Sparrow, and the House Crow (*Corvus splendens* Vieillot) cause damage to sunflower in India and Pakistan (Swindle, 1977, personal communication). Small (21) reported that species of parrots (cockatoos, corellas, and the galah) seriously damaged sunflower in North Victoria, Australia.

Fig. 2—Population reduction programs have been initiated in Argentina and Uruguay to relieve damage to sunflowers caused by the Monk parakeet (shown above). (Photo by Donald F. Mott.)

EXTENT OF LOSS

Only limited information is available on the amount of economic loss caused by birds damaging sunflower, but objective survey data are available for a substantial portion of the chief sunflower growing region of the USA for 1972. In 1972, 487 sunflower fields in North Dakota and Minnesota were sampled by biologists of the U.S. Fish and Wildlife Service. Birds had fed on 9.7% of the nearly 50,000 heads examined and had consumed about 1.2% of the standing crop (23). The loss was only about 13 kg/ha (12 lb/A) in 1972, which was a year of unusually late harvesting because of wet weather. Although this was a small loss for the total crop, birds had fed on more than 50% of the heads in 3.1% of the fields and on 20% of the heads in 13.4% of the fields—indications that birds can consume the profit margin for a considerable number of growers if their fields are located near favored roosting and loafing habitats for blackbirds (Fig. 3).

In a more intensive survey in 1972 in an area selected because of high bird damage, Schafer (1972, unpublished report) surveyed 94 sunflower fields, all within 16 km (10 mi) of a large roost of blackbirds in Steele County, North Dakota, and found that birds had consumed 4.7% of the

Fig. 3—Birds can consume the profit margin from some or many fields, such as this one with about 30% of the seed eaten by blackbirds. Note the attack from birds standing atop the head and some damage also from birds perching on the leaves. (Photo by the author.)

crop in the 298 km^2 (115 mi^2) surveyed. He found that blackbirds had consumed 21.1% of the crop in 12 fields in the 73 km^2 (28 mi^2) within 6.4 km (4 mi) of the roost.

The capability of blackbirds to damage sunflower can be estimated from studies of their food habits. Bird and Smith (5) found that sunflower seed made up about 20% of the diet of blackbirds at the height of the damage season in a study near Manitoba. In 1972 in Steele County, North Dakota, Royall and Mott (1973, unpublished report) found that sunflower seeds made up nearly 70% of the diet of Red-winged Blackbirds at the height of the sunflower damage season.

Even less information is available on the extent of loss in commercial sunflower plantings on other continents. Small (21) reported that in Australia as many as 600 sunflower kernels were found in the crop of an individual cockatoo that had just fed on sunflower. He calculated that the loss amounted to as much as 40 g of seed per feeding. Most of the seed on several hectares of large plantings is lost to flocks of 200 to 500 Monk Para-

Fig. 4—A head with 48% of the seeds removed by blackbirds. Note the involucral bracts where birds have fed and the intact bracts where they have not. The dropped ray flowers, and attached florets, and green involucral bracts indicate this head is at the stage of maximum vulnerability to blackbird attack. (Photo by the author.)

keets in some Uruguayan fields (Besser and De Grazio, 1975, unpublished report). These investigators found that a caged parakeet could consume about 48 g (1.7 oz) of seed daily or about 2 kg (4.4 lb) of seed for the 6 weeks when the sunflower crop is vulnerable to parakeets. Camprag (7) estimated that about 3% of the sunflower crop was lost to birds in the Voivodina region of Yugoslavia. Loss was estimated to be 8,000 metric tons annually for this region from 1970 to 1973.

Factors Affecting Loss

The severity of blackbird damage to sunflower in a specific field or year often is related directly to the weather. Early harvest may minimize damage in dry years but more blackbirds assemble at the fewer suitable roosting marshes also. In dry years, fewer growers suffer more severe damage for short periods of time. Growers suffer lighter damage for longer periods in wet years because a greater number of suitable roosting locations for blackbirds are available, the number of birds at each roost is smaller, but harvesting is likely to be delayed.

The blackbird damage suffered by an individual sunflower grower also is related to the number of hectares planted in the vicinity of a roost of a specific size, as many pioneering sunflower growers and experiment station researchers woefully have discovered. When there are only a few sunflower fields near a roost of a specific size, these fields suffer proportionately more damage than when there are many fields in the area. For example, a single 16 ha (40 A) field may take nearly the same total amount of feeding pressure from a roost of 20,000 blackbirds as twenty 16 ha (40 A) fields. Farmers are wise to plant sunflower in areas or years of high sunflower production to minimize bird damage.

The yield of fields generally is not correlated to the amount of damage growers sustain. All seed is seldom taken in a commercial field. Blackbirds feed as readily in fields that yield 500 kg/ha as in fields that yield 2,000 kg/ha. They are more likely, however, to consume nearly all of the crop adjacent to suitable loafing cover from 4 ha of a large low-yielding field compared to nearly all of the crop from 1 ha of a high-yielding field.

The maturity of the sunflower seeds in heads affects the amount of damage a flock of blackbirds of specific size inflicts. Blackbirds prefer the dough stage of maturity and select the green heads containing dough stage seed for feeding in preference to more mature heads (Fig. 3 and 4). Apparently the birds are able to extract the kernels from the hulls more easily at the dough stage and have little difficulty in removing the involucral bracts and florets to reach the seed (Fig. 4). A flock of blackbirds of a specific size generally causes more damage in August when kernels are not fully developed than a flock of similar size in September feeding on fully developed seeds. Fortunately for growers, flocks usually are smaller in August than in September.

The planting rate and density of plants also is related to damage sustained. Low planting rates and the resulting large heads with well filled seed to the centers are less vulnerable to birds. Heads 25 cm (10 in) in diam or larger turn downwards earlier, making it difficult for the bird to obtain seeds. The attack on large heads nearly always proceeds from the edge to the center of the head (Fig. 3 and 4). The birds usually use the bowl of the downturned head as a feeding platform, so feeding becomes more difficult for them after the seed on the outer 2.5 to 5 cm (1 or 2 in) has been eaten. The 1972 survey of bird damage by the U.S. Fish and Wildlife Service showed such large heads also are less vulnerable to wind damage. In 1972, 0.9% of the crop was lost to shattering of seed or rubbing of heads on adjacent plants compared to the 1.2% loss to birds (23).

All of the preceding factors, however, have less influence on seed loss to birds than the location of sunflower crops in relation to the best blackbird habitats. In the 1972 survey reported by Stone (23) the heaviest damage by blackbirds in North Dakota occurred in the western portions of Richland, Cass, Traill, and Grand Forks counties, areas on the eastern fringe of the Drift Plains physiographic regions described by Colton et al. (8). Less damage by blackbirds was found in fields on the flood plains of the Red River Valley (the Agassiz Lake Plain physiographic region). The difference was related to the greater number of permanent water areas which roosting blackbirds use heavily in the Drift Plains regions.

Since the 1972 survey, farmers on the Drift Plains and Coteau physiographic regions of central North Dakota have found that production of sunflower is profitable, and have steadily increased areas in sunflower production. Meanwhile, farmers in the Red River flood plain, having a number of profitable alternative crops, have increased sunflower plantings more slowly. The concentration of sunflower production in North Dakota in the Drift Plains and Coteau regions indicates that blackbird damage will be heavier in these expanding sunflower-producing regions, because the Drift Plains and Coteau regions have the highest number of water areas per square mile of any region in the USA. In some portions of Stutsman County in North Dakota, for example, there are more than 40 water basins per km². Agricultural statistics (16) show that in 1970, only 202 ha (500 A) of sunflower were planted in Stutsman County, whereas in 1976 more than 6,000 ha (15,000 A) of sunflower were planted (17). Other counties of North Dakota in the Drift Plains and Coteau regions showed similar increases in sunflower plantings. Clearly, protection of fields from blackbird damage will be a major and increasing concern of sunflower growers in the Northern Great Plains in the future. In contrast, the new sunflower plantings on the Southern Great Plains in Texas have suffered little damage from blackbirds, because far fewer blackbirds breed there and migrating blackbirds arrive after sunflower has been harvested.

PROTECTING SUNFLOWERS FROM BIRD DAMAGE

Sunflower growers protect sunflower fields from bird damage by two principal methods: (a) cultural practices and (b) use of mechanical frightening devices and chemical frightening agents. As the latter is employed only when the former fails to alleviate bird damage sufficiently, cultural practices will be discussed first.

Cultural Practices

The previous discussions on bird damage to sunflower and the ecology of the birds provide hints on effectively avoiding bird damage. Sunflower fields should not be planted within 400 m (0.25 mi) of small marshes or sloughs that consistently harbor 5,000 or more roosting or loafing blackbirds in late summer (18). Growers should avoid planting sunflower adjacent to shelterbelts, groves, or other woody areas regularly used by blackbirds for loafing cover and instead plant a minimum 90 m (about 100 yd) buffer strip of a crop not as attractive to blackbirds. Bird and Smith (4) list flax (*Linum usitatissimum* L.), forage, pasture, sugarbeets (*Beta vulgaris* L.) and potatoes (*Solanum tuberosum* L.) as crops that blackbirds do not harm in Canada. Soybeans [*Glycine max* (L.) Merr.], pinto beans (*Phaseolus vulgaris* L.) or other legumes can be added to this list for the USA. Most cereal crops, especially hard red spring wheat (*Triticum aestivum* L.), also sustain less damage than sunflower.

If farmers cannot select an alternate crop, or select a site to grow sunflower far enough from marshes and trees to provide a degree of bird protection, they should at least plant at lower planting rates to produce large heads. Farmers should plan their plantings to assure uniform ripening of the crop in the community, and plan with neighbors so that a single or a few growers will not have the only fields within the feeding range of a large roost of birds. Production of nonoilseed cultivars rather than oilseed and cultivars having heads that turn downward more rapidly also should offer some protection. Sunflower growers also should consider the use of desiccants and seed dryers to shorten the period that their crop is vulnerable to birds.

The most underrated cultural practice for relieving bird damage is both effective and simple. A major portion of harvested cropland near blackbird roosts should not be plowed or tilled until most or all sunflower fields have been harvested. Several authors (4, 18) and many of the wiser farmers recommend delaying tillage of nearby cropland. Birds, especially blackbirds, consume great quantities of weed seeds and some insects from crop stubbles. The benefit of blackbirds in controlling weeds and insects in crop stubbles is not established but leaving crop stubbles near blackbird roosts reduces crop damage. It is pertinent that Mott et al. (15) found that the diet of blackbirds at the height of the corn (*Zea mays* L.) damage season in South Dakota corn fields consisted of 41% foxtail or pigeon grass (*Setaria* spp.) and only 27% corn. Unfortunately, farmers living near marshes often are unwilling to delay fall tillage of part of their cropland to accommodate farmers growing sunflower. In some areas these farmers near marshland were forced to cease growing sunflower because of bird problems.

Further, when farmers are harvesting sunflower, the harvested stubble with shattered seed should be left as alternate bird feeding areas until all sunflower in the area is harvested. Leaving the sunflower stubble as alternate feeding areas is especially important if all grain stubble in the area is plowed or tilled. The discussion on this topic can be concluded with optimism. Perhaps as the sunflower belt moves slightly westward on the Northern Great Plains, farmers will leave more alternate field feeding sites for wind erosion control rather than fall plow to avoid wet fields in the spring as is common farther east, especially on soils of high clay content in the Red River Valley.

Mechanical Means of Crop Protection

Nearly all farmers are aware of the value of frightening birds from sunflower fields by patrolling them with rifles and shotguns, or using exploders and pyrotechnics. They also know this activity requires several work hours each morning and evening, and occasionally all day. Sunflower is vulnerable to birds for approximately 6 weeks—longer than any other crop grown in North America—so protection of sunflower fields by mechanical means is an especially formidable task and one likely to discourage the protector long before harvest.

Nevertheless, persevering farmers can nearly fully protect their crops by patrol. A farmer familiar with blackbirds can move birds effectively and economically by firing a 0.22 rifle round above the birds to cause the feeding flock to flush and follow this with a series of round behind the departing flock. The 0.22 rifle can be hazardous up to 2,000 m and must be used carefully.

Use of "shell crackers," a 12-gauge shotgun shell containing a powerful firecracker instead of shot, is a second effective means of moving flocks of birds. A 12-gauge shotgun launches the firecracker, which explodes above the birds about 100 yards from the gunner. As blackbirds, from their long association with hawks, instinctively dislike anything above them, the shell cracker is far more effective than regular shells in moving birds, and greatly increases the range and versatility of a person patrolling the field. Two or more gunners complement one another in protecting large fields.

Patrol by aircraft has not been as effective as the foregoing statement about hawks would indicate. Birds often take refuge from the plane in the crop instead of leaving the field, and they are extremely difficult to move from such loafing cover as shelterbelts with a plane. In addition, the cost of flying has spiralled upward faster than crop value, making this approach steadily less attractive.

A gunner attempting to kill birds with a shotgun, where killing is permitted, is one of the least effective of all methods of patrolling.

Gas-powered propane or acetylene exploders (Fig. 5) probably have saved as much sunflower seed from birds to date as all other means combined. The exploder produces a much louder sound than a shotgun blast. Some exploders have solar cells that turn them on at daylight and off at dark, and thus minimize maintenance time. However, for best results, exploders should be moved frequently so that birds do not become accustomed to them as quickly. One exploder can protect about 4 ha (10 A) of crop (4). Mounting the exploder on a stand above the crop gives the best deterrent effect. Mounting a small, thin steel drum, with the ends removed, just beyond the barrel of the exploder, enables the drum to serve as an amplifier and increases effectiveness.

Amplification of recorded tapes of distress, alarm, and escape calls of blackbirds and their natural enemies such as hawks have been used to protect crops. Electronically generated amplified sounds also have been used by some growers to protect crops. The equipment to broadcast calls is expensive, however, and often requires the knowledge of an electronic technician to keep it in proper running order under the weather conditions prevailing in crop fields. Because of attendant problems, deterrent calls to protect sunflower fields have not been widely used. Broadcasting calls from planes also has been attempted in other crops and found effective, but costs, always high, are becoming prohibitive.

Fig. 5—Gas-powered exploders probably have saved more sunflower from blackbirds than all other means combined to date. This propane powered exploder is mounted properly on a barrel and fires slightly above the crop. The adjacent heads were damaged before the exploder was installed. (Photo by the author.)

Chemical Frightening Methods

Mechanical frightening methods are being replaced by chemical frightening methods. The discovery of chemicals that cause some species of birds to emit cries while ascending in circular distress-display flights often to heights of 500 feet or more led to a new method of protecting crops—the use of chemical frightening agents (13). The ability of one of these agents, 4-aminopyridine (4-AP), to protect crop fields was first demonstrated in corn fields in South Dakota by biologists of the U.S. Fish and Wildlife Service in 1964 (11). They sprayed partially husked ears of corn with a 4-AP solution at the rate of five ears per each 0.8 ha (2 A) of corn per treatment on a grid pattern, and obtained a 78% reduction in blackbird damage in the 410 ha (1,013 A) treated. In 1965 in this same area, they broadcast cracked corn baits treated with 3% 4-AP that had been diluted with 29 parts of untreated corn at the rate of 1.12 kg of the diluted bait mixture per treated hectare (1 lb/A) and obtained on 85% reduction in damage (10). The ratio of treated to untreated particles was increased subsequently to 1 to 99 and this product was federally registered to protect corn fields from blackbird damage under the trade name of Avitrol® FC Corn Chops 99. In 1967, investigators of the U.S. Fish and Wildlife Service began using 4-AP in North Dakota to protect sunflower (Fig. 6). Nine years later (August 1976) 4-AP was federal-

Fig. 6—A spray plane baiting a sunflower field in Traill County, North Dakota with a chemical frightening agent to protect it against blackbird attack. Only 1 gram of the active ingredient, 4-aminopyridine, is used to treat 3 ha (3,333 A/lb) of sunflower field. (Photo by Donald R. Henne.)

ly registered to protect sunflower from blackbirds as the product Avitrol® FC Corn Chops-99S (AFCC-99S).

Most of the efficacy tests that permitted the registration of AFCC-99S were conducted in North Dakota. R. V. Hansen and W. K. Pfeifer (unpublished reports of the U.S. Fish and Wildlife Service, North Dakota) reported satisfactory to excellent results by spreading AFCC-99S in disked strips between standing sunflower in the initial Rolette County test in 1967, and in tests in Ransom County in 1968 to 1970, in Cass and Ransom counties in 1971, and in Cass County in 1972.

In 1973, an experimental permit was obtained and the first tests were conducted wherein AFCC-99S baits were broadcast within standing sunflower. AFCC-99S was broadcast at 1.1 kg/ha (1 lb/A) in all sunflower fields treated in a 73 km² (28 mi²) area of Steele County, North Dakota. In the year that fields were treated, only 3.4% damage occurred in this area compared with 23.6% in the previous year despite the presence of a peak population of 7 times as many birds in the test year (2). About $9 worth of sunflower seed was saved for each $1 spent in the baiting program. In 1974, blackbirds consumed about 2.5 times as much sunflower seed in 12 untreated fields in Traill and Grand Forks counties, North Dakota, as in 12 fields treated with AFCC-99S, and results were generally satisfactory over the 7,284 ha (18,000 A) of sunflower treated in Minnesota and the Dakotas under a permit granted that year (2).

Despite the effectiveness demonstrated in tests that secured registration of AFCC-99S in sunflower in the USA, several problems prevent growers from making the most effective use of this product. These problems involve professional applicators, lack of concerted and sustained baiting efforts, and improvement of use directions. Solutions to these problems are discussed in detail by Besser and Pfeifer (3). Most of the problems in application of AFCC-99S should be corrected within the next few years. Then sunflower growing areas with perennial bird damage problems should receive the kind of protection against blackbird attack indicated in the 1973 sunflower study in Steele County, North Dakota (2) and that demonstrated in Brown County, South Dakota, corn fields over a 7-year period from 1964 to 1971 (1). In the latter study, damage in a 243 km² (94 mi²) study area decreased by 90% (from 162 to 13 kg/ha or from 2.43 to 0.19 bu/A), roosting blackbird populations in the roosting marshes declined 80% (from 1.6 million to 300,000), and baiting efforts decreased by 50% from the peak use of 4,082 kg to 2,041 kg (9,000 lb to 4,500 lb) during the 7 years of study.

Chemical frightening agents are more promising than attempts to reduce populations of blackbirds by poisons. In order to obtain 80% protection of crops by poisons—a figure that would be a realistic goal—it would be necessary for 80% of the offending birds to ingest bait. The competition with other foods, such as ripening sunflower, and the short stay of blackbird populations during migration probably makes this figure unattainable, even by persons with the most knowledge of blackbird feeding habits and preferences. However, 80% protection has been obtained with the chemical frightening agent 4-AP—by affecting less than 1% of the offending blackbird population—a figure readily attainable by persons with little knowledge of blackbird feeding habits and preferences.

Other chemical methods to protect crops also have many disadvantages. Repellent sprays which show good potential for protecting other kinds of crops (9) appear to have little use in sunflower. When sprayed on sunflower heads in the early dough stage, the repellent falls on the florets, which are sloughed off as the head ripens. At later stages of maturity, the heads turn downward, effectively preventing the repellent chemical from reaching the hulls which birds must open to feed on the kernels.

Chemosterilants—birth control agents for blackbirds—which could effectively reduce production of young and largely negate breeding habitats, might be employed profitably. However, sterilization of the bird populations involved in a major portion of the damage to sunflower will require a better knowledge of bird movements than anyone presently possesses. Birds probably would have to be sterilized on their winter ranges where food is less plentiful, or in spring roosting assemblages of male blackbirds, such as the one described near Mound City, Missouri (24). This task would be formidable and expensive, and would require the federal registration of a suitable chemical. In laboratory and small-scale studies, Schafer et al. (19) have shown that certain chemicals permanently sterilize male blackbirds but do not affect their ability to successfully hold territories, and Bray et al. (6) have shown that vasectomization of Red-winged Blackbird males lowers production on territories held. Chemosterilant research on birds now has low priority by the U.S. Fish and Wildlife Service, pending the outcome of programs using more direct methods of protecting crops.

SUMMARY

Birds, especially blackbirds, cause locally serious problems in growing sunflower in North America, and birds are serious problems in sunflower production on several other continents. Wise selection of planting sites and cultural practices can circumvent some of the problems. For the most serious problems, moving the birds from fields by mechanical means and chemical frightening agents has been most successful. Chemical frightening agents now appear to be the best single answer for controlling bird damage to sunflower, if growers are willing to unite and persevere in their control efforts.

LITERATURE CITED

1. Besser, J. F., J. W. De Grazio, and J. L. Guarino. 1973. Decline of a blackbird population during seven years of baiting with a chemical frightening agent. p. 12–14. *In* H. N. Cones and W. B. Jackson (eds.) Proc. 6th Bird Control Seminar. Environmental Studies Center, Bowling Green State University, Bowling Green, Ohio.
2. ————, and J. L. Guarino. 1976. Protection of ripening sunflowers from blackbird damage by baiting with Avitrol® FC Corn Chops-99S. p. 200–203. *In* W. B. Jackson (ed.) Proc. 7th Bird Control Seminar. Environmental Studies Center, Bowling Green State University, Bowling Green, Ohio.

3. ————, and W. K. Pfeifer. 1977. Improvement of baiting methods for protecting ripening sunflowers from blackbirds. Proc. Sunflower Forum 2:11–12.

4. Bird, R. D., and L. B. Smith. 1963. Blackbirds in field crops. Canada Dep. of Agric. Publ. 1184. 4 p.

5. ————, and ————. 1964. The food habits of the redwinged blackbird, *Agelaius phoeniceus,* in Manitoba. Can. Field Nat. 78(3):179–186.

6. Bray, O. E., J. J. Kennelly, and J. L. Guarino. 1975. Fertility of eggs produced on territories of vasectomized red-winged blackbirds. Wilson Bull. 87(2):187–195.

7. Camprag, D. 1974. Harmfulness of birds on maturing sunflower plants in northeast regions of Yugoslavia. p. 701–705. *In* A. V. Vranceanu (ed.) Proc. Int. Sunflower Conf. (Bucharest, Romania). Research Institute for Cereal and Industrial Crops, Fundulea, Ilfov, Romania.

8. Colton, R. B., R. W. Lemke, and R. M. Lindvall. 1963. Preliminary glacial map of North Dakota. U.S. Geol. Surv., Misc. Geol. Invest. Map No. I-331.

9. Crase, F. T., and R. W. DeHaven. 1976. Methiocarb: Its current status as a bird repellent. p. 46–50. *In* C. C. Siebe (ed.) Proc. 7th Vertebrate Pest Control Conference (Monterey, Calif.). University of California, Davis.

10. De Grazio, J. W., J. F. Besser, T. J. DeCino, J. L. Guarino, and E. W. Schafer, Jr. 1972. Protecting ripening corn from blackbirds by broadcasting 4-aminopyridine baits. J. Wildl. Manage. 36(4):1316–1320.

11. ————, ————, ————, ————, ————, and R. I. Starr. 1971. Protecting ripening corn from blackbirds with 4-aminopyridine. J. Wildl. Manage. 35(3):565–569.

12. Dementev, G. P., and N. A. Gladkov. Birds of the Soviet Union. Vol. V. Smithsonian Institution and the National Science Foundation, Washington, D.C. p. 416.

13. Goodhue, L. D., and F. M. Baumgartner. 1965. Application of new bird control chemicals. J. Wildl. Manage. 29(4):830–837.

14. Martinez, O. E. 1971. Aves daninas en agricultura en el Uruguay. Spec. Rep. Dep. of Plant Health, Direccionde Sanidad Vegetal, Ministerio de Agricultura, Montevideo, Uruguay. 122 p.

15. Mott, D. F., R. R. West, J. W. De Grazio, and J. L. Guarino. 1972. Foods of the red-winged blackbird in Brown County, South Dakota. J. Wildl. Manage. 36(3):983–987.

16. North Dakota Crop and Livestock Reporting Service. 1971. North Dakota crop and livestock statistics. 1970 annual summary. U.S. Dep. Agric., Fargo, N.D. 65 p.

17. North Dakota Crop and Livestock Reporting Service. 1977. North Dakota crop and livestock statistics. 1976 annual summary. U.S. Dep. Agric., Fargo, N.D. 79 p.

18. Pfeifer, W. K. 1975. Blackbirds and sunflowers. National Sunflower Grower 2(5):12.

19. Schafer, E. W., Jr., R. B. Brunton, and N. F. Lockyer. 1976. Evaluation of 45 chemicals as chemosterilants in adult male quail (*Coturnix coturnix*). J. Reprod. Fert. 48:371–375.

20. Schaffner, L. W., and F. R. Taylor. 1965. Marketing outlets and marketing methods for sunflower seed and tame mustard seed. Agric. Econ. Rep. No. 40. Dep. of Agric. Economics, North Dakota State Univ., Fargo, N.D. 22 p.

21. Small, D. 1975. Sunflowers for the birds. J. Agric. (Melbourne, Australia) 73(6):217.

22. Stewart, R. E., and H. A. Kantrud. 1972. Population estimates of breeding birds in North Dakota. Auk 89(4):766–788.

23. Stone, C. P. 1973. Bird damage to agricultural crops in the U.S.—A current summary. p. 264–267. *In* H. N. Cones and W. B. Jackson (eds.) Proc. 6th Bird Control Seminar, Environmental Studies Center, Bowling Green State University, Bowling Green, Ohio.

24. West, R. R., and J. F. Besser. 1967. Dipnetting blackbirds in a marsh roost in Missouri. Inland Bird-Banding News 39(4):91.

JEROME F. BESSER: Research Biologist, U.S. Fish and Wildlife Service, Denver Wildlife Research Center, Denver, Colorado. Led research on methods to combat bird damage to agricultural crops in the USA, 1960–1976, including major research efforts in sunflowers 1972–1977; author and coauthor of over 40 scientific publications, mainly on birds, mammals, and agricultural crop protection; advisor and researcher on crop protection methods to combat bird damage in Central America, South America, and Africa through USAID and FAO Programs, 1968–1977.

Chapter 9

Breeding and Genetics

GERHARDT N. FICK

Breeding and selection for improvement of sunflower (*Helianthus annuus* L.) undoubtedly was practiced by the earliest growers. Monocephalic types with seeds quite similar to cultivars developed recently were grown by the early Ozark Bluff Dwellers in North America, and were known to occur in Europe as early as the 16th century (77, 78). Gundaev (70) states that peasants in the Voronezh and Saratov districts of the USSR practiced systematic selection of plants, heads, and seeds to improve yield and marketability of sunflower. Large numbers of local peasant cultivars were available by 1880.

Breeding and selection work to improve sunflower at experimental stations was initiated in the USSR at the Kharkhov station in 1910 and at the Kruglik and Saratov stations in 1912 and 1913, respectively (139). This early work included studies on the biology of the plant, and genetic and biochemical investigations, as well as studies on methods in breeding, selection, and seed production. Limited genetic studies on sunflower also were conducted in the USA about this time (25, 26), and probably in other countries also.

Except for limited research on improvement of sunflower for silage, the first significant breeding efforts to improve sunflower for North American agriculture occurred in Canada in 1937 (143). Breeding programs to improve sunflower were developed more recently in the USA. The cooperative USDA-Texas Agricultural Experiment Station program was initiated in 1950, primarily with the objective of developing a wide diversity of sunflower germplasm for possible future use in comprehensive sunflower research programs (Kinman, M. L., G. W. Rivers, E. J. Redman, J. T. Wade, and E. L. Flynt. 1963. Annual report of cooperative sesame, sunflower, and guar investigations. USDA and Texas Agric. Exp. Stn., College Station). Small scale sunflower breeding programs designed primarily to develop nonoilseed cultivars or hybrids also were conducted during the 1950's by the Minnesota and California Agricultural Experiment Stations, and Northrup King and Company in California. With the introduction of the high oil

From: Carter, Jack F. (ed.). 1978. *Sunflower Science and Technology*. Agronomy 19. Copyright © 1978 by the American Society of Agronomy, Crop Science Society of America, and Soil Science Society of America, 677 South Segoe Road, Madison, WI 53711 USA.

Soviet cultivars into the USA and Canada in the 1960's and the expanded area devoted to the nonoilseed types, numerous seed companies have become involved recently in sunflower breeding. The cooperative sunflower research program of the USDA-North Dakota Agricultural Experiment Station at Fargo, North Dakota, was started in 1970.

BREEDING

Objectives

Major objectives in sunflower breeding include improved seed yield, earlier maturity, shorter plant height, uniformity of plant type, and disease and insect resistance. High oil percentage is important in breeding oilseed types whereas large seed size, a high kernel-to-hull ratio, and uniformity in seed size, shape, and color are important objectives in breeding and selection of nonoilseed sunflower. The emphasis on specific objectives differs for various production areas and for different breeding programs. The objectives listed are those considered important for the current major production areas of the Great Plains in the USA and Canada. As consumer demands and economic factors change, other characteristics such as improved oil quality, protein percentage and protein quality may become additional and increasingly important breeding objectives.

Breeding Methods

Sunflower is a highly cross pollinated crop, pollination occurring primarily by insects (143). The plants have perfect flowers and show a wide range in self fertility (144). Most breeding methods for maize (*Zea mays* L.) and other cross pollinated crops are applicable to sunflower with certain modifications necessitated because of their flowering process and morphological characteristics. Methods used in early sunflower breeding as well as current procedures used to develop cultivars and hybrids will be reviewed in the following sections.

Methods for Improving Cultivars

Mass Selection

Mass selection generally refers to the phenotypic selection of individual plants for the improvement of a cultivar or population for some specific trait. Selected plants are harvested usually without control of pollination, and their seeds are mixed to form a new cultivar or source population.

Mass selection for improvement of sunflower was used commonly in the USSR during the early stages of cultivar improvement (70). The cultivars 'Fuksinka 3,' 'Chernianka 35,' 'Karlik,' 'Pioner Sibiri,' and 'Omskij' were developed using mass selection procedures.

The family selection method was one of the more widely used forms of mass selection (70). This method involved selection of individual open pollinated plants, and subsequent cross pollination within phenotypically similar families. Gundaev (70) listed 11 cultivars that were developed by the family selection method and regionalized for production in the USSR.

Mass selection also has been used extensively to produce improved sunflower cultivars in Argentina. The cultivars 'Manfredi INTA,' 'Guayacan INTA,' 'Cordobes INTA,' and 'Impira INTA' were developed by this method (117). According to Davreux et al. (29) the development of high yielding cultivars through mass selection procedures remains a part of their current program.

The availability of improved cultivars appears to be evidence that mass selection has been effective in improving selected characteristics. These characters include earlier maturity and resistance to broomrape (*Orobanche cumana* Wallr.), occurrence of the armor layer of the seed, disease resistance, lower hull percentage, and higher oil percentage (125). Mass selection also has been effective in improving seed size and uniformity in seed color in nonoilseed sunflower (165).

Pustovoit Method of Reserves

A widely used and highly successful method for improving sunflower cultivars was developed in the USSR by V. S. Pustovoit during the 1920's. The method is a form of recurrent selection that includes progeny evaluation and subsequent cross pollination among progenies with superior characteristics. The method has been especially effective in increasing oil percentage from about 30% in the early 1920's to over 50% in cultivars grown currently. All of the experimental stations involved in sunflower breeding in the Soviet Union and in most eastern European countries use the "method of reserves" (141).

A description of the "method of reserves" (Fig. 1), as outlined by Pustovoit (139), is as follows. Ten thousand to 15,000 plants with a large number of seeds per head (1,200 or more) are selected from a heterogeneous population which may trace to elite cultivars, populations, or the progeny of intercultivar or interspecific hybridization. The seed from these plants is analyzed for hull and oil percentage. Based on these analyses 1,000 to 1,200 heads are selected for progeny evaluation. Usually these are subdivided into two to four groups on the basis of agronomic characteristics such as maturity and height. Progenies from individual plants within each group are grown in single row plots with two replications. A check consisting of the best cultivar most similar to the lines being evaluated is included in every third plot as a control. The progenies are evaluated for maturity, height, head inclination, head diameter, seed yield, seed weight, seed type, hull percentage, and oil percentage. An evaluation for tolerance to broomrape or

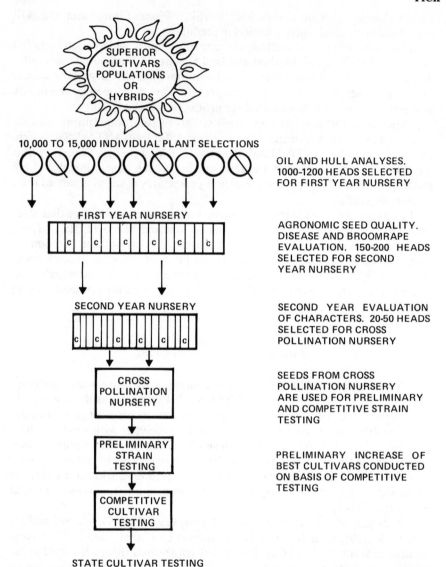

Fig. 1—Diagram of Pustovoit "Method of Reserves" for improvement of sunflower cultivars.

certain diseases may be conducted concurrently in a separate nursery planted specifically for this purpose. On the basis of the observations from the first year of testing, 15 to 20% of the best plants are evaluated for a second year using similar methods and techniques. Reserve seed from the original plants rather than the seed obtained from the first year of testing is used to plant the second year nursery.

After the second year the original seed of the best remaining 20 to 50 plants is planted in a replicated isolation nursery for cross pollination. Undesirable plants including disease infected, branched, extremely tall, or plants susceptible to shattering are removed during the season. Heads are harvested individually and a sample of the seed is used for laboratory analyses. On the basis of laboratory analyses, seed from the best plants within the 20 to 50 progenies is mixed for use in the next cycle of selection, for testing the following year in large plots, and for seed production. After 3 years of testing the best cultivars are submitted to the State for inclusion in regional performance trials.

The "method of reserves" has been successful primarily in increasing oil percentage and production of oil per hectare. Cultivars developed by this method were grown on 4.6 million ha or about 98% of the production area in the USSR in 1973 (141). Mean oil percentage on a dry weight basis was 50.8% in the USSR in 1973 with seed yields averaging 15.4 q/ha. In addition to increased oil percentage, the method has been useful in developing cultivars resistant to broomrape, Verticillium wilt (*Verticillium dahliae* Kleb.) and the sunflower moth (*Homoeosoma nebulella* Hb.) as well as improving adaptation to short growing seasons (70, 141).

The Pustovoit method of reserves has been successful in improving oil percentage because the procedure maintains the genetic variability necessary for effective selection. Precise progeny testing for one or more years involving replication and extensive use of control plots, necessary for detecting small differences, also has played an important role. In addition, a relatively large proportion of the genetic variance for oil percentage has been found to be additive (43, 153), a criterion that is necessary for the success of the method. Because oil percentage is approaching the biological limit and the genetic variability has been reduced as a result of intensive selection within a limited population, the method probably will be less effective in improving oil yield in future years (5). As a result hybrid sunflower breeding programs designed to take advantage of both additive and non-additive genetic variance have been initiated in the Soviet Union (204) and in most Eastern European countries previously using the Pustovoit method.

Inbreeding and Heterosis

Inbreeding as a method for improving sunflower was used as early as 1922. Cardon (18) described the variation occurring in the cultivar 'Mammoth Russian' and attempted to isolate different types by self pollination. Because of the high degree of self sterility, he believed that developing improved strains through inbreeding would be extremely difficult. Hamilton (74), however, reported that self sterility was not complete, and that many lines when self pollinated produced 15 to 50% as much seed as was obtained from cross pollination. Three generations of inbreeding increased uniformity and generally decreased the size of the plants. Certain lines retained excellent vigor for height and yield, and Hamilton (74) indicated that inbreeding was a promising method for sunflower improvement.

Inbreeding was used in the Soviet Union during the 1920's to develop lines with improved oil percentage, strong single stems, and disease and insect resistance (91, 204). As with other cross pollinated crops, breeders soon realized that the value of inbreeding was to develop lines with certain desirable characteristics for subsequent crossing to produce synthetic cultivars or interline hybrids. Some of the early results involving hybridization of inbred lines showed heterosis for plant height, head diameter, seed size, and seed yield. The increase in seed yields of early hybrids over open pollinated cultivars ranged as high as 60% (200).

Effects of Inbreeding

The most significant effect of inbreeding is on seed yield. Unrau and White (200) found that seed yields of two lines from the cultivar 'Mennonite' declined 35% in one generation of inbreeding by self pollination. Four generations of inbreeding decreased yields by 60% compared to the original population; thereafter inbreeding caused only a slight further decrease in seed yield. They also noted smaller but significant effects for plant height, head diameter, and 100-seed weight with the greatest reduction occurring in the first three or four generations of inbreeding.

Schuster (177), in an extensive study involving inbreeding by continuous selfing for 18 generations, reported that seed yields decreased an average of 40% relative to the initial population. The minimum decrease among individual lines was 29%. Inbreeding depression was considerably less for plant height, head diameter, and seed weight. Kernel and seed oil percentage showed only slight inbreeding depression, and inbreeding did not significantly affect the number of seeds occurring in the center portion of the head.

Kovacik and Skaloud (105) reported on the effects of six generations of inbreeding on plant and seed characteristics of lines selected from the cultivar 'Ruzynska No. 9.' The most significant inbreeding effects were on seed yield and seed weight. Minimum values for most of the characters occurred after four or five generations of inbreeding.

Putt (144) indicated that self fertility was affected by inbreeding. He observed a decrease in self fertility of lines from Mennonite after two generations of inbreeding. However, significant differences for self fertility occurred between and within types of different origin.

Development of Inbred Lines

Several different methods are used to develop inbred lines depending on such factors as the source populations available and specific program objectives. The most common procedure involves selection of individual plants within open pollinated cultivars or segregating generations of planned crosses. Plants that appear phenotypically desirable are self pollinated. At harvest many of those that exhibit disease susceptibility, unsuitable maturity, lodging, or other undesirable agronomic characteristics are discarded. Laboratory analyses for seed and kernel characteristics or greenhouse tests for disease reactions allow for further identification of superior plants. Progenies from the plants that are retained are grown the next season and again self pollinated. Further selection is practiced both within and among progeny rows. Only the better plants in the better progenies are chosen for further inbreeding. The process of selection and inbreeding often continues for 2 to 5 years before lines are tested for combining ability.

Although self pollination is used most commonly during the inbreeding process, sib pollinations also may have advantages in developing inbred lines. Aside from theoretical considerations involving the rate of approach to homozygosity, sib matings are especially useful when development or maintenance of lines with a high degree of self incompatability is desired. One example might be where resistance to a particular disease has been introduced by hybridization with a highly self incompatible genotype. Self pollination could lead to loss of the line during the developmental process.

Kinman et al. (Kinman, M. L., G. W. Rivers, E. J. Redman, J. T. Wade, and E. L. Flynt. 1963. Annual report of cooperative sesame, sunflower, and guar investigations. USDA and Texas Agric. Exp. Stn., College Station) suggested the use of alternate selfing and sibbing to develop and maintain self incompatible lines. They found that seed production of highly self incompatible lines could be increased up to 10-fold by sib matings. Alternate self and sib pollination also is used extensively in Romania to maintain inbred lines (215), and has been used in programs to develop cultivars in Germany (178). Evaluations of the effects of sib matings on various plant and seed characters during different stages of inbreeding were reported by Schuster (177) and Kovacik and Skaloud (106).

Recently, breeders in the USA have selected intensively for self fertility among lines during inbreeding, especially when the inbred lines are to be used in production of hybrids using the cytoplasmic male sterile and fertility restorer system. Inbred lines that produce large quantities of seed upon self pollination are not only easy to maintain and evaluate, but also produce highly self fertile hybrids which may result in high seed yields when insect pollinator populations are less than optimum (45, 54). Experience of the author has shown that inbred lines with nearly 100% seed set upon selfing can be developed.

Development of inbred lines through chromosome duplication of haploids also has received attention in sunflower (70). The main advantage of this method is that pure breeding lines may be developed without several years of inbreeding. Haploids occur among twin seedlings at a frequency of only 0.64 to 4.76%, however, and effective methods for producing haploids artificially are not currently available.

Correlation Between Characters of Inbred Lines and Hybrids

Numerous studies have been conducted to determine if associations exist among plant and seed characters that could aid in selection for high seed yield or oil percentage. In general these correlation studies show that characters that contribute to good vegetative plant growth are associated with high seed yield. Putt (145) found positive correlations between seed yield and days from sowing to maturity, height, head diameter, and stem diameter. These results were substantiated by recent investigations of Kovacik and Skaloud (105) and Pathak (131). Other characters reported to be correlated positively with yield include leaf area per plant (124, 187), photosynthetic activity (183), seed weight (131, 182), and disease resistance (52, 173). In contrast to the positive relationships of the aforementioned characters with seed yield, certain studies have shown a negative relationship between seed yield and days from sowing to blooming (170, 175) and leaf area per plant among certain genotypes (124). These results suggest that the correlation of individual characters with seed yield depends on the group of genotypes being evaluated, and that the correlations can be used only as a general guide in selection for high yield.

Individual characters that have been reported to be correlated with oil percentage and consequently with oil yield per hectare include kernel percentage of seeds and 100-seed weight (145, 173), disease reaction (173, 230), and the presence or absence of branching. Regarding the latter trait Fick et al. (58) found that seed from branched plants had about 2.5 percentage units higher oil than seed from single-headed types of similar genetic background, a fact believed to be caused by smaller seed size rather than by the branching trait itself. Correlations of oil percentage with agronomic characters such as seed yield, plant height, head diameter, maturity, and leaf area have not been consistent over the numerous studies that have been conducted.

If correlations are to be of value in selection and development of inbred lines with good combining ability, information is necessary on the relationships between various characters of inbred lines and the same characters in their hybrids. Russell (173) conducted one of the more extensive studies involving correlations between various plant and seed characters of inbred lines and hybrids. He found significant positive correlation coefficients between the characters of days to flowering, plant height, rust rating, and seed oil percentage of the inbred lines, and the same characters in their topcross hybrids in two years of study, and for head diameter and percent stalk lodging in 1 of the 2 years. However, he found no definite relationships between 10 plant characters of the inbred lines and their combining ability as expressed by seed yields of their topcrosses.

Results presented by Putt (153) indicated that in a diallel cross of 10 inbred lines the higher yielding lines generally produced the higher yielding hybrids and synthetics. Similarly Gundaev (70) and Kloczowski (101) indicated that seed yield of inbred lines was correlated positively with yields of the hybrids. A general correlation between the yields of inbred lines and their crosses agrees with results obtained in maize (11). Several investigations in maize also have shown that visual selection of inbred lines is effective in producing lines with high combining ability. Irrespective of the value of selection for seed yield and visible plant characters in the inbreeding process, most sunflower breeders practice this type of selection before evaluation in hybrid combinations.

Evaluation of Inbred Lines

A completely satisfactory decision on the value of an inbred line in a hybrid or synthetic can be made only by testing it in experimental hybrid combinations. Unrau (199) conducted some of the early studies on methods of testing for combining ability. He suggested that the best method to test inbred lines for combining ability involved production of single cross hybrids with two tester lines of good general combining ability. Because of problems associated with low crossing percentages, he further suggested that the use of polycross testing to evaluate inbred lines was limited to the testing of fairly self incompatible types. Russell (173) and Kinman et al. (Kinman, M. L., G. W. Rivers, E. J. Redman, J. T. Wade, C. H. Lednicky, and E. J. Flynt. 1964. Annual report of cooperative sesame, sunflower and guar investigations. USDA and Texas Agric. Exp. Station, College Station) used the topcross test for evaluation of inbred lines relatively early in breeding studies conducted in Canada and the USA, respectively. Gundaev (70) also refers to the use of the topcross test for evaluating lines in the Soviet Union.

An early problem associated with evaluating inbred lines in hybrid combinations was the low hybridization percentage that often occurred in crossing. Hand emasculation procedures generally are considered too tedious for producing large quantities of crossed seed. Thus most hybrid seed for testing was produced by hand pollinations without emasculation or by natural crossing in isolated plots. In crossing plots involving two or more lines, hybridization percentages ranged from 21 to 96% and from 4 to 100% in seed production studies conducted by Unrau (199) and Putt (148), respectively. Consistently higher crossing percentages were obtained by hand pollinations (148, 199), but often these were still less than desirable. Consequently the yields of the test cross hybrids were influenced by the crossing percentage as well as the combining ability of the inbred lines.

Current methods involving genetic or cytoplasmic male sterility, or induction of male sterility by gibberellic acid, allow for complete hybridization of lines and hence greater precision in estimating combining ability. Various tester parents and tester schemes are being used. Anaschenko (5) has conducted extensive testing for general combining ability by the topcross method with chemical emasculation of the female parent with gibberellic acid. He has used open pollinated cultivars, hybrids, and inbred lines as testers. Vranceanu (206) used a monogenic male sterile line as a female parent to test for general combining ability and subsequent diallel cross analysis with artificial emasculation to test for specific combining ability. Recent testing by breeders in the United States has included the rapid conversion of lines to cytoplasmic male sterility by using greenhouses and winter nurseries, and subsequent hybrid seed production in isolated crossing blocks using open pollinated cultivars, synthetics, composites, or inbred lines as testers.

The choice of tester parent or tester scheme depends on the specific program objectives as well as the genetic material available. Sprague and Tavčar (193) made general recommendations concerning choice of the tester parent in maize breeding, recommendations that likely are appropriate for sunflower also. These include:

a. The tester should differ in origin from the lines to be tested.
b. To identify an inbred line with the maximum specific combining ability, such as a replacement of a line in a single cross or three-way hybrid, the opposing line or lines would be the preferred tester.
c. Heterogeneous testers exhibit a minimum of line × tester interaction and also hybrid × location or hybrid × year interaction. Thus average effects can be evaluated with less expenditure of time and money.

Additional information in maize breeding on the relative level of tester performance suggests that a low yielding tester may have advantages in discriminating among lines for combining ability (115, 162). Recently, with increased emphasis on development of single cross maize hybrids, procedures also have been used which allow for testing lines in hybrid combination during the inbreeding process (72, 73). This procedure involves yield evaluation during each generation of inbreeding among pairs of lines derived from two different source populations. The method appears to have application in sunflower also (123).

Utilization of Inbred Lines

Hybridization of inbred lines to produce improved high yielding hybrids or synthetics is the primary objective in sunflower breeding programs that are designed to maximize heterosis. Gundaev (69) indicated that studies conducted in the USSR as early as 1930 showed that crosses of inbred lines yielded as much as 50% more than existing cultivars. Unrau and White (200) in 1944 in Canada reported a 61% greater seed yield and much greater uniformity in a hybrid than in unimproved Mennonite. Recently, Fick and Zimmer (50) and Putt and Dorrell (155) in the USA and Canada, respectively, reported that hybrids yielded up to twice as much as check cultivars. Heterosis has been observed for time to bloom, plant height, head diameter, seed weight, and seed oil percentage in addition to seed yield (102, 153).

Much of the current interest in sunflower breeding programs in the USA involves the development of inbred lines for use in single cross hybrids. The single cross hybrids have advantages over three-way hybrids and open pollinated or synthetic cultivars because of greater uniformity for agronomic, disease, and seed oil characters (46). Uniformity in flowering has been especially useful because fewer applications of insecticides are necessary for control of damaging insects such as the sunflower moth (*Homoeosoma electellum* Hulst). Uniformity in maturity, height, and head diameter also has facilitated harvesting procedures greatly.

Three-way hybrids are produced primarily to reduce seed costs. Experience has indicated that in producing hybrid seed using the cytoplasmic male sterile and fertility restorer system, seed yields from single-cross female parents are often 1.5 to 2 times greater than those obtained from inbred lines. Conversely, inbred lines that yield up to 84% of their crosses have been developed (153) and seed yields of 2,000 kg/ha have been obtained from hybrid seed production fields in North Dakota involving the cytoplasmic male sterile line cmsHA 89.

Three-way hybrids, because of greater genetic heterogeneity, generally have been considered more stable over environments than single crosses. Vulpe (219) further suggested using double-cross hybrids, produced by using a combination of the genetic male sterile and cytoplasmic male sterile restorer systems, to improve the adaptation and yield stability of hybrids. In studies conducted in North Dakota and Minnesota (54, 56), three-way hybrids or open pollinated cultivars, as a group, were slightly more stable than single cross hybrids. Stable single cross hybrids also were identified, however, indicating that yield stability can be attained with individual genotypes as well as with the genetic heterogeneity inherent in three-way hybrids and open pollinated cultivars.

The use of inbred lines to form synthetic cultivars also may have value. Putt (148) concluded that synthetic cultivars were a desirable alternative to single cross hybrids when the hybridization percentage was low and variable, as was often the case in hybrid seed produced by natural cross pollination. Putt (153) later reported that high yielding synthetics probably could be produced from relatively few lines. He predicted the seed yields and seed oil percentages for the highest ranking 100 synthetics that could be produced from 10 inbred lines assuming 50 and 60% outcrossing between the F_2 and F_6 generations. Fourteen of the highest ranking 100 synthetics in F_6 for yield and 30 for oil percentage consisted of two or three lines. Results presented by Kloczowski (101) supported Putt's suggestion that valuable synthetics could be developed from a small number of lines, perhaps as few as three to five.

Synthetics have been produced and tested in the Soviet Union (69, 204). Two synthetic cultivars, composed of 8 and 14 inbred lines respectively, showed as much as an 8 to 10% higher oil yield than the check VNIIMK 8931.

Kloczowski (100) evaluated parents, F_1, and F_2 generations of 20 interline hybrids to obtain information on the possibilities of developing synthetic cultivars from intercrossing only two or three inbred lines. Average yields in the F_2 were 48% less than in the F_1. The F_2 and F_1 were more similar for plant height, head diameter, plant maturity, and oil and hull percentage of seeds. The actual seed yields were significantly less than those predicted on a theoretical basis, a fact that he attributed to the lower than expected outcrossing percentages. These results suggest the importance of selecting lines for a synthetic that are relatively self incompatible with a strong tendency to cross pollinate, as well as selecting lines with good combining ability.

Synthetic cultivars may be especially useful in countries or programs where hybrid seed production is not feasible for technical or economic reasons. Ravagnan (161) suggested developing synthetics as a rapid and sufficiently reliable method to deal with immediate commercial seed production needs in Zambia. Putt (148) and Kloczowski (101) have discussed the value of synthetics as source populations for developing superior sunflower germplasm.

A disadvantage of the synthetics is less uniformity than in hybrids, especially the single cross hybrids. Further, all lines of a synthetic must possess resistance to the prevalent diseases, while resistance if dominant is required in only one of the parent lines of a hybrid.

Methods for Producing Hybrids

Natural Cross Pollination with Self Incompatibility

Efficient and economical production of hybrid sunflower seed for large scale commercial use was facilitated greatly by the discoveries of genetic and cytoplasmic male sterility. Procedures to utilize sterility have been developed only recently, however, primarily since 1970. Prior to 1970, hybrids were grown commercially from seed produced by natural crossing in fields planted with the two parents in alternating groups of rows. The hybrid Advance was introduced for commercial production in Canada in 1946, and was grown subsequently for most of the oilseed producton in that country in the early 1950's (173). Other hybrids, including 'Advent' and 'Admiral' were developed later, seed of which also was produced by natural cross pollination (148). Hybridization percentages often were less than 50% (148) because self and sib pollination occurred on the female parent in addition to the desired cross pollination. Thus the full yield potential of these "hybrids" was not realized.

More recently several nonoilseed hybrids, or partial hybrids, were produced by natural crossing using lines with relatively low inbreeding coefficients. Two of these hybrids, 'D693' and 'D694,' were grown extensively for commercial production in Minnesota and North Dakota in the early 1970's and produced significantly higher yields than most open pollinated cultivars available at that time (222).

Breeding methods to develop lines for use in producing hybrid seed by natural cross pollination require selection of a highly self incompatible female parent. Also, the male parent should be a profuse pollen producer. Such parents together with appropriate seed production procedures can result in seed lots with hybridization as high 90% (118, 148). Vranceanu and Stoenescu (207) reported production of hybrids with 71 to 87% hybrid plants using self incompatible lines. They further suggested using a seedling marker gene to identify and discard nonhybrid plants in commercial production fields.

Breeding procedures to produce hybrid seed by natural crossing have been replaced largely by schemes designed to use genetic and cytoplasmic male sterility. Nevertheless the method of natural crossing for producing hybrid seed may be promising enough to justify additional investigation should inherent problems develop with the male sterility systems. The procedure also may have value in programs designed to produce synthetic cultivars from hybrids composed of two or only a few lines.

Genetic Male Sterility

According to Gundaev (69), genetic male sterility first was reported in the Soviet Union by Kuptsov in 1934. Since then numerous investigators (108, 156, 205) have reported genetic male sterility in sunflower. Vranceanu (206) indicated isolation of more than 30 sources of male sterility in the Romanian program, most of which were controlled by a single recessive gene. Diallel cross analysis of 10 of these lines indicated the presence of five different genes.

Putt and Heiser (156) reported one of the first studies to assess the value of genetic male sterility to produce hybrid seed. They concluded that lines with partial male sterility might have the most immediate value in commercial production of hybrid seed. The partial male sterile lines hybridized well in crossing plots and also could be increased and maintained easily. The oilseed hybrid 'Valley,' which was produced for commercial production using the partially male sterile line CM 90RR as a female parent and the cultivar 'Peredovik' as a male parent, showed excellent seed yields (94). Hybridization percentages of this hybrid varied but seed lots with 60 to 80% hybrid were common.

The use of complete genetic male sterility to produce hybrid seed requires the maintenance of the male sterile locus in a heterozygous condition in the female parent. This is accomplished by sib pollinations of male sterile plants (*ms ms*) with heterozygous male fertile plants (*Ms ms*) within the female parent. The resultant progeny from the male sterile plants will segregate 1:1 for fertile and sterile plants. When such lines are used in hybrid seed production the fertile plants must be removed prior to flowering to obtain 100% hybridization with the male parent line.

Removal of the male fertile plants was facilitated greatly by the discovery of a close linkage between genes for genetic male sterility and anthocyanin pigment in the seedling leaves (108). This linkage allowed identification and removal of the male fertile plants during the seedling stage, and was used widely to produce hybrid seed in France and Romania in the early 1970's. Some of the best hybrids produced by this system yielded up to 24% more than open pollinated check cultivars (211). Production of hybrid seed by the genetic male sterile system has the advantage that fertile hybrid plants can be produced using any normal male fertile line as the male parent. Disadvantages include the requirement to incorporate and maintain the linked characters in the female parent, and the high labor costs required to remove the male fertile, anthocyanin pigmented plants from the female rows of seed production fields.

The system has been replaced largely by the cytoplasmic male sterile and fertility restorer system in most current hybrid sunflower breeding programs. The value of genetic male sterility now appears to be primarily an alternate method of hybrid seed production should problems develop with the use of cytoplasmic male sterility such as occurred in maize with susceptibility to southern corn leaf blight (*Helminthosporium maydis* Nishikado and Miyaka). The system also may have value for developing suitable testers for evaluating inbred lines, and subsequent production of hybrid seed for testing.

Cytoplasmic Male Sterility and Genetic Fertility Restoration

According to Gundaev (70) the first reference to cytoplasmic male sterility was by Shtubbe in 1958. The report by Durand (35) written in 1962 mentions investigations involving a source of cytoplasmic male sterility discovered by East German researchers. Other early discoveries were reported by Volf and Gundaev in the Soviet Union, Stoyanova in Bulgaria, and Vulpe in Romania (70, 217). These early reports indicated that most crosses of cytoplasmic male sterile plants with normal male fertile lines produced progeny with variable percentages of sterile plants. Gundaev (69) found, in crosses with a source from 'VNIIMK 8931,' that complete sterility might be expressed by as few as 10 or as many as 90% of the progeny. Generally, varying degrees of partial sterility also occurred. Through selection and test crossing, lines that produced 92 to 96% sterile progeny were developed and utilized in experimental production of hybrid seed (69, 70).

Leclercq (110) in France reported the discovery of cytoplasmic male sterility from an interspecific cross involving *H. petiolaris* Nutt. and *H. annuus* L. This source of cytoplasmic male sterility was shown to be very stable and is now the source used almost exclusively in breeding programs around the world. Genes for fertility restoration of the male sterile cytoplasm were identified subsequently by several investigators (40, 98, 111, 214) shortly after the widespread distribution of the sterility source by Leclercq. The first hybrids produced by the cytoplasmic male sterility and genetic fertility restorer system were made available for commercial production in the United States in 1972, and by 1976 it was estimated that these hybrids accounted for over 80% of the sunflower production in the USA (54). The best hybrids yielded over 20% more than the open pollinated cultivar Peredovik, and most were resistant to the three major diseases, of rust (*Puccinia helianthi* Schw.), downy mildew [*Plasmopara halstedii* (Farl.) Berl. et de Toni], and Verticillium wilt (*Verticillium dahliae* Kleb.).

Breeding Methods to Develop Cytoplasmic Male Sterile and Fertility Restorer Lines

Most breeding programs in the USA and throughout the world now place major emphasis on developing lines that can be used to produce hy-

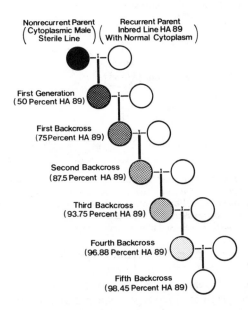

Fig. 2—Backcross method is used to develop and maintain a cytoplasmic male sterile line of HA 89. The final progeny will be identical with the recurrent HA 89 parent in most characteristics except that it will be male sterile instead of fully fertile. Theoretically the genetic content of the nonrecurrent male sterile parent will be reduced by half for each generation of backcrossing.

brids by the cytoplasmic male sterile and fertility restorer system. Cytoplasmic male sterile lines are developed by the backcrossing method (Fig. 2). Desirable lines that have undergone inbreeding and selection for several generations are crossed initially to a plant with cytoplasmic male sterility. Thereafter the inbred line to be converted is used as the recurrent parent in the backcrossing procedure. The final progeny should be genetically similar to the recurrent parent except that it will be male sterile.

If the inbred line has not been tested previously for combining ability, crosses with a cytoplasmic male sterile tester and subsequent evaluation of the sterile F_1 hybrid can provide valuable information on combining ability. If this procedure is used, plots of a normal pollen shedding line or cultivar must be interplanted in the trial to ensure adequate pollination. Conversion of an inbred line to cytoplasmic male sterility can be accomplished in a relatively short time, especially by using winter nurseries and greenhouses which allow for as many as three or four generations per year. Consequently several breeders in the United States have converted large numbers of inbred lines to cytoplasmic male sterility, and then tested for combining ability using various fertility restorer lines or populations as testers.

A few sunflower lines have proven difficult to sterilize, especially inbred lines selected from 'Chernianka 66' and 'Armavirec' (Fick, unpublished data) as well as several other important commercial cultivars. These

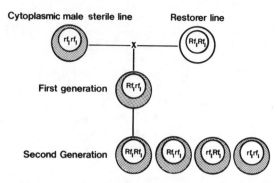

Fig. 3—A restorer line may be developed by crossing plants with dominant restorer genes (Rf) to a cytoplasmic male sterile line, and subsequent selection of male fertile plants in later generations. This procedure, unlike the backcross procedure using lines with normal cytoplasm, does not require test crossing to a cytoplasmic male sterile plant after each generation of selection to determine if the selected plants carry the restorer genes.

lines apparently possess genes for partial restoration of fertility, or perhaps lack genes that suppress expression of partial restoration.

Most sources of fertility restoration that are being used now in breeding programs originated from wild species of sunflower, including the widely used T66006-2 source discovered by Kinman (98). Occurrence of genes for complete fertility restoration of male sterile cytoplasm is rare in cultivated sunflower although it has been reported (48, 214).

Development of fertility restorer lines requires incorporation of the dominant restorer gene(s) in the male parent. This may be accomplished by backcrossing using established inbred lines as the recurrent parent. However, this procedure is tedious because the selected plants in each generation must be crossed to a cytoplasmic male sterile line to determine if the selected plants carry the restorer genes. The first generation plants from these test crosses will be fully male fertile if the selected plants carry the necessary fertility restoring genes.

New restorer lines can be developed by an easier procedure, as shown in Fig. 3, by crossing the line or population possessing the restorer genes to a cytoplasmic male sterile line. From the resultant male fertile hybrid, lines homozygous for the fertility restoring genes and with other desired characters may be isolated by continuous self pollination without test crossing. This procedure was used to develop the first downy mildew resistant restorer lines RHA 271, RHA 273, and RHA 274 (51). These lines possess genes for fertility restoration and resistance to downy mildew, and have recessive branching. The latter trait allows production of pollen over an extended period of time and is useful for producing hybrid seed with female lines that flower up to three weeks later than the fertility restorer lines.

No significant problems have been encountered using the cytoplasmic male sterile and fertility restorer system for production of hybrid seed. The cytoplasm controlling sterility has no apparent adverse effects on agronomic or seed oil characters when incorporated into inbred lines (203; Fick, unpublished data). Velkov and Stoyanova (203) have observed that bee visitation necessary for good seed set in production fields was similar for cytoplasmic male sterile lines and their normal male fertile counterparts. These results are in contrast to recent data obtained by Furgala et al. (61) which indicated that sterile lines had less nectar and possibly were less attractive to bees than normal male fertile types.

The male sterility source discovered by Leclercq (110) and the fertility restoring genes tracing to Kinman's T66006-2 source have been highly stable over a wide range of environments. Nevertheless, the search for new sources of male sterile cytoplasm and fertility restoring genes should be continued in order to reduce genetic vulnerability to diseases or other pests. Leclercq (112) presented preliminary evidence that suggested the occurrence of a second source of cytoplasmic male sterility from the cross of *H. petiolaris* Nutt. and Whelan (221) reported results that suggested the occurrence of cytoplasmic male sterility in crosses of *H. petiolaris, H. giganteus* L., and *H. maximiliani* Schrader. Fick et al. (55) and Dominguez-Gimenez and Fick (31) reported the widespread occurrence of fertility restoration of male sterile cytoplasm among wild species. In several collections fertility restoration appeared to be controlled by genes other than the original Rf_1 gene found by Kinman. Thus the possibilities for reducing potential problems of genetic vulnerability associated with using restricted germplasm exist for future investigations.

Recurrent Selection for Improving Source Populations

The success of isolating inbred lines with good combining ability or other desired characters by the standard procedure of inbreeding and selection depends on the frequency of superior S_0 plants in the source population. Recurrent selection appears to be one of the most promising methods to increase the frequency of desirable genotypes in a source population, and thus enhance the chances of success in isolating superior inbred lines. Two types of recurrent selection often are distinguished. These are (a) phenotypic recurrent selection in which the phenotype of the S_0 plant is the basis of selection, and (b) genotypic recurrent selection in which some type of progeny test forms the basis of selection. Each cycle of recurrent selection requires selfing and evaluation of selected plants, and subsequent intercrossing of the progenies of superior selfed plants to produce seed for the next cycle of selection.

Aside from the Pustovoit method for improvement of oil percentage, which is a modified form of recurrent selection not involving self pollination, relatively few recurrent selection studies have been conducted in sunflower. Kinman et al. (Kinman, M. L., G. W. Rivers, E. J. Redman, J. T. Wade, and E. L. Flynt. 1963. Annual report of cooperative sesame, sunflower, and guar investigations. USDA and Texas Agric. Exp. Stn., College Station; Kinman, M. L., G. W. Rivers, E. J. Redman, J. T. Wade, C. H. Lednicky, and E. J. Flynt. 1964. Annual report of cooperative sesame, sunflower and guar investigations. USDA and Texas Agric. Exp. Stn., College Station) used phenotypic recurrent selection to select for self incompatibili-

Table 1—Mean oil percentage and variance among self pollinated plants of two sunflower populations during three cycles of recurrent selection (45).

	Chernianka 66		12 cultivar composite	
Generation cycle	Mean	Variance	Mean	Variance
0	41.9	9.5	46.1	17.6
1	43.0	8.0	48.8	14.5
2	44.6	12.6	49.0	18.8
3	45.8	10.6	49.3	17.1

ty and plants with single stems. They found the number of seeds produced per plant from self pollination decreased from a mean of 106 to 32 after one cycle of selection within the inbred line S-37-388. Selection within the open pollinated cultivar Chernianka 66 was not effective for self incompatibility but resulted in a significant decrease in the percentage of plants with single stems.

Fick and Rehder (45) completed three cycles of recurrent selection for high oil percentage within two source populations. The results (Table 1) show that mean oil percentage increased from 41.9 to 45.8% in the cultivar 'Chernianka 66,' and from 46.1 to 49.3% in a composite of 12 cultivars. The variability for oil percentage among self pollinated plants in the selected populations was similar to that present in the original populations, suggesting that further selection likely would be effective.

Gundaev (70) presented limited results on the value of recurrent selection for high seed yield. He indicated that Karp in 1946 obtained a 13% increase in seed yield by intercrossing the best, or presumably the highest yielding, lines. Gundaev (69) reported that one cycle of recurrent selection for high yield, with test cross evaluation of progeny, increased seed yields by 6% over the original cultivar.

Several of the methods of recurrent selection used for hybrid development and improvement of maize breeding populations appear applicable to sunflower. These include the comprehensive breeding system outlined by Eberhart et al. (38), reciprocal full sib selection utilizing multiple-eared or multiple-headed populations proposed by Hallauer (72, 73), and several modified forms of recurrent selection reviewed by Sprague (192). Recurrent selection methods utilizing both S_1 progeny and test cross evaluation have been initiated in the USDA sunflower breeding program at Fargo, North Dakota. Preliminary results indicate that significant improvement in yield and combining ability of populations can be achieved by either S_1 or test cross evaluation (Fick, unpublished data).

The development of single cross hybrids utilizing the procedure described by Hallauer (72, 73) is a particularly intriguing method for use in sunflower. This procedure involves yield evaluations during each generation of inbreeding among pairs of lines derived from two source populations. Whereas Hallauer used two-eared or prolific maize populations, a composite of restorer lines with recessive branching and a normal single headed B-line composite can be used as the source populations in sunflower. Hybrid seed can be produced on one or several heads treated with gibberellic

Table 2—Range in flowering and height of wild sunflowers grown at Fargo, North Dakota, in 1973 (55).

Species	No. of populations	Days to 50% flowering†	Height
			cm
H. annuus	148	36–48	71–330
H. petiolaris	53	15–26	81–218
H. tuberosus	1	78	137
H. rigidus	1	75	142
H. grosseserratus	2	78–80	102–122
H. maximiliani	5	75–88	107–157
H. nuttallii	1	78	152

† Days after 1 June.

acid to cause sterility, and selfed seed can be obtained from untreated heads of the same plant. Populations to use this recurrent selection procedure have been developed in the North Dakota program (123).

Genetic Resources

The need for genetic variability and sources of resistance to diseases and pests has encouraged sunflower breeders to use a wide range of germplasm resources in their breeding programs. Wild species of sunflower, world collections, and mutagenically altered populations have all proven useful in the development of improved cultivars and hybrids.

Wild Species of Sunflower

The wild sunflowers, of which over 50 species are recognized (79), are extremely variable for agronomic type, disease reaction, and seed quality characters. Heiser and colleagues (79) presented extensive information on the variability for agronomic and morphological characteristics among the species occurring in North America. Although the data are primarily of taxonomic value the information also provides clues to the potential value of certain species in sunflower breeding programs.

Fick et al. (55) evaluated over 200 wild populations for agronomic traits that could be used to improve cultivated sunflower at Fargo, North Dakota, in 1973. A wide range in days to flowering and in height was observed both among and within species (Table 2). The variation was significantly greater than in the cultivated types grown in the area, and was considered to be of value in developing new cultivars adapted to specific environments and growing seasons. Variation among and within species also was observed for germination and seedling vigor, branching, seed shattering, and tolerance to fall frosts.

The wild species of sunflower are an excellent source of resistance to the major sunflower diseases and to several plant and insect pests. Putt and Sackston (157) first reported on the occurrence of rust resistance from natural crosses with wild annual sunflower near Renner, Texas. The 953-88 and 953-102 sources of resistance that were selected from these crosses have been used by sunflower breeders throughout the world in developing cultivars resistant to rust. Galina Pustovoit (140) and G. Pustovoit and Gubin (141) conducted extensive studies with more than 40 species of wild sunflower. They reported lack of infection or resistance to the pathogens causing rust, downy mildew, white rot (*Sclerotinia libertiana* Fuck.), gray rot (*Sclerotinia bataticola* Taub.), Verticillium wilt, Botrytis (*Botrytis cinerea* Pers.), Rhizopus head rot (*Rhizopus microsporus* Namm.), white rust (*Cystopus tragopogonis* Shred.), Fusarium wilt (*Fusarium* sp.), and to the plant and insect pests of broomrape, head moth and aphid (*Brachycaudus helichrysi* Kalt.). The autohexaploid species *H. tuberosus* L., *H. rigidus* (Cass.) Desf., *H. subcanescens* Wats., and *H. macrophyllus* Willd. were cited as especially valuable sources of resistance. Soviet scientists (141) have developed cultivars with immunity to several of the diseases and pests mentioned through interspecific hybridization, primarily with *H. tuberosus.*

Recent studies in the USA and Canada have confirmed the results of Soviet investigators and have shown that disease resistance occurs commonly both among and within large numbers of different collections of wild species. Zimmer and Rehder (229) found that resistance to rust was widespread among wild sunflower with plants free of rust occurring in 190 of 200 populations of seven species tracing to the U.S. North Central region. Resistance to Verticillium wilt occurred in all of 46 wild *H. annuus* and *H. petiolaris* populations collected from widely separated areas in Canada and the USA (85). Fick et al. (55) reported that resistance to downy mildew was common among five perennial species, but less frequent among the annual species with only 9 of 83 populations of *H. annuus* and *H. petiolaris* showing possible resistance. Zimmer and Fick (226) also found resistance to several of the minor diseases including Alternaria leaf and stem spots (*Alternaria* sp.) and Phoma black stem (*Phoma oleracea* var. *helianthi tuberosi* Sacc.) in several wild species. It is apparent from these and other investigations that wild sunflower offers a vast reservoir of disease resistance for improving cultivated types.

Sunflower researchers also have evaluated wild species of sunflower for variation in seed quality characteristics. The wild species generally have lower oil percentages than cultivated types. Oil concentrations ranged from 18.2 to 31.5% among 47 populations of six species (55), from 21.7 to 27.8% for seven populations of seven species (64) and from 18.4 to 28.6% among 42 populations of wild *H. annuus* (33). Wide variation in fatty acid composition of the oil from the wild species has been observed (42, 104, 142). Wild types with high linoleic acid and others with high oleic acid have been identified.

Protein percentage in the seeds of the wild species was reported to be considerably higher than that of cultivated types exceeding 40% in *H. scaberrimus* Ell. and over 70% in fat-free kernels of *H. rigidus* (64, 86). Chlorogenic acid percentage of hulled and fat free seeds of 42 wild *H. annuus* populations ranged from 1.58 to 2.70%, which on the average is somewhat lower than the cultivated types (33).

Wild species of sunflower have provided the key to efficient production of hybrid sunflower. The widely used source of cytoplasmic male sterility identified by Leclercq (110) originated from a cross of cultivated sunflower with the wild annual *H. petiolaris.* Additional sources of cytoplasmic male sterility are believed to have been identified in crosses involving *H. petiolaris* (112), *H. annuus* ssp. *lenticularis* CKll. (6), and *H. petiolaris, H. giganteus,* and *H. maximiliani* (221). Most sources of fertility restoration of cytoplasmic male sterility come from wild species (31). A survey of the wild species *H. annuus* and *H. petiolaris* indicated that genes for fertility restoration occur commonly in both species with restoration observed in all but 4 of 85 crosses (31, 55). Thus, alternate cytoplasmic male sterile and restorer systems appear to be available in the wild species should they be needed.

Wild species are used in breeding programs depending on several factors including the need for expanding variability and the difficulties in using the wild species. Because the cultivated sunflower traces to a relatively restricted germplasm base, it is unlikely that it possesses the wealth of genetic variability present in its wild relatives. Consequently, new sources of variability, primarily for disease and pest resistance, probably will be needed as new pests or races of pests arise.

Difficulty in using the wild species in breeding programs varies among species. Several of the annuals, *H. argophyllus* Torrey and Gray, *H. bolanderi* Gray, *H. debilis* Nutt., *H. paradoxus* Heiser, *H. petiolaris,* and *H. praecox* Englem. and Gray, cross quite readily with *H. annuus* and produce relatively fertile hybrids (79). More difficulty is encountered in crosses with the perennials although artificial hybrids have been produced with *H. decapetalus* L., *H. giganteus, H. grosseserratus* Martens, *H. hirsutus* Raf., *H. rigidus, H. strumosus* L., and *H. tuberosus* (63, 79, 221). Successful transfer of desirable traits from both the annual and perennial species has been reported, and in several cases without extensive backcrossing or reselection to recover the cultivated genotype (113, 140, 157). In a study designed to transfer fertility restorer genes from wild populations of *H. annuus* and *H. petiolaris,* investigators at Fargo, North Dakota (31, 57) found that plants with acceptable oil percentage and agronomic type were recovered after only one or two backcrosses.

World Collections

World collections of sunflower serve basically the same purpose as the wild species, which is to provide additional sources of genetic variability for developing improved cultivars and hybrids. Two of the larger world collections are those maintained at the N.I. Vavilov All-Union Scientific Research Institute of Plant Industry (VIR) at Leningrad, USSR, and at the USDA's North Central Regional Plant Introduction Station at Ames, Iowa. The VIR collection in 1974 had about 1,200 entries of cultivated sunflower, of which about 60% had been collected in the USSR and the remainder from countries around the world (5). Soviet researchers are concentrating on

Table 3—Number of sunflower cultivars from the USDA World Collection in different classes for plant height, maturity, seed oil percentage, and reaction to rust from tests conducted at Fargo, North Dakota, in 1974 (Fick, unpublished data).

Country of origin	No. of cultivars	Plant height (cm)			No. of cultivars	Maturity (days)		
		Short 76–125	Med. 126–175	Tall 176–225		Early 75–100	Med. 101–125	Late 126–150
Argentina	5	0	4	1	5	0	1	4
Austria	3	0	0	3	3	0	0	3
Brazil	5	0	2	3	5	0	0	5
Canada	3	0	2	1	3	1	1	1
Chile	4	0	3	1	4	0	4	0
Egypt	3	0	1	2	3	1	0	2
Ethiopia	5	0	4	1	6	0	0	6
France	9	0	9	0	9	1	7	1
Hungary	10	3	7	0	12	44	4	3
Iraq	10	1	2	7	10	0	2	8
Iran	40	3	29	8	40	2	20	18
Israel	6	0	2	4	8	0	6	2
Jordan	7	0	6	1	8	0	5	3
Kenya	9	0	3	6	8	0	0	8
South Africa	8	0	7	1	7	0	5	2
Spain	12	2	9	1	12	2	8	2
Turkey	103	2	66	35	106	2	47	57
Uruguay	3	0	3	0	3	0	0	3
USA	22	3	17	2	23	1	14	8
USSR	42	4	30	8	44	9	29	6
Others†	13	0	8	5	14	2	4	8
Total	322	18	214	90	333	25	158	150

Country of origin	No. of cultivars	Oil percentage			No. of cultivars	Rust reaction		
		Low 20–29	Med. 30–39	High 40–49		S	MR-R	Mixed
Argentina	4	4	0	0	5	3	0	2
Austria	3	3	0	0	3	2	0	1
Brazil	4	4	0	0	5	3	0	2
Canada	3	2	11	0 0	4	0	2	2
Chile	2	0	1	1	4	2	0	2
Egypt	2	1	1	0	3	3	0	0
Ethiopia	4	4	0	0	6	6	0	0
France	9	3	2	4	13	7	0	6
Hungary	8	3	3	2	15	15	0	0
Iraq	5	4	1	0	10	10	0	0
Iran	30	10	15	5	48	35	0	13
Israel	5	2	3	0	10	8	0	2
Jordan	6	6	0	0	9	7	0	2
Kenya	4	3	1	0	10	6	0	4
South Africa	6	3	2	1	10	5	1	4
Spain	7	2	3	2	13	8	0	5
Turkey	82	51	30	1	123	117	0	6
Uruguay	2	2	0	0	3	2	0	1
USA	17	7	7	3	20	11	4	5
USSR	30	2	10	18	48	36	0	12
Others†	12	5	5	2	18	14	0	4
Total	245	121	85	39	380	300	7	73

† Includes countries from where only 1 or 2 cultivars originated, viz, Afghanistan, China, Columbia, England, Germany, Indonesia, Lebanon, Paksitan, Poland, Portugal, Syria, Tanganyika, and Yugoslavia.

evaluating individual entries in the collection for general combining ability, oil percentage, and marker genes that could be used in breeding programs (5, 171). It was obvious to the author in a visit to the Kuban Experiment Station of VIR in 1976 that the collection has a great amount of genetic diversity.

The USDA world sunflower collection has about 500 entries from over 30 countries (188). Individual entries have been characterized for agronomic traits, disease and insect resistance, and seed quality characteristics primarily from data obtained from plantings for seed increase purposes. Experience of the author suggests that many of the entries have rather good combining ability when crossed with adapted lines or cultivars, but often possess low oil percentage and susceptibility to diseases. Data for plant height, maturity, seed oil percentage, and rust reaction from a partial evaluation of the collection in 1974 appear in Table 3.

Some of the more recent introductions to the USDA collection, consisting primarily of newly released Soviet cultivars, have proven extremely valuable as source material for developing new high oil inbred lines of hybrids.

Mutagenesis

Mutagenesis has had only limited use for increasing variability in sunflower breeding programs. Gundaev (70) indicated that irradiation and chemical mutagens produced a range of mutants, the most practical of which were early ripening, short statured plants. Voskoboinik and Soldatov (204) also obtained mutants with useful characters for breeding through chemical mutagenesis, among them plants that combine early maturity with high oil percentage and thin hull, dwarf types with an increased number of leaves, plants with altered fatty acid composition, and plants with male sterility. Luczkiewicz (119) reported the occurrence of mutants induced by X-ray treatment, most of which involved alterations in gross plant or leaf characters. He also cited references to other investigations involving use of both chemical and physical mutagens in sunflower. Perhaps the most valuable mutant type produced to date is the high oleic acid population developed by Soldatov (189). This population is characterized by individual plants with oleic acid as high as 90%, high seed oil percentage, and desirable agronomic traits.

Preservation of Germplasm

As in all crops, the preservation of available sunflower germplasm for use in future breeding efforts is imperative. This is particularly true for the wild sunflower species which, as indicated, possess a wealth of genetic diversity for further improvement of cultivated types. Although wild species still occur commonly in most parts of North America, they are disappearing in some areas and certain species such as *H. paradoxus* and *H. exilis* Gray are apparently nearly extinct (79, 92). Currently only limited collections of wild sunflower have been made, and most are maintained by individual breeders under less than optimum conditions. The world collections contain very little wild germplasm (5, 188). High priority in the near future must be given to organized exploration, collection, and maintenance of the wild North American sunflower species.

Seed repositories have been developed at the Leningrad and Kuban stations of VIR in the Soviet Union as well as at Fort Collins, Colorado, USA, where seed may be kept at optimum storage temperature and humidity for long life.

Breeding Techniques

General descriptions of some of the more widely used techniques in the breeding and evaluation of field crops have been described thoroughly by Poehlman (136). Specific techniques and evaluation procedures that are used primarily in sunflower breeding, however, have not been reported widely. Some of the following information on selfing and crossing techniques, the use of certain chemicals to alter flowering or growth habits, and the evaluation for agronomic, seed quality, or disease characteristics may be useful in developing a sunflower breeding program.

Self and Cross Pollinating

A knowledge of the pollination and flowering processes of sunflower is necessary to develop appropriate selfing and crossing techniques essential to a sunflower breeding program. Sunflower is highly cross pollinated, primarily by insects and only to a limited extent by wind (70, 143). The flowers must be covered with bags or otherwise isolated to prevent natural cross pollination in selfing and crossing procedures.

Several studies have been conducted to investigate the relative efficiency of different bagging materials for use in self pollinating. Putt (144) found that heads covered with cotton and tiffany bags produced slightly more seed than heads covered with paper bags, although the latter were satisfactory for use in a breeding program. Kalton (96) investigated three types of cloth and two types of paper bags in selfing studies. He obtained the poorest seed set with Kraft pollen bags and 11.35 kg paper grocery bags, and found neither paper bag satisfactory for durability. Heads covered with cheese cloth bags with a density of 16 threads/cm^2 produced the highest seed set but some foreign pollen contamination occurred. Muslin bags woven at 34 and 50 threads/cm^2 appeared the most desirable considering relative seed set, durability, and exclusion of foreign pollen.

In the USSR parchment paper bags usually are used for collecting pollen, and cloth bags with a density of 20 threads/cm, and 40 × 45 cm in size, for isolating maternal heads (70). The USDA breeders in North

Dakota and Texas have found that cloth bags of a density of 24 threads/cm and 33 × 33 cm in size are suitable for most purposes, and often can be used for 2 or 3 years. Recently several breeders in the USA and Canada also have made extensive use of bags made of plastic netting, which have allowed satisfactory seed set and exclusion of foreign pollen, and generally cost less than cloth bags.

Seed set on self pollinated heads can be increased significantly through manipulation of the heads by drawing a cotton batting brush over the stigmas, or by rubbing the hand over the surface of the head without removing the bag (144). These procedures are widely used in the USSR and most Eastern European countries; however, breeders in the USA and Canada have not used manipulation of heads extensively primarily because selection of highly self-fertile types has been considered advantageous in the inbreeding process.

Crossing techniques in sunflower have been described by Putt (144). Emasculation is performed by removing the anther tubes with forceps early on the morning that the flowers open. Free pollen is blown off and flowers not wanted are removed from their ovaries before fertilization occurs. It is best to pollinate the morning after emasculation, although the stigmas remain highly receptive for up to 4 or 5 days (70, 144). The morning after pollination the stigmas of those flowers that have been fertilized have withered and receded, thus providing an indication of crossing success. The efficiency of the procedure, as measured by the proportion of crossed seed compared to selfed seed that was obtained, averaged over 90% (144).

Because hand emasculation is tedious many breeders make crosses without emasculation, relying on the greater competitive capacity of foreign pollen to effect fertilization. Hybrid plants generally can be distinguished from selfed plants on the basis of vigor or the presence of certain marker genes. Crossing without emasculation is most useful when only a few hybrid plants are desired but has limitations when a high percentage of crossing is required or the maternal parent is highly self-fertile. Perhaps the best procedure to obtain large quantities of high percentage hybrid seed or extreme accuracy in the crossing procedure, is to use genetic or cytoplasmic male sterile plants as female parents. Such parents allow 100% hybridization but are restricted to plants or lines in which sterility has been identified or incorporated.

Chemical induction of male sterility by gibberellic acid to facilitate crossing has proven effective and recently has been used widely by sunflower breeders to produce hybrid seed for combining ability tests. According to Anaschenko (3) the best results are obtained when the sunflower plants are sprayed during bud initiation with 0.5 to 1.5 mg of a 0.005% solution of gibberellic acid per plant. Up to 100% male sterile plants can be obtained with only slight reduction of female fertility and seed yields of treated plants. Cultivars and inbred lines may show different reactions to gibberellic acid treatments (176) although Anaschenko (3) and Piquemal (133) suggest that most genotypes can be sterilized with appropriate treatment. Specific recommendations involving different concentrations, application rates, number of applications, and inclusion of other growth regulators have been made in numerous reports (3, 4, 133, 180, 181, 191). Experience of the author with induction of sterility by gibberellic acid suggests that time of application is the most important factor in successful use of gibberellic acid, and that genotypic differences, dosage rates, and application schemes are of secondary importance.

Pollination in sunflower is conducted by collecting pollen from heads, usually isolated with bags prior to flowering to avoid contamination by foreign pollen. Pollen can be collected effectively from flowering heads into paper bags by a light tap of the hand on the back of the head. Mature, viable pollen generally can be collected throughout the day, although direct sunlight on pollen for several hours will reduce viability (70). Pollen can be applied by a small piece of cotton, a camel hair brush, the corner of the cloth bag isolator, a small section of leaf, paper, or other suitable material that is dipped in the pollen and gently drawn over the receptive surfaces of the stigmas. After each cross, care must be taken to avoid contamination by wiping the hands with alcohol and cleaning or discarding the pollen applicator. The highest percentage of fertilization is obtained by pollinating with freshly gathered pollen, or with pollen that has been stored not more than 2 or 3 days (70). Studies on the duration of pollen viability have suggested that pollen can be stored without serious loss of viability for 1 to 2 weeks in cork-stoppered vials at ordinary room temperature (144), and with some viability still remaining after storage in paper wrappers at room temperature for 11 months (7). The author has stored pollen in paper or plastic bags at cool temperatures in a household refrigerator with no apparent reduction in viability after 3 months (Fick, unpublished data).

Chemical Alteration of Germination and Growth

To facilitate the rapid development of new inbred lines and cultivars most sunflower breeders use greenhouses and winter nurseries to grow two to four generations per year. Postharvest dormancy is frequently a problem associated with the rapid turnover of generations. The dormancy period varies among genotypes ranging up to 45 to 50 days (198, 232). To overcome dormancy, seeds can be treated with Ethrel (2-chloroethyl-phosphonic acid, Amchem Products, Inc., Ambler, Pennsylvania). Udaya Kumar and Krishna Sastry (198) reported that germination percentages of whole seeds five days after harvest and treated with 25 ppm Ethrel were 76% compared to 8% for the untreated control. Ruud (172, and personal communication) increased germination of seed lots from 1 to 2% to more than 80% with Ethrel treatments.

The procedure for breaking dormancy with Ethrel that has been used successfully in the USDA program at Fargo, North Dakota, involves soaking the seeds for 16 hours in a 10^{-3} M solution (0.6 ml Ethrel/1 liter water), or in a 10^{-4} M solution buffered to a pH of 7.0 if the seeds are to remain in the solution until germination occurs. Removing the hull further enhances germination of many seed lots. Using this procedure seeds harvested at 25 to

30% moisture and treated immediately with Ethrel have germinated within 1 week.

Another chemical that may have value in a program requiring short generation time and especially for greenhouse plantings is the growth retardant succinic acid, 2,2-dimethylhydrazide (SADH). Sunflower plants grown in the greenhouse under suboptimal light intensity of winter are tall with weak stems and are difficult to manage. According to Dorrell (32) SADH in aqueous solutions of approximately 6,000 ppm applied between the seventh leaf and flower bud stage produced plants with short, thick stems that required no artificial support. Seed yields were reduced slightly by the treatment but seed size, germination, oil percentage, and fatty acid composition were not affected significantly. Other chemicals that may have value in reducing height of sunflower plants are Ethrel and CCC (2-chloro-ethyltrimethylammonium chloride) (62, 116).

Agronomic, Seed Quality, and Disease Evaluation

Efficient procedures for evaluating breeding lines, cultivars, and hybrids are essential to identify superior genotypes rapidly. Proper conduct of field trials is especially important if trials are to accurately measure differences among lines and cultivars for seed yield, flowering, maturity, height, lodging, resistance to diseases, insects, and other pests, and other characters. Techniques used in the USSR for evaluating sunflower in field trials have been reported by Pustovoit (139). Cultivar trials generally are planted by hand in hills spaced 70 × 70 cm apart with two plants per hill, and consist of six-row plots with the center four rows used for observation and yield determinations. Every third plot is a check cultivar and three to four replicates are used. The scheme has allowed precise estimation of small differences in seed and oil yield.

Researchers in the USA and Canada usually plant sunflower with cone seeders in rows 76 or 91 cm apart and thin by hand to recommended stands. The randomized block design is used commonly with plot size and number of replicates varied according to test and location. Fick and Swallers (47) investigated the value of single-row plots for preliminary evaluation of experimental cultivars and found highly significant competition effects. They concluded that single row plots were of limited value for evaluating seed yield, but accurately estimated days to 50% flowering, plant height, rust reaction, test weight, and seed oil percentage.

Evaluation of seed quality characters has been facilitated greatly by new instrumentation in recent years. One of the most useful developments has been analysis for oil percentage by nuclear magnetic resonance (NMR) spectroscopy. The NMR analysis is rapid, accurate, and nondestructive of seeds. Specific procedures for preparation and analysis of sunflower seed samples have been described by Granlund and Zimmerman (67).

If oil analysis by NMR is not feasible due to expense, lack of appropriate facilities, trained labor, or other reasons, several simple, rapid and reasonably accurate alternative methods of estimating oil percentage of sunflower seeds have been devised. One alternative method is to determine hull percentage which is closely correlated with oil percentage of the whole seed (145, 173). Estimating oil percentage by appearance of whole seed combined with cutting through a few seeds with a knife or thumbnail is effective in safflower (*Carthamus tinctorius* L.) (22), and some breeders use the procedure effectively in selecting for high oil in sunflower. The appearance and texture of a coarsely ground seed sample and comparisons with standards of known oil percentage also may have value in estimating oil percentage (22, 173). Shewfelt and Putt (184) found that placing a given number of seeds between layers of filter paper, pressing them with a laboratory hydraulic press, and then measuring the weight loss of the seed estimated the oil percentage of small seed samples relatively rapidly and accurately.

Breeding for different levels of other seed quality factors is feasible as indicated by the occurrence of genetic variability for fatty acid composition (154), amino acid composition, protein and vitamin content (9, 88), and chlorogenic acid percentage (33). Developing lines or cultivars unique for these traits generally requires close cooperation with a biochemist to furnish expertise in evaluation procedures. One of the more useful techniques in breeding for altered fatty acid composition may be the single seed analyses of fatty acids by gas-liquid chromatography (GLC). Putt et al. (154) described procedures using a portion of the seed for analysis with the remainder containing the embryo saved for growing into a plant. Because fatty acid composition is assumed to be controlled largely by the genotype of the seed rather than by the genotype of the maternal parent, this procedure may allow selection of individual seeds from an F_1 plant rather than in later generations on an individual plant basis. According to Zimmerman and Fick (231), position of the seed within the head has a significant effect on fatty acid composition, and consequently selective sampling of seeds may be required. Position effects also were noted for seed oil percentage (50) and for chlorogenic and caffeic acid levels (138), thus suggesting that adequate sampling of seeds in selection studies for these traits is essential.

Breeding of nonoilseed sunflower requires evaluation of several seed quality factors not generally considered highly important in oilseed types. Included are seed size, shape, and color, ease of hull removal, kernel size and volume weight, and suitable roasting characteristics. Evaluations for seed size commonly are conducted by determining the percentage of seeds passing over 7.1, 7.9, 8.7, or 9.5 mm (18/64, 20/64, 22/64, or 24/64 in) round hole screens (114, 152). Seed color and shape are evaluated primarily by visual observations, with specific criteria depending on available markets. Ease of hull removal can be determined effectively with a small laboratory huller as described by Lofgren (114). The huller consists of a rotating disc, powered by a variable speed motor, and enclosed in a circular drum. The seeds are thrown by centrifugal force against a rubber impact ring on the inside of the drum causing relatively complete hull removal. Kernel size generally is determined by weight. Lofgren (114) describes a laboratory roasting procedure that consists of roasting 15 g samples of kernels in vegetable oil at 177 C for 4 min. Roasted kernels should be relatively light and uniform in color with a minimum of dark kernels (114).

Useful techniques and evaluation procedures in breeding for disease resistance depend on a disease environment that will differentiate between resistant and susceptible plants. Researchers with the USDA sunflower program at Fargo, North Dakota, have developed effective evaluation procedures for the diseases of rust, downy mildew, Verticillium wilt, and Sclerotinia stalk rot [*Sclerotinia sclerotiorum* (Lib.) de Bary] which generally are considered the four major diseases of sunflower in the USA. Evaluations for rust are conducted by spraying 3 to 4-week-old susceptible border row plants with a water suspension of uredospores collected from commercial fields the preceding season and stored at 2 C until used. Spraying generally is done in the evening when temperatures are lower, and plants are covered overnight with metal or plastic containers to provide a high level of humidity. Infection of the susceptible plants provides natural inoculum and subsequent spread of rust throughout the nursery, often allowing selection of resistant plants by early flowering before selfing and crossing procedures start. Greenhouse evaluations for rust are conducted on seedlings after inoculating with a 10:1 mixture of talc and uredospores.

Downy mildew evaluations are made principally in the greenhouse although field evaluations on land heavily infested with downy mildew also have been conducted (223). The greenhouse inoculations as described by Zimmer (224) are as follows: Germinated seeds with radicals 10 to 20 mm long are dehulled and placed in 50 mm petri dishes containing a 15 ml suspension of 10,000 or more zoosporangia per ml in distilled water for 18 hours in a darkened cabinet at 20 C. Following inoculation, the seeds are planted in sterile soil in 12.7 cm pots, and maintained on greenhouse benches for 14 days at 20 to 25 C. Resistance or susceptibility is determined by the occurrence of typical downy mildew symptoms, and by sporulation of the fungus on the cotyledons or leaves after the seedlings are placed in a saturated humidity chamber for 18 hours. The procedure is highly effective. Few escapes occur, and when it is conducted during the winter it allows more efficient use of labor than field evaluations made during peak labor demands of spring and summer.

Evaluations for Verticillium wilt and Sclerotinia stalk rot can be conducted effectively in field tests on land heavily infested with the disease organisms. An effective field evaluation nursery for Verticillium wilt was established at Hunter, North Dakota, which produced infections of 100% on susceptible check cultivars (49, 228). A field evaluation nursery for Sclerotinia stalk rot established near Wyndmere, North Dakota, resulted in infections of over 70% in each of three years of testing (Zimmer, unpublished data). The effectiveness of these nurseries in evaluating genotypes for Verticillium wilt and Sclerotinia stalks rot supports earlier results of Canadian workers in use of naturally infected field sites for these diseases (146, 147). Moser and Sackston (126) and others have presented methods for evaluating for Verticillium wilt by artificial inoculation of seedling plants in greenhouse tests. Several authors including Pawlowski and Hawn

(132) and Adams (2) have described methods of artificial inoculation of plants for Sclerotinia evaluation, which may involve infesting untreated soil or inserting sclerotia or mycelial infected toothpicks into stem or head areas.

Effective procedures have been developed for evaluation for resistance to several sunflower insects (19, 20; Schulz, personal communication), and broomrape (139).

GENETICS

Introduction

Estimates of genetic variation and heritability, the type of gene action, and the number of genes associated with a trait provide useful guidelines to determine the value of source populations and appropriate procedures to use in a breeding program. The following section presents this type of genetic information for a wide range of sunflower characters.

Yield and Yield Components

Much genetic variability for seed yield occurs in sunflower. Yield in sunflower, however, as in all crops, depends on many traits and varies greatly with environment. Because of the importance of environmental effects, heritability of seed yield is relatively low compared to other agronomic traits. Kloczowski (102) found that heritability in the broad sense on a single plant basis was 18% for yield as compared to a range of 22 to 49% for plant height, weight of 1,000 seeds, seed oil percentage, hull percentage of seeds, and head diameter. Shabana (182) reported a relatively high broad sense heritability of 69% for seed yield among individual plants, an estimate that was similar to those for oil percentage and 1,000-seed weight, but considerably less than the estimates for days to flowering, leaf area, number of leaves per plant, plant height, and number of seeds per head. Pathak (131) obtained a broad sense estimate for yield of 57% compared to values ranging from 20 to 81% for plant height, head diameter, stem diameter, 100-seed weight, and kernel percentage. Heritabilities for the major yield components of number of seeds per head and seed weight generally were higher than those for seed yield in these studies, suggesting that selection for increased yield through selection of yield components may have value as a breeding procedure.

Most researchers involved in development of hybrid sunflower have noted differences among inbred lines and cultivars for combining ability for seed yield although relatively few reports have been published. Unrau and White (200), Unrau (199), and Putt (153) conducted some of the earliest studies designed specifically to evaluate inbred lines in hybrid combinations. They observed marked differences among lines for combining ability. Anaschenko (5) reported on combining ability tests on more than 100 cultivars from the VIR world collection. Among the high oil Soviet cultivars, 'Armavirsky 3497,' 'Sputnik,' 'VNIIMK 8931,' 'Armaviretz,' 'Zaria,' and 'Voshod' had good general combining ability. Voskoboinik and Soldatov (204) listed 30 Soviet cultivars as having good, medium, or poor general combining ability. The value in identifying the relative combining ability of cultivars is based on the assumption that the highest frequency of inbred lines with good general combining ability can be isolated from cultivars with good general combining ability. According to Anaschenko (5) the heritability of general combining ability is high, ranging from 60 to 86%, in studies where the relative combining ability of inbred lines was compared to the combining ability of the source populations.

Putt (153) reported estimates of general and specific combining ability for eight characters including seed yield in a diallel cross of 10 inbred lines. Specific combining ability was more important than general combining ability for seed yield, suggesting that non-additive genetic variance was more important than additive in influencing yield. Assuming a relatively large, non-additive genetic component, breeding procedures that involve some form of testcross evaluation might be most effective in selection for seed yield. Conversely, if yield is determined primarily by additive genetic variance, selection procedures that utilize S_1 or S_2 progeny evaluation might prove most effective. Preliminary data from the USDA breeding program in North Dakota suggest that both S_1 and testcross evaluation may be effective means to improve seed yield. After one cycle of recurrent selection using S_1 progeny evaluation and a selection intensity of 20%, seed yield of a synthetic population formed from 12 cultivars was increased 6.9%. Following one cycle of recurrent selection in a synthetic composed of five restorer lines, using testcross evaluation and a selection intensity of 22%, 31 of 45 testcross hybrids exceeded the mean yield of the five check hybrids. The average yield of the testcross hybrids was 3.5% greater than the checks (Fick, unpublished data).

Stability of yield performance over environments also is apparently under genetic control. Fick and Zimmer (54) evaluated 18 hybrids and cultivars in 14 environments and concluded that yield stability could be attained with genetically homogeneous single cross hybrids as well as with heterogeneous cultivars or populations. Hybrids produced with the fertility restorer line RHA 274 appeared to be especially stable, suggesting that yield stability may have a relatively high heritability.

Plant Growth Characteristics

Genetic variation among characters associated with plant growth and resultant morphological or physiological differences serves as the basis for development of lines and cultivars with improved agronomic traits. Variation among characters such as plant height, flowering, and maturity is particularly useful because it allows for development of types adapted to specific environments or agro-climatic regions. Variation among other traits such as leaf or stem color, leaf shape, and leaf number is of less apparent value, although certain of these traits may be correlated with traits of direct economic importance or may be useful in basic genetic studies.

Chlorophyll Deficiencies, Color Differences, and Aberrant Seedling Types

Among early growth characteristics, deficiencies in chlorophyll content of seedling plants are perhaps the most common and easily recognized traits under genetic control. Often deficiencies in chlorophyll content are observed following several generations of inbreeding, and generally are controlled by recessive genes. Leclercq (109) reported that seedlings entirely devoid of chlorophyll, and that die in the early seedling stage, were controlled by a single recessive gene in three different cultivars. Leclercq (109) also reported chlorophyll deficient types characterized by an absence of chlorophyll in the cotyledons but which developed some green color in the first true leaves. Some seedlings die, although others apparently develop to maturity. Their occurrence in six different cultivars or source populations was controlled by a single recessive gene.

Plants that show nearly normal chlorophyll development as seedlings but become chlorophyll deficient as adults with leaves of an obvious streaked or mottled appearance have been observed in the USDA program in North Dakota. Among plants of a segregating line closely related to the fertility restorer RHA 294, 120 plants with a normal phenotype and 46 with chlorophyll deficiencies were observed. The results suggest control by a single recessive gene (P > 0.05).

Hockett and Knowles (81) observed plants in a line selected from 'Black Manchurian' having a yellow growing point, apparently due to a chlorophyll deficiency in the young, newly developed leaves but with normal color in mature leaf tissue. The yellow color was visible from the seedling to shortly before the mature plant stage. Genetic analysis indicated that a single recessive gene designated y was responsible for the yellow color. Stoenescu (194) also reported plants with a yellow growing point in the inbred line V-1327, and the author has observed such plants in lines isolated from the cultivar Chernianka 66.

Plants with leaves of a light yellow-green color throughout their growth period probably have been observed by many researchers involved in sunflower breeding. A line homozygous for this trait, PI 343765, was used extensively as a source of short stature in crosses in the USDA program in North Dakota. In one cross with a line having leaves of normal green color the F_2 segregated 42 normal and 16 light green suggesting a single recessive gene was responsible (P > 0.05).

In contrast to light green or chlorophyll deficient plants, lines with leaves of an intensively dark green color also have been observed. Luczkiewicz (119) studied a line from 'Karlik 68' in crosses and concluded that the dark green color of leaves was controlled by two complementary dominant genes. The F_2 of a cross with a normal line segregated 83 plants with normal green leaves and 54 plants with dark green leaves, results that agree with a dihybrid 9:7 ratio (P > 0.05).

Yugoslavian researchers (28, 216) found that significant genetic variation for chlorophyll content of a more or less continuous nature exists among cultivars and inbred lines, and that total chlorophyll per plant is correlated positively with seed yield. A heritability estimate of 89.8% was obtained from studies involving inbred lines and their F_1 hybrids.

The presence of anthocyanin pigment in seedling and mature plant leaves and stems is controlled primarily by a single dominant gene (30, 194, 199). The gene for anthocyanin pigment in leaves and stems was designated T by Satsperov, cited in Stoenescu (194), and apparently differs from the gene(s) responsible for anthocyanin pigment in the seeds (109, 194). Stoenescu (194) reported the same gene controlled anthocyanin pigment in lines from seven different sources of cultivated sunflower. Results of crosses with seven different populations of wild sunflower also indicated a single dominant gene for anthocyanin pigment, although crosses were not made to determine whether the same gene was present in all cases (30).

Genetic variation for early growth characters other than pigmentation also has been observed. Vranceanu and Stoenescu (207) reported seedling plants characterized by atrophy of the midrib in the first leaf pair. The trait was recessive in crosses with normal selections. The author has observed seedling plants in which the cotyledons are cupped from fusion of the cotyledonary tissue. In one such line isolated from Chernianka 66 this trait was homozygous and acted as a recessive character in crosses with normal lines. Undoubtedly numerous other aberrant seedling types have been observed, some of which could have value as markers in breeding and genetic studies.

Shape, Size, and Number of Leaves

Leaves of cultivated sunflower generally are relatively flat, cordately to subcordately shaped, conspicuously serrated, and usually hispid (79). However, wide variation occurs, some of which is controlled genetically. Luczkiewicz (119) described two abnormal leaf shapes designated as "spoon-like" leaf roll and leaf-curl, both of which were recessive and under monogenic control in crosses of plants with normal leaves. Clement and Diehl (23) also found that curled leaves were determined by a single recessive gene.

According to Gundaev (70) pronounced serration of the leaf margin is dominant over the more usual weak or fine serration. Skaloud and Kovacik (186) reported that pronounced serration was controlled primarily by a single dominant gene, although some intermediate types occurred in crosses of pronounced × finely serrated lines. Skaloud and Kovacik (186) further reported that leaves lacking a serrated leaf edge were recessive to the finely serrated types and were controlled by a single recessive gene.

Table 4—Phenotypic variability for agronomic, morphologic and seed quality characters in cultivated sunflower (Ventslavovich, cited in Gundaev, 70).

Character	Variation or range in measurements
Vegetation period, days	65–165
Plant height, cm	50–400
Stem diameter, cm	1.5–9.0
Number of primary branches	0–35
Length of primary branches, cm	5–125
Number of leaves (non-branching plants)	8–70
Length of leaf blades, cm	8–50
Head diameter, cm	6–50
Number of flowers in head, $\times 10^3$	0.1–8.0
Length of achenes, mm	6–25
Width of achenes, mm	3–13
Percentage of hull in achenes	10–60
Oil percentage of kernel	26–72
Ratio of oleic to linoleic acid in oil	1:5–4:1

Leaf size, leaf number, and total leaf area per plant may be influenced by environmental factors although significant genetic variation has been reported. Results of Ventslavovich (Table 4) indicated a range in leaf size, as measured by length, of 8 to 50 cm and a range in leaf number from 8 to 70. Crosses among lines with different numbers of leaves per plant show largely a continuous distribution in the F_2 generation, which suggests primarily quantitative inheritance (119; Fick, unpublished data). Broad sense heritability estimates as high as 94.4% have been obtained, indicating that the number of leaves per plant is highly heritable in some crosses (182).

Total leaf area per plant, which depends on both leaf size and number of leaves per plant, varies widely among genotypes. Vrebalov and co-workers (216) found nearly a twofold difference with values from 4,524 to 8,577 cm^2/plant among seven cultivars and about a sevenfold difference with values ranging from 1,566 to 11,150 cm^2/plant among 47 inbred lines. The heritability as obtained from an evaluation of inbred lines and their F_1 hybrids was 57.8%. Shabana (182) reported a broad sense heritability estimate of 88.8% for leaf area per plant and a high expected genetic advance, thus suggesting the relative importance of additive gene effects.

Petiole Characters

Luczkiewicz (119) isolated a plant from irradiated seeds of the cultivar Karlik 68 having erect petioles growing at an acute angle of less than 45° with the stem. In a cross with a normal line having petioles growing at nearly right angles to the stem, the F_1 was normal. Segregation in the F_2 closely approximated a ratio of 3 normal:1 erect, indicating erect petioles were controlled by a single recessive gene. Luczkiewicz also described plants from Karlik 68 having thick, short petioles. In a cross with a normal line the F_1 was normal. A segregation ratio of 9 normal to 7 mutant types appeared in F_2 indicating that thick, short petioles were controlled by two complementary genes.

Branching

Wide variation for the type of branching, length, and number of branches occurs in sunflower. The inheritance of branching is complex although genes with major effects have been identified. Putt (143) identified a single dominant gene, which he designated *Br,* for the occurrence of branching over the entire length of stem. Putt (151) later reported that, in contrast to most types of branching, the profuse branching in the line 953-88-3 was recessive to non-branching and monogenically inherited. He designated the gene as *b*. Hockett and Knowles (81) identified four genes for branching in a study that classified branching types as follows: 0—no branching, 1—basally branched, 2—top branched, 3—fully branched with a central head, 4—wild types or fully branched without a central head. The four genes identified were the dominant gene Br_2 for top branching, the duplicate dominant genes Br_2 and Br_3 from the wild type parent, and the recessive genes b_2 and b_3 which gave a fully branched phenotype only when homozygous for both genes. The authors suggested that the Br_2 gene may be the same as the *Br* gene reported by Putt, whereas b_2 and b_3 probably are different from the *b* gene.

Plant Height

Plant height is reported to vary within a range of 0.5 to 4 m (Table 4), although gigantic types as tall as 7 and even 12 m were reported by Dodonaeus cited in Cockerell (26). Crosses of plants of different heights usually produce plants as tall or taller than the tall parent in the F_1 (153, 199), and an almost continuous distribution of heights in F_2 (70). Pathak (131), Kloczowski (102), and Shabana (182) reported broad sense heritability estimates of 20, 49 and 90%, respectively. Unpublished results of the author showed that broad sense heritability estimates ranged from 4.1 to 84.9% with an average of 60.5% in 9 crosses involving inbred lines with dwarf and normal phenotypes. Narrow sense heritability estimates ranged from 20.4 to 37.5% with an average of 29.6% for three crosses that showed broad sense heritabilities of over 80%. These results suggest that non-additive gene effects are relatively more important than additive effects in the control of plant height. Previous information presented by Putt (153) indicated that specific combining ability was more important than general combining ability for plant height, results that also suggest the importance of non-additive gene effects.

Although plant height generally has been regarded as a quantitative character, single gene control of height has been reported. Enns (39) found that the dwarf character of the inbred line 77AB was controlled by a single recessive gene. The author also has observed segregation indicating that a single recessive gene controls expression of dwarf plants in a cross of RHA 273 and a line isolated from the Romanian hybrid HS 90.

Flowering and Maturity

Limited genetic information is available on flowering and maturity of sunflower. Unrau (199) and Putt (153) reported that with few exceptions the flowering dates of F_1 hybrids were as early or earlier than either parent, thus suggesting that early flowering was dominant over late flowering. Several hybrids also had flowering dates intermediate to the two parents, results indicating that the type of gene action associated with date of flowering is dependent on the genotypes being studied. Heritability of flowering is relatively high as indicated by broad sense estimates of over 90% (182), and by correlation coefficients of 0.86 and 0.91 between days to flowering of inbred lines and their topcross hybrids (173).

In some genotypes the inheritance of days to flowering may be relatively simple. An inbred line isolated by the author from the cultivar 'Volgar' segregated for plants that flowered around modes of either 48 or 66 days after planting. In 2 years of testing this line produced 16 early plants and 52 late plants, results which indicate a single recessive gene controlled early flower (P > 0.05).

Sunflower plants generally are considered physiologically mature when the backs of the heads are yellow with the involucral bracts turning brown. Days to maturity often are correlated closely with days to flowering, although genetic differences in the time required from flowering to maturity exist. Thus, the genetic information relative to flowering date also may apply to maturity date or in some genotypes may be independent of maturity date. According to Gundaev (70) the F_1 hybrids are intermediate when early maturing genotypes are crossed with late maturing types. Putt (153) found that F_1 hybrids often were earlier than the mid-parent value in days to maturity although he also observed intermediate hybrids. The estimated components for general combining ability were greater than for specific combining ability, which suggested that additive gene effects were relatively important in control of maturity date.

Both flowering and maturity date may be influenced by photoperiod. Some cultivars or genotypes have a short day response to photoperiod, whereas others may have a neutral or long day response (36, 167, 179). Most of the Soviet cultivars apparently are heterogeneous for this character, although long day cultivars have been developed (35, 179). Experience of the author with inbred lines, hybrids, and segregating populations grown in summer nurseries in North Dakota and winter nurseries in Florida suggests the inheritance of photoperiodic response may be relatively simple.

Head Size, Shape, and Inclination

Head size is influenced greatly by environmental effects, especially by plant population, soil moisture, and soil fertility. Thus, the portion of the total variation in head size attributable to genetic effects often is less than it is for certain other agronomic traits. Kloczowski (102) and Pathak (131) reported broad sense heritability estimates of 22 and 44%, respectively. Russell (173) found a positive correlation coefficient of 0.40 between head size of inbred lines and their topcross hybrids. The values obtained by Kloczowski and Russell, in particular, were generally lower than those for plant height, days to flowering, and seed weight. In crosses among inbred lines, significant heterosis for head size occurred with both additive and non-additive genetic effects considered to be of importance (153).

Head shape varies from strongly convex to strongly concave. Experience of the author suggests that head shape often is intermediate in crosses of different types. Severe distortion of heads, characterized by a grotesque appearance, severe faciation, and convex shape, occurred at Fargo, North Dakota, in an inbred line RM 3319 from Romania. Limited data suggested that a single recessive gene was responsible.

In one inbred line isolated from VNIIMK 8931, plants without heads have been observed. Data collected by the author in two years at Fargo, North Dakota, showed 31 normal plants and 9 plants without heads. The results suggest control by a single recessive gene.

Head inclination, which may be important in harvesting and depredation from birds, depends on the strength of the upper stem. According to Gundaev (70) the erect position of the head is dominant over bending head positions in crosses of different types.

Lodging

Genetic control of lodging probably is complex because it depends on several characters including plant height, stem strength and diameter, and the type of root system. Russell (173) found that percent lodged plants of inbred lines was correlated significantly ($r = +0.51$) with lodging in their topcross hybrids, thus suggesting a medium heritability. According to Vranceanu and Stoenescu (212) resistance to lodging in the F_1 generation usually was intermediate to the lodging of the parents.

Flower Characteristics

Self Incompatibility and Self Fertility

Self incompatibility in sunflower is common, and in some populations or cultivars is nearly complete (18, 79). The self incompatibility is apparently sporophytic in nature, although there is not complete S-allele dominance (71, 87).

In contrast to genotypes that exhibit nearly complete self incompatibility, genotypes that are nearly 100% self fertile also can be recovered by inbreeding within some populations. Luciano et al. (118) conducted perhaps the most extensive study on the inheritance of self incompatibility. They reported broad sense heritability estimates of 84 and 43% in the F_2 generation of two crosses, thus indicating that a significant portion of the total variation was due to genetic causes. Narrow sense heritability estimates were generally much lower than the broad sense estimates, indicating that dominance, epistasis, or both, were of major importance in the segregating progenies of these crosses.

Kloczowski (101) reported a parent progeny estimate of heritability of 38.5% in a study involving inbred lines showing wide variation for self fertility. The results of these investigations suggest that the heritability of self incompatibility, or conversely of self fertility, may be of an intermediate level.

Genetic Male Sterility and Partial Male Sterility

Several different types of genetic male sterility in sunflower have been described (156). Some male sterile types produce no pollen, or pollen grains of uneven or smaller than normal size which do not stain and are sometimes clumped. Other types, referred to as partial or psuedo-male steriles, have varying proportions of normal pollen. According to Vulpe (218) the frequency with which male sterility occurred was about 0.04% following one generation of selfing within 14 cultivars of diverse type.

Genetic male sterility usually is controlled by a single recessive gene (108, 109, 156, 205), although control by duplicate recessive genes (14, 209) and duplicate genes with epistatic effects (156) also has been reported. On the basis of diallel crosses Vranceanu (206) identified at least five different genes for male sterility, designated ms_1 through ms_5.

Putt and Heiser (156) found that partial male sterility was recessive in crosses with normal types, and in at least one cross appeared to be polygenically inherited.

Fertility Restoration of Cytoplasmic Male Sterility

Genes for fertility restoration of male sterile cytoplasm occur infrequently among cultivated sunflower, but are common among wild sunflower (31, 214). The first reports on fertility restoration by Kinman (98), Enns et al. (40), Leclercq (111), and Vranceanu and Stoenescu (214) indicated genetic control by a single dominant gene. Later reports by Fick and Zimmer (48) and Dominguez-Gimenez and Fick (31) suggested that in some crosses the inheritance of fertility restoration was more complex. Limited data from crosses of nonoilseed sunflower suggested that two independent complementary dominant genes were required for restoration (48). Segregation ratios of male-fertile to male-sterile plants in the BC_1 and F_2 generations of seven crosses between the cytoplasmic male sterile line cmsHA 89 and wild sunflower populations (Table 5) suggested that four genes were involved, and that one gene was present in the female parent. Fertility was re-

Table 5—The observed male fertile and male sterile plants in the BC₁ and F₂ generations of crosses between cmsHA 89 and wild sunflower (*Helianthus annuus*) populations, and the goodness of fit of observed to expected ratios (31).

Cross	Generation	No. of plants		Expected ratio	Probability range
		Fertile	Sterile		
					%
1	BC₁	43	51	1:1	30–50
	F₂	30	10	3:1	95–100
2	BC₁	45	48	1:1	80–90
	F₂	74	15	3:1	5–10
3	BC₁	47	49	1:1	80–90
	F₂†	--	--	3:1	--
4	BC₁	38	56	1:1	5–10
	F₂	54	24	3:1	10–20
5	BC₁	66	28	3:1	30–50
	F₂	129	19	54:10	30–50
6	BC₁	68	17	3:1	30–50
	F₂	127	19	54:10	30–50
7	BC₁	75	14	7:1	30–50
	F₂†	--	--	243:13	--

† F₂ data not available.

stored when at least two dominant alleles were present at any of the four loci. Segregation occurred for one gene in four crosses, three genes in two crosses, and possibly four genes in one cross (31).

Sometimes partial restoration of fertility is observed in segregating populations of crosses of cytoplasmic male sterile and fertility restorer lines, suggesting the presence of modifying genes with an influence on fertility restoration (31). Observations by the author suggest that the inheritance of partial restoration may be complex, and highly dependent on environmental conditions.

Flower Color and Morphology

Ray flower petals of sunflower are usually yellow but may vary from red to deep orange to very pale yellow (lemon) or near white. Studies on the inheritance of flower color indicate that most differences are controlled by a few genes with major effects. Cockerell (25) suggested that a single dominant gene controlled red flower color. Fick (44) concluded that two independent complementary dominant genes controlled red flower color, but that orange, lemon, and some yellow-flowered lines may possess one of the genes for red color.

Leclercq (109) reported that yellow flower color was dominant to orange, and that segregation in F₂ and backcross populations showed monogenic control. Skaloud and Kovacik (186) concluded that lemon was inherited as a monogenic recessive trait in crosses of orange × lemon and yellow × lemon. Crosses by Fick (44) among yellow, orange, and lemon lines indicated that yellow was dominant to orange, lemon showed recessive epistasis to yellow and orange, and that two genes were involved. The interaction of these two genes in crosses of yellow × lemon produced F₂ segregation ratios of 9 yellow:3 orange:4 lemon. In crosses of yellow × orange and orange × lemon segregation in F_2 for one gene occurred.

Color variations due to the presence or absence of anthocyanin pigment also occur among the different parts of disc flowers. According to several reports cited in Stoenescu (194) and Fick (44), the occurrence of anthocyanin pigment in disc flowers is controlled by a single dominant gene. This gene is apparently the same gene that causes anthocyanin pigment in seedling leaves and other vegetative plant parts. In disc flowers the anthocyanin pigment is visible primarily on the calyx, the top of the tubular corolla, and the entire length of the stigma.

The stigma of the disc flowers also may show various intensities of anthocyanin pigment even though none appears in other reproductive or vegetative plant parts. Luczkiewicz (119) reported that the presence of anthocyanin pigment in the stigma was dominant over its absence, and postulated that three independent genes with cumulative effects controlled the variation in stigma color.

The anther column of the disc flowers may range in color from yellow to brown to different shades of black. Stoenescu (194) indicated that the yellow color was inherited as a single recessive gene in a cross of a line with the usual black anthers. Luczkiewicz (119) suggested that as many as four genes were responsible for the various shades of brown and black. The normal yellow color of the pollen from the anthers is controlled by a single dominant gene in crosses of lines with light yellow or near-white pollen (194).

Several morphological variations in flower type also appear to be inherited simply. Luczkiewicz (119) reported that the length of ligulate ray flower petals was controlled by two independent complementary dominant genes with the short petals showing recessive epistasis to the normal long petals. He also suggested that two complementary dominant genes produced types with a reduced number of ligulate flowers. Fick (44) identified plants from three different sources that had tubular ray flower petals, two types that had very short tubular flowers, and one that had tubular flowers of normal length. Each was controlled by a single recessive gene. Crosses of the "chrysanthemum" type of sunflower, in which the disc flowers are entirely ligulate, showed that this trait is dominant in the F_1. Luczkiewicz (119) indicated that a single dominant gene was involved, while Fick (44) suggested that a minimum of two genes controlled the "chrysanthemum" type.

Cirnu et al. (21) reported significant genetic variation in the length and diameter of the tubular corolla of disc flowers, a feature that may have an effect on attractiveness or accessability of nectar to insect pollinators.

Nectar Production

Significant genetic variation occurs in the nectar produced by different sunflower cultivars, a factor believed to be important in visitation by insect pollinators and resultant seed yield (70). Free (60) in his review of the relevant literature found threefold differences reported in nectar production among cultivars and that the average sugar production per flower of various cultivars ranged from 0.4 to 0.6 mg. He also found the sugar concentration of the nectar normally is about 50% but that genetic variation for sugar percentage may occur.

Furgala et al. (61) detected significant variation in quantity and quality of nectar from pistillate florets of over 70 lines and cultivars. Nonoilseed cultivars, genetic and cytoplasmic male sterile lines, and certain fertility restorer lines generally secreted less nectar with lower solids content than oilseed cultivars of recent Soviet origin. The F_1 hybrids from crosses of cytoplasmic male sterile and fertility restorer lines produced high levels of nectar, exceeding that of either parent, which suggests that dominant and complementary gene effects may be important in controlling this trait.

Seed and Kernel Characteristics

Oil Percentage

Oil percentage of whole sunflower seeds depends on both the percentage of hull and the percentage of oil in the kernel. Hull percentage among genotypes may vary from 10 to 60% while oil percentage in the kernel may vary from 26 to 72% (Table 4). Hull percentage of seeds and oil percentage in the kernels of present high oil cultivars or hybrids is in the range of 20 to 25 and 57 to 67%, respectively. Gundaev (70) indicated that about two-thirds of the increase in seed oil percentage from past breeding and selection resulted from a reduction in hull percentage and about one-third from an increase in kernel oil percentage.

Oil percentage generally is considered to be quantitatively inherited (70, 206). Broad sense heritability estimates ranging from 27 and 32% for hull percentage (102, 131) to 32% for oil percentage of kernels (102) to 65 and 72% for oil percentage of seeds (43, 182) have been reported. Narrow sense heritability estimates reported by Nikolic-Vig et al. (128) for two populations were 20 and 37% for hull percentage and 57 and 75% for oil percentage of seeds. Fick (43) reported narrow sense estimates of 52 and 61% for oil percentage of seeds. The wide variation that occurs among these heritability estimates probably is due to differences in the germplasm studied, environmental effects, and the methods of estimation. The estimates as a whole suggest an intermediate level of heritability for oil percentage with the heritability of oil percentage of whole seeds perhaps slightly higher than heritability of oil percentage of kernels or of hull percentage.

Putt (153) found that general combining ability was greater than specific combining ability for oil percentage of seeds, and suggested that additive gene effects were more important than non-additive in control of oil percentage. Fick (43) indicated that the ratio of narrow to broad sense heritability was high, thus also suggesting that genetic effects for oil percentage of seeds were largely additive. According to Vranceanu and

Stoenescu (208) hull percentage is controlled by genes with largely additive effects, whereas oil percentage of the kernel is controlled primarily by genes with dominant or heterotic effects. Heterosis for oil percentage of the kernel was also observed by Pustovoit according to Gundaev (70).

Fatty Acid Composition

Sunflower oil is comprised primarily of palmitic, stearic, oleic, and linoleic acids, with oleic and linoleic accounting for about 90% of the total (27, 99, 163). There is an inverse relationship between oleic and linoleic acid which is highly influenced by environment, especially temperature during the growing season (17, 99).

Putt et al. (154) were among the first to show conclusively that significant genetic variation exists for fatty acid composition. Among 56 inbred lines grown in one environment they found the following ranges in composition: palmitic 4.7 to 8.2%; stearic 1.7 to 9.1%; oleic 13.9 to 40.3%; and linoleic 47.9 to 76.4%. These results are typical of the variation reported in cultivated sunflower by several other investigators (27, 88, 90) although wider variation may occur among the wild species (42, 104). Soldatov (189) recently identified genotypes from mutagenically altered populations with oleic acid percentages as high as 90%.

To date single genes with a specific influence on fatty acid composition, such as occur in safflower (103) and rapeseed (*Brassica campestris* L.) (34), have not been identified in sunflower. The type of gene action associated with differences in fatty acid composition appears to depend on the genotypes being evaluated. Hristova and Georgieva-Todorova (86) found that high linoleic acid percentage was a dominant trait in some crosses, whereas in others it was either recessive or near the mid-point value. Similar results were obtained by Ermakov and Popova (41).

Protein Percentage

The protein percentage of cultivated sunflower seeds varies from 9 to 24% (37, 70) and of kernels from about 24 to 40% (16, 88), although large numbers of different cultivars or lines were not analyzed. Protein percentages of the wild species generally are higher than the cultivated types (64, 142). The protein percentage in the F_1 of crosses of different types is intermediate or tending toward the low protein parent (196).

Several authors (88, 166, 190) have reported significant cultivar differences in amino acid composition of the protein. Among several amino acids essential for human nutrition Ivanov (88) reported a range of values of 1.88 to 4.00% for lysine, 3.42 to 6.25% for methionine, and 0.79 to 1.32% for tryptophan in inbred lines selected from different cultivars. In a later study Ivanov (89) identified several lines with lysine content as high as 5.20% of the protein. Little is known about the inheritance of amino acid composition in crosses, although selection for types with altered amino acid composition has been effective (89).

Table 6—Means, variances, and estimates of heritability for seed and kernel weight in two crosses of sunflower (Fick, unpublished data).

Population	200-seed wt		25-kernel wt	
	X	s^2	X	s^2
		g		
cmsHA 286 × RHA 295				
cmsHA 286 (P$_1$)	26.2	3.70	3.28	0.46
RHA 295 (P$_2$)	20.5	2.42	2.56	0.30
P$_1$ × P$_2$ (F$_1$)	24.9	2.47	3.12	0.31
P$_1$ × P$_2$ (F$_2$)	21.5	4.09	2.68	0.51
Heritability (broad sense)		69.9		51.0
Heritability (narrow sense)†		68.9		65.1
cmsHA 288 × RHA 280				
cmsHA 288 (P$_1$)	24.9	2.45	3.11	0.31
RHA 280 (P$_2$)	19.8	1.72	2.47	0.21
P$_1$ × P$_2$ (F$_1$)	30.8	4.36	3.85	0.55
P$_1$ × P$_2$ (F$_2$)	24.8	4.76	3.10	0.59
Heritability (broad sense)		59.7		61.0
Heritability (narrow sense)†		37.5		25.4

† Heritability in narrow sense estimated as correlation between F$_3$ and F$_4$ lines.

Other Chemical Components

Significant genetic variation for chemical constituents aside from the lipid and protein fractions of sunflower kernels also has been reported. These include tocopherol content of the oil (107), chlorogenic, caffeic, and quinic acid levels (12, 138), sugar content (12), beta carotene content (12), nicotinic acid, thiamine, riboflavine and biotine vitamins (9), and several mineral elements (166). Genetic variation for chlorogenic acid, a phenolic compound that causes discoloration of sunflower meal, is of particular interest. Dorrell (33) found a range of 1.12 to 4.50% in chlorogenic acid percentage of meal among 387 inbred lines of diverse origin. Breeding to reduce chlorogenic acid levels is in progress.

Seed and Kernel Weight

Variation in seed size occurs largely within a range of 6 to 25 mm in length and 3 to 13 mm in width (Table 4). Seeds of most oilseed cultivars grown currently weigh 40 to 100 g/1,000 whereas seeds of nonoilseed types often weigh more than 100 g/1,000.

Heritability of seed weight among oilseed types is of an intermediate level as indicated by broad sense estimates ranging from 30 to 66% (102, 131, 182). Putt (153) found that the components for general and specific combining ability for weight per 1,000 seeds were about equal, suggesting that both additive and non-additive gene effects were important in the control of seed weight.

Heritability estimates for seed weight among nonoilseed types obtained by Fick (unpublished data) were also of an intermediate level (Table 6). The narrow sense estimate for the cross cmsHA 286 × RHA 295 suggested that the additive portion of the total genetic variance was relatively high and that selection for seed weight should be effective. In the cross cmsHA 288 × RHA 280 the narrow sense estimate was low compared to the broad sense estimate and suggested non-additive genetic effects were important. These results suggest that the type of gene action controlling seed weight depends upon the genotypes being evaluated. The data for kernel weight in these crosses (Table 6) and the interpretation of results are similar to those for seed weight.

Seed Color

The color of sunflower seeds is determined by the presence or absence of pigment in each of three different seed coat layers. Each layer may develop pigment independently of the other layers (143). The outer layer or epidermis may be free of pigment, fully pigmented dark brown or black, or possess striped patterns of dark brown or black. The second or corky layer may either lack pigment or contain anthocyanin so intense as to mask the pigments in the other layers. The third or innermost layer, often referred to as the armor layer, is either non-pigmented or black.

The inheritance of seed color is considered relatively complex because of the numerous combinations of pigment that may occur in the different layers. Several genes with major effects, however, have been identified. Satsperov and Ananieva, cited in Stoenescu (194), and Putt (143) reported that the black pigment in the armor layer was controlled primarily by a single dominant gene. Leclercq (109) found that anthocyanin of seeds was due to a single dominant gene *Tf* which differed from the *T* gene identified by Satsperov for anthocyanin coloration of the entire plant. According to Velkov, cited in Stoenescu (194), gene *T* was responsible for anthocyanin in vegetative plant parts, and a second complementary gene controlled anthocyanin in the seeds.

White color of the seeds results from a lack of pigment in all three layers and is dominant over dark brown or black color in the epidermis according to unpublished observations of the author and Putt (personal communication). Striping of the seed, caused by uneven pigmentation in the epidermis, is dominant over solid colored pigmentation (70; Fick, unpublished data). When the black coloration of the armor layer is present, the background color of both white and striped seeds appears as various shades of gray.

Disease Resistance

Over 30 diseases are known to attack cultivated sunflower, most are incited by soil or windborne fungi (226). Four of these diseases, rust, downy mildew, Verticillium wilt, and Sclerotinia stalk and head rot, are considered of major importance in the USA and Canada. Phoma black stem, Alternaria leaf and stem spot, Septoria leaf spot (*Septoria helianthi* Ell. at Kell.), Rhizopus head rot, charcoal stem rot (*Sclerotinia bataticola* Taub.), and powdery mildew (*Erysiphe cichoracearum* DC) may cause significant losses in some years. Resistance or tolerance to most of these diseases has been identified, and development of disease resistant cultivars appears to be the most feasible means of control.

Rust

Resistance to rust occurs commonly, especially among wild species of *Helianthus* (80, 229). In some of the earliest studies Putt and Sackston (157, 159) identified rust resistance among crosses of wild annual sunflower from Renner, Texas, and successfully incorporated this resistance into commercial cultivars. Two dominant genes, R_1 from the source 953-102 and R_2 from 953-88, were identified and have been used extensively by breeders throughout the world. A third gene R_3, non-allelic to R_1 and R_2, was reported later by Canadian workers (Research Report 1967. Research Station, Morden, Manitoba).

Many different races of rust occur (84, 121) and several recent studies suggest the existence of many different genes for resistance (122, 225). Although resistance often is controlled by single dominant genes, observations by the author suggest that genes occur with additive or intermediate effects and Soviet investigators have published several reports on resistance controlled by recessive genes (137, 157).

Downy Mildew

Resistance to downy mildew from wild species has been incorporated recently into cultivated sunflower. Resistance among the wild species is most common among the perennial types (55, 140). Sources of resistance used extensively to develop resistant lines or cultivars include AD 66 from Romania (213), HA 61 from the USA (227), HIR 34 from France (202), and the *H. tuberosus* × VNIIMK 8931 populations from the USSR (140). Pedigrees of this germplasm indicate that the resistance of AD 66 and HA 61 traces to wild *H. annuus* of Texas, whereas resistance of HIR 34 and the VNIIMK 8931 crosses is attributable to *H. tuberosus*.

Resistance to downy mildew is controlled by a single dominant gene and as many as four different genes have been identified (201). These genes were designated *Pl*, Pl_2, Pl_3 and Pl_4 (201, 210, 227). Only the Pl_2 gene provides effective resistance to the race of downy mildew that occurs in North America (224). Field tolerance to downy mildew, suggestive of additional genes contributing partial resistance, also has been observed in sunflower (66, 201).

Verticillium Wilt

Resistance to Verticillium wilt occurs among cultivated sunflower (147, 228) and commonly among wild species (85). Resistance may be controlled by a single dominant gene (49, 150) or by several genes that show dominance for susceptibility, a lack of dominance, or heterosis for resistance (85, 150).

Sources of resistance used commonly in breeding programs in the USA are HA 89, HA 124, and P-21 VR1 which were isolated from VNIIMK 8931, VNIIMK 8883, and Peredovik, respectively. Each of these inbred lines has a single dominant gene for resistance (49).

Sclerotinia Stalk and Head Rot

Variation in susceptibility of lines or cultivars to Sclerotinia is observed in sunflower, although complete resistance or specific genes for resistance have not been identified. Putt (146) reported significant differences in susceptibility among 44 cultivars and inbred lines in the first year of testing, and among the 10 entries with lowest infection and the 10 with highest infection in 3 subsequent years of evaluation. Although inconsistencies in the infection of various entries occurred from year to year, the 4-year average ratings showed that 9 of the 10 entries with lowest infection were among the 10 entries selected originally. Luka (120) also reported significant variation among lines and cultivars and cited several earlier investigations in which differences in susceptibility were detected. Among cultivars grown commercially, the French hybrid INRA 4701 and the USDA hybrid 896 have been reported to have some tolerance (120, 169).

Pustovoit and coworkers have conducted extensive investigations on resistance to Sclerotinia among selections from interspecific crosses of *H. tuberosus* × VNIIMK 8931. They recently reported the development of a population with resistance ranging from 70 to 90% against nine different races of Sclerotinia (141). This resistance is primarily against infection in the head and upper stem, the type of infection that occurs widely in some European countries. In the USA and Canada the predominant disease phase of Sclerotinia is basal stem infection. Although some resistance to infection of the upper stem and head was suggested among the wild species evaluated by Zimmer and Fick (226), little if any resistance was observed for basal stem infection.

It can be concluded from the preceding investigations that the differences among genotypes are heritable, and that selection for populations with low levels of infection can be effective. In a recent study by Hoes and Huang (83), 3 years of testing and of phenotypic recurrent selection reduced infection from Sclerotinia in the cultivar Peredovik to about one-half that of the original population.

Other Diseases

Resistance or differences in susceptibility among genotypes have been reported for other diseases, although in most cases information on the genetics or inheritance of resistance is not as well defined as for rust, downy mildew, or Verticillium wilt. Differences among species or cultivars have been reported for reaction to Phoma black stem (226), Alternaria leaf and stem spot (127, 226), Septoria leaf spot (Fick and Zimmer, unpublished data), charcoal stem rot (127, 141), powdery mildew (185), Phytophthora black stem rot (*Phytophthora* sp.) (8), Fusarium wilt (129), Botrytis (68), white rust [*Albugo tragopogi* (Pers.) Schroet] (P. L. Thompson, Pacific Seeds, Toowoomba, Queensland, Australia, personal communication), Phialophora yellows (*Phiolophora asteris* f. sp. helianthi) (82), and the virus disease aster yellows (158).

The number of genes involved and type of gene action controlling resistance have been determined for only a few of the aforementioned diseases. Hoes and Enns (82) suggested that two genes controlled resistance to Phialophora yellows, one of which was dominant. Fick and Zimmer (unpublished data) found that a single dominant gene controlled resistance to Septoria leaf spot in two crosses of resistant × susceptible lines. Putt and Sackston (158) reported that resistance to aster yellows was genetically dominant.

Insect Resistance

Over 100 species of insects have been reported on sunflower in North America, although most are not of economic importance (1, 220). Ten to 12 species are of current concern in the major production areas of Canada and the northern U.S., several of which have caused moderate to severe yield reductions in recent years (24). These include the sunflower moth (*Homoeosoma electellum* Hulst), the banded sunflower moth (*Phalonia hospes* Walsingham), the sunflower midge (*Contarinia schulzi* Gagne), the sunflower bettle (*Zygogramma exclamationis* Fab.), and several species of weevils of which the adults, larvae or both may cause injury to roots, stems, heads, or seeds. Unlike resistance to the major diseases, cultivar resistance to the major insect pests is not generally available, although in some cases field tolerance or differences in susceptibility have been reported.

Head or Seed Damaging Insects

Perhaps the best example of resistance to an insect pest in sunflower is the resistance to the sunflower moth (*H. nebulella* Hb.) occurring in the

USSR and Eastern Europe. Resistance or susceptibility to this moth is determined by the presence or absence of a dominant, darkly colored layer in the seed coat, often referred to as the "armor" layer or "phytomelan" layer (70). The armor layer prevents the larvae of the moth from penetrating the hull, thus affording protection from injury. The presence of the armor layer is controlled primarily by a single dominant gene (70, 95, 144).

Carlson et al. (19) and Carlson and Witt (20) found the armor layer was at least partially effective in reducing damage from the sunflower moth (*H. electellum*) occurring in North America, although severe infestations of the moth have caused seed yield losses of over 40% among cultivars possessing the armor layer (197). Assuming the armor layer offers some protection, apparently lines or cultivars with a greater degree of tolerance could be developed. Perestova, cited in Gundaev (70), found that thickness and uniformity of the armor layer varied widely among genotypes and concluded that cultivars with an armor layer 2 to 2.5 times stronger than that of current cultivars could be developed. In addition to the armor layer, Carlson and Witt (20) suggested that chemical factors also were involved in resistance and that differences among genotypes were present. Observations by Jarvis (93) indicated that a "chrysanthemum" type of sunflower (PI 204578) offered resistance to the sunflower moth.

Differences among lines or cultivars in susceptibility to damage from the sunflower midge have been observed at Fargo, North Dakota (Fick, unpublished data). Two highly susceptible lines HA 234 and HA 285 were crossed with tolerant lines RHA 273 and RHA 282 to investigate possible mechanisms of susceptibility. Preliminary results showed at least a partial relationship between early flowering and susceptibility, thus suggesting that some of the apparent tolerance may be due to escape rather than genetic resistance per se. Schulz (174) suggested that commercial fields of oilseed sunflower were damaged less than fields of nonoilseed sunflower, and that larva numbers generally were lower in the heads of oilseed cultivars.

Schulz and coworkers (J. T. Schulz. 1971–1976. Annual reports, Department of Entomology, North Dakota State University, Fargo) suggested possible differences in susceptibility in field trials among different lines and cultivars for damage caused by the banded sunflower moth, a head maggot (*Gymnocarena diffusa* Snow), a "head clipping" weevil (*Haplorhynchites aeneus* Boh.), and the seed weevils (*Smicronyx fulvus* LeC. and *S. sordidus* LeC.). The mechanisms influencing susceptibility to these insects were not determined.

Foliage Feeding Insects

Zimmer and Fick (unpublished data) observed differences in plant injury caused by the sunflower beetle among over 50 different lines and cultivars in a nursery near Hunter, North Dakota, in 1973. Although some of the differences appeared to be due to stage of growth, this relationship was not complete. Several accessions from an interspecific cross involving *H.*

tuberosus appeared to be especially tolerant. A promising method for control of the sunflower beetle may be the development of cultivars with the dense leaf and stem pubescence that occurs in the wild species *Helianthus argophyllus*. In a preliminary study conducted at Fargo, North Dakota, in 1976 involving plants from a cross of cultivated sunflower × *H. argophyllus*, oviposition, viability of eggs, and larval feeding all appeared to be affected adversely by density of pubescence. Although the genetics of pubescence have not yet been determined, observations to date suggest that the density and length of pubescence are intermediate in the F_1 and that the trait may be inherited relatively simply.

Ramakrishnan et al. (160) observed differences in susceptibility to the grasshopper (*Atractomorpha crenulata* F.). Among 102 cultivars evaluated, they classified 10 as highly susceptible, 50 as moderately susceptible, and 42 as less susceptible. Pustovoit and Gubin (141) reported resistance to aphids (*Brachycaudus helichrysi* Kalt.) among several wild species. The author observed several plants from the *H. tuberosus* introduction PI 357397 that were resistant to greenhouse whiteflies.

Stem Infesting Insects

Schulz et al. (J. T. Schulz. 1971–1976. Annual reports, Department of Entomology, North Dakota State University, Fargo) and Schulz, cited in Cobia and Zimmer (24), evaluated lines and cultivars for resistance to the sunflower budworm (*Suleima helianthana* Riley) and the sunflower maggot (*Strauzia longipennis* Wied.) and suggested differences in susceptibility. The mechanisms influencing susceptibility were not determined, although damage from the sunflower maggot apparently was not influenced by stalk diameter.

Root Feeding Insects.

Bottrell et al. (10) found eight cultivars of sunflower equally susceptible to damage from the carrot beetle (*Bothynus gibbosus* DeGeer). Rogers and Howell (168) reported that carrot beetles damaged wild annual species of sunflower heavily but that perennial species tended either to tolerate injury or possess nonpreference qualities that prevented beetle attacks. Their data suggested a genetic basis for the apparent resistance.

Oseto (130) evaluated sunflower cultivars and lines for tolerance to two species of weevils (*Cylindrocopturus adspersus* LeC. and *Baris strenua* LeC.) that attack the roots and lower stem area. No differences in susceptibility were recorded. Observations by the author in a commercial hybrid seed production field near Sheldon, North Dakota, in 1973 suggested that the parent line RHA 266 was more tolerant to damage from *Baris strenua* than the parent cmsHA 89.

Broomrape Resistance

Broomrape (*Orobanche cumana* Wallr.) is a parasitic plant that feeds on sunflower roots, causing severe economic damage in parts of the USSR and Eastern Europe. Cultivars resistant to broomrape were developed as early as the 1920's although numerous local races and evolution of new virulent races makes breeding for resistance a continuous process (13, 139). Resistance to individual races may be controlled by a single, dominant gene (15, 70), or may be more complex with at least two or more genes influencing resistance (70). Resistance is common among cultivars grown currently in the USSR (139) and also is present among selections from interspecific crosses with *H. tuberosus* (141).

Tolerance to Bird Depredation

Resistance to bird depredation has not been identified in sunflower, although field observations suggest that differences in head inclination, head shape, seed size, and ease of seed removal may be important agronomic traits influencing the degree of bird damage. Harada (75) found that seed losses from sunflower heads were directly proportional to head angle, with heads having the seed surface toward the soil surface at or near an angle of 180° showing the least damage. Some breeders believe that types with a strongly concave head type, such as exhibited by the parent line RHA 271 and some of its hybrids, offer significant protection from birds. Robinson (164, personal communication) observed that nonoilseed types suffered less bird damage than oilseed types in yield trials, possibly because of greater difficulties encountered by the birds in removing the seed from the head and also in removing the hull from the kernel. Recent observations by Harada (76) suggested that certain oilseed types were tolerant to bird damage showing less than 10% seed loss when other types of similar agronomic type suffered nearly 100% loss. Possible chemical differences causing distastefulness to birds were suggested as one explanation for the apparent tolerance.

Tolerance to Adverse Environments

Differences among sunflower lines or cultivars have been reported for tolerance to drought (134, 135), salt or alkaline soils (97), germination at cool temperatures (65), and frost (143). These traits probably are influenced by many different morphological and physiological characteristics, and thus the genetic control and inheritance of tolerance may be complex. In one instance, however, Putt (143) found that tolerance to light fall frosts was controlled by a single dominant gene.

Gene Linkages

Only a few linkage relationships among genes controlling qualitative traits have been reported in sunflower. Hockett and Knowles (81) found that the Y gene for green versus yellow color of the growing point was linked to the Br_3 gene for branching with 11.6 ± 1.0% recombination. Leclercq (108) reported a close linkage between the T gene for anthocyanin and the ms_1 gene for genetic male sterility. This linkage served as the basis for hybrid seed production in France and Romania prior to the discovery of cytoplasmic male sterility. The crossover percentage was estimated at 0.3 to 0.7% (206) and 1.3 ± 0.2% (195). Vranceanu and Stoenescu (210) postulated that the R_1 gene for resistance to rust was linked to the Pl gene for resistance to downy mildew, although conclusive linkage tests were not conducted. Fick and Zimmer (53) found a close repulsion linkage, or possible allelic relationship, of the Pl_2 gene for downy mildew resistance with a rust resistance gene, presumably R_1. No recombinant types were recovered from among 543 F_3 and 105 F_4 lines from two crosses. A linkage of the gene determining formation of the scar at the base of the seed with a gene determining absence of striation was found by Ananieva cited in Gundaev (70). Stoenescu and Vranceanu (195) reported a repulsion linkage between the ms_2 gene for genetic male sterility with the fl gene for small tubular ray flowers with a crossover value of 19.3 ± 5.3%.

PROGRESS IN THE UNITED STATES AND CANADA AND CHALLENGES FOR THE FUTURE

Extensive breeding and genetics programs to improve sunflower have developed only recently in the USA and Canada, primarily since the introduction of the high oil Soviet cultivars in the 1960's. Nevertheless considerable progress has been made. Hybrids produced by the cytoplasmic male sterility and genetic fertility restorer system first were introduced for commercial production in the USA in 1972, and within 5 years were grown on about 90% of the production area. Direct substitution of adapted and tested hybrids is estimated to have resulted in yield increases in excess of 25%. Progress also has been made in developing lines and hybrids that are resistant to the important diseases, highly self compatible, and more uniform than the former cultivars in height, flowering, and maturity. The rapidity of these developments is an outstanding achievement made possible by the widespread cooperation of plant breeders, pathologists, agronomists, and seed production personnel with the USDA, Agriculture Canada, State Agricultural Experiment Stations, seed companies, and the processing industry.

Continued improvement of sunflower in the future appears feasible. Most breeders are extremely optimistic, and present information suggests that large numbers of new lines are available that are superior in combining ability, agronomic type, oil percentage, and/or disease and insect tolerance.

Also, wild species of *Helianthus* offer tremendous genetic diversity for further improvement of sunflower, but to date they have been used only to a limited extent.

Normal sunflower plants when grown at a plant population of 50,000 plants/ha have 1,000 to 2,500 individual florets per head with an average of about 1,500. Assuming a weight of 70 g/1,000 seed, an average head under optimum conditions theoretically should produce 105 g of seed. Based on these assumptions, seed yields in excess of 5,000 kg/ha would be expected. Yet average yields in the USA and Canada often are less than 1,200 kg/ha or less than 25% of the indicated potential. Thus, improvement of yield is theoretically feasible and represents a significant challenge for researchers involved in sunflower investigations. With currently available germplasm, resources, and facilities, plant breeders almost certainly will develop improved and more productive cultivars in future years.

LITERATURE CITED

1. Adams, A. L., and J. C. Gaines. 1950. Sunflower insect control. J. Econ. Entomol. 43: 181–184.

2. Adams, P. B. 1975. Factors affecting survival of *Sclerotinia sclerotiorum* in soil. Plant Dis. Rep. 59:599–603.

3. Anaschenko, A. V. 1967. The chemical castration of sunflower plants. (In Russian) Dokl. Vses. Akad. Sel'skokhoz. Nauk im V.I. Lenina 2:17–18.

4. ————. 1972. Methods for developing hybrid sunflower involving chemical castration. p. 229–230. *In* Proc. 5th Int. Sunflower Conf. (Clermont-Ferrand, France).

5. ————. 1974. The initial material for sunflower heterosis breeding. p. 391–393. *In* Proc. 6th Int. Sunflower Conf. (Bucharest, Romania).

6. ————. 1976. *Helianthus annuus* L.—Interspecific classification and genetic resources. p. 69–71. *In* Abstr. of papers 7th Int. Sunflower Conf. (Krasnodar, USSR).

7. Arnoldova, O. N. 1926. To the biology of sunflower blooming in connection with the techniques of its crossing. (In Russian). J. Exp. Agron. Southeast 3:131–143.

8. Banihashemi, Z. 1975. Phytophthora black stem rot of sunflower. Plant Dis. Rep. 59: 721–724.

9. Borodulina, A. A., P. S. Popov, and I. N. Harchenko. 1974. Biochemical characteristics of the present sunflower varieties and hybrids. p. 239–245. *In* Proc. 6th Int. Sunflower Conf. (Bucharest, Romania).

10. Bottrell, D. G., R. D. Brigham, and L. B. Jordan. 1973. Carrot beetle: pest status and bionomics on cultivated sunflower. J. Econ. Entomol. 66:86–90.

11. Briggs, F. N., and P. F. Knowles. 1967. Introduction to plant breeding. Reinhold Publishing Corporation, New York.

12. Brummett, B. J., and E. E. Burns. 1972. Pigment and chromogen characteristics of sunflower seed, *Helianthus annuus*. J. Food Sci. 37:1–3.

13. Bukherovich, P. G. 1966. Study of aggressiveness of broomrape of different origins and resistance of a whole series of sunflower varieties to it. Sb. Rab. Maslichn. Kult. 3:22–25. [Transl. Indian Natl. Sci. Documentation Centre, New Delhi. 1968.]

14. Burlov, V. V. 1974. Utilization of male sterility in sunflower breeding for heterosis. p. 353–360. *In* Proc. 6th Int. Sunflower Conf. (Bucharest, Romania).

15. ————, and S. V. Kostiuk. 1976. Inheritance of the resistance to the local race of broomrape (*Orobanche cumana* Wallr.) in sunflower. (In Russian). Genetica 12(2):44–51.

16. Burns, E. E., L. J. Talley, and B. J. Brummett. 1972. Sunflower utilization in human foods. Cereal Sci. Today 17:287–289.

17. Canvin, D. T. 1965. The effect of temperature on the oil content and fatty acid composition of the oils from several oil seed crops. Can. J. Bot. 43:63–69.

18. Cardon, P. V. 1922. Sunflower studies. J. Am. Soc. Agron. 14:69–72.

19. Carlson, E. C., P. F. Knowles, and J. E. Dillé. 1972. Sunflower varietal resistance to sunflower moth larvae. Calif. Agric. 26(6):11–13.

20. –––––, and R. Witt. 1974. Moth resistance of armored-layer sunflower seeds. Calif. Agric. 28(11):12–14.

21. Cirnu, I., V. Dumitrache, and E. Hociota. 1974. La pollinsation du tournesol (*Helianthus annuus* L.) a l'aide des abeilles-un facteur important pour l'augmentation de la production. p. 695–700. *In* Proc. 6th Int. Sunflower Conf. (Bucharest, Romania).

22. Claassen, C. E., W. G. Ekdahl, and G. M. Severson. 1950. The estimation of oil percentage in safflower seed and the association of percentage with hull and nitrogen percentages, seed size, and degree of spininess of the plant. Agron. J. 42:478–482.

23. Clément, Y., and E. Diehl. 1968. Contribution to the study of some characteristics of sunflower. C. R. Acad. Agric. Fr. 54:853–859.

24. Cobia, D., and D. Zimmer (ed.). 1975. Sunflowers, production, pests, and marketing. Extension Bull. 25. North Dakota State Univ., Fargo.

25. Cockerell, T. D. A. 1912. The red sunflower. Pop. Sci. Monthly. p. 373–382.

26. –––––. 1915. Specific and varietal characters in annual sunflowers. Am. Nat. 49: 609–622.

27. Cummins, D. G., J. E. Marion, J. P. Craigmiles, and R. E. Burns. 1967. Oil content, fatty acid composition, and other agronomic characteristics of sunflower introductions. J. Am. Oil Chem. Soc. 44:581–582.

28. Cupina, T., and L. Vasiljevic. Inheritance of sunflower photosynthetic apparatus model. Acta Bot. Croat. 35:177–178.

29. Davreux, M., E. J. Fernandez, P. W. Luduena, and A. Orlowsky. 1974. Present situation of sunflower breeding in Argentina. p. 411–413. *In* Proc. 6th Int. Sunflower Conf. (Bucharest, Romania).

30. Dominguez-Gimenez, J. 1974. Fertility restoration and other characteristics in wild sunflowers. M.S. thesis, N. Dak. State Univ., Fargo.

31. –––––, and G. N. Fick. 1975. Fertility restoration of male-sterile cytoplasm in wild sunflowers. Crop Sci. 15:724–726.

32. Dorrell, D. G. 1973. Controlling plant height in sunflowers with growth retardants. Can. J. Plant Sci. 53:417–418.

33. –––––. 1976. Chlorogenic acid content of meal from cultivated and wild sunflowers. Crop Sci. 16:422–424.

34. Downey, R. K., B. M. Craig, and C. G. Youngs. 1969. Breeding rapeseed for oil and meal quality. J. Am. Oil Chem. Soc. 46:121–123.

35. Durand, Y. 1962. Rapport de mission en USSR sur la recherche agronomique et la culture du tournesol. Centre Technique Interprofessionnel des Oleagineaux Metropolitains, 174, Av Victor Hugo, Paris, France.

36. Dyer, H. J., J. Skok, and N. J. Scully. 1959. Photoperiodic behaviour of sunflower. Bot. Gaz. 121:50–55.

37. Earle, F. R., C. H. Vanetten, T. F. Clark, and I. A. Wolff. 1968. Compositional data on sunflower seed. J. Am. Oil Chem. Soc. 45:876–879.

38. Eberhart, S. A., M. N. Harrison, and F. Ogada. 1967. A comprehensive breeding system. Der Zuchter 37:169–174.

39. Enns, H. 1959. A study of the inheritance and expressivity of some seedling characters in the sunflower *Helianthus annuus* L. M.S. thesis, Univ. Manitoba, Winnipeg.

40. –––––, D. G. Dorrell, J. A. Hoes, and W. O. Chubb. 1970. Sunflower research, a progress report. p. 162–167. *In* Proc. 4th Int. Sunflower Conf. (Memphis, Tennessee).

41. Ermakov, A. I., and E. V. Popova. 1974. Variations resulting from crosses in the ratio of fat acids in oil of sunflower seeds. (In Russian). Tr. Prikl. Bot. Genet. Sel. 53:255–261.

42. Fernandez-Martinez, J. 1974. Variability in the fatty acid composition of the seed oil of *Helianthus species*. M.S. thesis, University of California, Davis.

43. Fick, G. N. 1975. Heritability of oil content in sunflowers. Crop Sci. 15:77–78.

44. –––––. 1976. Genetics of floral color and morphology in sunflowers. J. Hered. 67: 227–230.

45. –––––, and D. A. Rehder. 1977. Selection criteria in development of high oil sunflower hybrids. p. 26–27. *In* Proc. 2nd Sunflower Forum (Fargo, N.D.).

46. –––––, and C. M. Swallers. 1972. Higher yields and greater uniformity with hybrid sunflowers. N. Dak. Farm Res. 29(6):7–9.

47. ————, and ————. 1975. Intercultivar competition in sunflower test plots. Agron. J. 67:743–745.

48. ————, and D. E. Zimmer. 1974. Fertility restoration in confectionery sunflowers. Crop Sci. 14:603–605.

49. ————, and ————. 1974. Monogenic resistance to Verticillium wilt in sunflowers. Crop Sci. 14:895–896.

50. ————, and ————. 1974. Parental lines for production of confectionery sunflower hybrids. N. Dak. Farm Res. 31(5):15–16.

51. ————, and ————. 1974. RHA 271, RHA 273, and RHA 274—sunflower parental lines for producing downy mildew resistant hybrids. N. Dak. Farm Res. 32(2):7–9.

52. ————, and ————. 1975. Influence of rust on performance of near-isogenic sunflower hybrids. Plant Dis. Rep. 59:737–739.

53. ————, and ————. 1975. Linkage tests among genes for six qualitative characters in sunflowers. Crop Sci. 15:777–779.

54. ————, and ————. 1976. Yield stability of sunflower hybrids and open-pollinated varieties. p. 37–38. In Abstr. of papers 7th Int. Sunflower Conf. (Krasnodar, USSR).

55. ————, ————, J. Dominguez-Gimenez, and D. A. Rehder. 1974. Fertility restoration and variability for plant and seed characteristics in wild sunflowers. p. 333–338. In Proc. 6th Int. Sunflower Conf. (Bucharest, Romania).

56. ————, ————, and R. G. Robinson. 1974. Performance of sunflower hybrids produced by cytoplasmic male-sterility and genetic fertility restoration. Am. Soc. Agron. Abstr. p. 52.

57. ————, ————, and T. E. Thompson. 1976. Wild species of Helianthus as a source of variability in sunflower breeding. p. 4–5. In Proc. 1st Sunflower Forum (Fargo, N.D.).

58. ————, ————, and D. C. Zimmerman. 1974. Correlation of seed oil content in sunflowers with other plant and seed characteristics. Crop Sci. 14:755–757.

59. ————, and D. C. Zimmerman. 1973. Variablity in oil content among heads and seeds within heads of sunflowers (Helianthus annuus L.). J. Am. Oil Chem. Soc. 50:529–531.

60. Free, J. B. 1970. Insect pollination of crops. Academic Press, London.

61. Furgala, B., E. C. Mussen, D. M. Noetzel, and R. G. Robinson. 1976. Observations on nectar secretion in sunflowers. p. 11–12. In Proc. 1st Sunflower Forum (Fargo, N.D.).

62. Georgiev, T. M. 1972. Improving the form of the sunflower plant by means of the growth regulator ethrel. C. R. Acad. Agric. G. Dimitrov. 5(2):139–146. Sofia, Bulgaria.

63. Georgieva-Todorova, J. 1974. A new male sterile form of sunflower (Helianthus annuus L.). p. 343–347. In Proc. 6th Int. Sunflower Conf. (Bucharest, Romania).

64. ————, and A. Hristova. 1975. Studies on several wild-growing Helianthus species. C. R. Acad. Agric. G. Dimitrov. 8(4):51–55. Sofia, Bulgaria.

65. Gimeno-Ramirez, V. 1975. Seleccion de plantas de girasol basadas en la variacion de la velocidad de germinacion a bajas temperaturas. Comun. Ser. Prod. Veg. Inst. Nac. Invest. Agrar. 5:73–75.

66. Goossen, P. G., and W. E. Sackston. 1967. Transmission and biology of sunflower downy mildew. Can. J. Bot. 46:5–10.

67. Granlund, M., and D. C. Zimmerman. 1975. Oil content of sunflower seeds as determined by wide-line nuclear magnetic resonance (NMR). Proc. N.D. Acad. Sci. 27:128–133.

68. Guillaumin, J. J., J. Kurek, M. Lalande, and A. Ramirez. 1974. Inoculation de capitules de tournesol au laboratorie par Botrytis cinerea et essai de mise au point d'un test de sensibilite varietale. p. 655–659. In Proc. 6th Int. Sunflower Conf. (Bucharest, Romania).

69. Gundaev, A. I. 1966. Prospects of selection in sunflower for heterosis. Sb. Rab. Maslichn. Kult. 3:15–21. [Transl. Indian Natl. Sci. Documentation Centre, New Delhi, India. 1969.]

70. ————. 1971. Basic principles of sunflower selection. p. 417–465. In Genetic Principles of Plant Selection. Nauka, Moscow. [Transl. Dep. of the Secretary of State, Ottawa, Canada. 1972.]

71. Habura, E. C. H. 1957. Parasterilitat bei sonnenblumen. Z. Pflanzenzuecht. 37:280–298.

72. Hallauer, A. R. 1967. Development of single-cross hybrids from two-eared maize population. Crop Sci. 7:192–195.

73. ————. 1973. Hybrid development and population improvement in maize by reciprocal full-sib selection. Egypt. J. Genet. Cytol. 2:84–101.

74. Hamilton, R. I. 1926. Improving sunflowers by inbreeding. Sci. Agric. 6:190–192.

75. Harada, W. 1975. Sunflower cultivar trials and bird damage studies at CSUF. M.S. thesis. California State Univ., Fresno.

76. ————. 1977. A possible deterrent to bird damage in sunflower. Proc. 2nd Sunflower Forum (Fargo, North Dakota).

77. Heiser, C. B., Jr. 1951. The sunflower among the North American Indians. Proc. Am. Phil. Soc. 95:432–448.

78. ————. 1976. The sunflower. Univ. of Oklahoma Press, Norman.

79. ————, D. M. Smith, Sarah Clevenger, and W. C. Martin. 1969. The North American Sunflowers (*Helianthus*). Mem. Torrey Bot. Club 22(3):1–218.

80. Hennessy, C. M. R., and W. E. Sackston. 1972. Studies on sunflower rust. X. Specialization of *Puccinia helianthi* on wild sunflowers in Texas. Can. J. Bot. 50:1871–1877.

81. Hockett, E. A., and P. F. Knowles. 1970. Inheritance of branching in sunflowers, *Helianthus annuus* L. Crop Sci. 10:432–436.

82. Hoes, J. A., and H. Enns. 1974. Inheritance of resistance to *Phialophora* yellows of sunflower. p. 309–310. *In* Proc. 6th Int. Sunflower Conf. (Bucharest, Romania).

83. ————, and H. C. Huang. 1976. Control of Sclerotinia basal stem rot of sunflower: A progress report. p. 18–20. *In* Proc. 1st Sunflower Forum (Fargo, N.D.).

84. ————, and E. D. Putt. 1962. Races of *Puccinia helianthi*. Phytopathology 52:736–737.

85. ————, ————, and H. Enns. 1973. Resistance to Verticillium wilt in collections of wild *Helianthus* in North America. Phytopathology 63:1517–1520.

86. Hristova, A., and Y. Georgieva-Todorova. 1975. A study of the fatty acid composition and protein content in the seeds of *Helianthus annuus, Helianthus scaberrimus* and their F_1 hybrids. C. R. Acad. Agric. G. Dimitrov 8:55–58. Sofia, Bulgaria.

87. Ivanov, I. G. 1975. Study on compatibility and incompatibility display in crossing selfed sunflower lines. (In Bulgarian). Rastenievud Nauk. 12(9):30–35.

88. Ivanov, P. 1974. Biochemical differentiation of sunflower varieties as a result of breeding. p. 225–229. *In* Proc. 6th Int. Sunflower Conf. (Bucharest, Romania).

89. ————. 1975. Variation of the protein, lysine, and chlorogenic acid content in some sunflower selfed lines. (In Bulgarian). Rastenievud Nauk. 10:23–27.

90. ————, and V. Nikolova. 1975. Variability of the fatty acid composition of oil in several selfed sunflower lines. (In Bulgarian). Rastenievud Nauk. 12(9):36–40.

91. Jagodkin, I. G. 1937. Application of method of inbreeding and diallel crossing in sunflower growing. (In Russian). Selektsija i Semenovodstvo 1:21–27. (Plant Breed. Abstr. 8:65, 1937–1938).

92. Jain, S. K., A. M. Olivieri, and J. Fernandez-Martinez. 1977. Serpentine sunflower, *Helianthus exilis,* as a genetic resource. Crop Sci. 17:477–479.

93. Jarvis, J. J. 1976. Damage to sunflower introductions by sunflower moth and birds. USDA Special Report A-1. North Central Regional Plant Introduction Station, Ames, Iowa.

94. Jensen, L. A., C. M. Swallers, and F. K. Johnson. 1970. Sunflower production in North Dakota 1970. N. Dak. State Univ. Circ. A-538.

95. Johnson, A. L., and B. H. Beard. 1977. Sunflower moth damage and inheritance of the phytomelanin layer in sunflower achenes. Crop Sci. 17:369–372.

96. Kalton, R. R. 1951. Efficiency of various bagging materials for effecting self-fertilization of sunflowers. Agron. J. 43:328–331.

97. Karami, E. 1974. Emergence of nine varieties of sunflower (*Helianthus annuus* L.) in salinized soil cultures. p. 167–172. *In* Proc. 6th Int. Sunflower Conf. (Bucharest, Romania).

98. Kinman, M. L. 1970. New developments in the USDA and state experiment station sunflower breeding programs. p. 181–183. *In* Proc. 4th Int. Sunflower Conf. (Memphis, Tenn.).

99. ————, and F. R. Earle. 1964. Agronomic performance and chemical composition of the seed of sunflower hybrids and introduced varieties. Crop Sci. 4:417–420.

100. Kloczowski, Z. 1971. Comparison of the F_1 and F_2 of linear hybrids in oil sunflowers. Genet. Pol. 12:359–362.

101. ————. 1972. Breeding of oil sunflower in Poland. p. 258–261. *In* Proc. 5th Int. Sunflower Conf. (Clermont-Ferrand, France).

102. ————. 1975. Studies on some features of oil sunflower and their significance in breeding that plant in Poland. (In Polish). Hodowla Rosl. Aklim. Nasienn. 19(2):89–131.

103. Knowles, P. F., and A. B. Hill. 1964. Inheritance of fatty acid content in the seed oil of a safflower introduction from Iran. Crop Sci. 4:406–409.

104. ————, S. R. Temple, and F. Stolp. 1970. Variability in the fatty acid composition of sunflower seed oil. p. 215–218. In Proc. 4th Int. Sunflower Conf. (Memphis, Tenn.).

105. Kovacik, A., and V. Skaloud. 1972. The proportion of the variability component caused by the environment and the correlations of economically important properties and characters of the sunflower (Helianthus annuus L.). Sci. Agric. Bohemoslov. 4:249–261. (Plant Breed. Abstr. 43:8100. 1973).

106. ————, and V. Skaloud. 1974. The influence of inbreeding and sib crossing on the characters of productivity in sunflower (Helianthus annuus L.). p. 435–438. In Proc. 6th Int. Sunflower Conf. (Bucharest, Romania).

107. Kurnik, E., M. Jaky, J. Parragh, L. Meszaros, and R. Szabo. 1970. Aspects of quality in sunflower breeding in Hungary. Agrortudomanyi Kozlemenyek, Budapest 29:507–514.

108. Leclercq, P. 1966. Une sterilite male utilisable pour la production d'hybrides simples de tournesol. Ann. Amelior. Plant 16:135–144.

109. ————. 1968. Heredite de quelques caracteres qualitatifs chez le tournesol. Ann. Amelior. Plant. 18:307–315.

110. ————. 1969. Une sterilite cytoplasmique chez le tournesol. Ann. Amelior. Plant. 19: 99–106.

111. ————. 1971. La sterilite male cytoplasmique du tournesol. I. Premieres etudes sur la restauration de la fertilite. Ann. Amelior. Plant. 21:45–54.

112. ————. 1974. Recherche d'un noveau type de sterilite male cytoplasmique. p. 331. In Proc. 6th Int. Sunflower Conf. (Bucharest, Romania).

113. ————, Y. Cauderon, and M. Dauge. 1970. Selection pour la resistance au mildiou du tournesol a partir d'hybrides topinambour × tournesol. Ann. Amelior. Plant. 20:363–373.

114. Lofgren, J. R. 1976. Seed quality of confectionery sunflowers (Helianthus annuus L.). p. 2–3. In Proc. 1st Sunflower Forum (Fargo, N. D.).

115. Lonnquist, J. H., and M. F. Lindsey. 1970. Tester performance level for the evaluation of lines for hybrid performance. Crop Sci. 10:602–604.

116. Lovett, J. V., and P. W. Orchard. 1974. Influence of CCC on sunflower growth, development and yield under controlled environment and field conditions. p. 153–159. In Proc. 6th Int. Sunflower Conf. (Bucharest, Romania).

117. Luciano, A., and M. Davreux. 1967. Produccion de girasol in Argentina. Publication Technica 37. Instituto Nacional de Technologia Agropecuaria, Estacion Experimental Regional Agropecuaria, Pergamino, Argentina.

118. ————, M. L. Kinman, and J. D. Smith. 1965. Heritability of self-incompatibility in the sunflower (Helianthus annuus). Crop Sci. 5:529–532.

119. Luczkiewicz, T. 1975. Inheritance of some characters and properties in sunflower (Helianthus annuus L.). Genet. Pol. 16:167–184.

120. Luka, C. 1974. Research on the resistance of sunflower inbred lines and hybrids to Sclerotinia libertiana Fuck. p. 303–308. In Proc. 6th Int. Sunflower Conf. (Bucharest, Romania).

121. Miah, M. A. Jabbar, C. M. R. Hennessy, and W. E. Sackston. 1967. Origin of new physiologic races of sunflower rust (Puccinia helianthi) through selfing and hybridization of the "Canadian" races. Proc. Can. Phytopathol. Soc. 34:23.

122. ————, and W. E. Sackston. 1970. Genetics of rust resistance in sunflowers. Phytoprotection 51:1–16.

123. Miller, J. F., and G. N. Fick. 1978. Adaptation of reciprocal full-sib selection in sunflower breeding using gibberellic acid induced male sterility. Crop Sci. 18:161–162.

124. Moradi, A., and P. Vojdani. 1974. Study of the relationship between leaf surface, grain yield and oil percent in different varieties of sunflower. p. 189–196. In Proc. 6th Int. Sunflower Conf. (Bucharest, Romania).

125. Morozov, V. K. 1947. Sunflower breeding in the USSR. Moscow: Piscepromizdat. p. 272. (Extended summary prepared by Imperial Bureau of Plant Breeding and Genetics, Cambridge, England.)

126. Moser, P. E., and W. E. Sackston. 1973. Effect of concentration of inoculum and method of inoculation on development of Verticillium wilt of sunflowers. Phytopathology 63:1521–1523.

127. Nikolic-Vig, V., and D. Skoric. 1974. Sunflower breeding problems in Yugoslavia. p. 401–403. *In* Proc. 6th Int. Sunflower Conf. (Bucharest, Romania).

128. ————, D. Skoric, and S. Bedov. 1971. Variability of oil and husk content in the sunflower seed of Peredovik and VNIIMK 8931 varietal populations and their heritability. Savremena Poljoprivreda 19(3):23–32.

129. Orellana, R. G. 1971. Fusarium wilt of sunflower, *Helianthus annuus*: First report. Plant Dis. Rep. 55:1124–1125.

130. Oseto, C. Y. 1977. Biology and control of stem weevils, *Cylindrocopturus adspersus* and *Baris strenua*. P. 4. *In* Proc. 2nd Sunflower Forum (Fargo, N.D.).

131. Pathak, R. S. 1974. Yield components in sunflower. p. 271–281. *In* Proc. 6th Int. Sunflower Conf. (Bucharest, Romania).

132. Pawlowski, S. H., and E.J. Hawn. 1964. Host-parasite relationships in sunflower wilt incited by *Sclerotinia sclerotiorum* as determined by the twin technique. Phytopathology 54:33–35.

133. Piquemal, G. 1970. How to produce hybrid sunflower seeds by inducing male sterility with gibberellic acid. p. 127–135. *In* Proc. 4th Int. Sunflower Conf. (Memphis, Tenn.).

134. Pirjol, L., C. I. Milica, and V. Vranceanu. 1971. Rezistenta la seceta a florii-soarelui in diferite faze de vegetatie. An. Inst. Cercet. Cerale Plante Teh. Fundulea Acad Stiinte Agric. Silvice Ser. C 37:191–208.

135. Pirjol-Savulescu, L. 1974. Variability of sunflower resistance to drought. p. 133–143. *In* Proc. 6th Int. Sunflower Conf. (Bucharest, Romania).

136. Poehlman, J. M. 1959. Breeding field crops. Henry Holt and Company, Inc., New York.

137. Pogorletskii, B. K., and E. E. Geshele. 1975. The problem of the immunity to rust in sunflower. (In Russian). Genetica 11(8):12–18.

138. Pomenta, J. V., and E. E. Burns. 1971. Factors affecting chlorogenic, quinic and caffeic acid levels in sunflower kernels. J. Food Sci. 36:490–492.

139. Pustovoit, V. S. 1964. Conclusions of work on the selection and seed production of sunflowers. Agrobiology 5:672–697. [Transl. R. P. Knowles, Agriculture Canada, Saskatoon, Saskatchewan. 1965.]

140. Pustovoit, G. V. 1966. Distant (interspecific) hybridization of sunflowers in the U.S.S.R. p. 82–101. *In* Proc. 2nd Int. Sunflower Conf. (Morden, Manitoba).

141. ————, and I. A. Gubin. 1974. Results and prospects in sunflower breeding for group immunity by using the interspecific hybridization method. p. 373–381. *In* Proc. 6th Int. Sunflower Conf. (Bucharest, Romania).

142. ————, A. G. Malysheva, and V. P. Shvetsova. 1974. Biochemical characteristics of sunflower interspecific hybrids and their initial forms. (In Russian). Skh. Biol. 9:844–848.

143. Putt, E. D. 1940. Observations on morphological characters and flowering processes in the sunflower (*Helianthus annuus* L.). Sci. Agric. 21:167–169.

144. ————. 1941. Investigations of breeding technique for the sunflower (*Helianthus annuus* L.). Sci. Agric. 21:689–702.

145. ————. 1943. Association of seed yield and oil content with other characters in the sunflower. Sci. Agric. 23:377–383.

146. ————. 1958. Note on differences in susceptibility to Sclerotinia wilt in sunflowers. Can. J. Plant Sci. 38:380–381.

147. ————. 1958. Note on resistance of sunflowers to leaf mottle disease. Can. J. Plant Sci. 38:274–276.

148. ————. 1962. The value of hybrids and synthetics in sunflower seed production. Can. J. Plant Sci. 42:488–500.

149. ————. 1963. Sunflowers. Field Crop Abstr. 16(1):1–6.

150. ————. 1964. Breeding behaviour of resistance to leaf mottle or Verticillium wilt in sunflowers. Crop Sci. 4:177–179.

151. ————. 1964. Recessive branching in sunflowers. Crop Sci. 4:444–445.

152. ————. 1965. Breeding for large sunflower seed. Res. Farmer 10(2):10–11.

153. ————. 1966. Heterosis, combining ability, and predicted synthetics from a diallel cross in sunflowers (*Helianthus annuus* L.). Can. J. Plant Sci. 46:59–67.

154. ————, B. M. Craig, and R. B. Carson. 1969. Variation in composition of sunflower oil from composite samples and single seeds of varieties and inbred lines. J. Am. Oil Chem. Soc. 46:126–129.

155. ————, and D. G. Dorrell. 1975. Breeding for sunflower production. p. 185–201. *In* J. T. Harapiak (ed.) Oilseed and pulse crops in western Canada. Modern Press, Saskatoon, Saskatchewan, Canada.

156. ——————, and C. B. Heiser, Jr. 1966. Male sterility and partial male sterility in sunflowers. Crop Sci. 6:165–168.

157. ——————, and W. E. Sackston. 1957. Studies on sunflower rust. I. Some sources of rust resistance. Can. J. Plant Sci. 37:43–54.

158. ——————, and ——————. 1960. Resistance of inbred lines and hybrids of sunflowers (*Helianthus annuus* L.) in a natural epidemic of aster yellows. Can. J. Plant Sci. 40: 375–382.

159. ——————, and ——————. 1963. Studies on sunflower rust. IV. Two genes, R_1 and R_2 for resistance in the host. Can. J. Plant Sci. 43:490–496.

160. Ramakrishnan, C., N. Rajamohan, and T. R. Subramaniam. Note on the susceptibility of sunflower types to the grasshopper, *Atroctomorpha crenulata* F. Madras Agric. J. 61: 178–180.

161. Ravagnan, G. M. 1974. Methods of selection to improve the production of sunflower (*Helianthus annuus*) in Zambia. p. 415–420. *In* Proc. 6th Int. Sunflower Conf. (Bucharest, Romania).

162. Rawlings, J. O., and D. L. Thompson. 1962. Performance level as criterion for the choice of maize testers. Crop Sci. 2:217–220.

163. Robertson, J. A., J. K. Thomas, and D. Burdick. 1971. Chemical composition of the seed of sunflower hybrids and open pollinated varieties. J. Food Sci. 36:873–876.

164. Robinson, R. G. 1973. The sunflower crop in Minnesota. Univ. of Minnesota Extension Bull. 299 (Rev.).

165. ——————. 1967. Registration of Mingren sunflower. Crop Sci. 7:404.

166. ——————. 1975. Amino acid and elemental composition of sunflower and pumpkin seeds. Agron. J. 67:541–544.

167. ——————, L. A. Bernat, H. A. Geise, F. K. Johnson, M. L. Kinman, E. L. Mader, R. M. Oswalt, E. D. Putt, C. M. Swallers, and J. H. Williams. 1967. Sunflower development at latitudes ranging from 31 to 49 degrees. Crop Sci. 7:134–136.

168. Rogers, C. E., and G. R. Howell. 1973. Behaviour of the carrot beetle on native and treated commercial sunflower. Progress Rep., Texas Agric. Exp. Stn., College Station.

169. Rollier, M. 1975. Etudes varietales tournesol. Centre technique interprofessionnel des Oleagineux metropolitains (C.E.T.I.O.M.), Paris, France 43:1–38.

170. Ross, A. M. 1939. Some morphological characters of *Helianthus annuus* L., and their relations to the yield of seed and oil. Sci. Agric. 19:372–379.

171. Rozhkova, V. T., A. V. Anaschenko, and T. V. Mileyeva. 1976. Sunflower collection as a help for the breeder. p. 68–69. *In* Abstr. of papers 7th Int. Sunflower Conf. (Krasnodar, USSR).

172. Ruud, R. R. 1976. The use of ethrel to break dormancy of sunflower seeds in a germination test. Newsl. Assoc. Off. Seed Anal. 50(3):43–44.

173. Russell, W. A. 1953. A study of the inter-relationships of seed yield, oil content, and other agronomic characters with sunflower inbred lines and their top crosses. Can. J. Agric. Sci. 33:291–314.

174. Schulz, J. T. 1973. Damage to cultivated sunflower by *Contarinia schulzi*. J. Econ. Entomol. 66:282.

175. Schuster, W. 1964. Inbreeding and heterosis in sunflower (*Helianthus annuus* L.), Wilhelm Schmitz Verlag, Giessen. p. 135. (Plant Breed. Abstr. 37:1207, 1967).

176. ——————. 1969. Beobachtungen uber mannliche sterilitat bei der sonnenblume (*H. annuus*), ausgelost durch genetische, physiologische und induzierte chemische faktoren. Theor. Appl. Genet. 39:261–273.

177. ——————. 1970. Die auswirkungen der fortgesetzten inzuchtung von I_0 bis I_{18} auf verschiedene merkmale der sonnenblume. Z. Pflanzenzuecht. 64:310–334.

178. ——————. 1973. Untersuchungen uber die fixierung von heterosis durch kombinationszuchtung bie sonnenblumen. Z. Pflanzenzuecht. 70:214–222.

179. ——————, and R. Boye. 1971. Der einfluss von temperatur und tageslange auf verschiedene sonnenblumensorten unter kontrollierten klimabedingungen und im freiland. Z. Pflanzenzuecht. 65:151–176.

180. Seetharam, A., and P. Kusuma Kumari. 1974. GA induced male sterility in sunflower. Sci. Cult. 40:398–399.

181. ——————, and ——————. 1975. Induction of male sterility by gibberellic acid in sunflower. Indian J. Genet. Plant Breed. 35:136–138.

182. Shabana, R. 1974. Genetic variability of sunflower varieties and inbred lines. p. 263–269. *In* Proc. 6th Int. Sunflower Conf. (Bucharest, Romania).

183. Shcherbak, S. N., and V. V. Efremova. 1966. Photosynthetic force and productivity of some varieties of sunflower. Sb. Rab. Maslichn. Kult. 3:55–59. [Transl. Indian Natl. Scientific Documentation Centre, New Delhi, 1968.]

184. Shewfelt, A. L., and E. D. Putt. 1958. A rapid method for estimating the oil content of sunflower seeds. Can. J. Plant Sci. 38:419–423.

185. Shopov. T. 1976. Powdery mildew, a new disease of sunflower in Bulgaria. (In Bulgarian). Rastit. Zasht. 23(9):10–12.

186. Skaloud, V., and A. Kovacik. 1974. Inheritance of some heteromorphic characters in sunflowers (*Helianthus annuus* L.). p. 291–295. *In* Proc. 6th Int. Sunflower Conf. (Bucharest, Romania).

187. Skoric, D. 1974. Correlation among the most important characters of sunflower in F_1 generation. p. 283–289. *In* Proc. 6th Int. Sunflower Conf. (Bucharest, Romania).

188. Skrdla, W. H., R. L. Clark, and J. L. Jarvis. 1976. List of seed available at the North Central Regional Plant Introduction Station. Helianthus. USDA.

189. Soldatov, K. 1976. Chemical mutagenesis for sunflower breeding. p. 57–58. *In* Abstr. of papers 7th Int. Sunflower Conf. (Krasnodar, USSR).

190. Sosulski, F. W., and G. Sarwar. 1973. Amino acid composition of oilseed meals and protein isolates. Can. Inst. Food Sci. Tech. J. 6(1):1–5.

191. Spirova, M. 1975. New data on the male sterility in sunflower induced by gibberellic acid. (In Bulgarian). Rastenievud Nauk. 12(9):10–17.

192. Sprague, G. F. 1966. Quantitative genetics in plant breeding. p. 315–347. *In* K. J. Frey (ed.) Plant breeding. University Press, Ames, Iowa.

193. ————, and A. Tavcar. 1956. Mais. I. General considerations and American breeding work. Handbuch der Pflanzenzuchtung II. 2:103–143.

194. Stoenescu, F. 1974. Genetica. p. 93–120. *In* A. V. Vranceanu. Floarea-soarelui. Editura Academiei Republicii Socialiste, Romania, Bucuresti.

195. ————, and A. V. Vranceanu. 1976. Linkage studies between *ms* genes and four marker genes in sunflower. p. 49–50. *In* Abstr. of papers 7th Int. Sunflower Conf. (Krasnodar, USSR).

196. Stoyanova, Y., and P. Ivanov. 1975. Inheritance of oil and protein content in first hybrid progeny of sunflower. (In Bulgarian). Rastenievud Nauk. 12(9):30–35.

197. Teetes, G. L., M. L. Kinman, and N. M. Randolph. 1971. Differences in susceptibility of certain sunflower varieties and hybrids to the sunflower moth. J. Econ. Entomol. 64:1285–1287.

198. Udaya Kumar, M., and K. S. Krishna Sastry. 1975. Effect of growth regulators on germination of dormant sunflower seeds. Seed Res. 3(2):61–65.

199. Unrau, J. 1947. Heterosis in relation to sunflower breeding. Sci. Agric. 27:414–427.

200. ————, and W. J. White. 1944. The yield and other characters of inbred lines and single crosses of sunflowers. Sci. Agric. 24:516–528.

201. Vear, F. 1974. Studies on resistance to downy mildew in sunflowers (*Helianthus annuus* L.). p. 297–302. *In* Proc. 6th Int. Sunflower Conf. (Bucharest, Romania).

202. ————, and P. Leclercq. 1971. Deux nouveaux genes de resistance au mildiuo du tournesol. Ann. Amelior. Plant. 21:251–255.

203. Velkov, V., and Y. Stoyanova. 1974. Biological pecularities of cytoplasmic male sterility and schemes of its use. p. 361–365. *In* Proc. 6th Int. Sunflower Conf. (Bucharest, Romania).

204. Voskoboinik, L. K., and K. I. Soldatov. 1974. The research trends in the field of sunflower breeding for heterosis at the All-Union Research Institute for Oil Crops (VNIIMK). p. 383–389. *In* Proc. 6th Int. Sunflower Conf. (Bucharest, Romania).

205. Vranceanu, V. 1967. Ereditatea surselor de androsterilitate la floarea-soarelui. Probl. Agric. 12:4–11.

206. ————. 1970. Advances in sunflower breeding in Romania. p. 136–148. *In* Poc. 4th Int. Sunflower Conf. (Memphis, Tenn.).

207. ————, and F. Stoenescu. 1969. Folosirea liniilor autoincompatibile cu gene marcatoare pentru crearea hybrizilor simpli de floarea-soarelui. An. Inst. Cercet. Cerale Plante Teh. Fundulea Acad. Stiinte Agric. Silvice 35:551–557.

208. ————, and ————. 1969. Hibrizi simpli de floarea-soarelui, o perspectiva apropiata pentru productie. Probl. Agric. 10:21–32.

209. ————, and ————. 1969. Sterilitatea mascula la floarea-soarelui si perspectiva utilizarii ei in producerea semintelor hibride. An. Inst. Cercet. Cerale Plante Teh. Fundulea Acad. Stiinte Agric. Silvice 35:559–571.

210. ———, and ———. 1970. Imunitate la mana florii-soarelui, conditionate monogenic. Probl. Agric. 2:34–40.

211. ———, and ———. 1970. Obtinerea de hibrizi de floarea-soarelui pe baza androsterilitate genica. An. Inst. Cercet. Cerale Plante Teh. Fundulea Acad. Stiinte Agric. Silvice 36:281–290.

212. ———, and ———. 1971. Aspecte genetice privind rezistenta la cadere a florii-soarelui, An. Inst. Cercet. Cerale Plante Teh. Fundulea Acad. Stiinte Agric. Silvice 37: 183–190.

213. ———, and ———. 1971. Ereditatea imunitatii la mana a florii-soarelui. An. Inst. Cercet. Cerale Plante Teh. Fundulea Acad. Stiinte Agric. Silvice 37:209–217.

214. ———, and ———. 1971. Pollen fertility restorer gene from cultivated sunflower (*Helianthus annuus* L.). Euphytica 20:536–541.

215. ———, ———, and P. Caramangiu. 1974. Maintenance of the biological and gentic value of inbred lines in sunflower hybrid seed production. p. 427–433. *In* Proc. 6th Int. Sunflower Conf. (Bucharest, Romania).

216. Vrebalov, T. 1975. Studies of the ecological adaptability of the genetically stable inbred lines and the existing varieties of sunflower from the aspect of yield, oil content, and disease resistance. Research Report, Grant No. FG-YU-225. Institute of Agric. Res., Novi Sad, Yugoslavia.

217. Vulpe, V. V. 1968. Tipuri de androsterilitate citoplasmatica la floarea-soarelui. Commun. Bot. 7:121–133.

218. ———. 1972. Surse de androsterilitate la floarea-soarelui, An. Inst. Cercet. Cerale Plante Teh. Fundulea Acad. Stiinte Agric. Silvice 38:273–277.

219. ———. 1974. Single, three-way and double-crosses in sunflower. p. 443–449. *In* Proc. 6th Int. Sunflower Conf. (Bucharest, Romania).

220. Walker, F. H., Jr. 1936. Observations on sunflower insects in Kansas. J. Kans. Entomol. Soc. 9:16–25.

221. Whelan, E. D. P. 1976. Sterility problems in interspecific hybridization of *Helianthus* species. p. 5–6. *In* Proc. 1st Sunflower Forum (Fargo, N.D.).

222. Wilkins, H. D., and C. M. Swallers. 1972. Sunflower production in North Dakota. N. Dak. State Univ. Circ. A-538 (Rev.).

223. Zimmer, D. E. 1972. Field evaluation of sunflower for downy mildew resistance. Plant Dis. Rep. 56:428–431.

224. ———. 1974. Physiological specialization between races of *Plasmopara halstedii* in America and Europe. Phytopathology 64:1465–1467.

225. ———. 1977. Rust resistance of today's sunflower hybrids. p. 7. *In* Proc. 2nd Sunflower Forum (Fargo, N.D.).

226. ———, and G. N. Fick. 1974. Some diseases of sunflowers in the United States—their occurrence, biology, and control. p. 673–680. *In* Proc. 6th Int. Sunflower Conf. (Bucharest, Romania).

227. ———, and M. L. Kinman. 1972. Downy mildew resistance in cultivated sunflower and its inheritance. Crop Sci. 12:749–751.

228. ———, ———, and G. N. Fick. 1973. Evaluation of sunflowers for resistance to rust and Verticillium wilt. Plant Dis. Rep. 57:524–528.

229. ———, and D. A. Rehder. 1976. Rust resistance of wild *Helianthus* species of the North Central United States. Phytopathology 66:208–211.

230. ———, and D. C. Zimmerman. 1972. Influence of some diseases on achene and oil quality of sunflower. Crop Sci. 12:859–861.

231. Zimmerman, D. C., and G. N. Fick. 1973. Fatty acid composition of sunflower (*Helianthus annuus* L.) oil as influenced by seed position. J. Am. Oil Chem. Soc. 50: 273–275.

232. ———, and D. E. Zimmer. 1978. Influence of harvest date and freezing on sunflower seed germination. Crop Sci. 18:

GERHARDT N. FICK: B.S. and M.S., University of Minnesota, Ph.D., University of California, Davis. Sunflower breeder and geneticist, SEA-USDA, 1971–1977; sunflower breeder and research director, SIGCO Sun Products, Breckenridge, Minnesota, 1977—present; major contribution to breeding and development of sunflower hybrids using cytoplasmic male sterile and fertility restorer system; participated in development of numerous parental lines, including several fertility restorer lines that are used as male parents of most hybrids currently available to U.S. growers; author or coauthor of more than 30 scientific publications, primarily on sunflower breeding, genetics, and production practices.

Chapter 10

Cytology and Interspecific Hybridization

ERNEST D. P. WHELAN

Chromosome morphology, ploidy, microsporogenesis and megasporogenesis of sunflower are discussed in this chapter. Information also is given on the effects of certain mutagens on chromosomal behavior, and a description of meiosis both in genetic and cytoplasmic male steriles. Such information is becoming increasingly important to sunflower breeders, many of whom are using interspecific hybridization as a potential source of cytoplasmic male sterility, fertility restoration, and insect and disease resistance. Furthermore, the sunflower is well suited to cytological investigations, and meiosis can be studied relatively easily with standard techniques.

The chapter concludes with a presentation of data both in narrative and tabular form on fertility and chromosomal behavior of interspecific hybrids of *Helianthus* spp. Species classification is based on the taxonomic treatment of Heiser et al. (35), and some information is given on all 49 species occurring in North America.

CHROMOSOME MORPHOLOGY

The genus *Helianthus* has a basic chromosome number of $n = 17$ (10, 22, 25) and contains diploid ($2n = 34$), tetraploid ($2n = 68$), and hexaploid ($2n = 102$) species.

Relatively little is known about chromosome morphology of *Helianthus* despite the importance of the sunflower crop and the many interspecific crosses that have been made. The morphology of the pachytene chromosomes has yet to be described in the literature, and to date, detailed karyotype studies have been conducted on three species only, namely, *Helianthus annuus* L., *H. debilis* Nutt., and *H. mollis* Lam. by Georgieva-Todorova and co-workers (13, 17, 75). Vranceanu (75) illustrates the chromosomes and idiogram of the common sunflower, *H. annuus,* from

From: Carter, Jack F. (ed.). 1978. *Sunflower Science and Technology.* Agronomy 19. Copyright © 1978 by the American Society of Agronomy, Crop Science Society of America, and Soil Science Society of America, 677 South Segoe Road, Madison, WI 53711 USA.

one of the foregoing studies. The 17 chromosomes divide into four groups based on the position of the centromere and the presence or absence of chromosomes with satellites. The first group contains two pairs of satellited chromosomes with submedian centromeres; the second group, five pairs of chromosomes with median to submedian centromeres; the third group, six pairs with submedian centromeres; and the fourth group, four pairs with submedian to subterminal centromeres. Chromosome length of the 17 chromosomes varies from 3.05 to 6.20 μ. Heiser and Smith (33) illustrate the somatic chromosomes of the 'Mammoth Russian' cultivar of *H. annuus*. Their drawing shows three rather than two pair of satellited chromosomes. They state that the karyotypes of *H. annuus* and *H. debilis* are very similar.

H. mollis differs somewhat from *H. annuus* and *H. debilis* (17). The chromosomes are smaller and the mean total length varies only from 3.16 to 4.50 μ. The chromosomes of *H. mollis* are divided into three main groups. The first group contains the two pairs of chromosomes with submedian centromeres and a satellite in the short arm. One of these chromosomes is the longest in the total complement. The second group has 11 pairs of chromosomes with submetacentric centromeres and the third group has four pairs with subterminal centromeres. The short arms of this latter group of chromosomes are noticeably shorter than those of the other chromosomes.

The 34 somatic chromosomes of *H. giganteus* L., illustrated in Fig. 6F, appear to have only one pair of satellited chromosomes. Long (55) examined hybrids between *H. giganteus* and *H. salicifolius* A. Dietr. and found the somatic chromosomes varied in length from 3.2 to 6.5 μ with a mean of 4.5 μ. He did not mention satellited chromosomes.

EUPLOIDY AND ANEUPLOIDY

Diploid, tetraploid, and hexaploid species occur naturally in *Helianthus,* and in two species, namely, *H. ciliaris* DC. and *H. strumosus* L., both tetraploid and hexaploid races occur within the species. *H. decapetalus* L. also contains both diploid and tetraploid races.

Haploids also are found occasionally at a very low frequency, as mentioned by Fick in his chapter. Triploids, as would be expected, are completely sterile or nearly so and one, *H.* × *multiflorus* L., is used as an ornamental; this taxon is a hybrid between *H. annuus* and the tetraploid race of *H. decapetalus* (33).

Aneuploids have been reported in the genus. Gundaev (19) states that twin embryos may have intermediate numbers of chromosomes between the haploid and diploid count and a continuous aneuploid series could be developed from such material. The maximum frequency of aneuploids was found in the group with $2n = 28$. Many of the embryos studied displayed chimeral structure with cells having 24 to 28 chromosomes.

Aneuploids can result from hybridization. Leclercq et al. (47) obtained trisomic plants ($2n + 1$) in backcross progeny of *H. tuberosus* L. × *H.*

annuus hybrids. Such progeny were resistant to downy mildew (*Plasmopara halstedii* (Farl.) Berl. & de Toni) and they postulated that the extra chromosome came from the *H. tuberosus* genome. The author also has found trisomic plants in backcross progeny of *H. petiolaris* Nutt. × *H. annuus,* and *H. maximiliani* Schrad. × *H. annuus.*

MICROSPOROGENESIS

Helianthus spp. have good characteristics for meiotic studies. The disc flowers usually are numerous and develop sequentially from the periphery of the head towards the center. Consequently, meiotic stages extending from early prophase to initial pollen development occur in a single head. When meiosis is occurring, the flower heads are quite large, exposed, and easily accessible. In the monocephalic cultivated types, a wedge containing most stages of meiosis can be removed from the head and the remainder left to develop normally. Meiosis within a floret is relatively synchronous so that meiocytes in all five anthers are at relatively the same stage of meiosis. The anthers, which form a cone, are 1.5 to 2.0 mm long when meiosis occurs. They can be removed readily from the flower to make squash preparations. If a specific meiotic stage is required, a single anther can be used to observe the developmental stage prior to detailed examination. Standard cytological techniques involving acid-alcohol fixatives and carmine stains can be used to study meiosis.

Despite the foregoing characteristics, no detailed meiotic studies of the common sunflower, *H. annuus,* have been published. The following description of meiosis is based on observations from smear preparations by the author. Samples were fixed in Newcomer's solution (61), the meiocytes stained in propiono-carmine, and the slides examined with phase contrast optics.

The first indication of the onset of meiosis is the development of the callose wall, an unbranched β-1, 3-linked glucan, surrounding each meiocyte during meiosis. When observed with phase contrast optics, these walls appear as grey, opaque, spherical masses with apparent holes, evenly distributed throughout the wall. Whelan (76) examined six different species of *Helianthus* and suggested that this configuration was characteristic of the genus. Subsequent studies of additional species supported this postulation. Examination of the callose wall with the scanning electron microscope (79) confirmed the presence of actual holes and intermeiocyte connections in four genera but they were not detected in *Helianthus.* Therefore, in *Helianthus* the apparent holes appear to be a surface patterning rather than actual discontinuities in the callose wall. In addition to the presence of the callose wall, the meiocytes develop a prominent nucleolus and the chromosomes are evident as chromatic threads throughout the cytoplasm. In zygotene, pairing of the chromosomes can be seen and synapsis appears to start near the centromere (Fig. 1A). Unfortunately, chromosome clumping occurs in both zygotene and pachytene, so individual chromosomes rarely can be identified. This type of chromosome behavior is encountered in many

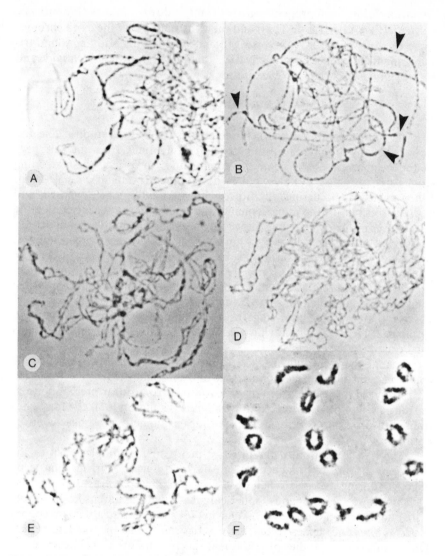

Fig. 1—Photomicrographs of microsporogenesis of *H. annuus*. Propiono-carmine smear preparations. A. Zygotene with partial pairing of the homologous chromosomes. Synapsis appears to initiate near the centromere and proceed distally (×1010). B. Pachytene. The bivalents tend to clump and few bivalents can be distinguished. The centromere (arrow) of two bivalents with submedian centromeres (top and bottom) and two with subterminal centromeres (center left and right) are indicated (×1010). C,D,E. The diffuse stages occurring during diplotene. Repulsion occurs between the homologous chromosomes (Fig. C) until a diffuse stage rather similar to leptotene-zygotene develops (Fig. D). With further contraction in late diplotene (Fig. E), bivalents stain readily and are more visible but bivalent counts are not possible (×880, 760, 760, respectively). F. Diakinesis with 12 to 13 ring and 4 to 5 rod bivalents (×1010).

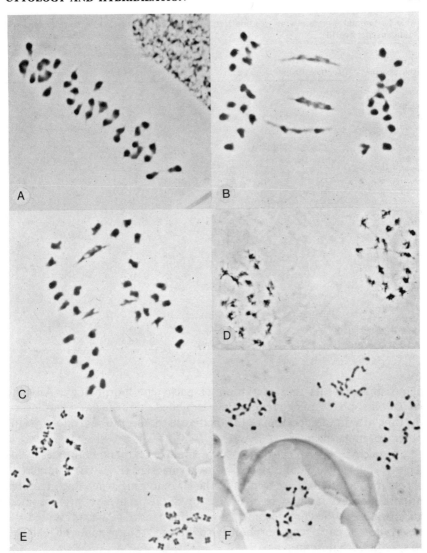

Fig. 2—Photomicrographs of microsporogenesis of *H. annuus*. Propiono-carmine smear preparations. A. Metaphase I with 13 ring bivalents and 4 rod bivalents (No. 1, 10, 13 and 16 from right hand side). Rod bivalents have started to undergo disjunction (× 975). B. Anaphase I with 14:14 segregation and three bivalents still in disjunction (× 975). C. Anaphase I with the last bivalent starting disjunction (× 975). D. Interphase. Chromatids have become diffuse and no longer can be counted (× 685). E. Metaphase II. Individual chromatids of each pair are evident and one pair with a median centromere has divided (× 685). F. Anaphase II. Seventeen chromatids have segregated to each of the four poles (× 400).

Table 1—Estimates of minimum chiasma frequency per cell for several *Helianthus* spp. and interspecific hybrids.

Female parent	Male parent	No. of cells	Chiasma frequency Range	Chiasma frequency Mean	Pollen fertility	Source
					%	
mollis		50	20–29	28.7	97.0	40
mollis		50	20–26	22.5	98.7	40
mollis	*divaricatus*	50	19–24	20.8	49.4	40
mollis	*atrorubens*	72	17–22	18.9	34.0	40
mollis		50	20–29	23.8	97.0	41
mollis	*occidentalis*	47	17–24	20.2	49.5	41
occidentalis		50	35–30	27.4	98.6	41
grosseserratus		50	21–28	23.8	99.7	40
grosseserratus	*mollis*	50	20–27	22.4	94.4	40
grosseserratus		12		23.9		51
grosseserratus	*salicifolius*	12		27.5		51
salicifolius		12		25.2		51
divaricatus		50	20–26		100.0	40
giganteus		50	20–30	22.0	99.4	40
giganteus		50	20–28	23.5	99.7	40
giganteus	*mollis*	50	20–26	22.9	59.7	40
atrorubens		50	29–32	31.0	92.4	40
annuus		48	24–30	28.1	95.0	†
annuus				23.9		12
argophyllus				20.8		12
argophyllus	*annuus*			19.9		12
decapetalus (4n)		120		39.3	96.0	16
rigidus				66.3	75.0	13

† Whelan, unpublished data.

plants and prevents the description of pachytene morphology. All nine *Helianthus* spp. examined to date have exhibited this behavior regardless of the time of year of collection or the fixative used. Georgieva-Todorova (13) observed similar behavior.

In pachytene, the centromeres are delimited clearly on both sides by intensely stained chromatic regions of varying size (Fig. 1B). Numerous chromatic regions also are evident in the chromosome arms. Such features usually allow easy and accurate description of the pachytene chromosomes. Unfortunately, the chromosome clumping in pachytene has prevented the description of pachytene morphology so linkage groups cannot be assigned to specific chromosomes.

Chromosome behavior in diplotene (Fig. 1C,D,E) appears to be unusually complex and similar to that observed for tomato, *Lycopersicon esculentum* Mill. (57), sweet cherry, *Prunus avium* L. (78), and *Rosa* spp. (42). The homologous chromosomes undergo a marked repulsion and chiasma frequently are evident. The bivalents become relatively diffuse and this stage may be identified incorrectly as zygotene. As diplotene progresses, the clarity of the bivalents increases. Bivalent counts cannot be made nor chiasma frequency estimated, however, until late diplotene or diakinesis (Fig. 1F) because a tendency for several of the bivalents to remain clumped gives the false impression that multivalents are present. A typical meiocyte in diakinesis contains 11 to 15 ring bivalents and 2 to 6 rod bivalents. Two of the bivalents are associated with the nucleolus. The observations on 48 such meiocytes of the cultivar 'Saturn' showed a mean chiasma frequency per cell of 28.1 with a range of 24.0 to 30.0. Georgieva-Todorova (12) estimated 23.9 chiasmata in the cultivar 'Peredovik.' Estimates for some other species appear in Table 1.

By metaphase I (Fig. 2A), the nucleolus, which has been evident throughout meiotic prophase, has disappeared and the intensely stained, highly contracted bivalents have become oriented on the metaphase plate. Sequential, not simultaneous, disjunction of the bivalents occurs in anaphase I (Fig. 2B,C). One to three of the bivalents, usually the rod bivalents, rapidly undergo disjunction and the resulting chromatids begin moving to the poles. Three of the bivalents frequently can be seen under-

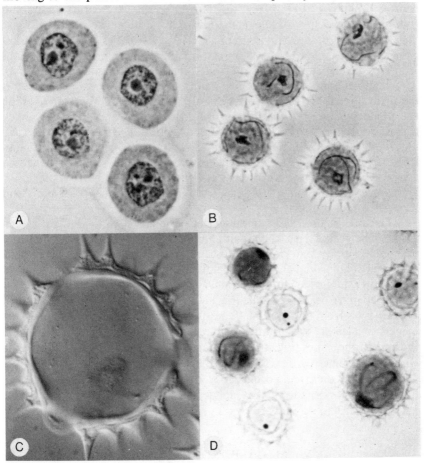

Fig. 3—Photomicrographs of microsporogenesis of *H. annuus* and its hybrids. Propiono-carmine smear preparations. A. Cytokinesis with the formation of four, uninucleate spores (× 960). B. Mature pollen containing the two long, sperm-like generative nuclei and the oval vegetative nucleus (× 295). C. Mature pollen grain (Nomarski interference phase optics) illustrating the spiny exine and three germ pores (× 900). D. Pollen grains from *H. rigidus* × *H. annuus* F₁ hybrid. Three grains are trinucleate, one is largest. Three aborted grains contain degenerative nuclei and little cytoplasm (× 275).

going disjunction after the other chromatids have migrated to the poles. Some meiotic abnormalities usually may occur in fully fertile cultivars, and up to 5% of the meiocytes contain such abnormalities as bridges and fragments or lagging chromosomes. Georgieva-Todorova (15) reported similar observations. Probably such abnormalities are due to spontaneous breakage and exchange rather than the presence of a paracentric inversion (48).

During telophase, a nucleolus forms, the chromatids lose their regular structure, and by interphase (Fig. 2D) chromosome counts no longer are possible. In metaphase II (Fig. 2E), the chromatids again can be counted and in those with median centromeres, the four arms of the diad frequently are evident. In anaphase II (Fig. 2F), disjunction of the chromatids occurs and four groups develop with 17 chromatids in each group. The chromatids lose their regular structure and nucleoli develop. Cytokinesis of the simultaneous type occurs at the end of the second meiotic division, peripheral constrictions develop, and walls form from these constrictions inward. Normally, four potential pollen grains form and their arrangement in the tetrad is tetrahedral (Fig. 3A). Following cytokinesis, the callose wall, surrounding each meiocyte since the onset of meiosis disintegrates, and the young pollen grains are released into the anther locule. As the pollen grain develops, two mitotic divisions of the nucleus occur to produce a trinucleate pollen grain (Fig. 3B) at anthesis. The vegetative or tube nucleus is oval, and the two generative or sperm nuclei are linear and sperm-like. At maturity, the pollen grains are yellow-orange, spherical, covered with spines (echinate), and have three apertures (Fig. 3C).

The author also used scanning electron microscopy to view the pollen grains. The grains were coated with a coat of gold about 250 Å thick and examined with a Cambridge Stereoscan MK IIa. In polar view, the grains show three equidistant furrows or colpi in the wall (Fig. 4B). The equatorial view shows the furrow to extend almost from pole to pole with a recessed area or aperture near the middle (Fig. 4A). The germinating pollen tube (Fig. 4 C) emerges from one of these "apertures", regions of the outer wall layer or exine which are very thin. Possible perforations do occur, however, at the base of the spines (Fig. 4E). The diameter of the body of the pollen grain without the spines, varies from 33 to 39 μ. Presumably aborted grains (Fig. 4D) appear smaller than normal, may lack colpi, and have more spines per unit area than normal grains. Gundaev (19) reports pollen grain diameter varying from 30 to 45 μ.

MEGASPOROGENESIS

Detail on the development and structure of the sunflower megagametophyte was lacking until this decade. In 1973, Newcomb (59) described development from the megaspore mother cell (MMC) to the mature megagametophyte. He used the cultivar 'Peredovik' with both light and electron microscopy techniques, and reported apparent nutritional interrelationships of the tissues within and adjacent to the embryo sac.

Newcomb's work showed that the ovule is tenuinucellar, the large MMC being surrounded by a single layer of nucellar cells. The MMC undergoes two meiotic divisions. The cell first divides transversely and unequally so that the chalazal cell is the larger. The chalazal cell again divides unequal-

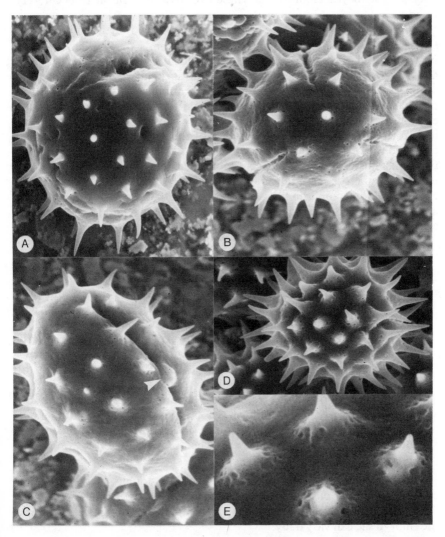

Fig. 4—Scanning electron micrographs of *Helianthus* pollen grains. A. Cultivated *H. annuus* illustrating the colpi or furrow with its central aperture through which the pollen tube emerges (× 1300). B. A polar view of a *H. maximilliani* pollen grain. Note the tricolpate nature of the grain (× 1585). C. The start of emergence of the pollen tube from the aperture of a *H. maximilliani* pollen grain (× 1510). D. A sterile pollen grain from the F₁ hybrid *H. petiolaris* × *H. annuus*. The frequency of spines per unit area appears to be much greater than in normal grains (× 1300). E. Possible perforations in the exine at the base of the spines in *H. annuus* (× 2950).

ly, finally producing a linear tetrad of haploid megaspores with the largest cell at the chalazal end. Following these divisions, the three micropylar megaspores and the nucellus degenerate and the functional megaspore at the chalazal end of the tetrad enlarges. Two mitotic divisions without wall formation occur in the chalazal cell to produce first a two-, then a four-nucleate coenocytic embryo sac. These two divisions appear to occur very rapidly.

Following these divisions, cell walls form and two further mitotic divisions occur to produce a seven- or eight-nucleate, six-celled, embryo sac. Newcomb (59) observed neither actual wall formation nor mitotic divisions and suggested that they occurred rapidly or at night when he collected no samples. The six-celled embryo sac usually contains two antipodal cells in the chalazal end rather than the usual three of the normal *Polygonum*-type embryo sac. Further coenocytic division may occur in these cells following their formation and prior to maturation of the embryo sac. The antipodal cells begin to degenerate before fertilization. The central cell containing the two polar nuclei, occupies the center of the embryo sac. The polar nuclei usually fuse before fertilization to form the single fusion nucleus. This cell contains all the cytoplasm that was not included in the antipodal, synergid, and egg cells. The two synergid cells, which appear to be haploid, and the egg cell are located at the micropylar end of the embryo sac.

Newcomb (60) also describes embryo sac development after fertilization:

"The degeneration of one synergid denotes the initiation of embryo and endosperm development. The other synergid is present until the late globular stage of embryogenesis. The primary endosperm nucleus divides before the zygote nucleus forming a coenocytic nuclear endosperm. When about eight endosperm nuclei are present during the early globular stage of embryogenesis, wall formation starts at the micropylar end of the embryo sac. The suspensor of the embryo consists of a large basal cell during the proembryo stages, a single row of cells during the early globular stages, and at the later globular stage a double tier of cells near the radicle of the embryo and a single row at the micropylar end of the embryo sac. Occasional embryo development ocurs in the absence of endosperm when only single fertilization takes place."

Polyspermy also can occur (73). Following double fertilization, the ovule may receive one or several pairs of sperm cells. Even when pollen from the same male parent is used daily, the number of sperm cells entering the embryo sac can vary daily.

MALE STERILITY

Two main types of sterility occur, namely, genetic male sterility (GMS) and cytoplasmic male sterility (CMS). The latter type only has been known for about a decade. The author is aware of only one publication by Paun (62) comparing cytological behavior of both GMS and CMS germplasm.

Paun (62) used GMS lines AS-110 and AS-116, both containing the ms_1 gene and behaving similarly. Diplotene-diakinesis was characterized by two to four univalents arising from either asynapsis or desynapsis. In subsequent stages, she observed unequal disjunction, equational division, and elimination of the univalents. Chromosome clumping and "agglutinations" also were evident along with chromatic bridges and fragments. Tetrad stages frequently were abnormal and contained micronuclei. At anthesis, 4 to 6% of the pollen stained with acetocarmine but it was only about 75% of the size of fertile pollen and lacked the spiny exine characteristic of *Helianthus* spp. Paun questioned the viability of it and doubted that it could compete with normal pollen. Paun (62) also investigated four CMS A-lines with the corresponding fertile B-lines. All originated from materials developed by Leclercq through hybridization of *H. petiolaris* and *H. annuus*. Two types of meiotic behavior were observed. The sporogenous cells in two of the four CMS lines degenerated before meiosis and produced no pollen. Only necrotic anthers were present. Meiosis in the other two lines was completely normal to tetrad stages, except that the frequency of sporogenous tissues undergoing meiosis was much less than normal. Pollen degeneration occurred following release of the spores from the tetrad. Meiosis and pollen development were normal in all four fertile B-lines.

Recently, Horner (36) used both light and electron microscopy techniques to compare microsporogenesis in a normal, fertile B-line, HA 232, and a CMS A-line of HA 232, developed by backcrossing. The study primarily involved the behavior of the anther and its sporogenous tissues rather than that of the chromosomes. Horner divided microsporogenesis into 11 stages, with stages 1 to 4 ranging from the first sporogenous tissue to the initiation of tetrads, and stages 5 to 11 from late tetrad to mature pollen. Both the A- and B-lines developed similarly until stage 4 but development within the A-line did not go beyond stage 5. By stage 5, disintegration of the tapetum and tetrads had occurred, resulting in sterility. Although a rudimentary exine formed around each microspore prior to their release from the tetrad, the characteristic spines of *Helianthus* pollen never developed. By the end of stage 5, the enlarged tapetal cells had degenerated and the middle layer cells of the anther wall had enlarged, flattening and crushing the microspore tetrads. The tapetum and tetrads rapidly became grossly disorganized, resulting in complete abortion.

The author also has studied meiosis in several CMS A-lines, and found it completely normal prior to tetrad formation in all but one line. The one exception was a sterile form of the cultivar 'Krasnodarets' produced by backcrossing to a cytoplasmic male sterile accession from Leclercq. In this material, 15 to 20% of the meiocytes at anaphase I or later division stages contained evidence of a single bridge and fragment, indicating a paracentric inversion. The author has observed similar phenomena in F_1 hybrids and backcross progeny of *H. petiolaris* × *H. annuus*. Chiasma frequency was 30.1 per cell in one normal A-line examined compared to 28.1 for the cultivar 'Saturn'.

Other studies (15, 58, 64) have involved genetic male steriles. Nakashima and Hosokawa (58) found that meiosis in the sterile line P21 *ms*, proceeded normally prior to the tetrad stage. The tapetal cells remained intact and enlarged after spore formation in contrast to their disintegration in normal anthers. As the anther matured, a sudden disintegration of the tapetal cells and also undeveloped pollen grains occurred. Pirev (65) also observed normal meiosis in another genetic sterile prior to tetrad formation. Pollen development stopped completely, however, at the uninucleate and binucleate stages. Sterile pollen grains were low in cytoplasmic and nuclear contents. As anthesis approached, the exine or outer wall of the pollen grain degenerated, the grains clumped, and dehiscence failed to occur. Georgieva-Todorova (15) obtained genetic male steriles from hybrids of *H. grosseserratus* Martens × *H. annuus*. One type had normal meiosis, and pollen abortion occurred after spore release from the tetrad. In the other, sporogenous tissue degenerated prior to meiosis.

In conclusion, male sterility in sunflower due to either genetic or cytoplasmic factors may be expressed from early meiosis to development of the pollen grain.

Biochemical differences also are evident between fertile and sterile lines. Pirev (64) could differentiate fertile and sterile forms by starch concentration in the pollen grains four to five days before flowering. Their protein concentration also differed. Protein increased throughout pollen grain development in fertile forms but was irregular and always less in sterile forms. Differences were most obvious shortly before anthesis. Kovacik and Kryzanek (44) found lower sucrose concentration in sterile pollen but the concentration of other saccharides was similar in sterile and fertile pollen. Amino acid content also varied and they found that the concentration of proline was high and asparagine low in fertile pollen; the reverse was true with sterile pollen. Phospholipids and nucleic acids were lower in sterile pollen (44), as might be expected.

Atanasoff (3) has suggested that cytoplasmic male sterility in sunflower, and indeed in most plants, is due to viral infection. Attempts to verify this hypothesis for sunflower were unsuccessful (J. A. Hoes and H. Enns, personal communication).

In addition to male sterility due to genetic and cytoplasmic factors, chemically induced male sterility has been reported. Piquemal (63) found that a solution of 20 ppm gibberellic acid (GA), applied directly to flower heads when the young heads were first visible, induced complete male sterility. Anashchenko (1, 2) also obtained complete male sterility in sunflower using GA. Although subsequent seed had perfect germination, it weighed 15 to 25% less than normal. Treatment of wild species of *Helianthus* with gibberellin was ineffective. Beglaryan (5) has reported mutagenic effects attributed to GA treatment. He found a mutant with cytological and morphological abnormalities in the M_4 progeny of 'Saratov Dwarf' following initial double treatment of the seed with 0.01% GA. He also observed an increase in the number of nucleoli per cell, some triploid and tetraploid cells, and several cells containing chromosome bridges, fragments, and lagging chromosomes. Initial studies by Campos (6) show that ethrel (2-chloroethylphosphonic acid) also may induce male sterility.

MUTAGENESIS

Mutagenesis already has been mentioned by Fick in his chapter, so only chromosomal effects will be discussed here.

Work by Gundaev (19) indicates that a radiation does of 16 kr to dry sunflower seed is virtually lethal. He reports that Beletskiy treated air-dried sunflower seed with 4 kr of X-rays, and moist, swelled seed with 1 kr and obtained about equal frequencies of chromosomal aberrations. The frequency of abnormal anaphase and telophase divisions was about 20%. Georgieva-Todorova (11) also observed cytological aberrations in anaphase following treatment of seed with X-rays. Chromosome bridges with one or two fragments were the most common defect. In contrast to Gundaev (19), however, she found seed germinated normally following irradiation with 8, 12, 16, 32, or 40 kr of X-rays. Gamma radiation from a cobalt-60 source has been reported to affect length and stainability of sunflower chromosomes (45). A dose of 2 kr applied to air-dried seeds immediately before sowing shortened the chromosomes and improved stainability with acetocarmine. The chromosomes became thinner with doses of 10 kr or more, reaching a limiting value at 20 to 30 kr. At 60 kr, stainability was reduced.

Chemical mutagens can affect sunflower, and Gundaev (19) cites the following critical concentrations: ethylenimine = 0.05%; dimethyl sulphate = 0.15%; diethyl sulphate = 0.2%; 1,4-bis-diazoacetylbutane = 0.1%; N-nitrosomethyl urea = 0.001%; N-nitrosoethyl urea = 0.001%. Treatment with 8-ethoxycaffeine has induced chromosome breakage (18).

Induction of polyploidy has been accomplished successfully using colchicine. Heiser and Smith (34) grew young seedlings for 8 hr on filter paper saturated with a 0.2% solution of colchicine and obtained some chromosome doubling. Using this technique, they obtained tetraploids from the perennial hybrids *H. giganteus* × *H. microcephalus* T. and G. One plant came from this cross, two from the reciprocal cross, and two others from *H. maximiliani* × 2n *H. decapetalus*. All five plants showed over 90% pollen stainability but somewhat reduced seed set. Most meiocytes had one to three quadrivalents, but an occasional one contained 34 bivalents. *H. decapetalus* was the only diploid species doubled successfully. Meiocytes contained from one to five quadrivalents. The most frequent configuration was 28 bivalents and three quadrivalents. Unfortunately, the plant died before pollen fertility could be determined. Colchicine-induced tetraploids of *H. annuus* also have been obtained (9). In diakinesis, 71% of the meiocytes contained 34 bivalents. The remaining 29% contained quadrivalents, trivalents, and univalents. About 67% of anaphase I meiocytes showed unequal division. Micronuclei and lagging chromosomes also were seen.

Despite these meiotic abnormalities, pollen fertility was estimated to be 83.8%. The tetraploids had a greater range in pollen grain size with the average similar to diploids.

INTERSPECIFIC HYBRIDIZATION

Interspecific hybridization of *Helianthus* has been conducted for many years. The main emphasis in the USSR was to obtain resistance to pests. The early interest in North America was more theoretical, and Heiser and coworkers published many reports on the taxonomy, morphology, and hybridization of *Helianthus* spp. In 1969, Leclercq (46) reported cytoplasmic male sterility (CMS) in backcross progeny of the hybrid *H. petiolaris* × *H. annuus*. All North American hybrids, and probably all sunflower hybrids developed to date using the CMS-fertility restorer system, are based on the Leclercq source of cytoplasm. There is now renewed interest in many countries in interspecific hybridization for insect and disease resistance and to find other sources of CMS and fertility restoration.

Hybridization techniques are discussed in the following section and information is presented in tabular form on interspecific hybrids and the behavior of their chromosomes.

Hybridization Techniques

The literature from the USSR discusses irradiation of pollen (72), temperature shocks (66, 67, 74), and grafting (66, 67) as aids in interspecific hybridization. These reports usually involve hybridization between *H. tuberosus* and *H. annuus*. They mention both direct grafting and "vegetative reapproachment". The latter is presumed to be approach grafting where two plants are grafted together while both components retain tissue systems above and below the point of union. The author also has hybridized *H. annuus* with *H. tuberosus, H. rigidus* (Cass.) Desf., and their putative hybrid *H.* × *laetiflorus* Pers. Initial F_1 hybrids, confirmed by chromosome counts of $2n = 68$, were obtained readily. Backcross seed was obtained with difficulty merely by repeated pollinations with *H. annuus* pollen. Researchers in France apparently have used similar methods with *H. tuberosus* × *H. annuus* hybrids (7, 47).

Whelan (77) used wild *H. annuus* as an "intermediate" parent or "bridge" to procure the first hybrids to be obtained between this species and both *H. giganteus* and *H. maximiliani*. Direct hybridization with the cultivar 'Krasnodarets' gave a single, highly sterile hybrid with *H. giganteus*. Repeated pollinations with or without subsequent embryo culture failed to give backcross progeny. Pollinating both *H. giganteus* and *H. maximiliani* with wild *H. annuus,* however, produced three and four hybrids, respectively. These hybrids subsequently gave small quantities of seed when pollinated with 'Krasnodarets' pollen. Continued backcrossing by commercial cultivars has yielded apparent cytoplasmic male sterile segregants from both the *H. giganteus* and *H. maximiliani* cytoplasm.

Interspecific Hybrids

This discourse and the accompanying tables are based on the taxonomic treatment of the genus by Heiser et al. (35), who recognized 49 species and divided them into three sections: namely, Annui (Table 2) with 13 species; Ciliares (Table 3) with two subgroups, *Pumali* and *Ciliares,* each containing three species; and Divaricati. Heiser et al. (35) further divided Divaricati into five subgroups: *Divaricati* (Table 4) with nine species, *Gigantei* (Table 5) with eight, *Microcephali* (Table 6) with five, *Angustifolii* (Table 7) with three, and *Atrorubentes* (Table 8) with five species. In addition to the tables, information is given for some species. The reader should consult this information as well as the chapter by Heiser. Only a few tables show crosses at the subspecies level, because most authors do not report the subspecies used. The tables also include reports of successful hybridizations, but these reports did not state the number of hybrids obtained, their pollen fertility, or meiotic behavior.

The following comments apply to Table 2 through 8.

(a) Parents: Female parent followed by the basic chromosome number, *n,* in parenthesis is listed first and italicized, followed by male parents inset two spaces. If the male parent has different possible ploidy levels, the appropriate level is shown in parentheses after the species name. A reciprocal cross is indicated by " ♀ " following inset species name.

(b) Pollen stainability: Where the report gives no numerical data, 0 = sterile, + = low fertility, + + = intermediate fertility, and + + + = high fertility.

(c) Chromosome pairing: R = regular or normal pairing, M = multivalents, U = univalents, B + F = bridges and fragments, X = defect was observed during meiosis.

(d) Type hybrid: A = artificial, N = natural, S = spontaneous garden hybrid.

(e) Source: numbers correspond to those in "Literature Cited."

Section Annui (Table 2)

H. agrestis Pollard. A highly self-compatible species. Only two successful hybridizations have been obtained and the progeny were completely sterile (35). Crosses with eight annual and eight perennial species failed to produce seed (30).

H. deserticola Heiser. A few seeds were obtained from crosses with *H. annuus, H. anomalus* Blake, and *H. petiolaris* but they failed to germinate. Reciprocal crosses failed (28).

H. anomalus. The seed is very difficult to germinate. Crosses with *H. annuus, H. deserticola, H. debilis* ssp. *debilis,* and *H. petiolaris* ssp. *fallax* Heiser, produced filled achenes but only those from the latter two crosses

Table 2—Summary of interspecific hybrids involving *Helianthus* section Annui with other
Helianthus spp.

Parents†	No. hybrids	Pollen stainability Range	Avg.	R	M	U	B+F	Type hybrid	Source
agrestis (17)									35
simulans	14	No pollen						A	35
floridanus (♀)	10	No pollen						A	30
similis (unknown)									35
deserticola (17)									28
anomalus (17)									35
debilis (♀)	5	30–60						A	35
petiolaris	1		2					A	35
niveus ssp. *niveus* (17)									32
ssp. *tephrodes*	5	50–90		X				A	35
laciniatus				X				A	35
microcephalus	6		10	X	X			A	30
nuttallii	2		10	X	X			A	30
occidentalis	15		9	X	X			A	30
debilis		20–88	61	X	X			A	35
praecox	2		5					A	35
spp. *canescens* (♀)			+ +					A	35
debilis	2		5	X	X			A	30
pumilis	1		13	X	X			A	30
ssp. *canescens* (17)									35
spp. *niveus*			+ + +					A	35
spp. *tephrodes*	4		99					A	30
angustifolius	8		10		X	X	X	A	30,31
annuus			+		X	X		A	35
argophyllus			+		X	X		A	35
bolanderi			+		X	X		A	35
neglectus			+		X	X		A	35
praecox			+		X	X		A	35
petiolaris			+					A	35
debilis	3		22	X	X			A	30
ssp. *tephrodes* (17)									35
ssp. *niveus*	1		90	X				A	35
spp. *canescens* (♀)	4		99					A	30
debilis (17)									35
niveus spp. *niveus* (♀)	2	34–88		X	X			A	30
niveus spp. *canescens* (♀)	3		18		X			A	30
neglectus	8	16–21						A	27
neglectus	3	46–68			X			A	27
petiolaris		9–73			X			A	29
praecox			+ +		X			A	35
annuus (♀)	17	0–22	6.7		X	X		A,N	23
anomalus	5	30–60	+					A	35
argophyllus			+	X	X			A	26,35
bolanderi			+	X	X			A	26,35
floridanus	2		5		X	X		A	30,31
paradoxus (♀)	6		25		X			A	30
paradoxus	6		4		X			A	30
praecox (17)									26
niveus spp. *niveus* (♀)			+ +	X				A	35
niveus spp. *canescens* (♀			+ +	X	X			A	35
debilis (♀)			+		X			A	35
annuus								A,N	23,35
argophyllus								A	35
bolanderi								A	35
petiolaris								A	35
paradoxus (♀)			+					A	35
occidentalis	2		4					A	30
petiolaris (17)									20,29

(continued on next page)

Table 2—Continued.

Parents†	No. hybrids	Pollen stainability		Chromosome pairing				Type hybrid	Source
		Range	Avg.	R	M	U	B+F		
niveus spp. canescens (♀)			+					A	21,35
annuus		0–30	14	X	X		X	A,N	20,29‡
argophyllus		15–28		X				A	29
bolanderi	3	1–14		X				A	29
debilis (♀)		9–73		X				A	29
neglectus	11	14–80		X				A	27
praecox									35
anomalus (♀)	1		2					A	35
paradoxus (♀)	7		14	X				A	30
neglectus (17)									27
niveus spp. canescens (♀)	1		9	X	X			A	29
petiolaris	16	9–80		X				A	27
annuus	2	2–3		X				A	27
debilis ssp. cucumerifolius	7	12–21		X				A	27
praecox	3	60–81		X				A	27
argophyllus (♀)	1		10					A	30
bolanderi (17)									22
niveus spp. canescens			+	X	X			A	35
petiolaris (♀)	3	1–14		X				A	29
debilis (♀)			+	X	X			A	26,35
praecox								A	35
annuus		2–35	10	X	X		X	A,N	22
paradoxus (17)									27
annuus	7		8	X				A,N	30
annuus	3		12					A,N	30
praecox			+					A	35
debilis	6		25	X				A	30
debilis (♀)	6		4	X				A	30
argophyllus	4		6	X				A	20
petiolaris	7		14	X				A	30
argophyllus (17)									35
niveus spp. canescens (♀)			+	X	X			A	35
petiolaris (♀)		15–28		X				A	29
neglectus	1		10					A	30
debilis (♀)			+	X	X			A	26,35
praecox (♀)			+					A	35
annuus			+ +					A,N	24,35
hirsutus			+					A	35
paradoxus (♀)	4		6	X				A	30
annuus (17)									22,25
niveus ssp. canescens (♀)			+	X	X			A	35
petiolaris (♀)		0–30	14	X	X		X	A,N	20,29‡
neglectus (♀)	2	2–3		X				A	27
debilis	17	0–22	6.7	X	X			A,N	23
praecox								A,N	35
argophyllus (♀)	5	30–36	33	X	X	X		A,N	24
bolanderi		2–35	10	X	X		X	A,N	22
paradoxus (♀)	7		8	X				A,N	30
decapetalus (4n)			0	X		X		A	33
hirsutus								A	35
strumosus								A	35
tuberosus (♀)		0–10	+	X	X		X	A	34,35‡
rigidus (♀)		0–10	+	X	X		X	A	‡
giganteus (♀)	3+1	0–5	0.4	X		X		A	77
maximiliani (♀)	4	0–5	+	X		X		A	77
grosseserratus (♀)	4		0	X				A	15

† See "Interspecific Hybrid" section of narrative for detailed explanation.
‡ Whelan, unpublished data.

germinated (35).

H. niveus ssp. *niveus* (Benth.) Brandegee. It is the only member of the genus known to hybridize with species in all three sections of the genus, i.e. Annui, Ciliares, and Divaricati (35).

H. debilis. This species originally was considered to contain eight subspecies (26) but three of the eight now are included as subspecific races of *H. praecox* (35).

H. praecox Engelm. and Gray. There are three subspecies, which produce fertile hybrids when intercrossed. They originally were included with *H. debilis* subspecies (26). Morphology and chromosome behavior indicate that *H. praecox* is related closely to *H. debilis* and *H. petiolaris.* Chromosome pairing suggests this species is related closely to *H. neglectus* Heiser (35). The cross *H. praecox* × *H. occidentalis* Riddell originally was reported as *H. debilis* ssp. *hirtus* × *H. occidentalis* (30).

H. petiolaris. Heiser (29) studied the cytology and morphology of this species and identified three cytological races. Hybrids among members of any cytological race are fertile but hybrids between races show some reduction in fertility. There are two subspecies, ssp. *petiolaris* and ssp. *fallax,* in addition to the cytological races. Crosses of *H. petiolaris* with *H. carnosus* Small, *H. decapetalus, H. mollis, H. pumilus* Nutt., *H. silphioides* Nutt., and *H. agrestis* were unsuccessful (29).

The presence of a paracentric inversion in *H. petiolaris* × *H. annuus* hybrids (20) has been questioned (29), but the present author frequently noted a bridge and fragment (Fig. 5B) in such hybrids, which is evidence of this chromosomal abnormality.

H. neglectus. This species appears to differ from *H. petiolaris* by a single translocation and from *H. annuus* by several translocations (35).

H. annuus. In Chapter 2, Heiser indicates six possible groups of wild and two of cultivated sunflower. As most publications fail to distinguish the type used, no classification of possible *H. annuus* subspecies is included in Table 2.

By hybridization between *H. petiolaris* and *H. annuus* and backcrossing with *H. annuus,* Leclercq (46) transferred the *H. annuus* genome into *H. petiolaris* cytoplasm and obtained the first cytoplasmic male steriles. The present author has used the same technique to obtain what appears to be cytoplasmic male sterility conditioned by the cytoplasm of the three species, *H. petiolaris, H. giganteus,* and *H. maximiliani.*

Heiser and Smith (33) suggest that *H.* × *multiflorus* is the hybrid between *H. annuus* and *H. decapetalus* (4n). The hybrid is a sterile triploid and 17 bivalents and 17 univalents usually occur in meiosis.

Section Ciliares Subsections *Pumili* and *Ciliaris* (Table 3)

H. pumilus. Artificial hybridization with the sympatric species *H. nuttallii* T. and G. was unsuccessful (35).

H. cusickii A. Gray. No seed set was obtained following pollinations with *H. bolanderi* A. Gray, *H. divaricatus* L., and *H. nuttallii* (35).

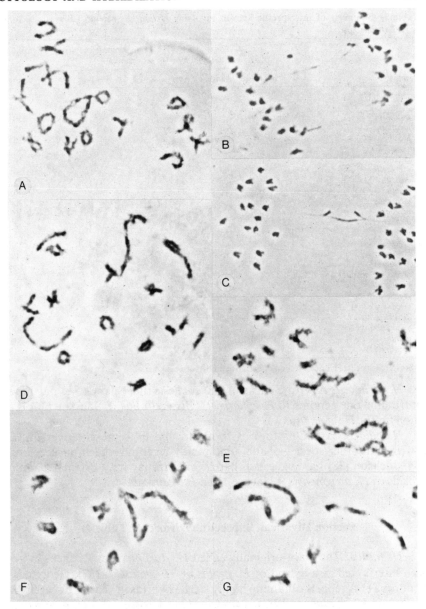

Fig. 5—Photomicrographs of meiosis from interspecific F_1 hybrids of *Helianthus*. Propiono-
carmine smear preparations. A. *H. argophyllus* × *H. annuus*. Diplotene with a circle of
four chromosomes and 15 bivalents. A typical configuration for these hybrids (× 880). B,C.
H. petiolaris × *H. annuus* second backcross hybrids. Anaphase I with (B) and without (C) a
single dicentric bridge and free acentric fragment, indicating paracentric inversion hybridity,
a frequent defect in these hybrids. The meiocytes also illustrate trisomic ($2n + 1$) segregates
(× 685). D,E. *H. maximiliani* × *H. annuus*. Diplotene with 11 bivalents, two chains of four
chromosomes, a trivalent, and a univalent (× 660), and (E) a circle of 10 chromosomes from
a meiocyte containing 10 bivalents and four univalents (× 1045). F,G. *H. giganteus* × *H.
annuus*. A chain of eight chromosomes (F) from a diplotene meiocyte containing 13 bivalents, and (G) a chain of six and of four chromosomes from one containing 12 II (both ×
1045).

Table 3—Summary of interspecific hybrids involving *Helianthus* section Ciliares and other *Helianthus* spp.

Parents†	No. hybrids	Pollen stainability		Chromosome pairing				Type hybrid	Source
		Range	Avg.	R	M	U	B+F		
gracilentus (17)									32
pumilus	5		28	X	X			A	30
pumilus (♀)	1		15					A	30
laciniatus			++					A	30
laciniatus (♀)	2	65,68			X			A	30
pumilus (17)									35
gracilentus	1		15					A	30
gracilentus (♀)	5		28	X	X			A	30
cusickii (♀)	3		30	X				A	35
niveus (♀)	1		13	X	X			A	30
laciniatus	1; died							A	30
niveus (♀)	1		13	X	X			A	30
cusickii (17)									32
pumilus	3		30	X				A	35
arizonensis (17)									38
laciniatus	4		91	X				A	30
laciniatus (17)									35,38
niveus ssp. niveus (♀)			+					A	35
gracilentus	2	35,63						A	30
gracilentus ♀)	2	65,85			X			A	30
arizonensis (♀)	4		91	X				A	30
dissectifolius	3		79					A	30
dissectifolius (♀)	5		93	X				A	30
pumilus (♀)	1; died							A	30
ciliaris (34,51)					X				32,38

† See "Interspecific Hybrids" section of narrative for detailed explanation.

H. laciniatus A. Gray. Jackson (38) suggests that this species may have contributed one genome to *H. ciliaris*. *H. dissectifolius* R. Jackson (30) is a synonym for *H. laciniatus*.

H. ciliaris. Attempts to cross both 4*n* and 6*n* representatives of *H. ciliaris* with other species of the same ploidy level have been unsuccessful (35). Jackson (38) has suggested that *H. ciliaris* may have originated from hybrids of *H. arizonensis* R. Jackson and *H. laciniatus*.

Section Divaricati Subsection *Divaricati* (Table 4)

H. mollis. The plant originally called *H. doronicoides* R. Jackson has been interpreted as a hybrid of *H. mollis* × *H. giganteus* (37). Georgieva-Todorova (14) reports obtaining highly sterile hybrids in crosses between *H. mollis* and *H. rigidus,* and *H. decapetalus* (2*n*). Crosses with *H. divaricatus* were partially fertile. Georgieva-Todorova and others (16) also report crosses with *H. scaberrimus* (4*n*). No tetraploid with this synonym is listed by Heiser et al. (35).

H. occidentalis. There are two subspecies (35), namely, ssp. *occidentalis* (*n* = 17) and ssp. *plantagineus* of unknown chromosome number. All crosses in Table 4 involve the first subspecies only.

H. divaricatus. Crosses with *H. hirsutus* Raf. and *H. tuberosus* produced seed, but it failed to germinate (35). This species is believed to have

Table 4—Summary of interspecific hybrids involving *Helianthus* section Divaricati, subsection *Divaricati*.

Parents†	No. hybrids	Pollen stainability Range	Avg.	R	M	U	B+F	Type hybrid	Source
mollis (17)									32
atrorubens			34.0	X		X		A,N	4,31,40
silphioides								N	4
divaricatus			49.4	X	X			A,N	39,40
grosseserratus (♀)			26	X	X			A,N	39,49
grosseserratus			94.4	X				A	40
giganteus (♀)			59.7	X		X		A,N	37,39,40
occidentalis			49.5	X	X			A,N	39,41
occidentalis (♀)			68	X				A,N	31
decapetalus (2n)	3		82	X				A	30
longifolius (♀)	3		43	X				A	30
maximiliani			77	X				A	31
microcephalus			85	X				A	31
occidentalis (17)									35
niveus spp. niveus (♀)	15		9	X	X			A	30
mollis (♀)			49.5	A	X			A,N	39,41
mollis			68	X				A,N	31
divaricatus								N	31,39
giganteus								N	35
grosseserratus (♀)	8	85-90		X	X			A,N	39,49
longifolius (♀)			34					N	31
atrorubens (♀)			48					A,N	31
maximiliani (♀)	2	78,81		X	X			A	49
maximiliani			80	X				A	31
praecox (♀)	2		4					A	30
nuttallii (♀)		40-70						A	56
divaricatus (17)									32
mollis (♀)			49.4	X	X			A,N	39,40
occidentalis								N	31,39
giganteus (♀)	7		42	X	X			A,N	49
grosseserratus (♀)			78	X				A,N	31
microcephalus (♀)	2		+++	X				A,N	70
decapetalus (2n)			75	X				A	31
giganteus (♀)	5		73	X	X			A	31
salicifolius								N	31
silphioides	1; failed to flower							A	31
hirsutus (34)				X					32,69
annuus			+	X		X		A	34
argophyllus			+					A	35
strumosus	7	81-95	+++	X	X			A	69
smithii	5	80-99	92	X	X			A	34
tuberosus	3	34-40		X				A	34
decapetalus (2n)	1		93	X				A	34
debilis			+	X		X		A	34
decapetalus (17, 34)				X					32,69
annuus			0	X		X		A	33,35
mollis			8			X		A	14
divaricatus (♀)			75	X				A	31
carnosus (♀)	1		40	X		X		A	30,31
giganteus (♀)	1		98	X				A	31
laevigatus	4		90	X	X			A	34
smithii (♀)			87					A	34
strumosus (4n)	2	90,98						A	69
maximiliani (♀)	7		84	X				A	30
debilis			+	X		X		A	34
praecox			+	X		X		A	34
resinosus	1	No pollen						A	34

(continued on next page)

Table 4—Continued.

Parents†	No. hybrids	Pollen stainability Range	Pollen stainability Avg.	R	M	U	B+F	Type hybrid	Source
eggertii (51)									35
strumosus (34, 51)				X					69
annuus			0	X	X			A	34
hirsutus (♀)	7	81–95		X	X			A	69
decapetalus (♀)(2n)	2	90,98						A	69
smithii (♀)	8	83–99	87		X	X		A	34
tuberosus			+++					A	34
rigidus								A	35
resinosus			+++					A	34
maximiliani	2		74					A	49
tuberosus (51)									35
annuus					X	X	X	A,N	34,35‡
hirsutus (♀)(2n)	3	34–40			X	X		A	34
strumosus (6n)			+++					A	30,34
rigidus			+++					A	8,35
resinosus			+++					A	34
schweinitzii			+++					A	34
rigidus (51)									35
annuus		0–10	+	X	X	X		A	‡
strumosus			+++					A	34
tuberosus			+++					A	8,35

† See "Interspecific Hybrids" section of narrative for detailed explanation.
‡ Whelan, unpublished data.

contributed one genome to *H. hirsutus*.

H. hirsutus. The cross *H. hirsutus* × *H. decapetalus* (2n) produced a hybrid with 34 bivalents; apparently an unreduced gamete from the diploid pollen parent was involved (34). The other two tetraploid × diploid crosses, *H. annuus* and *H. debilis,* gave almost sterile triploids.

H. decapetalus. This is the only known species of *Helianthus* which comprises both diploid and tetraploid races. No hybrids have been obtained between the two races (35). The cross *H. annuus* × *H. decapetalus* (4n) is the sterile triploid *H.* × *multiflorus* L., a cultivated ornamental (33). Crosses with *H. debilis* ssp. *cucumerifolius* and *H. praecox* (synonym *H. debilis* ssp. *hirtus*) were 4n × 2n, and gave almost sterile triploids (34). Georgieva-Todorova (16) reported 96% normal pollen in the tetraploid race but only 10 to 12% seed set. In diakinesis, 31% of the cells contained a quadrivalent or univalents. Bridges were seen in 24% of anaphase I meiocytes, and 33% of telophase II meiocytes contained micronuclei. Hybridization with *H. mollis* (14) gave completely female sterile plants with 8% normal pollen. Meiocytes contained numerous univalents and related defects.

H. eggertii Small. This species is related closely to *H. laevigatus* T. and G., which may be involved in its parentage. Heiser et al. (35) suggest that *H. eggertii* should hybridize readily with other hexaploids.

H. strumosus. This is the most variable perennial species. Heiser et al. (35) suggest that the most likely origin of the 4n forms is from hybridization of *H. divaricatus* × *H. grosseserratus,* followed by chromosome doubling.

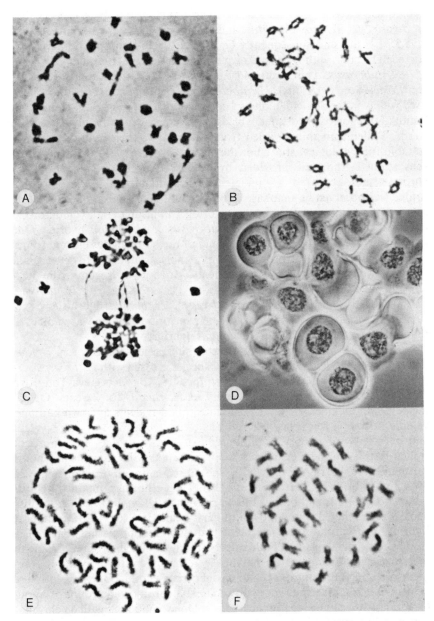

Fig. 6—Photomicrographs of meiosis from interspecific F₁ hybrids of *Helianthus*. Propiono-carmine smear preparations. A. *H. tuberosus* × *H. annuus*. Diplotene with 34 bivalents. Pairing is unusually good for this $6n \times 2n$ hybrid (× 575). B. *H. rigidus* × *H. annuus*. Diplotene with 29 bivalents, two quadrivalents, and two univalents. Pairing is more abnormal in this $6n \times 2n$ hybrid (× 640). C. *H. rigidus* × *H. annuus*. Anaphase I with bridges and lagging chromosomes (× 1010). D. *H. rigidus* × *H. annuus*. Abnormal tetrad formation including diads (× 400). E. The 68 somatic chromosomes from the F₁ hybrid *H. rigidus* × *H. annuus* (× 1010). F. The 34 somatic chromosomes from *H. giganteus*. Only one pair of chromosomes appears to have satellites (× 1010).

Subsequent crossing between this 4n hybrid and various 2n species such as *H. divaricatus, H. grosseserratus,* and *H. glaucophyllus* D.M. Smith should give 3n hybrids and subsequently 6n hybrids if chromosome doubling occurred. The cross with *H. annuus* gave an almost sterile triploid hybrid (34).

H. tuberosus. This hexaploid is one of the putative parents of *H.* × *laetiflorus* (See *H. rigidus*). *H. tuberosus* × *H. annuus* hybrids have been used widely in the Soviet Union as a source of disease resistance. Heiser and Smith (34) reported producing hybrids that were vigorous but almost female sterile. Pollen stainability varied from 12 to 53%. Their hybrids were not suitable for cytological study but they found a mean of 31 bivalents in 20 cells studied from one provided by the USDA. The remainder of the chromosomes appeared as univalents or multivalents. A chromosome bridge was observed in anaphase in only one of 50 cells examined. Kostoff (43) frequently observed chromosome bridges in his material. Cauderon (7) observed two types of meiosis in *H. tuberosus* × *H. annuus* hybrids. One type was strongly asyndetic, that is, many of the homologous chromosomes were not paired, and the other type was weakly asyndetic. However, both types showed meiocytes with 34 bivalents. Aneuploid progeny from such hybrids have been reported (47) following backcrossing with *H. annuus.* The author also has obtained *H. tuberosus* × *H. annuus* hybrids (see section on Hybridization Techniques). Their meiotic behavior was similar to that described for *H. rigidus* × *H. annuus* hybrids.

H. rigidus. There are two interfertile subspecies, namely, ssp. *rigidus* and ssp. *subrhomboideus.* The first subspecies crossed with *H. tuberosus* is thought to have given rise to *H.* × *laetiflorus* (8, 35). Georgieva-Todorova (13) studied meiosis and pollen fertility of *H. rigidus* for 2 years. Of 25 meiocytes in diakinesis, only 12 contained quadrivalents and two had univalents. Chiasma frequency per cell was estimated as 66.3 or 1.3/bivalent and over 95% of meiocytes in subsequent division stages were normal. Pollen fertility in the 2 years was 76.8 and 73.8%. She suggests a genome formulae of *Ar*1 *Ar*1 *Ar*2 *Ar*2 *BrBr.* She subsequently reported difficulty in obtaining hybrids with *H. tuberosus* (14). The author has obtained hybrids between *H. rigidus* and both wild and cultivated *H. annuus.* Pollen fertility usually is less than 5% and the grains vary greatly in size (Fig. 3D). In most hybrids, the chromosomes are clumped and do not permit cytological studies, but univalents and multivalents frequently occur from diplotene to metaphase I. Subsequent to metaphase I, bridges with or without fragments, lagging chromosomes, and chromosome elimination frequently are evident (Fig. 6C). Despite these difficulties, a chromosome count of about 68 and not 102 can be seen in the hybrids. One somatic cell with 68 chromosomes is illustrated (Fig. 6E).

Section Divaricati Subsection *Gigantei* (Table 5)

H. giganteus. This species has been investigated by Long (53, 54), and Jackson (37). Jackson believed it to be one of the parents of *H. doronicoides.* Whelan (77) obtained hybrids by pollination with cultivated and wild *H. annuus.* Their meiotic behavior differed.

H. grosseserratus. Long (55) suggests that this species may have origi-

Table 5—Summary of interspecific hybrids involving *Helianthus* section Divaricati, subsection *Gigantei.*

Parents†	No. hybrids	Pollen stainability Range	Pollen stainability Avg.	R	M	U	B+F	Type hybrid	Source
giganteus (17)									10
annuus—wild	3	0–5	+	X		X		A	77
cult.	1		3	X	X			A	77
mollis			59.7	X		X		A,N	37,39,40
occidentalis								N	35
divaricatus	7		42	X	X			A,N	49
divaricatus	5		73	X				A	31
decapetalus (2n)	1		98	X				A	31
grosseserratus (♀)	17	94–98		X	X			A,N	49,53,54
grosseserratus (♀)			90	X				A	31
atrorubens (♀)	1		40	X	X			A	31
maximiliani	4		90	X	X			A	49
maximiliani (♀)			90	X				A,N	31
nuttallii	4		91	X				A	49
salicifolius (♀)	5		39	X	X			A	49
salicifolius (♀)			70	X				A	31
microcephalus (♀)	6	55–97		X				A	30
nuttallii	4		91	X				A	31,49
grosseserratus (17)									10
mollis	1		26	X	X			A,N	39,49
mollis (♀)			94.4	X				A	40
occidentalis	8	85–90		X	X			A,N	39,49
divaricatus			78	X				A,N	31
giganteus	17	94–98		X	X			A,N	49,53,54
maximiliani (♀)	24	37–99	76	X	X	X		A,N	49,52
maximiliani (♀)			90	X				A	31
salicifolius (♀)			90	X				A	31
salicifolius		41–64		X	X	X		A	49,50,51
atrorubens		50–80		X				A	31
nuttallii	12	91–96		X	X			A	49,56
nuttallii (♀)			80	X				A	31,56
giganteus			90	X				A	31
annuus	4		0	X				A	15
nuttallii (17)									32
niveus ssp. *niveus* (♀)	2		10		X	X		A	30
grosseserratus (♀)	12	91–96		X	X			A	49,56
grosseserratus			80	X				A	31,56
giganteus (♀)	4		91	X				A	49
maximiliani			70	X				A	31,56
occidentalis		40–70						A	56
maximiliani (17)									10
annuus	4	0–5			X	X	X	A	77
mollis (♀)			77	X				A	31
occidentalis	2		78,81	X	X			A	49
occidentalis (♀)			80	X				A	31
giganteus (♀)	4		90	X	X			A	49
grosseserratus (♀)	24	37–99	76	X	X	X		A,N	49,52
grosseserratus			90	X				A	31
nuttallii (♀)			70	X				A	31,56
angustifolius		34–85		X		X		A,N	31
decapetalus (♀)(2n)	7		84	X				A	30
longifolius (♀)	1		55	X				S	31
salicifolius								A	35
strumosus	2		74					A	49
giganteus			90	X				A,N	31
salicifolius (17)									10
giganteus (♀)	5		39	X	X			A	49
giganteus			70	X				A	31

(continued on next page)

Table 5—Continued.

Parents†	No. hybrids	Pollen stainability		Chromosome pairing				Type hybrid	Source
		Range	Avg.	R	M	U	B+F		
grosseserratus			90	X				A	31
grosseserratus (♀)		41–64		X	X	X		A	49,50,51
maximiliani								A	35
angustifolius			65	X		X		A,N	31
carnosus	3	28–65		X	X	X		A	31
atrorubens		Abnormal flowers						A	31
divaricatus								N	31
californicus (51)									32
resinosus (♀)	5		98	X				A	30
grosseserratus (♀)	12	91–96						A	49
resinosus (51)									35
strumosus			+ + +					A	34
tuberosus			+ + +					A	34
californicus	5		98	X				A	30
schweinitzii			+ + +					A	34
smithii	1		79					A	34
decapetalus (4n)	1		62		X	X		A	34
schweinitzii (51)									35
tuberosus			+ + +					A	34
resinosus			+ + +					A	34
smithii	1		75					A	34

† See "Interspecific Hybrids" section of narrative for detailed explanation.

nated from hybridization between *H. giganteus* and *H. salicifolius*. Georgieva-Todorova (15) obtained four hybrids from the cross *H. grosseserratus* × *H. annuus* which had normal meiosis but were completely male sterile. Sterility was due to genetic rather than cytoplasmic factors. After backcrossing with 'Peredovik', two types of meiotic behavior were encountered in the steriles. One type had normal meiosis with occasional meiotic defects of similar frequency to that observed in 'Peredovik'. Pollen abortion occurred after spore release from the tetrad. The sporogenous tissue broke down prior to meiosis in the second type.

H. nuttallii. There are three subspecies, namely, ssp. *nuttallii* ($n = 17$), ssp. *parishii* ($n = $ unknown), and ssp. *rydbergii* ($n = 17$). This latter subspecies originally was classified as a species (56).

H. californicus DC. Some of the specimens from southern California that Long (56) treated as *H. californicus* were regarded as *H. nuttallii* by Heiser et al. (35).

H. resinosus Small. This species frequently is called *H. tomentosus*. It is suggested (35) that *H. giganteus* and *H. mollis* contributed two of the three possible genomes for this hexaploid.

Section Divaricati Subsection *Microcephali* (Table 6)

H. microcephalus. In the cross with *H. divaricatus,* two of the three hybrids were fertile with 17 bivalents; the third was partially sterile with a quadrivalent. Both hybrids obtained from the reciprocal cross had 17 bivalents and were fertile (70).

H. smithii Heiser. This species has been crossed successfully with both tetraploid and hexaploid species.

Table 6—Summary of interspecific hybrids involving *Helianthus* section Divaricati, subsection *Microcephali*.

Parents†	No. hybrids	Pollen stainability Range	Pollen stainability Avg.	R	M	U	B+F	Type hybrid	Source
microcephalus (17)									10
niveus ssp. *niveus* (♀)	6		10	X	X			A	30
mollis (♀)			85	X				A	31
divaricatus	3		++	X	X			A,N	70
atrorubens (♀)	1		42					A	31
giganteus	6	55–97		X				A	30
glaucophyllus		60–90		X				A	31
longifolius (♀)			46	X				A	31
glaucophyllus (17)									68
microcephalus (♀)		60–90		X				A	31
laevigatus (34)									32
decapetalus (♀)(2n)	4		90	X	X			A	34
smithii (34)									34
hirsutus (♀)	5	80–99	92	X	X			A	34
decapetalus (4n)			87					A	34
strumosus (4n)	8	83–99	87			X	X	A	34
strumosus (6n)								A	35
resinosus (♀)	1		79					A	34
schweinitzii (♀)	1		75					A	34
longifolius (17)									71
mollis	3		43	X				A	30
occidentalis			34					A,N	31
maximiliani	1		55	X				S	31
microcephalus			46	X				A	31
angustifolius (♀)			13	X		X		A	31
carnosus	1		42					S	31
floridanus (♀)			40	X	X			A	31
silphioides								N	35
atrorubens (♀)		6–69		X				A,N	31,71

† See "Interspecific Hybrids" section of narrative for detailed explanation.

H. longifolius Pursh. Although many of the artificial hybrids showed less than 50% stainable pollen, most meiocytes contained 17 bivalents (31).

Section Divaricati Subsection *Angustifolii* (Table 7)

H. angustifolius L. Attempted crosses with *H. agrestis, H. argophyllus* T. and G., *H. carnosus, H. debilis, H. heterophyllus* Nutt., and *H. petiolaris* were unsuccessful (31).

H. simulans E. E. Wats. This species, given as *H. floridanus* A. Gray, produced sterile hybrids in crosses with *H. agrestis* (30).

Section Divaricati Subsection *Atrorubentes* (Table 8)

H. silphioides. The only putative natural hybrids known are from Tennessee and involve *H. atrorubens* and *H. mollis* (35).

H. atrorubens L. Jackson and Guard (40) reported a quadrivalent in some cells in early diakinesis. Fragments were observed in late diakinesis, metaphase I, and anaphase I. A bridge with or without a fragment appeared in 5 of 54 anaphase I cells. Pollen stainability, however, was 92.4%.

Despite the reduction in pollen fertility of most hybrids, meiosis usually was regular with 17 bivalents. Similar behavior was observed with *H. longifolius* (35).

Table 7—Summary of interspecific hybrids involving *Helianthus* section Divaricati, subsection *Angustifolii*.

Parents†	No. hybrids	Pollen stainability		Chromosome pairing				Type hybrid	Source
		Range	Avg.	R	M	U	B+F		
angustifolius (17)									35
niveus ssp. *canescens* (♀)	8		10	X	X	X		A	30,31
maximiliani (♀)		34–85		X	X			A,N	31
salicifolius (♀)			65					A	31
longifolius			13	X	X			A	31
floridanus (♀)		65–100		X				A,N	31
simulans			90					A	35
atrorubens (♀)		20–55		X	X			A,N	31
radula	3		25	X	X			A	31
simulans (17)									35
angustifolius (♀)			90					A	35
atrorubens (♀)		75–85						A	35
agrestis	14	No pollen						A	30
floridanus (17)									35
debilis (♀)	2		5		X			A	30
longifolius			40	X	X			A	31
angustifolius (♀)		65–100		X				A,N	31
atrorubens (♀)	5		76					A	30
agrestis	10	No pollen						A	30

† See "Interspecific Hybrids" section of narrative for detailed explanation.

H. heterophyllus. Initial studies (31) indicated abnormal meiosis and possible triploidy but meiosis in subsequent collections was normal. The species appears to be unadapted to greenhouse studies. Artificial hybridization with *H. radula* (Pursh) T. and G. gave six seedlings but all died before flowering (35). Natural hybrids also have been found recently (C. B. Heiser, personal communication).

LITERATURE CITED

1. Anashchenko, A. V. 1972. Methods for developing hybrid sunflower involving chemical castration. p. 229–230. *In* Proc. 5th Int. Sunflower Conf. (Clermont-Ferrand, France).

2. ————. 1972. A study of male sterility in the sunflower. (In Russian). Tr. Prikl. Bot. Genet. Sel. 46(3):120–131.

3. Atanasoff, D. 1971. The viral nature of cytoplasmic male sterility in plants. Phytopathol. Z. 70:306–322.

4. Beatley, J. C. 1963. The sunflowers (genus *Helianthus*) in Tennessee. J. Tenn. Acad. Sci. 38(4):135–154.

5. Beglaryan, N. P. 1970. Some cytological characteristics of a sunflower mutant induced with gibberellic acid. (In Russian). Tsitol. Genetika 4(6):547–552.

6. Campos, F. A. 1974. The use of ethrel in the induction of male sterility in sunflower. p. 349–351. *In* A. V. Vranceanu (ed.) Proc. 6th Int. Sunflower Conf. (Bucharest, Romania). Research Institute for Cereal and Industrial Crops, Fundulea, Ilfov, Romania.

7. Cauderon, Y. 1965. Analyse cytogénétique d'hybrides entre *Helianthus tuberosus* et *H. annuus*. Conséquences en matière de sélection. Ann. Amélior. Plant. 15(3):243–261.

8. Clevenger, S., and C. B. Heiser. 1963. *Helianthus laetiflorus* and *Helianthus rigidus*— hybrids or species? Rhodora 65:124–133.

9. Dhesi, J. S., and R. G. Saini. 1973. Cytology of induced polyploids in sunflower. Nucleus 16(1):49–52.

10. Geisler, F. 1931. Chromosome numbers in certain species of *Helianthus*. Butler Univ. Bot. Stad. 11:53–62.

11. Georgieva-Todorova, Y. 1969. The effect of X-rays on chromosome aberrations in *Helianthus annuus* L. (In Bulgarian). Genet. Sel. 2:469–476.

Table 8—Summary of interspecific hybrids involving *Helianthus* section Divaricati, subsection *Atrorubentes*.

Parents†	No. hybrids	Pollen stainability Range	Pollen stainability Avg.	R	M	U	B+F	Type hybrid	Source
silphioides (17)									32
mollis (♀)								N	35
longifolius (♀)								N	35
atrorubens	4	61–65		X	X			A	31
divaricatus (♀)	1	Failed to flower						A	31
atrorubens (17)									32
mollis (♀)			34	X			X	A,N	4,31,40
occidentalis			48					A,N	31
giganteus	1		40	X	X			A	31
grosseserratus (♀)		50–80		X				A	31
salicifolius (♀)		Abnormal flower						A	31
microcephalus	1		42					A	31
longifolius		6–69		X				A,N	31,71
angustifolius		20–55		X		X		A,N	31
silphioides (♀)	4	61–65		X	X			A	31
heterophyllus (♀)	4		71	X	X			A	30
radula		Failed to flower						A	31
simulans		75–85						A	35
carnosus	2		49					A	30
floridanus	5		76					A	30
heterophyllus (17)									32
atrorubens	4		71	X	X			A	30
radula	6	All died before flowering						A	35
radula (17)									32
angustifolius (♀)	3		25	X		X		A	31
atrorubens (♀)		Failed to flower						A	35
heterophyllus (♀)	6	All died before flowering						A	35
carnosus (♀)	9	8–45	23	X	X	X		A	31
carnosus (17)									32
decapetalus (♀)2n)	1		40	X		X		A	30,31
salicifolius (♀)	3	28–65		X	X	X		A	31
longifolius (♀)	1		42					S	35
radula	9	8–45	23	X	X	X		A	31
atrorubens (♀)	2		49					A	30

† See "Interspecific Hybrids" section of narrative for detailed explanation.

12. ————. 1970. Zytogenetische Untersuchung der Hybriden *Helianthus argophyllus* × *H. annuus*. Z. Pflanzenzuecht. 64:357–366.

13. ————. 1971. Meiosis in *Helianthus rigidus* Desf. C. R. Acad. Seances Agric. Bulg. 4(4):407–411.

14. ————. 1972. A study of the compatibility between some species in the genus *Helianthus* (section "Divaricati"). C. R. Acad. Agric. G. Dimitrov 5(4):283–289.

15. ————. 1974. A new male sterile form of sunflower (*Helianthus annuus* L.). p. 343–347. *In* A. V. Vranceanu (ed.) Proc. 6th Int. Sunflower Conf. (Bucharest, Romania). Research Institute for Cereal and Industrial Crops, Fundulea, Ilfov, Romania.

16. ————. 1974. Cytological study of two tetraploid *Helianthus* species (2n = 68). C. R. Acad. Agric. G. Dimitrov 7(3):43–48.

17. ————, M. Lakova, and D. Spirkov. 1974. Karyologic study of *Helianthus mollis* L. C. R. Acad. Agric. G. Dimitrov 7(1):59–62.

18. ————, and V. Nuti. 1969. Chromosome aberrations in *Helianthus annuus* L. induced by 8-ethoxycaffeine. C. R. Acad. Seances Agric. Bulg. 2:225–230.

19. Gundaev, A. I. 1971. Basic principles of sunflower selections. (In Russian). p. 417–465. *In* Genetic principles of plant selection. Nauka, Moscow.

20. Heiser, C. B. 1947. Hybridization between the sunflower species *Helianthus annuus* and *H. petiolaris*. Evolution 1:249–262.

21. ————. 1948. Taxonomic and cytological notes on the annual species of *Helianthus*. Bull. Torrey Bot. Club 75(5):512–515.

22. ————. 1949. Study in the evolution of the sunflower species *Helianthus annuus* and *H. bolanderi*. Univ. California Publ. Bot. 23:157–196.

23. ————. 1951. Hybridization in the annual sunflowers: *Helianthus annuus* and *H. debilis* var. *cucumerifolius*. Evolution 5:42–51.

24. ————. 1951. Hybridization in the annual sunflowers: *Helianthus annuus* and *H. argophyllus*. Am. Nat. 85:65–72.

25. ————. 1954. Variation and subspeciation in the common sunflower, *Helianthus annuus*. Am. Midl. Nat. 51(1):287–305.

26. ————. 1956. Biosystematics of *Helianthus debilis*. Madrono. 13(5):145–176.

27. ————. 1958. Three new annual sunflowers (*Helianthus*) from the southwestern United States. Rhodora 60:272–283.

28. ————. 1960. A new annual sunflower, *Helianthus deserticola,* from the southwestern United States. Proc. Indiana Acad. Sci. 70:209–211.

29. ————. 1961. Morphological and cytological variation in *Helianthus petiolaris* with notes on related species. Evolution 15(2):247–258.

30. ————. 1965. Species crosses in *Helianthus*. III. Delimitation of "Sections". Ann. Missouri Bot. Gard. 52(3):364–370.

31. ————, W. C. Martin, and D. M. Smith. 1962. Species crosses in *Helianthus*. I. Diploid species. Brittonia 14(2):137–147.

32. ————, and D. M. Smith. 1955. New chromosome numbers in *Helianthus* and related genera (Compositae). Proc. Indiana Acad. Sci. 64:250–253.

33. ————, and ————. 1960. The origin of *Helianthus multiflorus*. Am. J. Bot. 47(10): 860–865.

34. ————, and ————. 1964. Species crosses in *Helianthus*: II. Polyploid species. Rhodora 66:344–358.

35. ————, ————, S. B. Clevenger, and W. C. Martin. 1969. The North American sunflowers (*Helianthus*). Mem. Torrey Bot. Club. 22(2):1–218.

36. Horner, H. T. 1977. A comparative light- and electron-microscopic study of microsporogenesis in male-fertile and cytoplasmic male-sterile sunflower (*Helianthus annuus*). Am. J. Bot. 64(6):745–759.

37. Jackson, R. C. 1956. The hybrid origin of *Helianthus doronicoides*. J. N. Engl. Bot. Club 58:97–100.

38. ————. 1963. Cytotaxonomy of *Helianthus ciliaris* and related species of the southwestern U.S. and Mexico. Brittonia 15:260–271.

39. ————, and A. T. Guard. 1955. Hybridization of perennial sunflowers in Indiana. Proc. Indiana Acad. Sci. 65:212–217.

40. ————, and ————. 1957a. Analysis of some natural and artificial interspecific hybrids in *Helianthus*. Proc. Indiana Acad. Sci. 66:306–317.

41. ————, and ————. 1957b. Natural and artificial hybridization between *Helianthus mollis* and *H. occidentalis*. Am. Midl. Nat. 58:422–433.

42. Klasterska, I., and A. T. Natarajan. 1974. The role of the diffuse stage in the cytological problems of meiosis in *Rosa*. Hereditas 76:109–116.

43. Kostoff, D. 1939. Autosyndesis and structural hybridity in F₁ hybrid *Helianthus tuberosus* L. × *H. annuus* L. and their sequences. Genetica 21:285–300.

44. Kovacik, A., and R. Kryzanek. 1969. Biological and biochemical analysis of sterile and fertile pollen of sunflower. Genet. Slechteni 5:179–184.

45. Kurnik, E., J. Parragh, and B. Pozsar. 1971. Effect of γ-ray dose (⁶⁶Co) on the size and stainability of root chromosomes of sunflower. (In Hungarin). Bot. Kozl. 58(4):235–237.

46. Leclercq, P. 1969. Une stérilité male cytoplasmique chez le tournesol. Ann. Amelior. Plant. 19:99–106.

47. ————, Y. Cauderon, and M. Dauge. 1970. Sélection pour la résistance au mildiour du tournesol à partir d'hybrides topinambour ✗ tournesol. Ann. Amélior. Plant. 20:363–373.

48. Lewis, K. R., and B. John. 1966. The meiotic consequences of spontaneous chromosome breakage. Chromosoma 18:287–304.

49. Long, R. W. 1955. Hybridization in perennial sunflowers. Am. J. Bot. 42:769–777.

50. ————. 1955. Hybridization between the perennial sunflowers *Helianthus salicifolius* A. Dietr. and *H. grosseserratus* Martens. Am. Midl. Nat. 54:61–64.

51. ————. 1957. A cytological analysis of a naturally occurring *Helianthus* hybrid. Ohio J. Sci. 57(2):63–69.

52. ————. 1959. Natural and artificial hybrids of *Helianthus maximiliani* × *H. grosseserratus*. Am. J. Bot. 46:687–692.

53. ———. 1960. Biosystematics of two perennial species of *Helianthus* (Compositae). I. Crossing relationships and transplant studies. Am. J. Bot. 47:729–735.

54. ———. 1961. Biosystematics of two perennial species of *Helianthus* (Compositae). II. Natural populations and taxonomy. Brittonia 13:129–141.

55. ———. 1963. Cytogenetic investigations of artificial *Helianthus giganteus* × *H. salicifolius* hybrids (Compositae). Ohio J. Sci. 63:273–281.

56. ———. 1966. Biosystematics of the *Helianthus nuttallii* complex (Compositae). Brittonia 18:64–79.

57. Moens, P. B. 1964. A new interpretation of meiotic prophase in *Lycopersicon esculentum* (tomato). Chromosoma 15:231–242.

58. Nakashima, H., and S. Hosokawa. 1974. Studies on histological features of male sterility in sunflower (*Helianthus annuus* L.). Proc. Crop Sci. Soc. Japan 43(4):475–481.

59. Necomb, W. 1973. The development of the embryo sac of sunflower *Helianthus annuus* before fertilization. Can. J. Bot. 51:863–878.

60. ———. 1973. The development of the embryo sac of sunflower *Helianthus annuus* after fertilization. Can. J. Bot. 51:879–890.

61. Newcomer, E. H. 1953. A new cytological and histological fixing fluid. Science 118:161.

62. Paun, L. 1974. The cytologic mechanism of male sterility in sunflower. p. 249–257. *In* A. V. Vranceanu (ed.) Proc. 6th Int. Sunflower Conf. (Bucharest, Romania). Research Institute for Cereal and Industrial Crops, Fundulea, Ilfov, Romania.

63. Piquemal, G. 1970. How to produce hybrid sunflower seeds by inducing male sterility with gibberellic acid. p. 127–135. *In* Proc. 4th Int. Sunflower Conf. (Memphis, Tennessee).

64. Pirev, M. N. 1966. A histochemical study of microsporogenesis in pollen-fertile and pollen-sterile forms of sunflower. (In Russian). Bull. Acad. Sci. Moldav. SSR 11:91–94.

65. ———. 1968. A study of the male reproductive organs of sunflowers with normal and sterile pollen. (In Russian). Bull. Acad. Sci. Moldav. SSR 14:62–71.

66. Pogorlets'kii, B. K., and V. F. Kukinn. 1971. New sources of male sterility in the sunflower. (In Ukrainian). Visnik Silskogospod. Nauki 1:39–42.

67. Pustovoit, G. V. 1969. Breeding sunflower for group immunity by the method of interspecific hybridization. (In Russian). Agric. Biol. 4:803–812.

68. Smith, D. M. 1957. A new species of *Helianthus* from North Carolina and Tennessee. Brittonia 10:192–194.

69. ———. 1961. Variation in the tetraploid sunflowers, *Helianthus decapetalus, H. hirsutus* and *H. strumosus.* Recent Adv. Bot. 1:878–881.

70. ———, and A. T. Guard. 1958. Hybridization between *Helianthus divaricatus* and *H. microcephalus.* Brittonia 10:137–145.

71. ———, W. C. Martin. 1959. Natural hybridization of *Helianthus longifolius* with *H. atrorubens* and *H. occidentalis.* Rhodora 61:140–147.

72. Tsvetkova, F. 1974. A study of hybridization between *Helianthus scaberrimus* and *H. annuus.* Abstr. Bulgarian Scientific Literature A19(3):Abstr. 623.

73. Vigfusson, E. 1970. On polyspermy in the sunflower. Hereditas 64:1–52.

74. Vinitskaya, O. P. 1973. Breeding sunflowers for group immunity by interspecific hybridization. Mikol. Fitopatol. 7(1):76–77.

75. Vranceanu, A. V., F. Stoenescu, A. Ulinici, H. Iliescu, and Fl. Paulian. 1974. Sunflower. (In Romanian). Editura Academiei Republicii Socialiste, Bucharest, Romania. p. 322

76. Whelan, E. D. P. 1974. Discontinuities in the callose wall, intermeiocyte connections, and cytomixis in angiosperm meiocytes. Can. J. Bot. 52:1219–1224.

77. ———. 1976. Sterility problems in interspecific hybridization of *Helianthus* species. p. 5–6. *In* Proc. 1st Sunflower Forum (Fargo, N.D.). North Dakota State University, Fargo.

78. ———, and C. A. Hornby. 1969. Meiotic prophase in *Prunus avium* cultivar Lambert. Can. J. Bot. 47:1813–1815.

79. ———, G. H. Haggis, E. J. Ford, and B. Dronzek. 1974. Scanning electron microscopy of meiotic chromosomes of plants *in situ.* Can. J. Bot. 52:1438–1440.

ERNEST D. P. WHELAN: B.S.A., M.Sc., Ph.D., Cytogeneticist, Research Station, Agriculture Canada, Morden, Manitoba, Canada (Present address, Research Station, Agriculture Canada, Lethbridge, Alberta, Canada). Breeding, cytology and interspecific hybridization of *Helianthus* spp.; author and co-author of over 20 scientific publications on genetics and cytology.

Chapter 11

Planting Seed Production

DONALD L. SMITH

The goal of a sunflower (*Helianthus annuus* L.) planting seed production program is to produce high quality seed genetically identical to the cultivar released by the breeder or developer. The production and seed management procedures to achieve this goal, for hybrid and open-pollinated sunflower cultivars, are covered in this chapter.

Attempts were made to establish a sunflower industry in North America during the decade between 1880 and 1890 (5). Not until the 1960–70 period, however, did two significant events stimulate the development of the sunflower crop and seed production. The first was the introduction about 1960 of the open-pollinated, high oil cultivars from the USSR. The second was the discovery in 1968 (7) of cytoplasmic male sterility and the subsequent reports (1, 3, 6) of fertility restoring genes. The introduction of high oil cultivars resulted in limited breeding and seed production by a few private companies. Subsequently attempts were made in 1968 and 1969 to incorporate the high oil characteristic into lines suitable for producing F_1 hybrids by using genetic male sterility. Discovery of cytoplasmic male sterility and fertility restoring genes substantially enhanced expectations for the successful production of hybrid sunflower seed. These discoveries provided evidence that the same biological system used to produce hybrid cultivars of corn (*Zea mays* L.) and grain sorghum [*Sorghum bicolor* (L.) Moench] on a field scale now was possible with sunflower. The release in 1971 by the ARS, USDA, and the Texas Agricultural Experiment Station of seven cytoplasmic male sterile and two fertility restorer inbred lines of sunflower provided the source of basic seedstocks for a hybrid sunflower seed industry.

Since most of the planting seed now being produced in North America is of hybrid cultivars, this chapter will emphasize procedures related to the production of hybrids using cytoplasmic male sterility and genetic fertility restoration. The procedures commonly used with open-pollinated cultivars, however, are reported with reference to differences.

From: Carter, Jack F. (ed.). 1978. *Sunflower Science and Technology*. Agronomy 19. Copyright © 1978 by the American Society of Agronomy, Crop Science Society of America, and Soil Science Society of America, 677 South Segoe Road, Madison, WI 53711 USA.

The word "seed" will refer to the achene, in deference to common usage. "Seed production" refers to production of planting seed.

A literature search revealed few reports on seed production. The author made a written inquiry of persons involved in sunflower seed production in the USA and Canada in March, 1977, to review current procedures. The responses to this inquiry are incorporated and acknowledged in this chapter.

HYBRID SEED

The hybrid sunflower seed production industry began in 1972 and developed in North Dakota, Minnesota, Texas, and California. Procedures that are common to most production programs have evolved, and they are often based upon limited data. Additional research is needed before some procedures can be employed with assurance of successful production of quality seed.

Production Procedures

Isolation

The origin and nature of the genus *Helianthus* presents seed producers with advantages and disadvantages. Wild populations of *Helianthus annuus* L. occur as weeds beside roads and streams, in ditches, native pastures, and cultivated fields in the Central, Western, and Southwestern U.S. These populations readily cross naturally with domesticated sunflower. Most of the wild sunflowers are branched with several inflorescences, and in some environments, the plants may produce pollen for a month or more. This characteristic increases the potential for contamination of hybrid seed production fields but is a beneficial character when incorporated in male parents of hybrids.

Volunteer sunflower plants from fields where sunflower was grown the previous season often create isolation problems. In small grains this problem normally is avoided by usual methods of weed control; whereas crops such as sugarbeet (*Beta vulgaris* L.) or soybean [*Glycine max* (L.) Merrill], may require special weed control to assure the elimination of volunteer sunflower plants.

Studies on the effect of spatial isolation on outcrossing of male sterile sunflower lines have been inconclusive. Gundaev (4) reported that Anashchenko in the USSR found 18.7% outcrossing in a plot of male sterile sunflower situated 1.05 km from a pollen source with a stand of trees as a barrier. Enns et al. (3) reported 13.75% and 18.0% outcrossing in small plots in Manitoba located 0.8 km and 1.2 km, respectively, from a large field of commercial sunflower. Bolson (2) used the ARS, USDA-developed male sterile line cmsHA89 in studies in 1976 in North Dakota and Minnesota. Seed set did not vary significantly among three plots located 0.8, 1.2, and 1.6 km from a pollinator field. The average outcrossing in each was approximately 50%. Outcrossing of 15% was reported in a 200 m² plot 2 km from a commercial field. An average of 10.5% outcrossing occurred in a 7 ha field of cmsHA89 located 2.4 km from any pollen source.

The field standards for the production of certified hybrid and open-pollinated seed were adopted in 1969 by the Association of Official Seed Certifying Agencies. The isolation requirement if 0.8 km from other varieties, strains, hybrids, noncertified crops of the same variety, volunteer sunflower, or wild annual *Helianthus annuus* L. The seed certifying agencies of California, Minnesota, and South Dakota adopted this same minimum isolation requirement. North Dakota, however, requires a minimum of 1.6 km isolation for the production of hybrid seed. The limited research results available appear to indicate that these official isolation requirements are inadequate for production of pure parental stocks in small plots but probably are adequate for production of quality hybrid seed in most regions of North America.

Sunflower seed producers of the Sacramento Valley of California agreed in 1976 upon standards of isolation for production of hybrid seed. These standards are as follows: 4.8 km from fields of commercial oilseed sunflower, 3.2 km from F_1 seed production of oilseed hybrid, 6.4 km from a field of any type of nonoilseed sunflower, and time isolation achieved by differential planting dates. Variable time of planting has been useful in achieving isolation between fields not having adequate distance isolation. A 30-day difference in planting dates has been the standard "time isolation." As the season progresses, however, time isolation is more difficult to achieve because many lines now in use mature more rapidly as planting dates are delayed. Late flowering plants in the first planted field may interfere with time isolation, especially on plants with multiple inflorescences. Each production organization has established its own standards for isolation from wild *Helianthus*. The author's experience has been that a distance of at least 4.8 km is the minimum isolation in the Sacramento Valley if the hybrid seed is to be nearly free of wild outcrosses.

Contracting

The production contract establishes the roles of the grower and the contractor and specifies production procedures. Contracts generally include agreements on the following: seedstocks to be planted, ratio of male to female rows, description of parental plants (including days from emergence to flowering), planting dates of the male and female rows, irrigation schedules if used, isolation requirements, control of wild sunflower, roguing procedures, pollination program, harmful insect control, disposition of male plants or seed, harvest procedures, quality specifications, storage arrangements, delivery requirements, field inspection, and service by the contractor.

Seedstocks

Some of the most important factors in planning for hybrid seed production are the specific characteristics of the inbred lines or single crosses to be used. Growth cycles must be documented over all production environments. The following characteristics must be known to plan the production procedures: influence of planting dates on the growth cycles, days from seeding to emergence, days from emergence to flowering, duration of receptivity of female flowers, duration of pollen production of restorer lines, amount and time of day pollen produced by restorer, attraction of honeybees to both male and female parents, disease resistance, and any genotypic or environmental factors influencing flowering cycles.

The following quality characteristics of parental seedstock also must be known: percentage germination, percentage outcrosses to other cultivars and to wild sunflower, percentage male fertile plants in the female stock, and seed size. These factors influence the planting rate and roguing procedure and resultant cost of production, quality of seed, and seed yield per hectare.

Planting

The choice of planting system is dependent upon equipment, parental seedstocks, field layout, and crop management system to be utilized. The main objective is to establish the desired population uniformly distributed and thus achieve the maximum yield of hybrid seed of desired sizes.

Planting procedures used successfully to produce hybrid corn seed have been adopted without modification for production of hybrid sunflower seed. Semiprecision and air planters designed for planting row crops, such as corn and soybean, are adapted readily for planting sunflower with only a change of seed plates and cut-off or drums. Most flat-plate planters require a change of the seed cut-off because the type commonly used for seeding corn may crack sunflower seed. A brush type cut-off or the type used for whole beet seed eliminates most of the cracking problem. The planting and harvesting machines should be matched regarding number and the distance between rows. The seedbed should be prepared to ensure uniform depth of seed placement, since a uniform and predictable date of emergence of seedlings is essential. Depth of seed placement, different planting dates, and split irrigation schedules aid in synchronizing inherently different flowering dates of male and female parents.

The optimum plant population depends on the characteristics of the female and male parents, soil type and condition, planting date, and the desired seed size. Sunflower plants tend to compensate for differences in population by producing larger heads and larger seeds at lower populations.

Plant populations ranging from 45,000 to 67,000/ha were reported by producers. The author has observed the optimum population for maximum yield of desired seed sizes for the USDA-developed female lines cmsHA89 and cmsHA290 to be approximately 60,000/ha under furrow irrigation in the Sacramento Valley of California.

The duration of pollen production also can be affected by planting rate. The time from emergence to flowering usually decreases as the plant spacing in the row decreases. The author has observed differences of as much as 1 week over a 3-year period in the USDA developed male line RHA274 planted at the rates of 30,000 and 60,000 plants/ha under furrow irrigation in the Sacramento Valley.

A north-south orientation of the rows improves the efficiency of roguing. The spacing between rows varies with the region and type of planter. Producers reported row spacings ranging from 56 to 102 cm with most production planted in rows spaced 76 cm apart. Row spacing influences the selection of appropriate plates for the desired distribution of the seed in the row, cultivation for weed control, ground application of sprays, and harvesting. Charts have been developed which show the plates to be used for various planters, seed sizes, seed viabilities, row spacing, and plant population/ha.

F_1 hybrid seed is produced on the female parent in alternating groups of rows with the male parent. Female to male row combinations ranging from 2:1 to 7:1 have provided adequate pollination under a number of environments according to producers. The pollination requirements of the various female stocks, available planter unit, width of header on the harvester, and disposition of the male rows are factors influencing the ratio of female to male rows.

Crop Management

Effective weed control with chemicals and cultivation is important in producing hybrid seed. Weeds of similar growth cycle and height as sunflower reduce yield and hinder roguing and harvest. Recommended fertilizer programs and timely irrigation especially during flowering and seed formation help assure a high proportion of desirable sized seed. The last irrigation can critically affect plumpness of seed. Generally an irrigation is recommended when the backs of the heads are beginning to change from dark to light green.

Harmful insects must be controlled. The sunflower moth, *Homoeosoma electellum* (Hulst.), is the most destructive. The interrelation of the insect control and roguing is discussed in the following section.

Seed losses from birds can be a serious problem. In California the most destructive species is the house finch (linnet), *Carpodacus mexicanus frontalis* (Say). The author has conducted various trials using explosive devices, pyrotechnics, chemical agents, such as Avitrol® and Mesurol®, and patrols. No effective control has been found.

Roguing

The objective of roguing is to remove atypical plants from parental stocks. The breeder's description of the parental stock is the basis for determination of atypical plants. The effects of plant spacing, temperature, and soil moisture on the development of the sunflower plant must be considered during roguing.

Successful roguing requires education of personnel and timely scheduling of field work. The educational phase includes learning the distinguishing features of the parental stocks and developing the ability to recognize off-type plants and male fertile plants in the female stocks.

A distinguishing characteristic of the male fertile plants among most inbreds is the darker brown color of the anthers as contrasted to the yellow color of the pollen sterile flower. Male fertile plants are identified readily by roguers after a few hours of experience even though only a few disk flowers may be developed fully and show pollen.

Off-types are those plants which are atypical morphologically including characteristics such as height, leaf shape, petiole angle, and wild type branching, or show red or purple stem or petiole, and disease especially downy mildew [*Plasmopara halstedii* (Farl.) Berl. & de Toni]. Roguing for off-type plants should begin at the bud stage. The time between bud formation and the first ray flowers generally is sufficient to remove most off-type plants.

Sunflower heads will continue to develop and shed viable pollen after removal from the stalk. This source of contaminating pollen can be eliminated by removing the head from the stalk and turning it face down on the soil.

The most critical period in roguing is during flowering. Most off-type plants, however, are detected easily on a morphological basis and removed prior to flowering. On the average a typical single-headed, male sterile line completes flowering in approximately 7 days. An individual head of a typical branching restorer under similar conditions completes pollen production in about 5 days although, due to multiple heads, pollen production by an individual plant may extend for 3 to 4 weeks. If the daytime temperature continually exceeds 38 C during the flowering period, the time lapse from the appearance of first to last anthers of a single head may be as short as 4 days. If the temperature remains below 22 C, the process may require 2 weeks. If the plant is male sterile, the stigmas usually remain intact and receptive, if they are not pollinated, for 3 to 5 days under optimum growing conditions.

When flowering begins, the male fertile plants should be removed each day. The stage of development of the heads as well as the time of day that roguing is done are key factors in assuring thorough roguing. Some off-type plants may be detected easily throughout the day, but male fertile plants in the female rows are detected more efficiently in the morning before bees have removed the pollen. As the ray flowers are first evident, the heads may be inspected by lightly pulling back the corollas of the ray flowers to expose the first ring of disk flowers.

Sunflower is phototropic until the early stage of flowering. After ray flowers are fully developed, the head generally faces the east. This feature makes roguing inefficient if the row direction is east-west. If the direction is north-south, however, roguing is done always looking westward at the heads. This procedure also avoids the discomfort of looking directly into the sun.

If a field must be furrow irrigated during roguing, only alternate furrows should be watered so that a dry row can be used for walking. Reducing the length and increasing the frequency of application of irrigation by circle sprinklers may be necessary during roguing.

Sometimes during roguing an insecticide must be applied to control sunflower moth or other damaging insects. Careful planning and scheduling is imperative to assure that each application is timed correctly and the safety of the roguers is maintained.

Pollination

In hybrid sunflower seed production, pollination involves the physical transfer by insects of pollen grains from the male parent to the male sterile female parent. After the female gamete is fertilized, an F_1 hybrid achene develops. Many pollinating insects collect pollen and nectar from sunflower, but the domesticated honeybee *Apis mellifera* (L.) is the only one obtainable in sufficient numbers to pollinate a field adequately.

Hives of honeybees are placed in the field at the beginning of flowering. The number of hives placed at any one time depends on the percentage and stage of flowering. Overstocking may force the bees to seek other sources of pollen and nectar. The total complement of hives usually is placed within 7 days. Seed producers reported placing hives at the rate of 0.5 to 2.5/ha. Hives should be located in groups on the field periphery or on blank strip areas within large fields at approximately 200 m intervals. Some producers suggested that bees prefer male fertile plants. Observations by the author in the Sacramento Valley of California over 3 years indicated that flight patterns are random and length of visit to a single head varied widely with no apparent preference for the male sterile (cmsHA89) or male fertile (RHA274) flowers. Hives at the rate of 2.5/ha were used. The principal criterion to determine the effectiveness of pollination was the percent of normal achenes in the receptacles.

Numerous species of potential pollinators other than honeybees have been observed upon sunflower florets. According to Torchio (see acknowledgement) these include species from the genera *Bombus, Diadasia, Melissodes,* and *Megachile.* The author has observed the genus *Bombus* in hybrid production fields in the Sacramento Valley of California. It is probable that some of these insects also are effective in the transfer of pollen.

Some other observations by the author relative to pollination and seed set have been made in the Sacramento Valley. In 1976 groups of rows 15 m wide of the male sterile line cmsHA89 were sampled in two fields in which RHA274 was the pollen source. Seed set was complete in all heads examined regardless of distance from the pollen source. Distinct rings, with only 60% seed set, appeared in another field which flowered during 4 days of temperature above 38 C and relative humidity of 20%. The low seed set in these rings may have been due to several factors. Obviously the high temperature reduced bee activity. Dehiscence, elongation of the style, and resulting discharge of the pollen occurred in approximately 2 hours, substantially reducing the time available for pollen transfer. Much of the pollen appeared to be bleached, suggesting that its viability was reduced. The length of time that the stigmas remained receptive was also likely reduced. It is probable that high temperature and low humidity during flowering will affect pollination and seed set adversely. Producers in northern areas also reported that pollination and seed set may be reduced when flowering occurs coincident with long periods of cool or wet weather.

Harvesting

Several specific procedures and harvester modifications are necessary for harvesting planting seed. The harvester is a grain combine with special modifications for collecting, threshing, and cleaning of sunflower seed. The objectives in harvesting are to minimize mechanical damage and loss of seed and to maintain high purity.

The harvester header attachment should have wide pans which match the spacing between rows. Elevators and augers are sources of mechanical damage to seed. Only bucket-type elevators should be used. Augers should be inspected thoroughly for wear, adjustment, and speed. High speed and excess spacing between the auger flite and housing cause damage to seed. The cylinder should be equipped with rasp bars as contrasted to spike teeth. The rasp bars create a rubbing action compared to the flailing action of the spike tooth.

Most harvesters have tailings returned to the front of the cylinder. This procedure may increase damage to the seed that passes twice between the cylinder and concave. The tailings return on most harvesters can be modified to discharge at the rear of the cylinder, thus reducing the hazard of seed injury.

The harvester must be cleaned thoroughly prior to harvest and between fields of different sunflower cultivars or inbred lines. Some areas in the harvester are impossible to clean without removing certain components. Additional access ports in the body of the harvester simplify cleaning. Places inside the machine where seed may lodge should be filled with a silicone rubber sealant. A checklist is an aid in the cleaning procedure. A source of air pressure (700 g/cm^2) and a vacuum source are necessary to clean the harvester thoroughly. The header should be removed to improve access to the machine. A thorough cleaning requires about 3 hours if the machine has been modified so that all interior sections are accessible with either air pressure or vacuum.

Preharvest adjustments and modifications allow maximum recovery of the crop and assure high purity and quality of seed. The value of planting seed is much greater than commercial oilseed so the efficiency of the harvest is based on different factors. The total weight of seed threshed daily must be a secondary consideration. Harvester adjustments may not allow utilization of maximum threshing and cleaning capacity. It may not be possible or advisable to separate the seed and foreign material as completely as in harvesting oilseed.

The cylinder speed should not exceed 300 rpm or 530 m/min (1,725 ft/min) peripheral speed for a 55.9 cm (22 in) cylinder diam. It should be reduced to approximately 200 rpm if the heads have 8% moisture or less. Generally the seeds dislodge easily from the receptacle, so wide spacing between the cylinder and concave at both the front and rear is recommended. As a guide, spacing of approximately 25 mm (1 in) at the front and 19 mm (0.75 in) at the rear of the concave usually gives good results. These settings should be adjusted to threshing conditions in the field. If the concave and cylinder spacing are adjusted properly and the cylinder speed set to remove all the seed, the receptacle should be discharged from the harvester with little breakage and there should be no seed in the tailings. Several achenes should be dissected to observe breaks and fractures which are evidence of mechanical damage to the embryo, cotyledon, and pericarp. The tetrazolium staining procedure is an aid in assessing damage.

If weather permits, harvest should not start until the sunflower seed has matured naturally to a moisture percentage of 11%. If waiting is not possible, however, a chemical desiccant may be applied to hasten field drying. Two desiccants, dimethyl bipyridium dichloride [Paraquat®] and sodium chlorate, are used successfully except when the air temperatures are less than 20 C after application. If the daytime temperatures remain above 20 C or higher after applying the desiccant, the harvest date is advanced 7 to 14 days. Germination percentages of seeds are not affected if the desiccants are applied at the recommended stage of plant development.

Sunflower seed is mature when the back of the head turns yellow, although the seed will contain 20 to 35% moisture at this stage of plant development. Air temperature, rainfall, and wind velocity determine the appropriate harvest procedure. In the Western U.S., when daytime temperatures are 25 C or above, percentage of seed moisture declines to 11% approximately 3 weeks after the backs of the heads turn yellow. In this region usually neither desiccation nor artificial drying of the seed is necessary. Both desiccation and artificial drying of seed may be necessary in the Northern Plains of North America. Frost accelerates the maturing process but may not cause a rapid loss in seed moisture, especially if followed by rainy weather or a long period of high relative humidity.

The time to start harvesting can be decided by hand threshing a random sample of 20 to 30 heads and testing the seed for moisture content. Harvesting may begin when seed can be stored safely, which is at a seed moisture of approximately 9.5% in the Northern Plains and 11% in the Western U.S.

Frequent adjustments of the harvester may be necessary, because temperature and humidity vary during the day.

As the harvest season progresses, the moisture percentage of the seed usually decreases and the seed becomes more vulnerable to mechanical damage in the threshing, separating, and handling processes. Frequent evaluation of seed moisture and harvester adjustments are imperative to maintain seed quality. Since the moisture in the head and seed varies in relation to the humidity of the air, in dry climates it may be necessary to restrict the harvesting to the periods of highest daily relative humidity.

If the pollinator rows are not removed after pollination, they should be harvested prior to the harvest of the female rows, preferably with a separate harvester. To avoid contamination, no pollinator heads should be left intermingled with the female rows. Under certain conditions, such as lodging, a person should follow the harvester to remove these sources of contamination.

Sunflower seed with moisture as high as 20% may be threshed and cleaned successfully, although drying is necessary to assure safe storage and maintain quality. If the harvest procedure includes drying, certain precautions are necessary to maintain seed quality. The dryer construction should allow thorough cleaning to prevent mixtures with other sunflower cultivars or other crops. The air temperature control must be accurate and easily adjusted. The air temperature must not exceed 43.3 C (110 F) as germination usually is reduced if the seed is dried at higher temperatures. Artificial drying may be necessary in some regions. A dryer should not be used, however, unless absolutely essential, because each handling of seed causes some additional mechanical damage. All transportation equipment and warehouse facilities must be cleaned to avoid seed mixtures. Loaded trucks or other containers must be covered in transit. Seed transfer should be by bucket-type elevator, portable auger, or belt conveyor. Regular grain dumping pits should be avoided since they are nearly impossible to clean thoroughly. It is advisable to clean seed with an air and screen cleaner as it arrives in the warehouse. This cleaning removes most foreign material, reduces the storage space required, and, by removing material of higher moisture content, helps assure safe storage.

Seed Management

Storage

Temperature of seed in storage should be checked regularly and more frequently immediately after placing in storage. Inspection frequency will vary with region, type of storage, and condition of seed when placed in storage. Various insects may attack sunflower seed in storage so inspections should include analysis for presence of larvae, webs, adults, or other typical evidence of injurious activity.

Inspection should include observation for symptoms of seed damage from rodent activity. Most small rodents which commonly inhabit grain storage facilities are especially attracted to sunflower seed. A thorough review of the storage area prior to harvest, and an eradication program for mice and rats will help to minimize losses in storage. Sealed storage bins eliminate rodent damage most effectively.

Quality Assessment

Samples for moisture percentage and quality determinations should be taken from every lot. One kilogram usually is adequate. Physical analysis of each seed lot as received at the processing plant is necessary to determine the required procedures for cleaning and sizing. This analysis includes determining percentage immature, shriveled, cracked, broken, insect, and disease-damaged seed, and foreign material. The cleaned seed is evaluated for percentage of various seed size classes. A germination test of the seed size classes is conducted according to the Rules of the Association of Official Seed Analysts.

The seed of some inbred lines and cultivars of sunflower produced in certain environments is physiologically immature at the optimum stage of plant development for harvesting. The author observed that seed lots of the male sterile inbred, cmsHA89, which matured under two environments and were harvested at approximately 11% moisture showed different levels of physiological maturity as determined by the germination test immediately after harvest. Seed matured in an average daily high temperature of 36 C germinated an average of 18% as compared to 68% for the lot matured in an average daily high temperature of 20 C. After storage for 2 months at an average daily temperature of 14 C, both lots germinated above 90%.

Freezing temperatures during plant maturation also may affect dormancy. Zimmerman and Zimmer (10) reported increased germination for sunflower seed subjected to natural freezes at −4 C for 3 to 4 hours in North Dakota at 46 and 53 days after flowering. Samples were collected at 1-week intervals beginning at 28 days after flowering and stored at −18 C for 16 hours followed by air drying in storage at 25 C. Germination percentage was near zero for samples harvested prior to 42 to 49 days after flowering. Zimmerman (9) found that seed treatment in a 145 ppm solution of Ethrel® for 16 hours, followed by drying, may accelerate seed maturation after harvest, reduce dormancy, and increase active germination to near maximum.

The genetic characteristics of a seed lot are determined by the process termed "grow-out" which is the planting of parental inbred seedstock or F_1 seed to observe the phenotype of plants produced. Plants are classified for proportions typical of the cultivar, atypical plants including outcrosses with wild sunflower or other cultivated sunflower, and for self or sib-pollinated plants. The sample to be used for planting the grow-out must be cleaned and sized comparably to the system used to clean and process the whole lot.

The germination percentage of each lot is determined before planting the grow-outs. If germination percentage is low, treatment with Ethrel® or freezing to relieve dormancy may be necessary to avoid delay in grow-out trials. The grow-out should include 1,000–2,000 F_1 plants/seed lot to adequately sample genetic variability. If the female line is variable, the sample size should be increased to approximately 3,000 plants. Subsequent research and information may permit reduction of sample size while obtaining adequate information on genetic variability of the seedstocks.

Grow-outs of sunflower have been conducted in Florida, Puerto Rico, Mexico, South Africa, and Hawaii. Tests for reaction to such diseases as rust and downy mildew usually are conducted in a greenhouse. A population of 50 to 100 seedlings generally is sufficient to evaluate disease reaction.

Processing

Most sunflower seed processing installations include indent cleaners to remove pieces of foreign material longer than the seed; air screen machines for preliminary grading and the removal of all the remaining foreign material, immature seed, and broken sunflower seed; a series of precision graders for separating seed of different sizes; a bagging unit; conveying equipment; and often a gravity separator. Bucket elevators of steel with a rolled lip on the bucket or plastic and belt conveyors are the most satisfactory means of seed conveyance. The capacity of the conveying systems should be designed to deliver the appropriate volume of seed when operated at slow speed to minimize seed damage.

Processing usually involves passing the seed through the following equipment, in sequence: air screen cleaner, indent cleaner, precision graders, and packaging. The screens used in the air screen cleaner depend upon the range in seed sizes of the lot, the kinds of foreign material and broken sunflower seed, and the sizes of seed desired in the final product. Screens with round holes ranging in diameter from 4.0 to 9.5 mm are quite effective in the cleaning and preliminary sizing. The precision grader separates the cleaned seed into desired sizes based upon thickness and width with cylinder selection based on variability in seed size. The seeds of certain weeds are nearly the same size as sunflower seed or have a surface which may require additional or special cleaning operations to separate from sunflower seed. One of the most difficult weed seeds to separate is cocklebur (*Xanthium* sp.) because the seed coat is composed of rigid, hooked appendages.

Grades of planting seed offered in North America range from approximately 10,000 to 17,600 seeds/kg. Most companies make available at least four grades with the following average specifications: Grade 2—11,000 seeds/kg, Grade 3—13,250 seeds/kg, Grade 4—15,500 seeds/kg, and Grade 5—17,600 seeds/kg. Some companies make seed available by seed counts of 80,000 to 120,000/bag. Much of the seed is packaged in multi-walled paper bags containing 11.4 kg (25 lb) or 22.7 kg (50 lb). The usual size of the 11.4 kg bag is 38 × 76 mm (15 × 30 in) and the 22.7 kg bag is 56 × 91 mm (22 × 36 in).

Seedstock Maintenance

The purpose of the seedstocks program is to increase the seed quantity and maintain precisely the specific characteristics of the parental stocks used in hybrid seed production. Thus, comprehensive botanical descriptions of each component line and the hybrid are necessary to establish bases of reference. A basic requirement in the increase of parental stocks is familiarity with their plant types so that atypical plants can be eliminated entirely.

One program for maintaining male fertility restoring lines involves the following: planting approximately 3,000 seeds at 40 cm intervals in rows 75 cm apart, examining all plants and removing prior to flowering those that do not fit the botanical description of the line, and selecting at least 500 identical plants for bagging to produce selfed seed when approximately 5% of the plants have commenced flowering. Each selfed head must be numbered to maintain its identity throughout the procedure. The seed from those plants that does not fit the botanical description for the line is discarded.

The second season of the program requires a location isolated at least 6.4 km from other sunflower. Seed from each retained head is planted in a head-to-row system. Rows that do not conform to the description of the line are discarded before anthesis along with remnant selfed seed of the discarded rows. At least two plants in each of the remaining rows are bagged and sib-pollinated. The remainder of the nursery is allowed to open-pollinate. The seed from the open-pollinated plants is used the following season to plant a foundation seed field. This field should be isolated from other sunflower by at least 4.8 km or 30 days difference in flowering date. The seed from sibbed heads is used to repeat the procedures followed in the second season. An optional procedure would be to return to 3,000 breeders seeds in alternate seasons and eliminate sib-pollination in the head-to-row nursery.

Maintenance of the B line, or maintainer line, is the primary consideration in a female parent stock program. An effective procedure is to plant at least 2,000 seeds of each of breeder seed of the A and B lines 40 cm apart in adjacent rows. Plants that do not conform to the description for the B line are removed from all rows before flowering. At least 1,000 plants of the male sterile A line are paired, bagged, and crossed with pollen from 1,000 B line plants. The B line plants used are selfed and their identity maintained. Atypical plants are discarded throughout the crossing program. All plants are harvested and threshed with individual identity maintained.

In the second season 20 seeds from each crossed A line plant are planted in a single row. All plants are classified for pollen production and any rows that contain male fertile or other atypical plants are discarded.

Based on the grow-out results, 10 to 15 of the paired combinations from the first cycle are selected and planted in adjacent row combinations to repeat the paired cross procedure. Remnant seed of the same selected 10 to 15 paired A and B line plants is combined to produce separate bulks of A line and B line stocks that are used to produce foundation seed of the A line.

The increase of foundation seed of the A line is similar to that used to produce hybrid seed, except that the field should be isolated from other sunflower by at least 8 km, or 30 days difference in flowering period, and the ratio of female to male rows should be no more than 3:1. The B line should be increased in a separate field isolated by 4.8 km, or 30 days difference in flowering date. Only one increase of the B line is recommended before returning to breeders seed.

OPEN-POLLINATED SEED

Sunflower seed production evolved as an open-pollinated process with varying degrees of manipulation by man. Until about 1940, seed production involved a substantial amount of phenotypic selection since each interested individual had his own concept of the desired plant and seed type. A selection program started by the Canada Department of Agriculture in 1937 (8) produced the cultivar 'Sunrise' in 1942. This was the beginning of the evolution of the present technology used in North America for the production of open-pollinated cultivars.

Many procedures used for open-pollinated cultivars have formed the basis for the procedures evolved for hybrid seed production. The significant differences between the two are the roguing, pollination, and seedstocks maintenance programs. The roguing program involves the removal of off-type plants which often are difficult to identify positively because of the normal variation in the cultivar. There are differences of opinion and inconsistent data on the value of honeybees in pollinating open-pollinated sunflower. These differences are principally due to the variation in self compatibility among cultivars. The recommendation in certain areas, however, is to place hives in and around the field to assure a high percentage of cross pollination and thereby reduce the possibility of undesirable effects of inbreeding depression that could occur if plants were primarily self-pollinated.

An effective seedstock maintenance program for an open-pollinated cultivar may involve the following procedures: selecting visually 2,000 typical seeds, planting, selecting and bagging at least 500 plants typical for the cultivar, rubbing the selfed head at least twice to distribute pollen during the flowering process, harvesting and threshing each selfed head, discarding those with off-type seed, planting seed from each of the selfed plants in an isolated location (6.4 km) in a head-to-row system, removing all the plants from any row that has off-type plants before flowering starts, sib-pollinating at least four plants within each row, harvesting the sibbed heads and maintaining the seed with a row identification, and harvesting the remainder of the open-pollinated seed from the block. The sibbed seed serves as the basis for the selection and sibbing program in the following season. The open-pollinated seed is the basic seedstock for starting the increase program to produce marketable quantities of seed.

The procedures for the production of quality seed of open-pollinated sunflower cultivars are well established, and those for hybrid cultivars have evolved to a high level of technology in a relatively short period. In this chapter the status of the procedures has been recorded. Since sunflower appears to excel all other annual crops for oil production per hectare in diverse environments the opportunities are considerable for further research and seed production development.

ACKNOWLEDGMENT

The author is indebted to the following persons for sharing their experience in sunflower seed production in response to my inquiry.

Anderson, V. D.	North Dakota State Seed Dep., Fargo, N.D.
Bergen, P.	CSP Foods, Ltd., Altoona, Manitoba, Canada
Bevis, R.	Interstate Seed & Grain, Fargo, N.D.
Claassen, C. E.	Pacific Oilseeds, Inc., Woodland, Calif.
Dunn, F. W.	Northrup, King & Co., Woodland, Calif.
Ebeltoft, D. C.	North Dakota State Univ., Fargo, N.D.
Ferguson, D. B.	David & Sons, Fresno, Calif.
Fick, G. N.	Sigco Sun Products, Breckenridge, Minn.
Lider, W. R.	Cal/West Seeds, Woodland, Calif.
Lofgren, J. R.	Dahlgren, Inc., Crookston, Minn.
Moses, C. A.	Interstate Seed & Grain, Fargo, N.D.
Palmer, D. S.	Northrup, King & Co., Yuba City, Calif.
Polson, D.	Agway, Inc., Grandin, N.D.
Putt, E. D.	Agriculture Canada, Research Station, Morden, Manitoba, Canada
Shein, S.	Northrup, King & Co., Woodland, Calif.
Shuler, J.	Sigco Sun Products, Breckenridge, Minn.
Stanton, C. R.	Pacific Oilseeds, Inc., Woodland, Calif.
Vaccaro, J. W.	Sacramento, Valley Milling Co., Ordbend, Calif.

For communicating his experiences with bees as sunflower pollinators, I am indebted to: P. F. Torchio, USDA, ARS, Western Region, Bee Biology and Systematics Laboratory, Utah State Univ., Logan.

LITERATURE CITED

1. Anonymous. 1970. Key to hybrid sunflowers found by USDA Scientist. Crops Soils 23(3):21.
2. Bolson, E. L. 1977. Effect of isolation distance on outcrossing in seed production of sunflower (*Helianthus annuus* L.). M.S. Thesis. North Dakota State Univ., Fargo.
3. Enns, H., D. G. Dorrell, J. A. Hoes, and W. O. Chubb. 1970. Sunflower research, a progress report. p. 162–167. *In* Proc. 4th Intl. Sunflower Conf. (Memphis, Tennessee). National Cottonseed Products Association, Memphis.
4. Gundaev, A. I. 1971. Basic principles of sunflower selection. p. 417–465. *In* Genetic Principles of Plant Selection. Nauka, Moscow. (Transl. Dep. of the Secretary of State, Ottawa, Canada. 1972).

5. Heiser, C. B., Jr. 1976. The sunflower. Univ. of Oklahoma Press, Norman.
6. Kinman, M. L. 1970. New developments in the USDA and state experiment station sunflower breeding programs. p. 181–183. *In* Proc. 4th Intl. Sunflower Conf. (Memphis, Tennessee). National Cottonseed Products Association, Memphis.
7. Leclercq, P. 1969. Une sterilite male cytoplasmique chez le tournesol. Ann. Amelior Plant. 19(2):99–106.
8. Putt, E. D. 1940. Observations on morphological characters and flowering processes in the sunflower (*Helianthus annuus* L.). Sci. Agric. 21:167–179.
9. Zimmerman, D. C. 1977. Sunflower seed germination as influenced by maturity and Ethrel treatment. *In* Proc. 2nd Sunflower Forum. (North Dakota State Univ., Fargo).
10. ————, and D. E. Zimmer. 1978. Influence of harvest date and freezing on sunflower seed germination. Crop Sci. 18:In press.

DONALD L. SMITH: B.S., Ph.D.; Director of Research, Cal/West Seeds, Woodland, California. Plant breeder and agronomist; involved in oilseeds research and seed production since 1958. Instrumental in establishment of sunflower breeding and seed production programs for Cal/West Seeds; Pacific Oilseeds Inc., Woodland, California; Semillas Pacifico, S.A., Sevilla, Spain; and Semillas Nacionales, Culican, Mexico. Participated in the introduction of oilseed sunflower into Spain.

Chapter 12

Production Costs and Marketing

DAVID W. COBIA

Production practices and costs for oilseed and nonoilseed sunflower (*Helianthus annuus* L.) are nearly identical. The two types of sunflower are sold in separate and unrelated markets; therefore, most of the marketing and processing companies, especially beyond the first handler, deal only in oilseed or nonoilseed sunflower or one or more of their end products. Market conditions occasionally encourage the flow of limited quantities of oilseed sunflower into the bird and petfood market. Some kernel chips, as well as rejects from the dehulling operation of nonoil sunflower to obtain kernels for human food, are sold to oilseed processors. In the USA and Canada oilseed sunflower is not used, in the whole seed or hulled form, for the human food market because of difficulties in removing the hull. In other parts of the world, however, oilseed sunflower is used directly for human food.

ECONOMICS OF FARM PRODUCTION

Production Patterns

Sunflower is gaining major crop status in the USA, while production in Canada has fluctuated with no perceptible trend. The crop has been grown primarily in the northern fringes of the Corn Belt and north into southern Manitoba where corn (*Zea mays* L.) and soybeans (*Glycine max* L. Merr) have not performed well either because of a short growing season or lack of rainfall during critical periods.

Commercial production of nonoilseed sunflower has occurred on a limited basis in the USA and Canada for several decades. California was the major producer of nonoil seed sunflower before the late 1950's. Since that

From: Carter, Jack F. (ed.). 1978. *Sunflower Science and Technology*. Agronomy 19, Copyright © 1978 by the American Society of Agronomy, Crop Science Society of America, and Soil Science Society of America, 677 South Segoe Road, Madison, WI 53711 USA.

time most of commercial production of nonoilseed sunflower has been in the Red River Valley of North Dakota and Minnesota, where most first line processors of nonoilseed sunflower are located. Nonoilseed sunflower production in the USA increased to over 100,000 metric tons in 1971 and has remained relatively stable since that time (Fig. 1). The two types of sunflower are not separated in Canadian statistics, but industry sources indicate that nonoilseed sunflower production in that country generally amounts to about 1,000 metric tons per year.

Sunflower has been crushed for oil in Canada since 1944. Soviet cultivars with higher oil percentage were first produced in Manitoba in 1963 (P. Bergen, personal communication).

The first sustained commercial production of oilseed sunflower in the USA occurred in 1966 when about 4,000 ha of Soviet cultivars were grown. Production of oilseed sunflower has increased more rapidly than nonoilseed types, principally because of the strong export demand for oilseed sunflower. This dramatic increase has averaged about 45% per year for the 1970–1977 period. Sunflower production in Canada during this same period has been erratic. Year-to-year changes have averaged 140%. Production reached 77,000 metric tons in 1972, declined to 8,300 metric tons in 1974, and then increased to an estimated 78,000 metric tons in 1977. Weather and relative prices of competing crops, especially wheat (*Triticum aestivum* L.) and rapeseed (*Brassica napus* L.), seem to be responsible for this variability in production.

Most of the sunflower production since the late 1950's has been in the North Dakota and Minnesota counties adjacent to the Red River and within 160 km of the U.S. border in Manitoba (Table 1). Geographic expansion in the northern production area began in 1972 and has been primarily into counties west of the Red River Valley in North Dakota and south into South Dakota. Sunflower is planted in the Northern Plains of North America in late May and early June and harvested in October.

Interest in sunflower production also has developed in other regions of the USA, especially in Texas. Sunflower has been introduced into Texas to provide additional supplies of oilseed so that excess cottonseed crushing capacity can be utilized better and as an alternative crop for farmers. As an alternative crop, sunflower is used as a second crop after wheat and as a replacement crop where cotton (*Gossypium hirsutum* L.) has been destroyed by wind or hail, competing in these uses primarily with grain sorghum [*Sorghum bicolor* (L.) Moench]. Planting in Texas can begin as early as April, but most second or replacement crop planting of sunflower occurs in July. Second crop sunflower usually is harvested in November and early December.

A large increase in sunflower production occurred in the USA in 1972 because of improved sunflower prices, government limitations on wheat and feed grain production, and the set-aside farm program. Under this program, participating farmers withdraw a specified area from the normal production of basic commodities (feed grains, wheat, and cotton) and transfer or set aside this land to an approved conservation use. The farmer receives a

Table 1—Planted sunflower, USA and Canada; 1969–1977.†

Year	ND	MN	SD	TX	CA	Other	Total USA	Manitoba, Canada
				thousand ha				
1969	44.5	34.4	‡	0.3	1.4	0.9	81.5	19.4
1970	51.4	37.2	0.2	‡	0.4	0.6	89.8	28.7
1971	98.3	65.6	6.1	‡	0.8	2.5	173.3	96.7
1972	169.2	121.8	17.0	0.2	1.6	38.7	348.5	87.8
1973	169.2	105.2	32.8	0.4	0.8	1.6	310.0	52.2
1974	154.6	78.1	35.8	2.8	0.8	1.6	273.7	8.5
1975	219.4	87.0	72.0	109.3	1.8	1.6	491.1	66.0
1976	250.9	86.6	40.5	74.4	2.7	2.6	457.7	51.2
1977§	485.6	186.2	72.9	101.2	3.6	60.7	910.2	66.8

† From 3, 4, 10, and F. G. Thomason, unpublished data.
‡ Less than 50 ha.
§ Preliminary estimates.

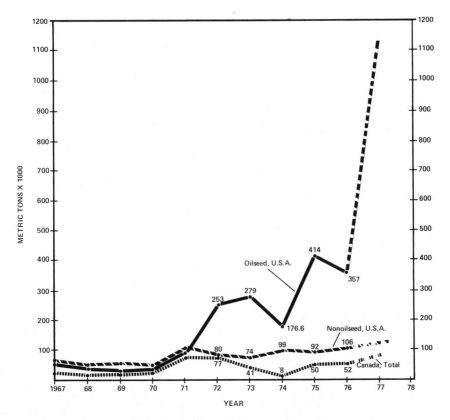

Fig. 1—United States oilseed and nonoilseed and total Canadian sunflower production, 1967–1976 and projected 1977 (From 3, 4, 10, and F. G. Thomason, unpublished data).

Table 2—Estimated 1978 per hectare production cost and yield for sunflower, wheat, barley, and flax on nonfallow land in east central North Dakota.†

Production item	Crop			
	Sunflower	Wheat	Barley	Flax
	$			
Variable costs				
Seed	11.68	11.74	10.87	13.34
Fertilizer	9.14	26.69	29.03	6.55
Pesticides	22.12	15.44	15.44	18.66
Machinery repair	16.43	13.71	13.96	13.34
Fuel and lubricants	10.38	7.54	7.91	7.17
Crop insurance	7.41	5.19	5.93	6.42
Drying	2.96			
Custom hire	1.98	2.97	2.97	
Interest on opr. capital	3.48	3.33	3.46	2.47
Total variable costs	85.58	86.61	89.57	67.95
Fixed costs				
Machinery depreciation	19.15	15.69	16.19	15.32
Interest on mchn. invest.	14.95	10.01	10.38	9.64
Labor	19.15	15.07	15.81	14.21
Land charge	74.13	74.13	74.13	74.13
Overhead	18.16	18.16	18.16	18.16
Total fixed costs	145.54	133.06	134.67	131.46
Total production cost/ha	231.12	219.67	224.24	199.41
Expected yields, metric ton/ha	1.23	2.29	2.74	0.94
Break-even price, $/metric ton	187.00	96.00	82.00	21.00

† From L. W. Schaffner, unpublished data.

payment from the federal government based on the area of land diverted from the production of basic commodities deemed to be in surplus. In some years farmers have had the option of planting approved alternative crops, such as sunflower on diverted land without forfeiting their entire set-aside payment. An estimated 176,000 ha of sunflower were grown under this program in 1972 and dropped to 61,000 ha in 1973. The program of allowing sunflower production on diverted land was discontinued in 1974 (14).

Total area planted to sunflower in the USA exceeded 400,000 ha (about 1 million A) for the first time in 1975. In 1977 favorable sunflower prices at planting time compared to small grains, and reputed drought tolerance of sunflower prompted a doubling of production to over 900,000 ha in the USA and 67,000 ha in Canada.

Average Yields

Average sunflower yields per harvested hectare in the USA have ranged from 0.962 to 1.132 metric tons for nonoilseed and from 1.033 to 1.240 for oilseed sunflower during the 1967–1976 period (Fig. 2). Until 1974 oilseed cultivars generally yielded about 100 kg/ha more than nonoilseed types. This higher yield was due, in part, to greater disease resistance in earlier oilseed types. Nonoilseed sunflower yields have been gaining relative to oilseed cultivars because of the recent improvements in the disease resistance of nonoilseed cultivars. In addition most of the increase in oilseed production in North Dakota has been on land with lower rainfall which tends to reduce average oilseed yields.

Production Costs

Generalizations on production costs and yields are difficult because of different management practices, land, and weather. Production costs given in Table 2 represent better-than-average management and may be used for comparisons with competing crops in North Dakota. Production costs for sunflower are slightly higher than those for small grains depending on use of fertilizer and pesticides. In the past, sunflower production costs have been lower than those for small grains. As production of sunflower increases, however, and farmers strive to increase yields, the use of hybrid seed and pesticides has and will continue to increase. The data in Table 2 reflect the use of nearly 100% hybrid seed for sunflower while the seed costs for the other crops reflect the common practice of buying certified seed about every 3 years. Farmers generally use less fertilizer on sunflower than on many other crops because they believe sunflower does not respond as well. The impact of inadequate fertilizer, however, frequently is noticed in subsequent crops.

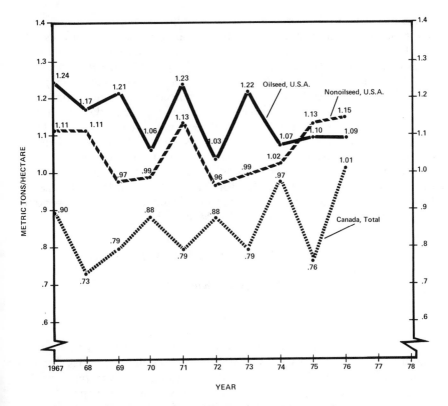

Fig. 2—Average yields of U.S. oilseed and nonoilseed and total Canadian sunflower; 1967–1977 (From 3, 4, and 10).

Higher herbicide costs for sunflower reflect the cost of selective broad-leaf and other weed control chemicals, which are more expensive than those commonly used for small grains. Machinery costs reflect double preplant incorporation of herbicides. The little postemergent cultivation required is about offset by the swathing expense of small grains and flax (*Linum usitatissimum* L.). Farmers frequently substitute postemergence cultivation for herbicides, a practice that would eliminate herbicide costs. Machinery expense is about the same for postemergent cultivation and offsets the cost of preplant incorporation of herbicides. Using cultivation rather than herbicides to control weeds is risky because rainy weather may prevent timely cultivation, and yields are reduced by early competition of weeds. Machinery and related expenses are higher than for small grains because of double incorporation of herbicides, and needs for a special header on the combine, for drying, and for transportation. Sunflower requires drying at harvest time more frequently than do small grains. Transportation costs are higher for sunflower because fewer elevators handle the crop in several production areas.

Relative Profitability

Sunflower production has been expanding primarily in response to favorable prices. In the northern production area, sunflower competes with small grains and flax and, to a lesser extent, with pinto bean (*Phaseolus vulgaris* L.), soybean [*Glycine max* (L.) Merr.], potato (*Solanum tuberosum* L.), corn, and sugarbeet (*Beta vulgaris* L.). Since 1973, sunflower prices generally have yielded higher net profit per hectare than most competing crops. Prices of wheat, barley (*Hordeum vulgare* L.), and flax required to achieve the same profit/hectare as sunflower are given in Table 3. These prices are based on estimated production costs and yields given in Table 2. For example, if $200/metric ton is estimated for sunflower, the net profit would be $15.60/ha. The price/metric ton of other crops required to yield the same profit level would be $103 for wheat, $87 for barley, and $228 for flax. Costs and yields vary from farm to farm and area to area, depending on land, cultural practices, costs, and yield potential.

THE MARKETING SYSTEM

Marketing Channel

Oilseed Sunflower

The flow of oilseed sunflower in the USA from the farmer, through country elevators, processors, and other marketing intermediaries, has become more complex and more like that of small grains with time and increased production. Most oilseed sunflower was grown under special grower agreements or contracts until 1974 and 1975. These contracts generally had several provisions, including price, area, delivery, quality, and payment

Table 3—Estimated prices of wheat, barley, and flax required to return the same profit as specified prices of sunflower in east central North Dakota for 1978.

Sunflower priced at	Crop		
	Wheat	Barley	Flax
	$/metric ton		
150	76	65	163
200	103	87	228
250	130	110	294
300	157	132	359
350	184	155	425
400	211	177	490
	est. 1978 prod. cost/ha		
$231	$220	$224	$199
	est. avg. yield, metric tons/ha		
1.233	2.287	2.744	0.942

schedule. While the crop was becoming established and the market was limited, growers needed a guaranteed buyer and price, high quality seed for seeding, and information on recommended agronomic practices provided by contractors (1). Contractors needed an assured supply to meet their processing requirements and to fulfill their forward commitments.

Special agreements for oilseed sunflower growers have been largely abandoned in the northern production area of the USA because of expanded market alternatives for growers and exporters. Processors also have a more stable, expanding supply from which to draw their requirements. Farmers in the USA have the option, as with most common small grains, of growing sunflower for the open market or to contract a portion of the crop with a standard cash-grain contract. Over 90% of the sunflower crop in Canada still is grown under contract. Contract activity also is prominent in Texas and California.

Most of the crop is trucked at harvest to country elevators to be cleaned and shipped by truck or rail to export or processing locations. A few cotton gin cooperatives in Texas receive sunflower from farmers as well. Some farmers truck their sunflower directly to export or processing locations.

Sunflower merchandisers, including country originators, originator exporters, and grower-marketing associations, and a few processors who also merchandise sunflower, acquire their supplies from country elevators. Exporters acquire their supplies at country origin points or from other intermediaries. Most oilseed sunflower produced in the USA moves through the Great Lakes via Duluth (Table 4) to Rotterdam, where it is transferred to barges for movement to inland processing points. The Duluth port is used because over 85% of the oilseed sunflower crop is grown in the north and a larger share of the oilseed sunflower grown in southern U.S. is crushed domestically. The increase in exports through New Orleans resulted from northern-grown sunflower moving down the Mississippi River on barges for export. Exports through Houston have been primarily sunflower grown in Texas. The proportion of the Canadian crop moving into export varies con-

Table 4—Sunflower exports by customs district, United States, 1974–1976.†

Customs district	1974	1975	1976
	% of total		
Duluth, Minn.	94.0	90.1	75.1
Pembina, N.D.	2.6	1.7	0.8
New Orleans, La.	3.0	6.8	9.7
Houston, Tex.	--	--	13.9
Others	0.4	1.4	0.5
Total	100.0	100.0	100.0
Total exports, metric tons	184,859	210,290	398,831

† Data supplied by F. G. Thomason from Bureau of Census, U.S. Dep. of Commerce. EM-522 monthly magnetic tapes.

siderably and has reached as high as 50%. Most of this sunflower is exported through Thunder Bay, Ontario.

Nearly all of the crop for export has been moved directly into shipping channels at harvest (Table 5) before the St. Lawrence Seaway freezes. From 63 to 84% of the crop was exported during October, November, and December for the years for which data are available. An additional surge of exports is made in May when the St. Lawrence Seaway reopens.

The USA has become the prime world source for wholeseed sunflower by supplying over 80% of wholeseed sunflower in world trade since 1975 (9). Very little sunflower oil is exported from North America to the Common Market countries because of a 10% ad valorem import tax on crude vegetable oil. No import tax is assessed against wholeseed sunflower (6).

The portion of oilseed sunflower processed domestically has been erratic in both the USA and Canada (4 and Thomason, unpublished data). This portion ranged from 10 to 37% in the USA during the 1973–1976 period. Most of the domestic crush of northern-grown sunflower in the USA is in modified flaxseed processing plants at Minneapolis. Small amounts also are processed at Gonvick, Minnesota, and Culbertson, Montana. Canadian sunflower is processed at Altona, Manitoba. Oil from sunflower grown in the Northern Great Plains is used primarily for margarine and cooking oils, with some being used in paints and varnishes. Southern grown sunflower is processed in cottonseed and castor bean processing plants. Sunflower oil from these plants has been used largely by snack food manufacturers for deep fat frying. Southern sunflower oil, when matured during hot summer months, is better adapted to this purpose than northern-grown oil because of its higher oleic acid percentage.

Sunflower oilseed meal produced in the USA has a relatively high fiber content of 24% because the hulls are not removed during processing as in most countries. Therefore, domestically produced sunflower oilseed meal is limited to use in feeds for ruminant animals, primarily dairy rations (11). In Canada, all seed processed for oil is hulled, and the fiber content of the meal from the crushing operation is in the range of 11 to 13%. This meal is used by feed manufacturers to produce rations for both ruminant and monogastric animals, including hogs and poultry (E. D. Putt, personal communication).

Table 5—Sunflower exports by month, USA, 1974–1976. †

Month	Year		
	1974	1975	1976
		% of total	
January	1.6	0.1	0.2
February	0.8	2.3	4.8
March	0.1	0.3	0.7
April	0.1	3.7	2.3
May	23.9	8.8	16.3
June	7.2	0.8	3.7
July	2.8	0.2	2.8
August	0.1	0.1	1.7
September	0.2	0.2	1.7
October	14.1	6.6	15.8
November	31.5	47.9	33.7
December	17.6	29.0	16.3
Total Exported (metric tons)	184,859	210,290	398,831

† Data supplied by F. G. Thomason from Bureau of Census, U.S. Dep. of Commerce, EM-522 monthly magnetic tapes.

Nonoilseed Sunflower

Fewer than 20 companies in the USA contract with farmers, either directly or through agents, and process nonoilseed sunflower in varying degrees for assorted markets. Over 90% nonoilseed sunflower is grown under contract with processors. About 60% of the nonoilseed crop is trucked directly from the farm to processor rather than via country elevators (1). Farmers provide storage for 20 to 30% of the crop. This amounts to a much higher portion than for oilseed sunflower. Several nonoilseed processors have either limited funds and/or limited storage and, therefore, require the grower to store the crop until needed.

From 3 to 5% of the U.S. nonoilseed sunflower is exported to Canada through Pembina, North Dakota (Table 4 and Fig. 1). Most nonoilseed sunflower receives its preliminary processing in the production area. The largest seed, normally those passing over a 7.9 mm (20/64 in) round-hole sieve, is sold to other processors for roasting whole and packaging, primarily for sale in competition with packaged nuts and snacks. Medium sized seed, usually those passing through 7.9 mm round-hole sieve but over a 7.1 mm (18/64 in) sieve, are dehulled to obtain kernels for roasting. Roasted sunflower kernels are prepared in vacuum-packed cans, small cellophane packages, and bottles. Large packages are prepared for bakeries and candy manufacturers. The smallest nonoilseeds, normally those that pass through a 7.1 mm sieve, are packaged either pure or in mixtures of grains and seeds as bird and petfood.

Sunflower hulls generally are ground and pelleted, either pure or as a mixture with other roughages, such as alfalfa (*Medicago sativa* L.) hay, and sold as livestock feed (3 and 11).

There are three relatively large and several smaller nonoilseed sunflower processors in Canada that process from about 4,000 to 5,000 metric tons of nonoilseed sunflower per year. Of this amount Canadian growers supply one-fourth to one-third. The balance is imported from the USA. Processing and utilization are similar to those described for the USA.

The Pricing Mechanism

Oilseed Sunflower Prices

Oilseed prices are determined by a complex economic system involving several factors, such as foreign trade, yield of oil and meal, substitution among competing oils and meals, other crops competing for farmland, and production costs (7, 8).

The prices of sunflower, in contrast to those of soybean, are related more closely to vegetable oil prices than to meal prices. The oil of sunflower accounts for about 75% of its value because sunflower contains approximately 40% recoverable oil that has sold for about four times the value of the meal. Soybeans contain only about 18% oil, and soybean meal is of a higher quality than sunflower meal.

Most vegetable oils can be substituted, within limits, for several edible and industrial uses. Each oil, however, has unique characteristics that determine its desirability for specific uses. Soybean oil, the largest source of vegetable oil in the world, is one of the lowest priced vegetable oils other than the highly saturated palm (*Elaeis quineensis* Jacq.) and coconut (*Cocos nucifera* L.) oils. Soybean oil contains about 9% linolenic acid, which oxidizes rapidly. Sunflower oil has a low level of saturated fatty acids with only a trace of linolenic acid. It is considered a premium oil for human consumption.

Differences in quality help explain price relationships of various vegetable oil prices. When sunflower oil is in short supply, the price differential between soybean oil and sunflower oil increases, and the sunflower oil price moves independently to soybean oil prices. When excess sunflower oil is available and the premium oil markets have been supplied, then sunflower oil prices decline and fluctuate with soybean oil prices (11).

The price for oilseed sunflower is export-demand oriented. Most North American production is exported. Prior to 1976 the North American production had a minimal influence on foreign and domestic prices because its production was about 3% of the total world supply. Now North America will have a more significant impact on world prices because its production is over 10% of the total world supply and it supplies over 80% of the wholeseed sunflower exports. Domestic prices are influenced temporarily by fluctuating foreign exchange and ocean freight rates, elevator handling margins, attitudes of dealers, and fluidity of the market.

Nonoilseed Sunflower Prices

Markets for nonoilseed sunflower are different from those for oilseed. Nonoilseed sunflower is sold to narrow, low-volume domestic markets with unrelated demand for end products. Sunflower roasted whole competes with snack foods, roasted kernels compete with nuts, and the smallest sunflower seeds compete with grain and seeds in bird and petfood markets.

Processors offer a contract price to farmers each spring which they believe will result in sufficient production to meet their forward commitments and other projected sales. These forward commitments are based on farm prices that processors believe will be required to attract land from other crops. Processors contract for most of their nonoilseed sunflower requirements because of the limited market. Some nonoilseed sunflower is grown without a contract. This uncommitted sunflower sells at the "open" or "cash" market price. The cash market price is at times higher than the contract price.

Farm Prices

The average sunflower price received by farmers ranged from $86/metric ton for oilseed to $122 for nonoilseed in the USA during 1966–1972. Nonoilseed sunflower required higher producer prices than the oilseed type partially because its yield was about 100 kg/ha lower (Fig. 3). Sunflower

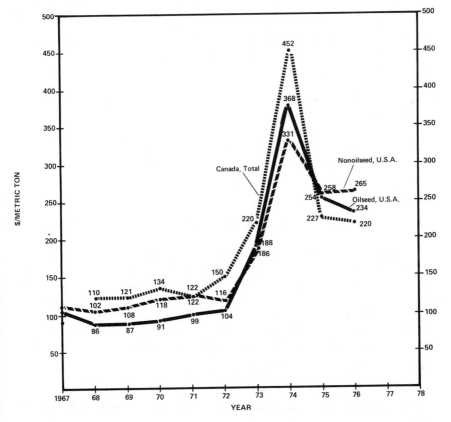

Fig. 3—Average prices received by farmers in the USA for oilseed and nonoilseed and in Canada for all sunflower; 1967-1977 (From 3, 4, and 10).

prices increased more than most other crops in 1973–1974. Oilseed reached a high of $485/metric ton and nonoilseed $374 during 1974 while averaging $378 and $331, respectively, for the year. Prices in Canada generally have been higher than in the USA. These higher prices can be explained by differences in foreign exchange rates and efforts to secure sufficient domestic supplies for processing.

Potential for a Futures Market in Oilseed Sunflower

No futures market is available for sunflower in Europe, Canada, or the USA. A futures market would facilitate and expand marketing alternatives for farmers, processors, end product buyers, and merchandisers. Hayenga (5) cites the following reasons for the present lack of a futures market in the USA:

Insufficient production. Past production of oilseed sunflower has been insufficient to support a futures market. A sustained production of more than 1 million metric tons annually in the USA should support a viable sunflower futures market. The 1977 U.S. production surpassed this volume, but sunflower production has fluctuated from year to year (see Table 1 and Fig. 1).

Lack of carry-over supplies. Many advantages of a futures market are related to a trader being able to make or accept delivery during contract months. To date, U.S. farmers have considered sunflower as a cash crop and sold most of their production at harvest, over 80% of which has been sold at harvest for prompt export (Table 5). To support a futures market, farmers and merchandisers must be willing to hold supplies so that the market would be fluid, at least during designated futures-market delivery months. It might be possible to use soybean, soybean oil, and soybean meal futures market as a proxy for a sunflower futures market until these conditions develop and a futures market is established for sunflower.

Grades and Standards

Grades and standards, which are carefully designed and objectively applied, make possible a high level of pricing and operational efficiency in the marketing channel. This tool reduces marketing costs and elicits integrity, trust, and credibility in the industry. Pricing efficiency means that, over long periods, prices accurately reflect the costs of production and marketing services through the marketing channel to consumers, and that quantity and quality desires of consumers are reflected back to the producer. Suppliers are rewarded appropriately for furnishing the quality and quantity desired and consumers pay for what they want. Operational efficiency refers to the reduced time and expense in processing and marketing functions associated with uniform quality and clearly identified product lots.

Table 6—Canadian grades and standards for sunflower.† From (4).

Grade name	Degree of soundness	Maximum cracked and hulled seed
		%
No. 1 Canada	Well matured, sound, sweet, and uniform in size	2
No. 2 Canada	Reasonably well matured, sweet; reasonably free from damaged kernels	5
No. 3 Canada	May be frosted or otherwise damaged; may have the natural odor associated with low quality seed, but shall not be distinctly sour, musty, rancid, nor have any odor that would indicate serious deterioration or contamination	10

† All grades require that sunflower seed be "commercially clean" and assume a 9.5% moisture.

Benefits from objectively applied grading systems are: the likelihood of misinformation and misrepresentation of quality is reduced; bargaining can be centered on price rather than on quality; price quotations are meaningful; buy/sell arrangements by description are enhanced, and costs of physical inspection are reduced; an increased demand for some producs may result; and pooling or comingling of the product is facilitated.

Existing Grades

Official Canadian sunflower grades provide for three grades of commercially clean seed at 9.5% moisture (Table 6). The only quantifiable criterion separating the three grades is the percentage of cracked or hulled seed. Subjective criteria are maturity, soundness, odor, uniformity of size, and presence of frosted or otherwise damaged seed. There are no separate standards for oilseed and nonoilseed sunflower but cultivar name is an option available on the grade certificate.

Official USDA grades for the USA are not yet established for sunflower. The sunflower industry, until 1976, functioned on loosely agreed upon standards—or by having each transaction contain its own specifications. Standards and associated discounts and premiums were specified in detail in most grower contracts. Increased volume and number of transactions have accelerated the need for grades and standards acceptable across the industry. The Sunflower Association of America (SAA) established a Grades and Standards Committee in 1975 to meet this need.

The USDA is in the process of developing standards similar to those recommended by the sunflower industry (P. E. Parker, personal communication). Standards developed by the SAA have been adopted by Minnesota, and are being used now for commercial transactions. These grades and standards for nonoilseed and oilseed sunflower are given in Tables 7 and 8.

400 COBIA

Table 7—Minnesota grades and grade standards for Class I-edible and bird or petfood culti-vars of sunflower. From (2).

Grade	Minimum test weight		Maximum moisture	Maximum limits of damaged seed		
	Large seed†	Small seed‡		Heat	Insect	Total
	——— kg/hl (lbs/bu) ———			——— % ———		
1	30.9 (24)	33.5 (26)	10	0.5	2	5
2	28.3 (22)	30.9 (24)	12§	1.0	5	8
3	27.0 (21)	29.6 (23)	14§	1.5	7	10

Sample grade—shall include sunflower seed that does not meet the requirements of a numerical grade or which contains more than 14% moisture; or which is musty, or sour, or heating, or which has any commercially objectionable foreign odor; or which is otherwise of distinctly low quality; or which shows evidence of chemicals not approved.

Class I mixed—are sunflower seeds that contain more than 2% of Class II. The percentage shall be shown as a factor.

† Sample containing 30% or more of large seeds shall be graded as large when using 7.9 mm (20/64 in) round-hole sieve.

‡ Sample containing more than 70% of seeds which pass through 7.9 mm round-hole sieve shall be graded small.

§ Tough shall be sunflower seed in grade number 2 and grade number 3 containing more than 12% moisture but not more than 14% moisture.

Unique Problems

Every agricultural commodity presents unique problems in establishing and implementing standards. The more prominent unique problems for sunflower follow.

Moisture. The outer shell or hull acquires and loses moisture more rapidly than the kernel. Therefore, seed tested during periods of changing moisture conditions, such as early morning dew in the field or drying, may give an erroneous value. This condition is corrected by tempering the sample in a moisture-proof container at 18 to 24 C for 24 hours. Electronic moisture testing equipment becomes increasingly inaccurate on sunflower seed as moisture levels move above 15% because calibration has assumed a linear relationship between moisture and electric current. As soon as non-linear calibration charts and/or dials become available, moisture can be tested accurately above 15% (12).

Dockage. Accurate adjustment of equipment to remove dockage is much more crucial for sunflower than small grains because the differences in the aerodynamic characteristics of sunflower seed and dockage are much smaller than for most other seed crops. Empty hulls and dehulled seed also cause special problems (10). Mechanical determination of dockage is not accurate above the 15% moisture level because the dockage is about as heavy as sunflower.

Hulls. Sunflower seed must be dehulled so that the kernel can be examined for some insect, weather, and heat damage.

Mixture of oilseed and nonoilseed sunflower. Oilseed and nonoilseed sunflower should not be mixed during harvest, storage, transportation, and handling operations. Nonoilseed sunflower mixed with oilseed sunflower

Table 8—Minnesota grades and grade standards for Class II-oilseed cultivars of sunflower. From (2).

Grade	Minimum test weight	Maximum moisture	Maximum limits of damaged seed		
			Heat	Dehulled	Total
	kg/hl (lbs/bu)		%		
1	34.7 (27)	10	0.5	5	5
2	32.2 (25)	12	1.5	5	8

Sample grade—shall include sunflower seed that does not meet the requirements of a numerical grade or which contains more than 12% moisture; or which is musty, or sour, or heating, or which has any commercially objectionable foreign odor; or which is otherwise of distinctly low quality; or which shows evidence of chemicals not approved.

Class II mixed—are sunflower seeds that contain more than 2% of Class I. The percentage shall be shown as a factor.

will reduce average oil percentage. Oilseed types cannot be dehulled readily for human consumption. Sunflower of either class may be mixed for bird or petfood.

Sampling. Representative samples must be obtained carefully. Empty hulls, light seeds, heavy seeds, and dockage of sunflower tend to classify or segregate and form layers during harvest, storage, transportation, and handling more in sunflower than with most common seed crops.

Needed developments. The creation of ideal standards is limited by available technology. For example, oilseed sunflower should receive a premium or be graded on the basis of oil percentage. The development of commonly acceptable procedures for inexpensive tests of oil percentage using nuclear magnetic resonance (NMR) is being pursued. Some disease and weather damage cause nonoilseed sunflower kernels to turn dark brown when roasted. Frequently these defective seeds must be roasted to be identified. Growers should be rewarded for high oil percentage and undamaged seed; but, until better testing equipment is available, proxy measures, such as test weight, must be used.

Frequently the linoleic-oleic fatty acid ratio is important in oilseed sunflower for the export market and for sunflower oil used in the "fast food" and "snack food" industry. This ratio is a function of temperature as the crop matures. Therefore, the region of production, or the percentage of oleic acid, may become part of future standards. Oleic acid percentage can be determined using a hand refractometer but results are inconsistent unless tests are conducted under carefully controlled conditions.

Because the testing of sunflower is relatively recent in the USA, industry-wide testing procedures are not in use. This has resulted in some problems in test results. A special committee has been organized by SAA to establish testing procedures which can be followed by testing laboratories and others to yield reproducible tests (D. C. Zimmerman, personal communication).

Another problem that hampers the sunflower trade, along with most other segments of the economy, is the confusing array of English units of weights, measures, and containers. Different units are used all along the marketing channel—bushels, pounds, hundredweights, short tons, and metric units of kilograms and tons or tonnes. The U.S. government finally gave its sanction and encouragement to a conversion to the metric system when the Metric Conversion Act was enacted in December, 1975. This act called for a voluntary industry-by-industry conversion and the establishment of a U.S. Metric Board to coordinate that conversion. No formal movement to adopt the metric system in the oilseed and grain trade in the USA is underway. The USDA Metric Action Plan, however, calls for agencies to implement conversion to metric as soon as the nonfederal sectors are ready (P. E. Parker, personal communication). Proposed revised standards for corn and soybeans include metric standards. A conversion to the metric system involving the entire agricultural industry, including the grain and oilseed trade, took place in Canada on 1 Feb. 1978.

MAJOR INDUSTRY AND RESEARCH ORGANIZATIONS

Trade associations and state and federal agencies conduct research and in other ways aid in the development and promotion of sunflower. They also provide and disseminate information about the crop. Private companies and farm organizations also carry out some or all of these functions but are not listed in this section. The following are brief descriptions of the functions and purposes of public agencies and trade associations involved with sunflower.

State and Federal Research and Extension Agencies

The Science and Education Administration (SEA), formerly the Agricultural Research Service, of the USDA, U.S. Department of Interior, and Research Branch of Agriculture Canada, working in conjunction with state and provincial experiment stations and cooperative extension personnel located in the major sunflower producing areas, have developed improved cultivars and parental lines for hybrids. They have identified and studied methods of control of insects, diseases, birds, and other sunflower pests. Agronomic practices related to weed control, irrigation, planting dates, plant population, fertilizer use, harvesting, drying, and storage have and are being investigated by public agencies. Such economic considerations as cost of production, processing feasibility, and marketing also receive attention. This research in North America supplements previous and contemporary investigations in other countries. Listed below are some of the more active institutions.

North Dakota State University, Fargo, ND 58102, Agricultural Experiment Station, Cooperative Extension Service, and attached SEA, USDA personnel (agronomic and economic).

Texas A&M, College Station, TX 77843 (and branch stations at Lubbock and Bushland), Agricultural Experiment Station, Cooperative Extension Service, and attached SEA, USDA personnel (agronomic).

University of Minnesota, St. Paul, MN 55108, and Crookston, Minnesota, Agricultural Experiment Station and Cooperative Extension Service (agronomic).

Richard B. Russell Agricultural Research Center, SEA, USDA, Athens, Georgia (end product).

University of California, Davis, CA 95616, Agricultural Experiment Station, Cooperative Extension Service, and attached SEA, USDA personnel (agronomic).

Research Branch, Agriculture Canada, Morden Research Station, Morden, Manitoba, and Saskatoon Research Station, Saskatoon, Saskatchewan (agronomic).

Government Statistical Reporting Services

Canadian sunflower statistics are released by Statistics Canada and published by the authority of the Minister of Industry, Trade, and Commerce. Statistics on sunflower in Canada have been published for over 25 years and include area planted, production, average yield, and farm prices.

In the USA, sunflower statistics are released by the Economics, Statistics, and Cooperative Service, formerly Statistical Reporting Service, of the USDA. Federal support for gathering sunflower statistics in the USA first became available for North Dakota and Minnesota in 1976, and South Dakota and Texas in 1977. Previously these four states had provided funds for estimating sunflower statistics. North Dakota has collected data on area planted and harvested, production, and prices received by farmers since 1962, and was the first state to separate oilseed and nonoilseed sunflower statistics in 1967. The other three states have released similar, though less comprehensive, data for shorter periods of time.

The Sunflower Association of America

The Sunflower Association of America was organized in 1975 to promote the sunflower industry and to solve common industry problems. This association is intended to represent the interest of growers, processors, seed companies, researchers, and marketing companies, including country elevators, exporters, and country originators. Its membership is composed of voting and associate members. Voting members must be engaged actively in the production, handling, processing, or marketing of sunflower or sunflower products in the USA. Associate members are individuals or organizations interested in the industry and the objectives of the association, but not engaged actively in commercial endeavors. The association has standing committees in such areas as research, grades and standards, and publications. In addition it publishes *The Sunflower* magazine and sponsors "The Sunflower Forum," a biennial meeting devoted to scientific reports and topics of general interest.

The International Sunflower Association

The objective of the International Sunflower Association is to improve international cooperation and to exchange information in the promotion of research of agronomics, processing techniques, and nutrition associated with the production, marketing, processing, and use of sunflower. The association is composed of a council whose membership is a representative from each country, individual members, affiliated members or organizations, and honorary members.

The objective of the association is accomplished primarily by sponsoring and publishing the proceedings of biennial international conferences. The first such conference in 1964 was organized by Dr. Murray L. Kinman and held at Texas A&M University. Since 1966, the conferences have been held at Morden, Manitoba; Crookston, Minnesota; Memphis, Tennessee; Clermont-Ferrand, France; Bucharest, Romania; and Krasnodar, USSR. Attending the Krasnodar conference were about 600 persons representing 40 countries. The 1978 conference was held in July at Minneapolis, Minnesota.

An executive committee, through its chairman, provides a point of continuing coordination between international conferences. A periodical, *Sunflower Newsletter*, is published quarterly by the association. It carries association news items, research results, and an annotated bibliography of publications dealing with sunflower.

The North Dakota State Sunflower Council

The North Dakota State Sunflower Council was created by the 1977 North Dakota Legislature to promote the production, development, marketing, and promotion of North Dakota sunflower. The Council's activities are financed by a $0.22/metric ton ($0.01/cwt) levy on all sunflower grown and sold to dealers in the state. A six-member council, elected by sunflower growers, administers the funds collected.

National Cottonseed Products Association, Inc.

Membership of the National Cottonseed Products Association is made up primarily of southern oilseed processors, cottonseed processors being dominant. Brokers, refiners, and related interests also hold membership. A few state processor associations also are members with some firms holding joint membership.

The fundamental interest in sunflower by the national and state associations, especially in Texas, is to provide farmers information about alternative crops while providing additional supplies of oilseed for their member processors. This association has been active in supporting and encouraging research and evaluation of the sunflower crop, assisting in the development and dissemination of information on sunflower, and generally promoting the crop.

LITERATURE CITED

1. Cobia, D. W. 1975. Sunflower production contracts: provisions and analysis. Agric. Economics, Rep. No. 104, North Dakota Agric. Exp. Stn., Fargo.

2. ————. 1977. The sunflower grades and standards picture. The Sunflower 3(6):20-21.

3. ————, and D. E. Zimmer (ed.). 1975. Sunflowers: production, pests and marketing. Bull. 25. North Dakota Coop. Extension Service, North Dakota State University, Fargo.

4. Foods Systems Branch. 1977. Oilseeds in Canada an oilseed reference. Agriculture Canada, Ottawa.

5. Hayenga, R. 1975. Sunflower futures market still in the planning stages. The Sunflower 1(4):12.

6. Helgeson, D. L., D. W. Cobia, R. C. Coon, W. C. Hardie, L. W. Schaffner, and D. F. Scott. 1977. The economic feasibility of establishing oil sunflower processing plants in North Dakota. North Dakota Agric. Exp. Bull. 503.

7. Houck, J. P. 1973. Domestic markets. *In* B. E. Caldwell (ed.) Soybeans: improvement, production, and uses. Agronomy 16:612-614. Am. Soc. Agron., Madison, Wis.

8. ————, M. E. Ryan, and A. Subotnik. 1972. Soybeans and their products, markets and policy. Univ. of Minnesota Press, Minneapolis.

9. Mielke, S. (ed.). 1977. Oil world weekly. pub. weekly. Hamburg, West Germany.

10. North Dakota Crop and Livestock Reporting Service. 1970-1977. Sunflower seed. Semi-annual mimeo. Fargo, N.D.

11. Schmidt, K. M. 1977. Estimated demand for sunflower oilseed meal and hulls as animal feed stuffs. M.S. thesis. North Dakota State Univ. Xerox Univ. Microfilms. Ann Arbor, Mich. pub. no. 13-11090.

12. Schuler, R. T. 1976. Sunflower dockage testing results. p. 15-16. *In* Proc. Sunflower Forum. Sunflower Association of America, Fargo, North Dakota.

13. Smith, J. R. 1976. About those prices. The Sunflower 2(2):18.

14. Thomason, F. G. 1974. The U.S. sunflower seed situation. ERS 590. U.S. Government Printing Office, Washington, D.C.

15. Zimmerman, D. C. 1976. Determination of moisture content in sunflowers with electronic moisture meters. J. Am. Oil Chem. Soc. 53(8):548-550.

DAVID W. COBIA: B.S., Brigham Young University, M.S. and Ph.D., Purdue University; Professor, Department of Agricultural Economics, North Dakota State University, 1976 to date; conducted research and prepared materials on marketing and processing feasibility of sunflower as part of overall duties in agribusiness and marketing.

Chapter 13

Processing and Utilization of Oilseed Sunflower

D. GORDON DORRELL

The sunflower (*Helianthus annuus* L.) has become an important oil-seed crop in North America, with an estimated 1 million ha grown in Canada and the USA in 1977. Commercial crushing began on a continuous basis in Manitoba, Canada, in 1944. The industry has subsequently progressed until today several plants in Canada and the USA are crushing a combined total of more than 70,000 metric tons of seed annually. This development of sunflower can be attributed to the introduction of cultivars from the USSR which contain more than 40% oil, the availability of surplus crushing capacity in flaxseed and cottonseed crushing mills, and the general need for crop diversification in the Red River Valley and parts of the Cotton Belt. Until recently, most of the crop has been exported as seed. With greater public awareness of the excellent quality of sunflower products and better identification on product labels, however, the domestic utilization of sunflower should increase significantly.

SEED

The sunflower seed is more correctly described as an achene, a specific type of indehiscent fruit. It is pointed at the base, rounded at the top, approximately 10 to 15 mm long, and in cross section appears to be four-sided. The outer portion of the pericarp or hull consists of elongated and pigmented cells, below which are several layers of sclerenchyma cells with pitted walls and bundles of fibers with heavily pitted walls. Immediately beneath these layers is the testa, or seedcoat, which is a white, papery layer of compressed epidermal cells and spongy parenchyma. In sunflower, the endosperm is only one or two cells thick as it is almost totally consumed dur-

From: Carter, Jack F. (ed.). 1978. *Sunflower Science and Technology.* Agronomy 19. Copyright © 1978 by the American Society of Agronomy, Crop Science Society of America, and Soil Science Society of America, 677 South Segoe Road, Madison, WI 53711 USA.

ing the formation of the embryo (101). The embryo consists of two cotyledons attached to a protruding radicle and is commonly referred to as the
kernel.

Chemical Composition

Some data are available on the proximate composition of sunflower
seeds; however, the seed sources frequently are not identified. If the cultivars are known, they were grown in different years or locations, making a
true comparison most difficult. Kinman and Earle (35) grew new introductions from the USSR and experimental hybrids from Texas in 1961 and reported major differences in chemical and physical composition. The mean
percentage of kernel in seeds from Soviet introductions was 67.5% compared with 58.9% for the North American F_1 hybrids. The oil and protein
concentrations of the introductions also were higher at 40.6 and 18.3%
compared with 33.6 and 17.0%, respectively. A comparison of more recent
cultivars from the USSR and hybrids from North America indicates a
further increase in oil concentration and considerable variability among
genotypes for size and density of seeds (Table 1). It appears that further
selection for both oil and protein levels is possible. Nikolic-Vig et al. (60)
noted that 'Peredovik,' with an average of 24.6% hull and 48.1% oil, had a
population range of 17 to 32% for hull and 36 to 55% for oil.

The chemical composition of the kernels and meal will be discussed in
detail later in this chapter. The pericarp is composed of approximately
equal proportions of lignin, pentosans, and cellulosic materials representing
82 to 86% of the total weight (16). Consequently, sunflower hulls contain
more lignin than many other fibrous seed products. The mineral content is
relatively constant among oilseed cultivars and within the range of most oilseeds.

OIL

Physical Characteristics

The ranges of certain physical characteristics and the maximum per-
missible levels for quality parameters and contaminants, as recommended
by the Codex Alimentarius Commission (1), are outlined in Table 2.

Chemical Composition

Triglycerides

Sunflower fits the general hypothesis for distribution of fatty acids on
the glyceride molecule. The saturated acids, palmitic and stearic, are esterified predominantly at the 1 and 3-positions and the unsaturated acids, oleic

Table 1—Physical and chemical characteristics of seeds from cultivars and experimental hybrids, Morden, Manitoba, 1975.

Entry	1,000 seed weight	500 ml weight	Bulk† density	Hull	Oil	Protein	EtOH soluble extract	Fiber residue	Ash
	g					%			
Cultivar									
Chernianka	50.2	218.3	0.436	24.8‡	48.9	13.8	7.5	27.1	2.7
Krasnodarets	57.2	228.8	0.458	23.7	50.6	16.1	6.7	24.2	2.4
Peredovik	55.6	214.0	0.428	22.8	49.9	16.3	8.9	22.4	2.5
Salyut	54.0	228.3	0.457	29.0	50.2	17.2	8.3	21.6	2.7
Sputnik	58.9	228.9	0.458	21.8	52.8	15.4	9.4	19.7	2.7
Hybrid									
Cargill 204	38.4	241.8	0.484	26.6	48.7	14.5	7.8	26.5	2.5
Morden 3	53.0	245.9	0.494	22.5	52.4	19.2	8.8	16.6	3.0
Morden 4	54.3	231.8	0.463	23.2	51.0	17.2	8.6	20.5	2.7

† g/cc.
‡ All components expressed on a dry weight basis.

Table 2—Recommended composition and quality factors for sunflower oil. †

Identity characteristics	
Density 20 C	0.918–0.923
Refractive index (^nD 40 C)	1.467–1.469
Saponification value (mg KOH/g oil)	188–194
Iodine value (Wijs)	110–143
Unsaponifiable matter	≤ 15 g/kg
Quality characteristics	
Acid value of virgin oil	≤ 4 mg KOH/g oil
Peroxide value	≤ 10 meq peroxide oxygen/kg oil
Contaminants	
Matter volatile at 105 C	≤ 0.2% m/m
Insoluble impurities	≤ 0.05% m/m
Soap content	≤ 0.005% m/m
Iron in virgin oil	≤ 5 mg/kg
Copper in virgin oil	≤ 0.4 mg/kg
Lead	≤ 0.1 mg/kg
Arsenic	≤ 0.1 mg/kg

† From Codex Alimentarius Commission. CAC/RS 23-1969 (11).

Table 3—Fatty acid composition of selected oilseeds.

Oil	12:0	14:0	16:0	18:0	20:0	18:1	18:2	18:3	20:1	22:1	Others
						%					
Sunflower			7	4		17	72				
Coconut†	44	18	11	6		7	2				12
Corn			12	2		29	56	1			
Cottonseed†		1	29	4		24	40				2
Olive†			14	2		64	16	2			2
Palm†		1	48	4		38	9				
Peanut†			6	5	2	61	22				4
Rapeseed			4	2		17	13	9	15	41	
Rapeseed (LEAR)‡			5	2		63	19	9	1		1
Safflower			7	2		13	78				
Safflower (High Oleic)			5	1		78	16				
Soybean†			11	4		25	51	9			

† From Manton (46).
‡ Low erucic acid rapeseed.

Table 5—Oil percentage and composition of sunflower kernels expressed on dry weight basis.†

Cultivar	Oil	Sterols	Phospho-lipids	Tocoph-erols	Carot-enoids	Phospholipid classes			
						Phosphatidyl-choline	Phosphatidyl-ethanolamine	Phosphatidyl-inositol	Phosphatidic acid
		—— % ——		—— mg/100 g ——		—— % of total ——			
Armavirski 3497	63.4	0.30	0.86	68.4	0.11	58.8	19.6	20.6	1.1
Peredovik	63.0	0.28	0.75	67.2	0.13	64.2	19.5	15.2	1.2
Salyut	61.8	0.26	0.72	62.8	0.15	56.1	17.0	22.4	4.5
VNIIMK 8931	62.2	0.30	0.82	69.8	0.16	55.4	18.2	24.0	2.2

† From Borodulina et al. (5).

Table 4—Oil percentage and fatty acid composition of seeds of *Helianthus* species native to Manitoba.

Species†	No. accessions	Oil	Fatty acids			
			16:0	18:0	18:1	18:2
				%		
H. annuus	30	14–29	2.6–6.1	0.7–2.6	12.3–24.0	69.5–82.6
H. petiolaris	18	23–40	3.0–3.7	0.3–2.4	11.8–27.5	67.7–84.3
H. maximiliani	18	20–36	3.5–5.5	0.7–2.6	6.9–14.3	80.5–87.4
H. rigidus	12	16–31	2.5–4.2	0.4–2.5	9.9–18.0	74.5–84.9

† Material collected by E. Whelan and H. Marshall.

and linoleic, randomly distributed at the positions not occupied by the saturated acids. The exception occurs with linoleic acid being preferentially esterified at the 2-position (48). Sunflower oil contains up to 72% triunsaturated glycerides. Mono-oleyl-dilinolein, which is one molecule of oleic acid and two of linoleic acid esterified with one molecule of glycerol, is the predominant form (39%), with lesser amounts of dioleyl-monolinolein (19%) and trilinolein (14%) (38).

Fatty Acids

Sunflower seed oil is characterized by a high concentration of linoleic acid, moderate level of oleic acid, very low level of linolenic acid, less than 15% of the saturated fatty acids, palmitic and stearic, and usually less than 1% of acids with fewer than 16 or more than 18 carbon atoms (Table 3). While lauric, arachidic, behenic, lignoceric, and eicosenoic acids may be present, they are of little practical concern. Traces of oxygenated fatty acids also have been found in some sunflower seeds stored for a prolonged period (49).

The effect of environment on fatty acid composition is well documented. Seeds produced in cool, northern climates normally contain 70% or more linoleic acid, whereas oil produced in more southerly latitudes may contain as little as 30% linoleic acid (35). There is a strong inverse relationship with oleic acid.

Breeding to modify the quality of sunflower oil has had little attention; therefore, within similar environments the fatty acid composition of cultivars is quite uniform. When cultivars are grown under controlled environments to minimize the strong effect of temperature, it becomes possible to identify considerable diversity in fatty acid composition. Seeds from inbred lines grown at Morden, Manitoba, contained from 5 to 78% oleic, 17 to 88% linoleic, and 4 to 25% saturated fatty acids. Soviet plant breeders are attempting to incorporate the high oleic-low linoleic acid characteristic into cultivars. The main difference in these genotypes is that the concentration of oleic acid in triglycerides, phospholipids, and sterols remains high throughout seed development and does not decrease in relation to linoleic acid, which is the trend in commercial cultivars (33). Apparently the dehydrogenase enzyme involved in the synthesis of linoleic from oleic acid is

Table 6—Distribution of sterols in different oils and refining byproducts. †

Material	Free sterols	Steryl esters	Steryl glycosides	Total
		%		
Sunflower oil				
Crude	0.22	0.07	0.03	0.32
Hydration residue	0.34	0.01	1.68	2.03
Refined oil	0.20	0.04	--	0.24
Soybean oil				
Crude	0.30	0.01	0.23	0.55
Hydration residue	0.74	0.01	1.93	2.68
Corn oil				
Crude	0.26	0.49	0.05	0.80
Hydration residue	0.20	0.07	0.54	0.81
Refined oil	0.20	0.45	--	0.65

† Reproduced from Popov et al. (63).

blocked. Canadian researchers (D. G. Dorrell, unpublished data) have found that high oleic acid genotypes are unstable and strongly influenced by environment.

Some accessions of native *H. annuus, H. petiolaris* (Nutt.), and *H. maximiliani* (Schrad.) maintained at the Morden Research Station (E. Whelan and H. Marshall, personal communication) contain very high concentrations of linoleic acid and low concentrations of saturated fatty acids (Table 4). Thus, breeders should not be concerned at lowering linoleic acid concentrations if these species are used as parents to introduce disease and insect resistance.

Phospholipids

Phospholipids are the main non-triglycerides or non-neutral oils in sunflower seeds. These lipids are composed of glycerol esterified with fatty acids and phosphoric acid. Phosphoric acid, in turn, is combined with a nitrogen-containing compound. While the non-triglyceride fraction may reach 5% of the total lipids, the phospholipids usually comprise only about 1% of the lipids.

Published values of the relative proportions of the three main phospholipids vary widely: phosphatidylcholine (14 to 64%), phosphatidylinositol (13 to 36%), and phosphatidylethanolamine (13 to 24%). Small quantities of phosphatidic acid also are present (Table 5) (5). Shcherbakov (85) observed that the concentration of total phospholipids decreased as the seed matured but increased during artificial drying. The quantitative and qualitative variation among Soviet cultivars is limited (5).

Waxes

The wax content of sunflower seeds usually is less than 1% of the total lipids. Approximately 83% of the wax and wax-like material is in the hull, 17% in the testa, and only traces in testa-free seed (37). Earle et al. (16)

Table 7—Individual tocopherols in vegetable oils. †

	α	β	γ	δ
		mg/kg		
Sunflower (R)‡	608	17	11	--
Cottonseed, crude	402	1.5	572	7.5
Miaze, germ (R)	134	18	412	39
Peanut	169	5.4	144	13
Rapeseed (R)	70	16	178	7.4
Safflower	223	7	33	3.9
Sesame	12	5.6	244	32
Soybean (R)	116	34	737	275

† Reproduced from Muller-Mulot (56).
‡ Refined.

noted that the wax content of hulls varied from 1.4 to 3.0% in cultivars grown in North America. Thus, the wax content of oil will depend more on the efficiency of dehulling than on the cultivar.

Popov et al. (61) isolated wax and found it to contain 25% hard and 75% liquid wax. The hard wax contained 9% unsaturated fatty acids (C11 to C24), 16% saturated fatty acids (C11 to C32), 9% hydrocarbons (C15 to C35), and 66% alcohols (C17 to C32). The predominant fatty acids were similar to those found in oil. Some arachidic, behenic, tricosanoic, and lignoceric acids also occurred. The settlings in holding tanks tend to contain a large portion of oil; therefore, the fatty acid composition of this material will be more similar to the oil than to pure wax.

Sterols

Sterols are a minor component in sunflower oil and cause few problems during extraction and refining. Stigmasterol, β-sitosterol, campesterol, Δ⁷-stigmasterol, and Δ⁷-avenasterol are present in varying concentrations (94). A stenol, Δ⁷-stigmastenol, also has been reported (51). Popov et al. (63) found the sterol content of crude sunflower oil was lower than either soybean or corn oils and that steryl glycosides predominated in the hydration residue following refining (Table 6).

Tocopherols

Tocopherols as a group also are known as the fat soluble vitamin E. They are important antioxidants, and the concentration in sunflower is somewhat lower than in soybean or cottonseed oil but higher than in peanut, rapeseed, or corn oil (Table 7). The α-form is the main tocopherol in sunflower oil, whereas the γ-form is more common in most other vegetable oils (56). The various tocopherols have somewhat similar antioxidant activities under normal conditions but at elevated temperatures the order of activity is γ, β, then α-tocopherol. Consequently, sunflower oil has less antioxidant activity during frying than soybean oil. The activity of tocopherols as vitamins is in the reverse order, thereby making sunflower oil more potent (57). Differences of up to 20 mg total tocopherols/100 g of oil have been reported among sunflower cultivars (39). The synthesis of tocopherols closely parallels that of lipids but is independent. While temperature does not influence synthesis directly, it does affect the tocopherol/linoleic acid ratio (14).

Other

The carotenoids reported by Borodulina et al. (5) in Table 5 are xanthophyll and carotene. These carotenoids are fat soluble and normally appear in extracted oil. The water soluble vitamins are discussed in the section dealing with the chemical composition of the meal.

Extraction and Processing of Oils

Seed Preparation

The method and effectiveness of drying seeds directly influences oil quality. Incomplete drying or the presence of extraneous material and debris, which tends to promote microbial activity, can lead to spoilage. This activity, in turn, increases the amount of soluble pigments and free fatty acids in the extracted oil. Such oil must be refined and bleached more extensively. Shcherbakov and Malyshev (87) observed during thermal drying that phospholipids increased, sterols and carotenoids were not affected, while diglycerides, hydrocarbons, and free fatty acids decreased proportionately to the drying temperature. Drying and storage is covered extensively in the chapter by Schuler.

Dehulling

Sunflower seed may or may not be dehulled prior to extraction, depending upon the design of the crushing plant. In older plants, especially those that crushed flaxseed, dehulling equipment was not available and extractors were successfully adapted to handle ground achenes. This system had the advantage of being simpler and more economical to operate, but produced a meal which contained less than 30% protein.

Plants specifically designed to crush sunflower seed contain extensive dehulling equipment. Dehulling reduces the movement of an unnecessary mass through the system; it reduces wear in the expeller; it reduces the wax content of the oil; and it reduces the fiber content of the meal. It is difficult to achieve an optimum balance between excessive dehulling, which leads to loss of oil, and insufficient dehulling, which reduces efficiency, because of differences in cultivars and source of seed. Some of the problems can be at-

Table 8—Effect of method of hull removal on the composition of sunflower seed products.†

Product	Fat	Cellulose‡	Protein‡
		%	
Whole seed	50.1	29.5	31.0
Hand separated kernels	60.9	8.4	52.0
Hand separated hulls	5.4	47.1	7.3
Meal from solvent extn.	1.2	13.5	46.0
Meal after 4 screenings	1.3	6.5	50.8

† From Prevot et al. (66).
‡ Values reported on a dry oil-free basis.

tributed to the use of dehullers designed specifically for other oilseeds and the general inefficiency of the best sunflower dehullers. Crushers in North America currently use Carver cottonseed, Entoleter or Specictrum impact dehullers, or the Bauer disc dehuller.

In many crushing plants, seeds are not segregated according to size prior to dehulling. Defromont (12) suggested that sizing, combined with well-adjusted and maintained separators and cyclones, would prevent loss of oil beyond the 1.9% normally present in the hull and would reduce the cellulose content of the meal to a minimum. Another approach is to assume that the dehulling process is incomplete and attempt to remove the remnant hulls and fibrous material after extracting the oil. Prevot et al. (66) sifted ground meal through 900 μm mesh screens and obtained an enriched meal similar to that from kernels dehulled by hand (Table 8).

Extraction

Extraction of oil from sunflower kernels poses no special problems; therefore, a broad range and combination of equipment and operating conditions can be employed successfully. Bailey's Industrial Oil and Fat Products (97) should be consulted for specific details.

Generally the kernels are crushed or rolled, then cooked to facilitate the disruption of oil-bearing tissue. Heating also inactivates phospholipases and lipases, thereby reducing the concentration of nonhydratible phosphatides and minimizing the acid number (44). The flaked kernels usually are prepressed in a screw press or expeller (98). With a dual speed press, the first stage has a shallow pitch which permits a greater capacity during the phase of maximum oil extraction. The second stage has a sharper pitch and operates at higher temperatures and pressure than is possible in a single stage press. The emerging pressed oil cake is granulated prior to solvent extraction.

The general operation of solvent extractors is similar whether they are constructed horizontally or vertically, or whether the oil cake moves or is stationary (50). The solvent and flaked oil cake move in counter flow. Fresh solvent is added just before the extracted meal leaves the extractor and flows as oil containing solvent or miscella toward the freshly introduced meal, which contains the highest oil concentration. This process usually removes all but about 1% of the total seed oil.

The main petroleum solvent used is hexane. Other solvents, such as petroleum ether, cyclohexane, benzene, and acetone, will extract oil satisfactorily but cost or toxicity preclude their widespread use. Solvents are recovered from the miscella by evaporation using excess heat from the meal desolventizer-toaster process. The miscella moves as a rising or falling film through an evaporator column; the vaporized solvent is drawn off, condensed, and then recycled. The oil goes through a stripping column where steam removes the final traces of solvent. This latter process is done under negative pressure. Up to 5% of the solvent may be lost through evaporation and entrainment.

The solvent must be removed completely from the extracted meal. The most common systems use steam, while newer systems use superheated solvent vapor combined with a short exposure to superheated steam to vaporize the solvent. Frequently, sunflower meal is toasted although this does not improve flavor and color or inactivate toxic factors. When toasting takes place, it is usually because the equipment was designed for soybeans [*Glycine max* (L.) Merr.], and modification is uneconomical.

Refining

Crude sunflower oil usually is degummed before refining to remove the hydrophilic phosphatides and reduce subsequent refining losses. Accurately monitored quantities of water, usually less than 1.5%, are mixed with the oil and the resulting hydrated gums removed by continuous centrifugation.

Sunflower oil should be pretreated with phosphoric acid or a phosphate salt before alkali refining to break phosphatidic linkages (36). The oil then is mixed with dilute solutions of sodium hydroxide to neutralize the free fatty acids and react with residual phosphatides, pigments, and other minor constituents. The strength of sodium hydroxide, temperature, length of pretreatment, etc., depend on the amount of non-neutral oil in the crude oil and the equipment being used (96). The mixture is heated to break the emulsion then separated by continuous centrifugation. The oil is washed with small quantities of water to remove traces of sodium hydroxide and soap. Crude oil with 0.25% wax, up to 1% phosphatides, and 1.5% free fatty acids can be refined to less than 100 ppm wax with a refining loss of 2.7% (36).

Bleaching and Deodorization

Sunflower oil contains moderate amounts of carotenoids and xanthophylls but no chlorophyll; therefore, it rarely poses serious problems in bleaching. In fact, many crushers add pigments to bring the oil color to established specifications. Occasionally colored products will develop during storage of poor quality seed but these are removed readily by refining and bleaching (27). Bleaching is accomplished by a batch or continuous operation using less than 1% activated clay or bleaching earth (26). Occasionally the problem may be serious enough to require the addition of activated carbon at rates up to 20% of the clay. Since as much as 4% oil can be lost through entrainment, bleaching is used only when absolutely necessary.

The process of deodorization is the last opportunity during refining to improve the flavor and odor of the oil. This is accomplished by injecting steam to strip the undesirable materials from the oil. Temperatures up to 270 C at reduced pressures are used, therefore, efficient heat transfer and recovery systems are essential (104). Deodorization removes volatile products such as free fatty acids, materials such as sterols, and some unsaponifiables which were not removed during refining. It also destroys peroxides which may have formed during refining.

Winterization

Sunflower oil is a premium salad oil and must be able to withstand prolonged storage under refrigeration without clouding or developing precipitates. According to American Oil Chemists' Society (AOCS) official methods, an oil is considered satisfactory if it remains clear when held at 0 C for 5.5 hours (1).

The process of winterization involves the slow chilling of oil to supersaturation, initiation of crystal development, a period of crystal growth, followed by separation of the crystals from the liquid oil. This process is difficult and slow because crystal development is gradual and somewhat unpredictable: the oil is viscous, filtration must be conducted at a low pressure to avoid disrupting the crystals, and the cake of filter clay on the filters must be removed frequently. In addition, oil losses reaching 4% can occur due to entrainment in the filter media.

Alternate methods have been proposed wherein miscella, rather than finished oil, is winterized (54). Miscella, containing from 40 to 60% hexane, has good flow characteristics and filters readily. Morrison and Thomas (12) were able to reduce the phosphatides approximately 10-fold, the waxes from 0.021% to less than 0.009%, and produced a finished oil that remained clear for more than 7 days. Crystal formation was not affected by the presence of the solvent.

Oil Stability

The oxidative stability of oil is an important consideration during all stages of processing from refining to storage. Although the crude oil contains sufficient tocopherols to confer some antioxidant activity, a large portion are removed during the refining. The presence of iron derivatives intensifies color development, increases the concentration of oxypolymers (81), and causes a deterioration in flavor (42). The presence of oxygen accelerates the accumulation of peroxides (64). Thus, the oil should be protected by using metal scavengers such as citric acid; storage tanks should be free of metal ions; and the tanks and valves should be blanketed under nitrogen.

A range of natural and artificial antioxidants have been tested in sunflower oil with varying degrees of success (83). Luckadoo and Sherwin (43) treated oil with 0.02% tertiary butyl hydroquinone (TBHQ) and effectively reduced the development of peroxides during storage. Untreated crude oil, stored 4 months at 24 C, then deodorized, had an active oxygen method (AOM) rating of 9 hours; the addition of TBHQ increased it to 51 hours.

Environment also directly influences the stability of sunflower oil. When the crop is grown over a wide range of latitudes and the seed enters a common marketing system, the fatty acid composition of the oil will be variable and subsequently influence stability. Morrison (53) observed that oil produced in the Southern U.S. had a higher concentration of oleic acid and a greater oxidative stability than that produced in the north (Table 9). Northern oil that had been refined and bleached, however, deteriorated at a slower rate during prolonged heating than equivalent Southern oil.

Disposal of Byproducts

Disposal of byproducts and wastes from sunflower crushing plants causes no unique problems. With increased environmental protection regulations and more stringent restrictions against adding fatty wastes to municipal sewerage systems, however, routine disposal of any wastes is a concern. The main byproducts of extraction and refining are gums, soapstock, washwater, bleaching and filtering clay, and distillates from the deodorization process. All of these "waste" materials have value, but they must compete with the same well-established soybean byproducts.

A satisfactory, commercial lecithin can be extracted from the gums; soapstock can be acidulated to produce an acid oil used in fatty acid and soap production or as a high energy livestock feed ingredient; bleaching and filtering clays can be reprocessed to extract residual oil then used as fillers in livestock feed; and distillates are utilized in tocopherol and fatty acid production (102). Disposal of wash water is a major problem to all vegetable oil processing plants. Although the water contains low levels of impurities, the biological oxygen demand (BOD) is high; therefore, the large volumes make treatment more difficult.

Margarine Production

The manufacturing of margarine requires that the physical characteristics of the oil be modified to raise the melting point. This process usually is achieved by a combination of selective hydrogenation and interesterifica-

Table 9—Composition and properties of crude sunflower seed oils from northern and southern United States.†

Evaluation		Northern	Southern
Iodine value‡	109.4	129.2	109.4
Peroxide value (meq/kg)		2.1	3.9
Free fatty acids (% of oleic)		2.0	1.5
Active oxygen method (hrs)		10.5	17.9
Fatty acid composition (%)			
16:0		5.9	5.2
18:0		4.7	4.4
18:1		26.4	50.9
18:2		61.5	37.9
20:0		0.5	0.5
22:0		0.7	0.7

† Reproduced from Morrison (53).
‡ Calculated from fatty acid composition.

tion. Selective hydrogenation generally reduces the high level of trilinolein and increases the proportion of monolinolein, diolein, and triolein forms. This process, of course, reduces the level of polyunsaturated fatty acids in the product. Interesterification is the process wherein fatty acids of different oils interact to produce new esters and a new distribution of fatty acids on the glyceride molecule. This change is achieved by either random or directed rearrangement. With the random process the different fats are heated in the presence of a catalyst and random rearrangement takes place. In the directed process, the temperatures are lower and the times are extended. During the course of crystallization some liquid portion may be withdrawn, creating a higher proportion of solid materials with high boiling points.

Sunflower oil is converted readily to a plastic product. For example, the melting point of an interesterified sunflower oil can be raised to 20 C (31). Nevertheless, a high quality margarine based solely on sunflower oil is difficult to produce. The low level of palmitic acid and the relative absence of low molecular weight fatty acids causes margarine, based on partially hydrogenated oil, to have an inadequate consistency due to a fine crystalline β-polymorphic structure. Such margarines are plastic in the 20 to 30 C range but are hard and brittle at lower temperatures (84). Freier et al. (23) suggested that this problem could be overcome by interesterifying sunflower oil with certain hydrogenated vegetable oils and using the resulting oil at 30 to 50% of the total fat level. Similarly, Hustedt (31) indicates that an interesterified sunflower oil blended with 5% hardened hydrogenated oil will produce an acceptable solid fat with greater than 60% essential fatty acids. Liquid sunflower oil also is being blended with highly saturated and hardened fats, such as palm oil, to raise the level of polyunsaturated fatty acids. The number of patents granted in the past 2 years indicate considerable progress in the production of semiliquid and soft sunflower oil margarines.

Utilization

Salad Oil

Sunflower oil is a premium salad oil because of its light color, bland flavor, and high concentration of the polyunsaturated fatty acid, linoleic acid. Approximately 80% of the deodorized sunflower oil produced in Canada is sold as salad oil. Most of this oil is marketed as pure oil rather than as a blended oil. The reverse is true in the USA where sunflower oil usually is blended with soybean and palm oils to increase the level of unsaturation. Recently, new blends have been marketed specifically advertising their high concentrations of polyunsaturated fatty acids. These consist mainly of sunflower oil. It is expected that pure sunflower salad oil will be widely available as consumers become more aware of differences in the quality of oils.

Cooking Oil

Sunflower oil has value as a cooking oil because of its high smoke point. Robertson et al. (72) concluded that partially hydrogenated sunflower oil had a flavor equal to and was as stable as the cottonseed-corn oil blends commonly used for frying potato chips. They also noted that while the viscosity of the sunflower oil was higher than the cottonseed-corn oil blend at the beginning of the test, viscosities were similar after 23 hours of frying (Table 10). During this same period the blend oxidized more than sunflower oil. The fatty acid composition of the sunflower oil was quite constant during frying. When oil was extracted from potato chips that had been stored for 11 weeks, the linoleic acid content was similar to or slightly higher than the original sunflower oil. When sunflower oil produced in the

Table 10—Chemical evaluation of potato chip frying oils. †

Iodine value		Free fatty acids		Viscosity		Peroxide value	
S‡	C-C§	S	C-C	S	C-C	S	C-C
		% as oleic		centistokes, 70 C		meq/kg	
0	109.0	117.0	0.06	0.04	16.0	14.5	3.5	2.0
2	105.8	114.5	0.04	0.04	16.2	14.7	4.4	3.3
4	104.7	115.2	0.05	0.04	16.3	14.8	1.8	2.9
6	104.7	114.8	0.07	0.07	16.6	15.0	2.5	3.9
8	103.7	113.9	0.09	0.08	16.9	15.4	2.2	3.2
10	103.6	112.6	0.08	0.09	17.0	15.7	1.6	2.8
12	102.8	112.2	0.11	0.10	17.0	15.8	1.4	2.6
14	102.7	111.7	0.14	0.11	17.2	16.3	1.7	3.7
16	102.5	110.4	0.15	0.15	17.4	16.8	1.6	2.6
18	102.8	110.7	0.18	0.16	17.4	16.9	1.6	3.1
20	102.5	109.9	0.15	0.16	17.4	17.5	1.8	3.0
23	103.4	109.7	0.19	0.17	17.4	17.6	1.8	2.9

† Reproduced from Robertson et al. (72).
‡ Hydrogenated sunflower oil.
§ Cottonseed-corn oil mixture (70:30).

Northern U.S. was lightly hydrogenated, it oxidized less after abuse and remained useful for a longer period than commercial shortening (71).

Lecithin

The lecithin (phosphatidylcholine) content of sludges from the degumming process is sufficiently high for the commercial production of this antioxidant from sunflower oil. Popov et al. (62) took partially dehydrated sludges containing 5% water, 39% oil, and 56% phosphatides and by simultaneously dehydrating and defatting this material, produced a powdered lecithin with sufficient stability and antioxidant properties to be of commercial interest. While this material is as stable as the common soybean, corn, or peanut lecithins, the market in North America is supplied almost totally by soybean lecithin (28).

Industrial Uses

Sunflower oil is not commonly used for industrial purposes because of limited supply and higher prices in relation to soybean oil. Nevertheless, its properties and advanced drying oil technology would make it possible to adapt sunflower oil to a wide range of applications (Fig. 1). The high concentration of linoleic acid and very low concentration of linolenic acid mean that despite the moderate iodine number of 125 to 140, it has good drying qualities without the yellowing associated with the high linolenic acid oils. Thus, sunflower oil could find a limited market in oil-based white and pastel paints where yellowing is a problem. Sunflower and linseed oil have

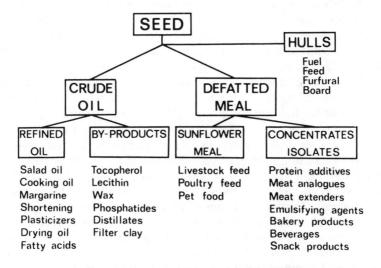

Fig. 1—Potential uses for sunflower seed products.

been mixed and transesterified to modify density and felxibility of the film. Various patents have been granted for producing modified resins, special lacquers, epoxy esters, polyester films, various copolymers, etc., based on sunflower oil (45). Trotter et al. (100) suggest that sunflower oil would compete with safflower oil, but substitution will depend largely on price and long term availability.

Nutrition

Human

Sunflower oil has excellent nutritional properties. It is practically free of significant toxic compounds and has a high concentration of linoleic acid. This polyunsaturated fatty acid is an essential fatty acid not synthesized by humans, and is a precursor of gamma linoleic and arachidonic acids.

The relationship between cholesterol concentration in blood plasma and the risk of coronary heart disease has been debated extensively. Generally, it is accepted that increasing the proportion of unsaturated to saturated fatty acids in the diet will lower the level of serum cholesterol. There is also an established relationship between cholesterol and arterial sclerosis. There is no agreement, however, that the consumption of a high proportion of polyunsaturated fatty acid will reduce the incidence of arterial sclerosis or prevent coronary heart disease (103). At this stage the consumption of sunflower oil can be considered only as an aid in controlling cholesterol levels.

The daily requirement for linoleic acid has been considered to be about 100 mg per kg of body weight per day (57). Press et al. (65), however, report that as little as 3 mg of linoleic acid per kg of body weight per day applied as a cutaneous application of sunflower oil, rapidly corrected fatty acid deficiencies. A recent patent suggests that certain nutritional disorders can be corrected with parenterally injected, sterilized emulsions of purified sunflower oil and gamma tocopherol (9).

The value of a vegetable oil depends on both the triglyceride and unsaponifiable components, because there is a relationship between the amounts of linoleic acid and tocopherols required by humans (57). An adequate ratio has been defined as 6 mg of α-tocopherol per g of linoleic acid. While there are disagreements as to the exact ratio, it is generally agreed that as the consumption of linoleic acid increases, the body requires more vitamin E. Table 11 shows that sunflower oil has a relatively high ratio due primarily to the high proportion of α-tocopherols in the total tocopherols of sunflower oil. In terms of vitamin activity, the α-form is the most potent, the reverse of its antioxidant properties (57). The temperature during seed development has a pronounced affect on the tocopherol/linoleic acid ratio. Dompert and Beringer (14) found that the ratio was 7.8 when sunflower was grown at 35 C but only 0.9 when grown at 10 C. While synthesis of tocopherols is independent of temperature, the concentration of linoleic acid in

Table 11—Tocopherol content and linoleic acid percentage in oils in relation to vitamin E/ polyunsaturated fatty acid ratios.

Oil	α-tocopherol†	Linoleic acid‡	α-tocopherol / Linoleic acid
	mg/kg	%	
Sunflower	608	72	0.84
Cottonseed	402	40	1.00
Peanut	169	22	0.77
Corn	134	56	0.24
Rapeseed	70	19	0.37
Safflower	223	78	0.28
Soybean	116	51	0.23

† Data from Table 7.
‡ Data from Table 3.

the oil increases as the temperature during synthesis decreases, thus the ratio declines. The method of processing affects the concentration of tocopherols in the oil; however, when normal refining and bleaching procedures are used, at least two-thirds of the tocopherols are retained (53).

Animal

It has been shown that diet can modify the proportion of fatty acids in milk and meat. It is now possible to feed highly unsaturated oils, such as safflower and sunflower oils, treated in such a way that linoleic acid is protected from hydrogenation in the rumen. Thus, a greater proportion of unaltered linoleic acid will be absorbed and deposited in meat tissue and milk. Work pioneered in Australia by Scott et al. (80) showed that the linoleic acid level of milk fat was increased significantly when cows were fed encapsulated safflower oil. Cows fed 1,480 g of safflower oil protected by a formaldehyde-protein matrix, produced milk containing up to 30% linoleic acid in the butter fat (17). This amount compares with a normal level of 2.1% linoleic acid. However, the milk with elevated levels of linoleic acid tended to develop an oxidized flavor. These researchers also noted that the linoleic acid concentration in meat fat could be increased fourfold by feeding similar rations. When a small number of adults were fed dairy products and beef obtained from animals fed protected vegetable oil, they showed a reduction in plasma cholesterol after 4 weeks (59).

MEAL

Chemical Composition

Effects of Processing

The chemical composition of sunflower seed meal compares favorably with most other oilseed meals. The main exceptions are the higher fiber and ash contents which tend to reduce the metabolizable energy (Table 12).

Table 12—Proximate composition of solvent extracted oilseed meals reported on a dry weight basis.†

	Protein	Crude fiber	N-free extract	Ash	Ether extract	DE cattle Mcal/kg
			%			
Sunflower	50.3	11.6	26.7	8.3	3.1	2.88
Cottonseed	46.0	12.5	34.9	6.8	2.3	3.60
Flaxseed	40.0	9.7	42.2	6.3	1.9	3.36
Rapeseed (B. napus)	44.0	10.1	36.8	7.8	1.2	3.13
Peanut	51.8	14.3	27.7	4.9	1.3	3.39
Soybean	52.4	5.9	33.8	6.6	1.3	3.63

† From Atlas of nutritional data on United States and Canadian feeds, 1971 (58).

Table 13—Changes in essential amino acid content of sunflower meal heat treated for 1 hour.†

Amino acid	Heat treatment, C				
	0	75	100	115	127
	g amino acid/16 g N				
Lysine	4.03	3.77	3.94	3.88	3.83
Histidine	2.67	2.63	2.72	2.63	2.66
Arginine	10.38	10.35	10.19	10.14	9.92
Threonine	5.33	5.10	5.02	4.88	4.97
Valine	5.21	5.15	5.21	5.08	5.26
Methionine	2.16	2.11	2.14	2.31	2.20
Isoleucine	4.28	4.16	4.23	4.17	4.32
Leucine	7.11	6.90	7.07	7.05	7.11
Phenylalanine	5.25	4.90	5.14	5.05	5.09
Total	46.42	45.07	45.66	45.19	45.36

† Reproduced from Amos et al. (2).

Composition of meal tends to vary directly with the efficiency of the dehulling and extraction procedures. Where dehulling is a standard procedure, it is biased toward minimum loss of oil-bearing tissue; therefore, the meal contains more fiber than desired for optimum quality. Crushing plants that do not dehull produce a meal with high concentrations of fiber. When processing operations are harsh or when excess heat is used to desolventize the meal, some decline in biological value occurs, apparently due to the destruction of lysine, arginine, and tryptophan (10). Heating meal above 75 C selectively altered the concentration of some amino acids (2). Lysine was lowered from 4.03 to 3.83 g, whereas methionine increased from 2.16 to 2.20 g of amino acid per 16 g nitrogen, when sunflower meal was exposed to 127 C (Table 13). Shcherbakov et al. (86) found that high temperature-water treatments prior to solvent extraction improved extraction of oil but denatured the protein and caused losses of up to two-thirds of the water soluble proteins. Under good processing conditions, meal of equivalent nutritive value to solvent extracted soybean meal is produced readily.

Amino Acids

The values reported for protein concentration and amino acid composition vary greatly with the source of seed. Analyses usually were conducted on nonoilseed types and older, low oil cultivars, or unidentified

Table 14—Amino acid composition of oilseed meals.†

Amino acid	Sunflower	Soybean	Flax	Rapeseed
		g amino acid/100 g N		
Essential				
Arginine	56.2	42.2	55.6	38.5
Histidine	14.4	17.0	13.6	16.7
Isoleucine	29.2	29.6	28.0	27.1
Leucine	38.4	44.0	36.9	42.3
Lysine	19.5	38.3	23.1	36.6
Methionine	11.4	8.1	10.9	13.2
Cysteine + cystine	13.4	12.9	11.0	17.1
Tryptophan	10.0	11.1	12.2	10.4
Phenylalanine	29.4	28.8	28.0	24.4
Valine	34.9	33.3	36.5	33.9
Threonine	22.3	22.1	23.9	27.2
Total	279.1	287.4	279.8	287.4
Non-Essential				
Alanine	24.5	26.1	27.6	27.1
Aspartic acid	54.9	70.8	58.3	45.9
Glutamic acid	143.0	113.0	131.0	113.0
Glycine	33.9	26.3	37.7	31.4
Proline	31.1	32.1	22.6	38.3
Serine	26.3	29.0	27.4	27.6
Tyrosine	14.2	19.0	13.8	16.2
Total	327.9	316.3	318.4	299.5
N recovered, %	97.3	91.3	95.0	92.1

† Reproduced from Tkachuk et al. (99).

samples. Tkachuk and Irving (99) compared the cultivar Peredovik with other oilseed meals and noted a lysine deficiency in an otherwise relatively well balanced amino acid composition (Table 14). Sunflower meal has an essential amino acid index of 68, compared with 79 for soybean meal and 100 for whole egg. Similarly, the protein score, based on the most-limiting amino acid, relative to FAO nutritional requirements, was 63 for sunflower and 67 for soybean meal (93).

Protein Classes

Sunflower seed proteins are characterized by a moderately low level of albumin and high level of globulin proteins (Table 15). The globulins represent 55 to 60%, albumins 17 to 23%, glutelins 11 to 17%, prolamines 1 to 4% and the combined nonprotein nitrogen and insoluble residue less than 11% of the total nitrogen in the meal (25, 91). Care must be exercised in interpreting results from an Osborne classification since salts normally present in meal will solubilize some globulins, thereby inflating the albumin concentration. In addition, globulins show differential solubility in dilute and 5% salt solutions.

Recent research in the USSR indicates that a redistribution of protein fractions has occurred during the breeding of high oil lines, a change resulting in an increase of the albumin fraction (6). Apparently the water soluble

Table 16—Amino acid composition of sunflower meal and its protein fractions.†

	Defatted meal	Nonprotein N	Low mol wt albumin	High mol wt albumin	Fractions of high mol wt albumin			Total globulin	Residual protein
					a	b	c		
N/total N%	100	5.9	6.8	6.2	2.4	2.2	1.4	61.4	17.6
					g/16 g N				
Glycine	5.4	3.8	3.9	7.0	9.3	6.1	4.6	6.2	5.6
Alanine	4.1	3.0	3.0	5.1	6.6	5.0	3.9	4.1	5.0
Valine	4.8	2.5	4.2	5.4	5.5	5.1	4.6	5.0	5.5
Leucine	6.2	3.2	5.9	6.1	6.2	5.0	5.2	6.4	7.7
Isoleucine	3.9	2.4	3.9	3.7	3.8	3.3	3.8	4.2	4.0
Serine	4.2	2.5	3.4	5.1	7.8	4.8	4.8	4.8	4.5
Threonine	3.8	1.8	2.4	5.4	6.1	5.2	3.6	3.7	5.2
Tyrosine	2.9	1.8	2.4	3.2	2.2	2.6	2.4	2.9	3.4
Phenylalanine	4.6	1.2	2.3	3.4	3.8	3.1	3.0	5.7	5.2
Proline	5.1	4.4	4.9	4.2	4.9	4.4	5.6	4.6	6.5
Methionine	2.6	0.5	5.0	3.1	1.6	1.8	2.6	1.2	1.1
Cystine	1.8	2.8	6.2	1.6	0.9	1.8	2.5	0.9	0.5
Lysine	3.6	6.6	4.2	6.7	9.6	7.8	4.6	3.0	3.7
Histidine	2.4	3.0	1.8	2.2	2.8	2.2	2.1	2.4	2.8
Arginine	8.5	13.5	10.5	7.0	4.8	7.0	8.2	9.0	7.3
Aspartic acid	9.6	7.7	7.8	9.7	10.1	8.8	9.3	10.8	8.6
Glutamic acid	21.4	18.6	29.8	19.8	15.8	19.8	22.9	23.3	16.1
Total	95.0	79.3	101.6	98.8	101.4	93.8	93.8	98.0	92.7

† From Mosse and Baudet (55).

Table 15—Protein classification based on solubility. †

| Crop‡ | N content of meal | Solubility of N in | | | | | N solubility |
		H₂O	5% NaCl	10% EtOH	0.2% NaOH	Residue	
				%			
Sunflower	9.9	19.7	57.0	3.6	11.7	8.1	94.5
Flax	6.9	45.8	41.5	1.5	3.2	8.1	94.0
Rapeseed	6.7	50.1	21.1	3.7	8.6	16.5	83.8
Soybean	8.0	72.4	7.1	4.2	4.8	11.6	83.7

† Reproduced from Sosulski et al. (91).
‡ Average of at least two cultivars.

proteins are active during oil synthesis, and their level may predetermine oil concentration. The salt-soluble globulins function as reserve proteins. Borodulina and Suprunova (6) indicate that growing sunflower under dryland conditions tends to increase the salt-soluble fraction and lower protein quality. Mineral nutrition also influences quality. Mosse and Baudet (55) found that the amino acid composition of the albumin and globulin fractions differed significantly. Globulins represented 61.4% of the total nitrogen, had higher levels of phenylalanine and aspartic acid, but lower concentrations of methionine, cystine, and lysine than the albumins (Table 16). When the albumins were partitioned into low and high molecular weight fractions, the former fraction contained 5.0 g of methionine and 6.2 g of cystine compared with 2.6 g and 1.8 g/16 g nitrogen in the whole meal. Conversely, the high molecular weight fraction had appreciably more lysine. When this fraction was subdivided further, the portion with a molecular weight of more than 48,000 contained up to 9.6 g of lysine per 16 g nitrogen. An investigation of three genotypes revealed that differences in the amino acid composition of the protein classes were relatively small (3). It is clear, however, that any increase in the ratio of albumin to globulin will improve protein quality greatly.

Carbohydrates

Cegla and Bell (8) isolated 8.3% sugars from dehulled and defatted sunflower meal. These sugars included 0.6% glucose, 2.3% sucrose, 3.2% raffinose, and 0.8% trehalose, indicating that the majority of the soluble carbohydrates can be hydrolyzed and digested easily. The available carbohydrates in commercial sunflower meal will be somewhat different due to the presence of variable amounts of hull fragments. Pure hull material contains up to 26% reducing sugars which include 14% xylose, 7% arabinose, and 2% galactose (7). These carbohydrates are less readily available and would alter the composition of meal.

Minerals

Sunflower meal compares favorably with most oilseed meals as a source of calcium and phoshorus (Table 17). No other mineral in either a deficient or toxic proportion has been noted.

Table 17—Mineral content of some solvent extracted oilseed meals.†

Meal	Ca	P	K	Na	Cl	S	Mg	Fe
					%			
Sunflower	0.26	1.22	1.08	--	0.19	--	--	--
Cottonseed	--	1.19	1.47	0.05	0.04	0.21	0.58	0.015
Linseed	0.40	0.83	1.38	0.14	0.04	--	0.60	0.033
Peanut‡	0.16	0.54	1.15	0.42	0.03	0.18	0.24	--
Soybean	0.29	0.64	1.92	0.34	--	0.43	0.27	0.013

† From Morrison (52).
‡ Mechanical extraction.

Vitamins

Sunflower meal is an excellent source of the water soluble B-complex vitamins, namely nicotinic acid, thiamine, pantothenic acid, riboflavin, and biotin (10). The concentration of nicotinic acid in sunflower meal is about 170% higher than that found in peanut meal, which is perhaps the best alternative source (Table 18). Similarly, the thiamine concentration is 360% higher than in linseed meal. Pantothenic acid and riboflavin concentrations are somewhat lower than found in peanut meal. Borodulina et al. (5) reported a range of 79.7 to 148.3 μg nicotinic acid and 16.0 to 29.0 μg thiamine/g of seeds in 11 Soviet cultivars. These values are lower than reported in Table 18 because the analysis was conducted on full-fat seeds. The cultivar 'Salyut' gave the best combination of 144.7 μg/g nicotinic acid and 21.9 μg/g thiamine. The high oil cultivars had much higher concentrations of nicotinic acid but somewhat lower concentrations of thiamine than low oil cultivars.

Other Components

Arginase and trypsin inhibitors have been observed in sunflower seeds (76). These components are heat labile and inactivated easily, however, and in the case of trypsin inhibitors, the activity is at an exceedingly low level (74).

The polyphenolic compound, chlorogenic acid, has been considered an impediment to the broad use of sunflower meal for human consumption. While chlorogenic acid is not considered to be a toxic compound, it is responsible for the production of yellow-green coloration following oxidation in the presence of alkali. Delic et al. (13) found that when chlorogenic acid was fed to mice at a level of 2% of the total diet, food consumption declined by 33% and weight gain declined by 66%. Similar but reduced effects were observed at levels of chlorogenic acid as low as 0.3%. Methionine and choline chloride partially offset these effects. When 3-week-old rats were fed a diet of casein containing a 1% chlorogenic acid extract from green coffee beans, however, the protein efficiency ratio, biological value, and

Table 18—B-complex vitamin concentration in solvent extracted meals.†

Meal	Nicotinic acid	Thiamine	Pantothenic acid	Riboflavin
			mg/kg	
Sunflower	318.7	37.8	44.8	3.6
Cottonseed	46.5	8.3	15.1	4.8
Linseed	33.3	10.4	13.4	3.2
Peanut	190.1	(7.9)‡	56.7	5.1
Rapeseed‡	177.7	2.0	10.5	4.5
Soybean	30.1	7.4	17.7	3.5

† From Atlas of nutritional data on United States and Canadian feeds, 1971 (58).
‡ Mechanical extraction.

digestibility of the diets were not affected significantly (18). In addition, there were no changes in the nitrogen balance or in hematological conditions.

Both genotype of the seed and environment during seed maturation have a direct bearing on the concentration of chlorogenic acid in the seed. Dorrell (15) analyzed 387 inbred lines and found that the concentration of chlorogenic acid ranged from 1.4 to 4.0%. Early seeding and warm temperatures during seed maturation favored higher levels of chlorogenic acid. The synthesis of chlorogenic acid and oil are associated closely; therefore, factors which reduce the final level of chlorogenic acid tend to reduce oil percentage (D. G. Dorrell, unpublished data).

When sunflower seed production reaches the level where it becomes economically feasible to produce protein concentrates and isolates for use in food products, the inactivation or removal of chlorogenic acid will be necessary. Several methods have been proposed, including the use of antioxidants (24), water diffusion, and alcohol extraction (76); however, these methods are deficient because of incomplete extraction of chlorogenic acid, loss of protein, or the use of costly reagents.

Utilization as Animal Feed

Ruminants

Sunflower meal is fed successfully to ruminants whenever there is an assured supply. It is availability of the meal, not necessarily the higher fiber, lower lysine, or lower metabolizable energy that has limited its use in the past (4). Yearling steers fed for 139 days did equally well when fed sunflower and soybean meal as protein supplements (32). Similarly, sunflower meal was equivalent to cottonseed meal (34). It is suggested that sunflower meal could be included in any feeding regime if care is taken to assure adequate levels of lysine.

Table 19—Growth, feed conversion, and carcass trait of pigs fed two levels of sunflower meal.†

	Soybean meal	25% Sunflower	100% Sunflower
Avg. daily gain, kg	0.66	0.66	0.38
Avg. daily feed consump., kg	2.20	2.37	1.68
Feed:gain ratio	3.30	3.61	4.54
Dressing percentage	71.4	72.6	71.8
Avg. backfat, cm	3.15	3.40	3.02
Longissimus muscle (cross section area, cm²)	32.5	30.7	19.9

† Reproduced from Seerley et al. (82).

Poultry

Many researchers have found that up to 50% of the protein concentrates in rations for laying chickens could be replaced with sunflower meal without significantly reducing egg production (89). Net protein value was increased from 44 to 65% by supplementing the meal with lysine, indicating that even more sunflower meal could be used if adequate levels of lysine were available (68). At these higher levels of substitution, however, staining of the eggshells occurred due to the presence of chlorogenic acid in the meal (4). Up to 30% sunflower meal could be used in broiler rations if the feed were pelleted, although no more than 20% is recommended in the mash form (34).

Swine

Rations with up to 25% of the protein from sunflower meal produced rates of gain similar to soybean meal, but required larger quantities of meal (Table 19). These additional feed requirements were lowered by adding lysine (82). After the animals reached about 35 kg, additional sunflower meal could be fed due to a decreased requirement for essential amino acids (34).

It is concluded that sunflower meal is deficient in lysine and has a lower energy value (79) than several oilseed meals (Table 20). These deficiencies can be overcome with supplementation and the addition of acidulated soapstock or other neutralized high lipid refining wastes.

Concentrates and Isolates

Chemical Composition

The composition of concentrates and isolates will be somewhat similar to that noted under the previous section, Chemical Composition of Meal, with a few pertinent exceptions.

There are few antimetabolic or toxic compounds that survive normal heat treatment during seed conditioning or meal desolventizing; however,

Table 20—Effects of protein sources on feed consumption, weight gain, energy, and protein utilization by mice.†

Protein source	Feed intake	Weight gain	Energy gain	PER NX6.25	Apparent digestibility	Apparent BV
	g			Kcal	%	
Casein	55	10.4	21.5	1.89	84.4	30
Flaxseed	66	10.1	19.7	1.53	72.2	32
Rapeseed	47	7.4	13.5	1.57	81.7	28
Soybean	54	10.1	22.2	1.87	78.2	34
Sunflower	52	9.1	17.3	1.75	80.2	32

† Reproduced from Sarwar et al. (79).

the process of producing more concentrated products always has the potential of concentrating undesirable compounds. For example, chlorogenic acid is considered to be a possible arginase inhibitor by causing oxidative changes in arginase (69). Sabir et al. (77) found that about one-half of the phenolics were extracted with the dilute salt soluble proteins. Chlorogenic acid was bound to those proteins with a molecular weight of less than 5,000. Thus, unless special extraction procedures are used, substantial quantities of chlorogenic acid will appear in protein isolates.

Sunflower flour may contain up to 29% carbohydrate material. Simple sugars constitute about 38% of the total carbohydrates, with arabinose being the predominant monosaccharide (78). In concentrates, the level of carbohydrates dropped to 21% with less than 10% as simple sugars.

Production of Concentrates

The level of proteins in sunflower concentrates varies with the effectiveness of removing hulls, lipids, and alcohol soluble materials from the meal. Commercial sunflower meal usually contains 38 to 40% protein, which can be increased to 40 to 42% by improved dehulling. Prevot et al. (66) suggested that the concentration can be further increased to as high as 51% by efficient screening of the meal. As noted in Table 8, seed dehulled by hand will contain 52% protein. When such material is extracted exhaustively to remove residual free and bound lipids, a concentration of up to 62% protein can be achieved (47). By further extraction with ethanol and dilute acid to remove soluble carbohydrates, minerals, etc., the concentration can be increased to about 73% to be close to a maximum practical level.

Production of Isolates

The protein isolates currently available are produced almost exclusively from soybean or cottonseed meal. Much of the existing soybean technology, however, could be applied directly to sunflower meal if an adequate supply of meal was assured (29).

Table 21—The influence of extraction conditions on the yield, composition, and properties of protein concentrates and extracts prepared by batch and countercurrent procedures.†

Extraction (Ext.) and diffusion (Diff.) procedures	No. of extractions	Total solvent-to-flour ratio	Extract protein loss	Protein concentrate		N solubility index		Chlorogenic acid in product	Color of slurry pH 9
				Yield	Protein content	pH 7	pH 9		
		v/w		%	%			g/100 g	
Control flour	--	--	--	--	59.2	26	90	3.9	Green
Batch extraction procedures									
Acid ext. of flour, 24 C	3	15	23.1	60.5	75.2	53	98	1.0	Brown
Water ext. of flour, 24 C	3	15	21.3	67.5	69.0	2	18	0.7	Yellow
Alcohol ext. of flour, 24 C	3	15	5.1	77.3	72.7	23	70	0.4	Creamy-white
Acid ext. of flour, 24 C	5	25	25.2	58.5	75.7	53	89	0.5	Yellow
Acid ext. of flour, 80 C	5	25	30.0	60.1	69.0	27	44	0.3	Creamy-white
Acid diff. of seed, 80 C	4	24	17.3	71.4	68.6	7	47	0.4	Creamy-white
Countercurrent extraction procedures									
Acid ext. of flour, 24 C	5	6	24.9	60.6	73.4	46	90	1.2	Green
Acid ext. of flour, 80 C	5	6	26.5	60.0	72.5	--	--	0.8	Light green
Acid ext. of flour, 24 C	6	6	26.1	57.2	76.5	57	94	0.5	Yellow
Alcohol ext. of flour, 24 C	5	6	2.8	78.0	73.8	18	76	0.4	Creamy-white
Acid diff. of seed, 80 C	5	6	12.1	73.8	70.5	9	41	0.4	Creamy-white
Acid diff. of seed, 65 C	6	6	17.0	69.6	70.6	13	69	0.8	Light green

† Reproduced from Fan et al. (21).

In 1948, Smith and Johnsen (90) outlined a procedure for solubilizing sunflower protein at pH 10 followed by precipitation at the isoelectric point of pH 5. Their yields were only 32% of the total nitrogen, and the product was dark green due to chlorogenic acid. Gheyasuddin et al. (24) were able to extract more than 75% of the total nitrogen using 1 M NaCl in a 10:1 solvent to meal ratio; however, color was still a problem. Subsequently, pre-extraction of flour and seeds with isopropanol (24) and continuous water diffusion (92) was used to extract chlorogenic acid prior to solubilizing the protein. These, along with methods of decolorizing with alcohol after the protein had been precipitated (75), did not solve the problem completely. Recently, Fan et al. (21), using either batch or countercurrent extraction procedures, extracted flour with 70% ethanol at 24 C and were able to produce a 73% protein concentrate, a yield of 78%, a protein loss of less than 6%, and a product that did not color seriously at pH 9 (Table 21). The main disadvantage of their procedure was that it reduced nitrogen solubility from 90% in the control to less than 76% in the extracted flour.

It appears that if the chlorogenic acid level of sunflower concentrates can be reduced to less than 0.5%, then flour can be extracted under alkaline conditions, the solubilized protein can be precipitated at pH 4 to 6, and an isolate can be produced which contains approximately 93% protein.

Functional Properties

Sunflower meal and meal products have very good functional properties, particularly emulsion capacity, water absorption and retention, and aeration. Huffman et al. (30) found that a 6% suspension of meal in water at pH 7, mixed in a Waring blender at 4,500 rpm, was capable of forming an emulsion containing up to 190 ml of oil per g of meal. Sunflower flour will absorb 107% of its weight in water or 208% of its weight in fat (40). This amount compares with 130 and 84%, respectively, for soy flour (Table 22). These values are increased somewhat in concentrates and isolates, but the relationship between sunflower and soybean flour remains the same. Sunflower concentrates and isolates have foaming and whipping characteristics that approach those of fresh egg whites (40).

Fleming et al. (22) reported that concentrates ranged in viscosity from 50 to more than 83,300 centipoise (cps), depending upon the method of preparation, properties indicating excellent potential for a variety of uses in the food industry. Isolated sunflower proteins can be spun successfully to produce textured products using existing technology (20).

Utilization in Food Products

Sunflower flour and concentrates can be used in most products where color and fiber are not of serious concern. The protein concentration of bread can be increased and the amino acid balance improved by adding 5 to

Table 22—Some functional properties of soy and sunflower meal products.†

Sample	Protein solubility index	Water absorption	Fat absorption	Oil emulsion	Vol. increase on whipping	Vol. of foam	
						1 min	1 hour
		——— % ———				——— ml ———	
Wheat flour	21.7	60.2	84.2	11.7			
Soy flour	21.4	130.0	84.4	18.0	70.0	160	61
Concentrate (Isopro)	2.3	227.3	133.0	2.8	170.0	400	8
Concentrate (Promosoy)	6.0	196.1	92.0	18.7	135.0	370	30
Isolate (Supro 610)	17.4	447.6	154.5	25.2	235.0	670	545
Isolate (Promine D)	71.1	416.7	119.2	22.2	230.0	660	535
Sunflower flour	16.1	107.1	207.8	95.1	230.0	600	467
Concentrate (DE 60)	3.3	137.8	254.9	14.0	220.0	610	380
Concentrate (DE 90)	2.1	203.0	226.5	10.1	225.0	540	140
Isolate (DE 60)	18.2	155.1	256.7	25.6	230.0	630	493

† Reproduced from Lin et al. (40).

Table 23—Properties of flour and breads baked with oilseed-wheat flour blends at 17.5% protein.†

	Farinograph absorption		Loaf volume		Reflectance value	
	No heat	Heat	No heat	Heat	No heat	Heat
	%		cc		%	
Control	62.0		3,035		59	
Cottonseed	64.4	65.8	2,600	3,000	49	45
Peanut	63.5	65.2	2,900	2,900	56	51
Sesame	63.0	65.5	3,000	2,900	56	50
Sunflower	64.7	65.4	2,600	2,840	44	45

† Reproduced from Rooney et al. (73).

Table 24—Effect of protein additives on physical and organoleptic properties of cooked wieners.†

Protein additive	Shrink-age‡	Cooking loss	Firm-ness§	Flavor¶	Texture¶
	%		mm		
Control	18.6	3.3	208.7	8.2	7.6
Soy flour	15.7	3.7	195.3	7.4	7.0
Concentrate	16.3	4.0	191.2	6.9	7.6
Sunflower flour	16.7	2.9	218.2	4.4	3.9
Concentrate (DE 60)	15.7	3.1	195.6	7.0	7.2
Concentrate (activated DE 60)	14.7	2.8	192.1	7.0	7.0

† Reproduced from Lin et al. (41).
‡ Uncooked loss after 16 hours.
§ Penetrometer measurement.
¶ 1 to 9 scale (9 liked most).

10% sunflower concentrate. This balance can be achieved without changing crumb color (88). While various oilseed flours are used to fortify wheat flours, they usually are not interchangeable because of differences in functional properties. Rooney et al. (73) noted that the properties of sunflower flour were improved with heating, whereas sesame and peanut flours were impaired (Table 23). Bread fortified with sunflower flour tended to be somewhat darker. Increasing the protein content of the blend from 17.5 to 20% weakened the dough structure, indicating limitations to the use of sunflower flour or concentrates. The general use of sunflower products in the baking industry will depend on the need for specific functional properties and the absence of color in the flour.

Lin et al. (41) concluded that wieners could be produced using both soybean and sunflower flours and concentrates as extenders. Sunflower flour, however, was judged somewhat less satisfactory than soybean flour because the wieners were softer and had an unusual flavor when heated (Table 24).

Nutrition

Sunflower flour and concentrates have considerable nutritional potential because they lack toxic factors which must be inactivated and they are a particularly good source of calcium, phosphorus, and nicotinic acid.

Deficiencies of lysine may be aggravated, however, during the production of isolates. Sarwar et al. (79) reported that while the total essential amino acid content of sunflower meal and isolates was similar, the loss of lysine during the production of the isolates reduced the protein score. This reduction may be due to the use of alkali and heat during protein extraction. Provansal et al. (67) observed a reduction in the concentrations of cystine, arginine, threonine, serine, isoleucine, and lysine when protein isolates were treated with increasing concentrations of sodium hydroxide. They also noted the formation of unusual amino acids such as lysinoalanine, lanthionine, alloisoleucine, and ornithine (67). The formation of crosslinked proteins will influence digestibility directly. Lysinoalanine has been implicated as a renal toxic factor in rats, and reduces protein utilization (95). These compounds appear widely in processed foods and at this time are not considered toxic to humans but rather indicate excessive processing of the protein. Additional nutritional data are contained in the review by Robertson (70).

OTHER PRODUCTS

The major byproduct of sunflower seed-crushing is hull material. At present, most hulls are ground and sold as filler or roughage for livestock rations. The hulls have at least 1% natural lipids, predominantly waxes, plus an additional 1 to 3% lipids that are absorbed during dehulling or picked up as fine kernel material or from small immature seeds. Some plants may add soapstock or refinery wastes to increase the energy value of the hulls further. The protein content of such material may reach 20%. Hulls can be treated with sulfuric acid at 120 C in a two-step process wherein the hydrolysis of pentosans is nearly quantitative and the hydrolysis of cellulose gives yields of up to 79%. When the products of hydrolysis are neutralized, they provide good substrates for yeasts. *Paecilomyces varioti* Bainier produced dry mycelium yields up to 94 g/100 g of available reducing sugar (19). Zhil'nikov (105) suggested that up to 40% of the polysaccharides in hulls could be hydrolyzed with cellulases, thereby further increasing its value as a feed. Hulls have also been used in the production of xylose, furfural, fertilizer, and fiberboard (7).

AUTHOR'S NOTE

The reader's attention is drawn to the following reviews: "Use of sunflower seed in food products" by Robertson (70), "Sunflower seed and oil: a review" by Mantha and Subrahmanyam (45), and "Sunflower seed oil meal" by Clandinin (10). In addition, the author has a collection of several hundred papers covering central and eastern European lipid research, prepared and generously provided by Prof. A. Popov. Many of these are in the original language.

LITERATURE CITED

1. American Oil Chemists' Society. 1974. Official and tentative methods of analysis. American Oil Chemists Society, Champaign, Ill.

2. Amos, H. E., D. Burdick, and R. W. Seerley. 1975. Effect of processing temperature and L-lysine supplementation on utilization of sunflower meal by the growing rat. J. Anim. Sci. 40:90-95.

3. Baudet, J., and J. Mosse. 1977. Fractionation of sunflower seed proteins. J. Am. Oil Chem. Soc. 54:82A-86A.

4. Bell, J. M. 1975. Utilization of protein supplements in animal feed. p. 633-656. In J. T. Harapiak (ed.) Oilseed and pulse crops in Western Canada. Western Cooperative Fertilizers Ltd., Calgary, Alberta.

5. Borodulina, A. A., P. S. Popov, and I. N. Harchenko. 1974. Biochemical characteristics of the present sunflower varieties and hybrids. p. 239-245. In A. V. Vranceanu (ed.) Proc. 6th Int. Sunflower Conf. Research Institute for cereal and Industrial Crops, Fundulea, Ilfov, Romania.

6. ————, and L. V. Suprunova. 1976. Change of sunflower seed protein complex qualitative composition depending on genotype and ecological factors. p. 173-175. In Proc. 7th Int. Sunflower Conf. (Krasnodar, USSR).

7. Cancalon, P. 1971. Chemical composition of sunflower seed hulls. J. Am. Oil Chem. Soc. 48:629-632.

8. Cegla, G. F., and K. R. Bell. 1977. High pressure liquid chromatography for the analysis of soluble carbohydrates in defatted oilseed flours. J. Am. Oil Chem. Soc. 54:150-152.

9. Chang, S. S. 1976. Parenterally dispensable nutrient oil. Ger. Offen. 2533612 (Sweden) 4 March. (See Chem. Abstr. 84:163187).

10. Clandinin, D. R. 1958. Sunflower seed oil meal. p. 557-575. In A. M. Attschul (ed.) Processed plant protein foodstuffs. Academic Press, New York.

11. Codex Alimentarius Commission, Recommended International Standard for Edible Sunflower Oil. 1969. Bulletin CAC/RS 23. FAO and WHO, Rome Italy.

12. Defromont, C. 1972. Dehulling of sunflower seeds. p. 353-361. In Proc. 5th Int. Sunflower Conf. (Clermont-Ferrand).

13. Delic, I., N. Vucurevic, and S. Stojanovic. 1975. Investigation of inactivation of chlorogenic acid from sunflower meal under in vitro conditions in mice. Acta Vet. (Belgrade) 25(3):115-119.

14. Dompert, W. U., and H. Beringer. 1976. Effect of ripening temperature and oxygen supply on the synthesis of unsaturated fatty acids and tocopherols in sunflower seeds. Z. Pflanzenernaehr. Bodenkd. 2:157-167.

15. Dorrell, D. G. 1976. Chlorogenic acid content of meal from cultivated and wild sunflowers. Crop Sci. 16:422-424.

16. Earle, F. R., C. H. VanEtten, T. F. Clark, and I. A. Wolff. 1968. Compositional data on sunflower seed. J. Am. Oil Chem. Soc. 45:876-879.

17. Edmondson, L. F., R. A. Yoncoskie, N. H. Rainey, F. W. Douglas, and J. Bitman. 1974. Feeding encapsulated oils to increase the polyunsaturation in milk and meal fat. J. Am. Oil Chem. Soc. 51:72-76.

18. Eklund, A. 1975. Effect of chlorogenic acid in a casein diet for rats: nutritional and pathological observations. Nutr. Metab. 18:258-264.

19. Eklund, E., A. Hatakka, A. Mustranta, and P. Nybergh. 1976. Acid hydrolysis of sunflower seed husks for production of single cell protein. Eur. J. Appl. Microbiol. 2:143-152.

20. Fabre, M. 1975. Spinning of sunflower proteins. Physical properties in relation to the organoleptic qualities of the spun products. Rev. Fr. Corps Gras 22:593-598.

21. Fan, T. Y., F. W. Sosulski, and N. W. Hamon. 1976. New techniques for preparation of improved sunflower protein concentrates. Cereal Chem. 53:118-125.

22. Fleming, S. E., F. W. Sosulski, A. Kilara, and E. S. Humbert. 1974. Viscosity and water absorption characteristics of slurries of sunflower and soybean flours, concentrates and isolates. J. Food Sci. 39:188-191.

23. Freier, B., O. Popescu, A. M. Ille, and H. Antoni. 1973. Sunflower oil interesterification. Method for improving the quality of margarine. Ind. Aliment. (Bucharest) 24: 604-607.

24. Gheyasuddin, S., C. M. Cater, and K. F. Mattil. 1970a. Preparation of a colorless sunflower protein isolate. Food Technol. 24:242-243.

25. ———, ———, and ———. 1970b. Effect of several variables on the extractability of sunflower seed proteins. J. Food Sci. 35:453-456.

26. Goebel, E. H. 1976. Bleaching practices in the U.S. J. Am. Oil Chem. Soc. 53:342-343.

27. Goldovskii, A. M., and V. N. Nikitinskaya. 1973. Formation of oil coloration during seed treatment. IV. Coloration of vegetable oils. Maslo-Zhar Prom. 1973(6):9-14.

28. Graf, T., and L. Meyer. 1976. Use of lecithin in food. Flavors p. 218-221 (Sept./Oct.).

29. Horan, F. E. 1974. Soy protein products and their production. J. Am. Oil Chem. Soc. 51:67A-73A.

30. Huffman, V. L., C. K. Lee, and E. E. Burns. 1975. Selected functional properties of sunflower meal (*Helianthus annuus*). J. Food Sci. 40:70-74.

31. Hustedt, H. H. 1976. Interesterification of edible oils. J. Am. Oil Chem. Soc. 53:390-392.

32. Kercher, C. J., S. Maxfield, L. Paules, W. Smith, and G. Costel. 1974. Sunflowers as a protein source for ruminants. Proc., Western Section, Am. Soc. Anim. Sci. 25:328-330.

33. Kharchenko, L. N., and A. A. Borodulina. 1976. Accumulation and metabolism of fatty acids in seeds of high oleic mutant sunflowers. p. 178-180. *In* Proc. 7th Int. Sunflower Conf. (Krasnodar, USSR).

34. Kinard, D. H. 1975. Feeding value of sunflower meal and hulls. Feedstuffs 47(45):26-31.

35. Kinman, M. L., and F. R. Earle. 1964. Agronomic performance and chemical composition of the seed of sunflower hybrids and introduced varieties. Crop Sci. 4:417-420.

36. Klein, K., and L. S. Crauer. 1974. Further developments in crude oil processing. J. Am. Oil Chem. Soc. 51:382A-385A.

37. Krasil'nilov, V. N., V. P. Rzehin, and N. A. Nedacina. 1972. Cerous substances in sunflower seeds—their composition and distribution on the morphological components of the seed—their distribution in various oil products. p. 409-412. *In* Proc. 5th Int. Sunflower Conf. (Clermont-Ferrand).

38. Kuksis, A. 1972. Newer developments in determination of structure of glycerides and phosphoglycerides. p. 5-163. *In* R. T. Holman (ed.) Progress in the chemistry of fats and other lipids. Pergamon Press, Oxford, England. p. 5-163.

39. Kurnik, E. 1967. Tocopherol content in sunflower varieties. Takarmarybazis 7:7-12 (Bull. Agric. Res. Inst., Iregszemcse, Hungary).

40. Lin, M. J. Y., E. S. Humbert, and F. W. Sosulski. 1974. Certain functional properties of sunflower meal products. J. Food Sci. 39:368-370.

41. ———, ———, ———, and J. W. Card. 1975. Quality of wieners supplemented with sunflower and soy products. Can. Inst. Food Sci. Technol. J. 8:97-101.

42. List, G. R., C. D. Evans, and H. A. Moser. 1972. Flavor and oxidative stability of northern-grown sunflower seed oil. J. Am. Oil Chem. Soc. 49:287-292.

43. Luckadoo, B. M., and E. R. Sherwin. 1972. Tertiary butylhydroquinone as antioxidant for crude sunflower seed oil. J. Am. Oil Chem. Soc. 49:95-97.

44. Macuk, J. P. 1971. Obtaining an easily hydratable sunflower oil. Tr. Vses. Nauchno-Issled. Inst. Zhirov. 28:41-48.

45. Mantha, K. S., and V. V. R. Subrahmanyam. 1973. Sunflower seed and oil: a review. J. Oil Technol. Assoc. India. p. 11-17, 26-34 (Jan.-Mar.).

46. Manton, S. D. 1971. Industrial oils, fats and waxes. Soap Perfum. Cosmet. 44:705-714.

47. Marinchevski, I., P. Petrov, and L. Georgieva. 1974. Possible production of protein concentrates from sunflower seeds for nutritional purposes. I. Model formation of a protein product with a minimal cellulose concentration. Maslo-Sapunena Prom. St. 11(1):16-23.

48. Mattson, F. H., and R. A. Volpenhein. 1963. The specific distribution of unsaturated fatty acids in the triglycerides of plants. J. Lipid Res. 4:392-396.

49. Mikolajczak, K. L., R. M. Freidinger, C. R. Smith, Jr., and I. A. Wolff. 1968. Oxygenated fatty acids of oil from sunflower seeds after prolonged storage. Lipids 3:489-494.

50. Milligan, E. D. 1976. Survey of current solvent extraction equipment. J. Am. Oil Chem. Soc. 53:286-290.

51. Mordret, F. 1972. Some characteristics of the unsaponifiable portion of sunflower seed oil. p. 399-408. *In* Proc. 5th Int. Sunflower Conf. (Clermont-Ferrand).

52. Morrison, F. B. 1957. Feeds and feeding. 22nd ed. Morrison Publishing Co., New York.

53. Morrison, W. H. 1975. Effect of refining and bleaching on oxidative stability of sunflowerseed oil. J. Am. Oil Chem. Soc. 52:522-525.

54. ————, and J. K. Thomas. 1976. Removal of waxes from sunflower seed oil by miscella refining and winterization. J. Am. Oil Chem. Soc. 53:485–486.

55. Mosse, J., and J. Baudet. 1972. Biochemical aspects of quality and content of lysine in sunflower seed proteins. p. 437–440. In Proc. 5th Int. Sunflower Conf. (Clermont-Ferrand).

56. Muller-Mulot, W. 1976. Rapid method for the quantitative determination of individual tocopherols in oils and fats. J. Am. Oil Chem. Soc. 53:732–736.

57. Mullor, J. B. 1968. Improvement of the nutritional value of food oils. Rev. Fac. Ing. Quim. Univ. Nac. Litoral 37:183–210.

58. National Academy of Science. 1971. Atlas of nutritional data on United States and Canadian Feeds. National Academy of Sciences, Washington, D.C.

59. Nestel, P. J., N. Havenstein, H. M. Whyte, T. J. Scott, and L. J. Cook. 1973. Lowering of plasma cholesterol and enhanced sterol excretion with the consumption of polyunsaturated ruminant fats. N. Engl. J. Med. 288:379–382.

60. Nikolic-Vig, V., D. Skoric, and S. Bedov. 1971. Variability in the percentages of oil and husk in the seeds of populations of sunflower varieties Peredovik and VNIIMK 8931 and their heritability. Savremena Poijoprivreda 19:23–32. (Savez Zemljoradnickih zadruga A. P. Voljvodine, Novi Sad, Yugoslavia.)

61. Popov, A., M. Dodova-Anghelova, Ch. P. Ivanov, and K. Stefanov. 1970. Further investigations on the composition of the wax isolated from sunflower husks. Riv. Ital. Sostanze Grasse 47:254–256.

62. ————, M. Gardev, N. Yanishlieva, and L. Hirstova. 1971. Method of obtaining native powdered sunflower lecithin. J. Am. Oil Chem. Soc. 48:305–306.

63. ————, Ts. Milkova, and N. Marekov. 1975. Free and bound sterols in sunflower, soya and maize oils. Nahrung 19:547–549.

64. ————, and N. Yanishlieva. 1969. The influence of some factors on the autooxidation of lipid systems. Izv. Otd. Khim. Nauki. Bulg. Akad. Nauk. 11:549–557.

65. Press, M., P. J. Hartop, and C. Prottey. 1974. Correction of essential fatty-acid deficiency in man by the cutaneous application of sunflower-seed oil. Lancet 1(7858):597–598.

66. Prevot, A., C. Bloch, and C. DeFromont. 1972. Enrichment of the protein in sunflower meal by mechanical means. p. 368–384. In Proc. 5th Int. Sunflower Conf. (Clermont-Ferrand).

67. Provansal, M. M. P., J-L. A. Cuq, and J-C. Cheftel. 1975. Chemical and nutritional modifications of sunflower proteins due to alkaline processing: Formation of amino acid cross-links and isomerization of lysine residues. J. Agric. Food Chem. 23:938–943.

68. Rad, F. H., and K. Keshavarz. 1976. Evaluation of the nutritional value of sunflower meal and the possibility of substitution of sunflower meal for soybean meal in poultry diets. Poultry Sci. 55:1757–1765.

69. Reifer, I., and H. Augustyniak. 1968. Preliminary identification of the arginase inhibitor from sunflower seeds. Bull. Acad. Pol. Sci. Ser. Sci. Biol. 16:139–144.

70. Robertson, J. A. 1975. Use of sunflower seed in food products. Crit. Rev. Food Sci. Nutr. 6:201–240.

71. ————, and W. H. Morrison. 1977. Effect of heat and frying on sunflower oil stability. J. Am. Oil Chem. Soc. 54:77A–81A.

72. ————, ————, D. Burdick, and R. Shaw. 1972. Flavor and chemical evaluation of partially hydrogenated sunflower oil as a potato chip frying oil. Am. Potato J. 49:444–450.

73. Rooney, L. W., C. B. Gustafson, S. P. Clark, and C. M. Cater. 1972. Comparison of the baking properties of several oilseed flours. J. Food Sci. 37:14–18.

74. Roy, N. D., and R. V. Bhat. 1974. Trypsin inhibitor content in some varieties of soybean (Glycine max) and sunflower seed (Helianthus annuus). J. Sci. Food Agric. 25:265–269.

75. Rucci, A. O., and M. H. Bertoni. 1974. Proteins from sunflower seed by-products. II. Colorless isolates from methyl alcohol extracted meal. Biological evaluation and preparation trials in salt and acid media. An. Asoc. Quim. Argent. 62:347–356.

76. Rutkowski, A. 1972. Oilseed proteins and their characteristics. Riv. Ital. Sostanze Grasse 49:416–427.

77. Sabir, M. A., F. W. Sosulski, and A. J. Finlayson. 1974. Chlorogenic acid-protein interactions in sunflower. J. Agric. Food Chem. 22:575–578.

78. ————, ————, and N. W. Hamon. 1975. Sunflower carbohydrates. J. Agric. Food Chem. 23:16–19.

79. Sarwar, G., F. W. Sosulski, and J. M. Bell. 1973. Nutritional evaluation of oilseed meals and protein isolates by mice. Can. Inst. Food Technol. J. 6:17–21.

80. Scott, T. W., L. J. Cook, K. A. Ferguson, I. W. McDonald, R. A. Buchanan, and G. L. Hills. 1970. Production of poly-unsaturated milk fat in domestic ruminants. Aust. J. Sci. 32:291-293.

81. Sedlacek, B. A. J. 1972. Study of the UV spectra of heated fats. 9. The effect of the addition of iron derivatives on changes in sunflower oil during heating. Nahrung 16: 167-177.

82. Seerley, R. W., D. Burdick, W. C. Russom, R. S. Lowrey, H. C. McCampbell, and H. E. Amos. 1974. Sunflower meal as a replacement for soybean meal in growing swine and rat diets. J. Anim. Sci. 38:947-953.

83. Segal, B. 1970. Effect of some antioxidants on the thermal degradation of sunflower oil. Igiena 19:595-600.

84. Sergeyev, A. G., N. L. Melamud, R. L. Perkel, G. V. Chebotaryova, I. V. Mikhailova, and A. V. Statsenko. 1976. Obtaining of food modified fats on the base of hydrogenated and re-esterified sunflower oil. p. 194-196. *In* Proc. 7th Int. Sunflower Conf. (Krasnodar, USSR).

85. Shcherbakov, V. G. 1969. Formation of a lipid complex of sunflower seeds during ripening and postharvest treatment. Byull. Nauchno-Tekh. Inf. Maslichn. Kult 1969:40-43.

86. ————, L. M. Gorshkova, N. P. Kovalenko, E. N. Shkurupii, Z. A. Chaika, and E. I. Polyakova. 1976. Effect of conditions for extracting oil from sunflower seeds on the yield and quality of protein substances. Izv. Vyssh. Uchebn. Zaved. Pishch. Tekhnol. 1976(2): 45-47.

87. ————, and A. M. Malyshev. 1968. Alteration of sunflower seed lipids in thermal drying. Izv. Vyssh. Uchebn. Zaved. Pishch. Teknhol. 1968(1):34-36.

88. Sirko, V. N., and V. A. Bukhantsov. 1973. Use of sunflower oil proteins in bread baking. Izv. Vyssh. Uchebn. Zaved. Pishch. Tekhnol. 1973(2):129-130.

89. Smith, K. J. 1968. A review of the nutritional value of sunflower meal. Feedstuffs 40: 20-23.

90. Smith, A. K., and V. L. Johnsen. 1948. Sunflower seed protein. Cereal Chem. 25:399-406.

91. Sosulski, F. W., and A. Bakal. 1969. Isolated proteins from rapeseed, flax and sunflower meals. Can. Inst. Food Technol. J. 2:28-32.

92. ————, M. A. Sabir, and S. E. Fleming. 1973. Continuous diffusion of chlorogenic acid from sunflower kernels. J. Food Sci. 38:468-470.

93. ————, and G. Sarwar. 1973. Amino acid composition of oilseed meals and protein isolates. Can. Inst. Food Sci. Technol. J. 6:1-5.

94. Stefanov, K. L., and A. D. Popov. 1976. Esterified sterol glycosides from sunflower seed lipids. Dokl. Bolg. Akad. Nauk. 29:1289-1291.

95. Sternberg, M., C. Y. Kim, and F. J. Schwende. 1975. Lysinoalanine: presence in food and food ingredients. Science 190:992-994.

96. Sullivan, F. E. 1968. Refining of fats and oils. J. Am. Oil Chem. Soc. 45:564A-615A.

97. Swern, D. 1964. Bailey's industrial oil and fat products. Interscience Publishers, New York. 1103 p.

98. Tindale, L. H., and S. R. Hill-Haas. 1976. Current equipment for mechanical oil extraction. J. Am. Oil Chem. Soc. 53:265-270.

99. Tkachuk, R., and G. N. Irvine. 1969. Amino acid compositions of cereals and oilseed meals. Cereal Chem. 46:206-218.

100. Trotter, W. K., H. O. Doty, W. D. Givan, and J. V. Lawler. 1973. Potential for oilseed sunflowers in the United States. USDA/ERS, AER-237.

101. Vaughan, J. G. 1970. The structure and utilization of oilseeds. Chapman and Hall Ltd., London.

102. Watson, K. S., and C. H. Meierhoefer. 1976. Use or disposal of byproducts and spent material from the vegetable oil processing industry in the U.S. J. Am. Oil Chem. Soc. 53: 437-442.

103. West, C. E., and T. G. Redgrave. 1974. Reservations on the use of polyunsaturated fats in human nutrition. Search 5:90-94.

104. Zehnder, C. T. 1975. Deodorization 1975. J. Am. Oil Chem. Soc. 53:364-369.

105. Zhil'nikov, V. L. 1971. Processing sunflower husks. Tr. Voronezh. Tekhnol. Inst. 19: 158-161.

D. GORDON DORRELL: B.S.A., M.Sc., Ph.D., Head, Crop Science Section, Research Station, Agriculture Canada, Morden, Manitoba, Canada. Sunflower oilseed quality investigations, Agriculture Canada 1968 to date.

Chapter 14

Sunflower for Confectionery Food, Birdfood, and Petfood

JAMES R. LOFGREN

Nonoilseed sunflower, (*Helianthus annuus* L.) is grown in gardens of many countries of the world to be consumed as a raw or roasted and salted snack. Large quantities of seeds also are used for feeding birds and small animal pets in the USA and Canada. Field production of nonoilseed sunflower in North America is concentrated in North Dakota and Minnesota. Lesser quantities are grown in other areas, primarily the states of South Dakota, California, and Texas, and the Canadian provinces of Manitoba, Saskatchewan, and Alberta.

Seed used as a confection or for snack food is typically of large size, i.e., those seeds passing over a 7.9 mm (20/64 in) round-hole sieve. The outer layer of the hull is striped black and white. A dark inner layer, characteristic of many genotypes, may give a grey appearance to the white stripes of the outer layer (20). Nonoilseed sunflower hybridizes readily with oilseed and wild *H. annuus,* thus requiring space or seasonal isolation to prevent cross pollination when seed is grown for planting purposes.

This chapter will discuss the processing, quality determination, preservation, and uses of nonoilseed sunflower.

PROCESSING PROCEDURES

Country elevators and processing plants purchase nonoilseed sunflower from the farmer according to the Grading Standards found in Chapter 12. Processing plants usually remove "dockage" as the first step in cleaning the seed. Dockage consists of petiole and stalk pieces, portions of the head, cracked hulls, and other foreign material. Seeds generally are separated into three groups based on size. The large size, which is often that seed going over a 8.7 mm (22/64 in) round-hole sieve, is used as "in shell" product. This seed is salted and roasted whole and usually constitutes 15 to

From: Carter, Jack F. (ed.). 1978. Sunflower Science and Technology. Agronomy 19. Copyright © 1978 by the American Society of Agronomy, Crop Science Society of America, and Soil Science Society of America, 677 South Segoe Road, Madison, WI 53711 USA.

Table 1—Effect of temperature, moisture percentage, and sample depth on the percentage of surface-disinfected sunflower seeds producing *Alternaria* spp. and *Aspergillus glaucus* when plated on T-6 agar media. Adapted from Christensen (5).

Observation	Surface†		90 cm (3 ft)†		180 cm (6 ft)†		270 cm (9 ft)†	
	Avg.	Range	Avg.	Range	Avg.	Range	Avg.	Range
Temperature at sampling time, C				Temperature, C				
	−6.0	−6 to −7	−5.0	−5 to −6	−5.0	−2 to −6	−2.0	−6 to +2
Moisture at sampling time, %				Seed moisture, %				
	13.7	10.9–17.2	11.8	10.3–13.1	12.9	10.3–14.8	11.0	10.6–11.5
% of surface disinfected seeds yielding				% of seeds infected				
Alternaria (field fungi)	98	92–100	100		89	72–100	99	96–100
A. glaucus (storage fungi)	9	0–36	0		28	0–56	0	

† Mean of four samples collected on March 10, 1970 from four positions and held in air tight containers at 22 to 25 C for 10 days before being tested.

25% of the seed crop. The medium or "hulling" size, which constitutes 40 to 60% of the crop, is that seed passing through a 8.7 mm (22/64 in) but over a 7.1 mm (18/64 in) round-hole sieve. The hull is removed from the "kernels" or mature ovules, and the kernels are then used with or without roasting as a snack or in various confectionery or bakery products. The remaining 15 to 20% which passes through a 7.1 mm (18/64 in) sieve normally is used as "birdfeed." Birdfeed is packaged as 100% sunflower seed or in blends with grain sorghum [*Sorghum bicolor* (L.) Moench.], corn (*Zea mays* L.), proso millet (*Panicum miliaceum* L.), wheat (*Triticum aestivum* L. em Thell.), and/or oat (*Avena sativa* L.) for sale to those persons who feed pet or wild birds and to householders feeding seed-consuming pets such as rodents.

Seed may be stored on the farm or in a country elevator for an indefinite time prior to processing. Moisture percentage and temperature affect the quality of sunflower seed in storage (5). Samples collected in the winter from storage bins at various positions and depths ranged from −6.0 to +2.0 C, and 10.3 to 15.9% moisture (Table 1). No deterioration had occurred at −6.0 to +2.0 C although the seed surface was contaminated with spores of *Alternaria* spp. and *Aspergillus glaucus* (Link) (5), but storage fungi rapidly invaded such samples at laboratory temperatures of 22 to 25 C. This study shows the importance of periodically testing the temperature and moisture of sunflower seed in storage to avoid loss, especially in warmer weather comparable to the laboratory temperatures. *Aspergillus* spp. were ubiquitous on seed from field infestations (Christensen, personal communication). Storage fungi invaded sunflower seed stored at temperatures of 20 C and above, especially above 7% seed moisture.

The percentage of the crop in different seed sizes (Table 2) varies greatly among cultivars according to Lofgren (14). He found that 'D747' (D = Dahlgren and Company designation), 'D509', and 'Sundak' had 84, 79, and 65%, respectively, while 'D508' had only 43% seeds passing over a 7.1 mm (18/64 in) round-hole sieve. In addition to these apparent genetic effects, seed size depends on plant density as shown by the author (Table 3). Highly significant differences in seed size occurred for the cultivar 'Mingren' at six plant densities. The lowest plant density gave the highest percentage of seed passing over a 7.9 mm (20/64 in) round-hole sieve. Highly significant positive correlations (0.72 and 0.36, in 144 samples) were found between large seed size and seed weight, and between large seed size and head diameter, respectively. A highly significant negative correlation (−0.66) was observed between large seed size and oil concentration of whole seed.

QUALITY OF NONOILSEED SUNFLOWER

In-shell Sunflower Seed

"In-shell" sunflower seed is considered of high quality if it meets the following criteria: (a) relatively large, uniform seed; (b) a large kernel surrounded by a smooth, loose hull with bright white stripes on black back-

Table 2—Means and standard errors of six quality characteristics of open-pollinated Sundak and three hybrid nonoilseed sunflower cultivars as reported by Lofgren (14).

Cultivar	Seed size†	% of crop‡	Hullability§		Kernel¶		Kernel weights				Roast			
							Twenty		Volume#		Dark count††		Color‡‡	
			score	S.E.	%	S.E.	g	S.E.	g	S.E.	score	S.E.	score	S.E.
Sundak	1(19)	39	15	3.7	50.3	1.14	1.51	0.08	5.84	0.13	0.00	0.00	2.6	0.51
	2(7)	20	12	2.7	51.0	1.98	1.60	0.13	5.73	0.09	0.00	0.00	3.2	0.46
	3(5)	5	10	4.1	46.8	2.65	1.79	0.12	5.79	0.08	0.25	0.50	3.5	0.58
	4(1)	1	10	–	44.4	–	1.55	–	5.96	–	1.00	–	4.0	–
D747	1(18)	29	13	3.1	49.6	2.95	1.44	0.07	5.80	0.15	0.33	0.58	2.9	0.41
	2(7)	32	11	2.4	48.4	1.49	1.57	0.08	5.72	0.26	0.20	0.41	3.0	0.00
	3(6)	16	9	3.8	46.6	0.47	1.71	0.07	5.70	0.10	0.00	0.00	2.5	0.55
	4(4)	7	9	2.5	43.1	2.02	1.83	0.04	5.69	0.17	0.00	0.00	2.7	0.96
D508	1(19)	37	16	3.1	54.5	2.12	1.69	0.35	5.85	0.12	0.00	0.00	2.9	0.46
	2(4)	5	10	4.1	51.6	1.44	1.77	0.28	5.79	0.11	0.00	0.00	3.0	0.00
D509	1(18)	35	14	3.7	52.8	1.73	1.43	0.06	5.78	0.09	0.00	0.00	2.7	0.46
	2(8)	32	9	2.3	51.7	1.53	1.52	0.05	5.65	0.11	0.10	0.35	3.0	0.00
	3(7)	11	7	2.7	49.4	3.87	1.69	0.07	5.76	0.14	0.10	0.35	2.7	0.46

† Seed size as determined on round-hole sieve, 1 = over 7.1, 2 = 7.9, 3 = 8.7, and 4 = 9.5 mm; numbers in parentheses following seed size indicate number of samples (50 g minimum) to determine statistics; Sundak size 4 one sample no standard error; D508 no size 3 and 4; D509 no size 4.
‡ Crop grown in 1975 at Crookston, Minnesota. Portion through 7.1 mm sieve not included.
§ Hullability—see text for explanation.
¶ Kernel percentages adjusted on the basis of hullability score.
# Volume weight of kernels in a 12.0 cc (20 × 38 mm) cylinder.
†† Dark count scale—0 = none, to 9 = all dark.
‡‡ Color score—1 = light tan, to 5 = very dark brown.

Table 3—Percentage of seed produced by the Mingren cultivar grown at variable plant populations and row spacings at Crookston, Minnesota, in 1968–70 that passed over a 7.9 mm round-hole sieve, as reported by Lofgren.

Population†	Row spacing, cm†			Avg.
	50	75	100	
plants/ha			%	
37,500	30.0	34.9	34.9	33.3 a*
50,000	28.1	23.1	22.1	24.4 b
62,500	17.5	20.9	23.1	20.5 bc
75,000	17.0	18.0	19.4	18.2 bc
87,500	12.6	17.4	13.5	14.5 c
Avg.	27.4 a*	29.4 a	26.6 a	

* Means among populations or row spacing followed by the same letter are not statistically different at the 0.05 level according to Duncan's multiple range test.
† Plant populations 15, 20, 25, 30, and 35 thousand/acre, respectively, and spacings between rows of 20, 30, and 40 in, respectively.

ground color; (c) freedom from insect, rodent, fungal, or weather damage; (d) freedom from particles sloughed off the outer layer of the hull; and (e) freedom from extraneous sunflower plant parts and other foreign matter. The consumer prefers large seed. The "in-shell" seed is sold to processing companies for roasting or processed to their specifications. One processor may prefer seed passing over a 8.7 mm (22/64 in) sieve while another processor will use seed passing through a 8.7 mm but over a 7.9 mm (20/64 in) round-hole sieve.

The hull of "in-shell" sunflower seed sometimes is stained by weathering due to alternate wetting and drying or to frost before maturity. Frost before maturity reduces seed quality severely by (a) lowering the volume weight substantially, (b) creating an undesirable "in shell" roasted product, and (c) causing kernels to be dark brown or black when roasted. Putt (20) observed three layers in the hull (pericarp). The innermost layer may be an amorphous dark color under striping which may occur in the outermost layer or color may be absent in the outermost and second layers. Frequent late fall rains wet the hull and subsequent drying may loosen the outermost layers of the hull so that they separate partially in processing. This condition produces an unattractive salted and roasted product, i.e., a white hull or a white hull with irregular black patches.

Diseased plants produced smaller seed than healthy plants according to Zimmer and Zimmerman (25). Plants of the Mingren cultivar infected with downy mildew [*Plasmopara halstedii* (Farl.) Berl. et de Toni] had 24 and 0% seed over 7.1 and 8.7 mm round-hole sieves, respectively, while healthy plants had 90 and 30%, respectively, over 7.1 and 8.7 mm sieves. Seed of the 'Commander' cultivar from plants affected with rust (*Puccinia helianthi* Schw.) had 75 and 0%, respectively, while seed from healthy plants had 92 and 37%, respectively, passing over 7.1 and 8.7 mm round-hole sieves. Plants of an oilseed cultivar infected with *Verticillium* wilt [*Verticillium albo-atrum* R.&B. (*V. dahliae* Kleb.)] produced smaller seed than healthy plants. Plants infected with *Rhizopus* spp. near physiological maturity produced seed of the same size as healthy plants.

Damage to sunflower seed by the sunflower moth (*Homoeosoma electellum* Hulst), by the banded sunflower moth (*Phalonia hospes* Wlshm.), and by the seed weevils (*Smicronyx fulvus* LeC. and *S. sordidus* LeC.) may occur in the field, and by the Indian meal moth (*Plodia interpunctella* Hübner) and other insects in storage. Insect resistance dependent on the armor layer (third layer) of the hull may reduce field damage due to *H. electellum* and *P. hospes* according to Carlson et al. (4). Most nonoilseed cultivars now grown lack the armor layer. Fumigating and aerating storage structures to maintain temperature and humidity at safe levels should lessen damage from storage insects.

Hulling Sunflower Seed

Medium size seed is hulled and the resulting kernel used in confectionery and bakery products. In this size seed quality requirements are: (a) freedom from other plant parts, soil, stones, weed seed, and seed of other crops; (b) uniformity of size; (c) relatively high kernel to hull ratio; (d) uniform size and color of kernels weighing 8.0 g/100 or more; and (e) uniformly light brown kernels when roasted.

Prior to hulling, seed is passed through an air separation chamber to remove foreign matter of similar size, shape, and specific gravity to sunflower seed. The seed is hulled in a drum-shaped chamber. Seed is fed into the top and drops onto a revolving plate that impels it against the smooth inner wall of the drum by centrifugal force (22). Feed rate and plate speed are varied to obtain the maximum percentage of undamaged kernels. A moisture percentage of 10 to 11% will result in the greatest recovery of undamaged kernels. Lower seed moisture leads to broken kernels, and higher seed moisture increases hull resistance to cracking.

According to Lofgren (14) the genetic constitution of sunflower cultivars also affects ease of hulling and percentage recovery of kernels (Table 2). The author studied four cultivars: Sundak, D747, D508, and D509. They showed 54.5, 52.8, 50.3, and 49.6% kernel recovery, respectively, for the seed size over the 7.1 mm sieve.

The hulls and the kernels are separated by air and gravity. Kernels then pass over mechanical color sorters equipped with electric eyes to remove dark kernels or kernels with adhering fragments of the hull. Next, kernels go over a vibrating table for visual inspection and removal of any unacceptable kernels with small vacuum hoses. The final raw product is stored at 10 C until roasted.

Storage fungi attack hulled sunflower kernels, according to Talley et al. (24). The hulled kernels are perishable, especially if the storage temperature exceeds 10 C (6) and the moisture percentage exceeds 5.2%. Other requirements to maintain high quality of kernels in storage are: (a) separation from other food or non-food items that may transfer odors or flavors to the kernels; (b) protection from rodents which could contaminate the product; and (c) prevention of invasion by storage insects by using approved sanitation methods.

Table 4—Effect of moisture percentage and storage time at 22 to 35 C, after previous storage for 6 months at 10 C, on percentage of sunflower kernels infected by storage fungi *Aspergillus glaucus* (Link) or *A. restrictus* (Smith). Adapted from Christensen (6).

Sample no.	0 days		46 days		79 days		113 days	
	Moisture	Kernels infected†	Moisture	Kernels infected†	Moisture	Kernels infected†	Moisture	Kernels infected†
					%			
1	5.53	4	5.99	0	5.15	0	5.00	0
4	6.02	4	5.74	2	5.68	0	5.32	2
6	6.25	2	5.84	12	5.60	8	5.49	14
9	6.46	0	6.87	10	6.34	8	5.84	4
11	6.74	2	7.01	18	6.22	6	6.19	10
12	6.99	6	6.81	6	5.96	22	5.84	26
15	7.26	4	7.45	84	7.09	100	6.42	96
17	7.47	8	7.77	100	6.81	96	6.86	100
20	7.64	0	7.47	100	6.73	90	6.84	98
Mean‡	6.70	3	6.78	37	6.17	36	5.83	37
S.E.‡	0.68	3	0.80	42	0.61	43	0.62	42

† Percentage of kernels infected by storage fungi *Aspergillus glaucus* (Link) or *A. restrictus* (Smith) after surface disinfection using 1 min of shaking in 2% NaOCl, rinsing with distilled water, placing on T6 agar, and incubating at 27 C.
‡ Mean and standard error (S.E.) calculated for 20 samples.

Christensen (6) sampled sunflower kernels from twenty 22.7 kg (50 lb) bags that had been stored at a constant 10 C for 6 months. Kernel moisture ranged from 5.53 to 7.64% (Table 4). When the kernels were surface-disinfected immediately after sampling, colonies of *Aspergillis glaucus* (Link) developed on 0 to 12% of the kernels plated on T-6 agar (Difco tomato juice agar, 25 g; Difco agar, 15 g; NaCl, technical grade, 60 g; distilled water, 900 ml) incubated at 27 C. Under storage in air-tight containers for 46 to 113 days at laboratory temperatures of 22 to 35 C, *A. glaucus* and *A. restrictus* (Smith) formed mycelium on kernels, especially when moisture exceeded 6.5% (Table 4). This indicated that surface-disinfection failed to destroy mycelia within kernel tissue, and when storage conditions were optimum for fungal development, additional growth of the fungi occurred. Further, when 12 samples were drawn from each of two additional bags stored at a constant 10 C, the range in kernel moisture was 6.88 to 7.90%, or about one-half the range found in the other 20 bags, indicating the variation in kernel moisture percentage was greater among bags than within individual bags. Equilibrium moisture percentage of kernels was 7 to 8, 8 to 9, and 10 to 11% when held at 22 to 25 C and 75, 80, and 85% relative humidity, respectively. Christensen (6) concluded that the moisture percentage of stored sunflower kernels should not exceed 6.5%, and relative humidity of storage containers should not exceed 70% to minimize hazards from storage fungi.

Aflatoxin B_1 and other aflatoxins are highly toxic and carcinogenic metabolites produced by *Aspergillus* spp. and other fungi according to Jones (11). Nagarajan et al. (19) inoculated rehydrated whole or broken seed with two isolates of *A. flavus* (Link) and three of *A. parasiticus* (Speare) and incubated at 28 C for 7 days. They found aflatoxin B_1 production differed for both cultivar of sunflower and fungal isolate. Isolate

NRRL (Northern Regional Research Laboratory, Peoria, Illinois) 2999 of *A. parasiticus* and NIN (National Institute of Nutrition, Hyderbad, India) 195 of *A. flavus* produced maximum levels of 15.6 and 0.1 ppm aflatoxin, respectively. Isolate NRRL 2999 produced only 6 to 16 ppm aflatoxin on sunflower kernels compared to 20 to 30 ppm on soybeans [*Glycine max* (L.) Merr.] and 25 to 50 ppm on hulled peanut (*Arachis hypogaea* L.). Broken seed and whole seed of sunflower produced 33 and 11 ppm alfatoxin, respectively, indicating that the armor layer of the hull or the hull itself inhibits the fungus.

The storage conditions for sunflower kernels influenced their acid value, color, flavor, odor, and their invasion by microflora according to Robertson and Thomas (23). Acid value, as measured by the milligrams of potassium hydroxide necessary to neutralize 1 g of extracted oil, increased at kernel moisture percentage of 5.2, 10.5, and 14.7%, especially when kernels were stored for 12 weeks at 24 and 35 C (Fig. 1). Kernels stored for 6 weeks at the two higher moisture percentages had poor flavor, dark color, and also musty odor in some samples. Kernels with an acid value of 4 or more had a distinct acidic or sour flavor. Kernels stored at 10.5% moisture and 24 and 35 C were dark and inedible after 2 weeks. Microflora counts generally decreased during storage because oxygen was depleted and free fatty acids increased. These acids with carbon chains ranging from 9 to 12 are bactericidal. The predominant microorganisms on the surface of sunflower kernels after longer storage periods were facultative anaerobic yeast rather than bacteria. The data presented by Robertson and Thomas (23) indicate the importance of storing kernels at low moisture of 5.2% and low temperature of 2 C to maintain their flavor and color.

Robertson and Thomas (23) compared sunflower seed from three processors and found that the plate counts of fungi of the kernels were only 2 to 8% of those of whole seed. They concluded that the hull has high surface contamination of microorganisms but hulling apparently fails to transfer these microorganisms to the kernel.

Laboratory Methods to Evaluate Hulling Quality

Hulling quality is assessed by passing 100 g of seed at the feed rate of about 600 g/min through a laboratory huller which is a miniature model of a commercial huller. Next, the hulls and kernels are separated with a bench cleaner using a 3.1 mm (8/64 in) round-hole sieve below to allow small fragments of the hull to pass through and forced air to raise and remove the large hull fragments. Then the kernels are collected and scored visually for percentage of whole seed remaining. This percentage is called the "hullability" rating. Lofgren (14) used this method on four sizes of seed from a yield trial and found large seed hulled easier than small seed (Table 2).

Lofgren (14) in a further study of the same yield trial found the percentage yield of marketable kernels depends on the genotype and the specific size of the seed being hulled (Table 2). Yield of kernels among cultivars varied from 49.6 to 54.5%, for seed passing through a 7.9 mm and over a 7.1 mm sieve. As seed size increased the percentage yield of kernels decreased, e.g., the range in D747 was from 43.1% for the largest seed to 49.6% for the smallest seed. Hybrid D508 has a thinner hull than D747 and a higher kernel recovery, 54.5 and 49.6%, respectively, for the same size of seed. Although cultivars with a thin hull are generally desirable, the hull can be so thin or fragile that it breaks in normal handling of whole seed. Kernels from hulled seed of such cultivars may be lost or damaged during harvest and processing and the yield of marketable, high quality seed reduced. Special handling to recover these kernels prior to microbial contamination would add to the cost of processing.

Environment influences the percentage of hull of sunflower seed. Clandinin (7) noted that sunflower seed produced on soils of high clay content in northern Europe had a higher percentage of hull than those grown

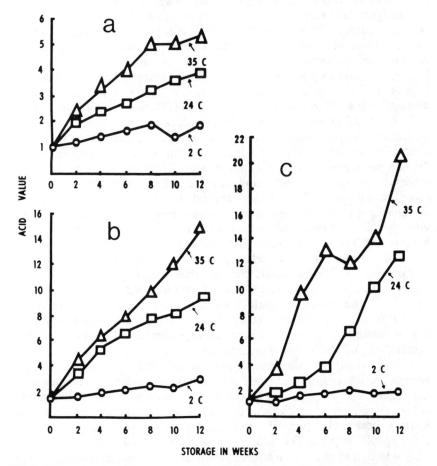

Fig. 1—Effect of storage temperature and duration on the acid value (mg of KOH required to neutralize 1 g of oil of the oil extracted from dehulled raw sunflower kernels with (a) 5.2, (b) 10.5, and (c) 14.7% moisture. Adapted from Robertson and Thomas (23).

on sandy soils in southern Europe.

Generally, kernel weight increases as seed size increases (14) (Table 2). Volume weight of kernels showed no significant association with whole seed size. Volume weight of kernels might be expected to decrease from premature frost or severe disease because kernels would lack complete development and therefore be lower in density.

Zimmer and Zimmerman (25) reported that the kernel percentages from healthy plants were significantly higher than those from diseased plants for the cultivars Commander and Mingren in the field (Fig. 2). Sunflower rust (*Puccinia helianthi* Schw.), head rot (*Rhizopus* spp.), or downy mildew [*Plasmopara halstedii* (Farl.) Berl. et de Toni] reduced kernel percentage from 54.0 to 51.0%, 52.5 to 49.5%, and 49.5 to 47.8%, respectively. Achene (seed) weights were reduced significantly by disease as shown in Fig. 2. Infection by *Rhizopus* spp. influenced seed weight greatly but affected seed size slightly according to the authors (25).

Lofgren (14) scored roasted kernels for dark count and color as reported in Table 2. Dark count was rated visually from 0 to 9, with 0 indicating no dark kernels and 9 indicating all dark kernels. Kernel samples of 15 g were roasted in vegetable oil at 177 C for 4 min and then cooled to 20 C. High dark count may be associated with frost damage before seed matures, heat damage during seed drying or storage, or effect of seed infection by diseases such as *Rhizopus* spp. (25). Lofgren (14) rated kernel color from 1 to 5 with 1 indicating light tan and 5 very dark brown (Table 2). His data indicated differences among sizes of seed and among cultivars as well as an interaction between cultivar and size. The color score of roasted kernels of Sundak was higher than of D747 and increased with increasing seed size but tended to decrease with increasing seed size in D747. Lofgren (14) also observed variation in color in all cultivars and all sizes which may be caused by (a) genetic impurity of the cultivar, (b) variation among heads due to environmental differences, (c) lack of uniformity in moisture content of the kernels, or (d) lack of uniformity in kernel size.

Color of dry roasted kernels at 177 C, determined by a Gardner Automatic Color Difference Meter, changed over roasting time, according to Talley et al. (24). The measure of color brightness, *Rd,* declined from 22.3 to 7.3 for 5 and 30 min of roasting, respectively. Values of *a,* or redness, increased from 3.2 at 5 min to 8.0 at 15 min and then decreased to 6.3 at 30 min of roasting. Yellowness, or *b,* values were highest at 19.9 for 10 min but declined to 13.8 for 30 min of roasting time.

Talley et al. (24) also investigated the physical changes of sunflower kernels during roasting and when used in bakery products. They measured the "chewiness" of sunflower kernels by determining the shear value. Shear value of 1 g samples of dry roasted kernels was determined using an Allo-Kramer shear press equipped with a standard shear compression cell and a 226.2 kg (500 lb) ring. The shear value decreased as roasting time increased from 0 to 20 min at 177 C in a forced draft oven. Beyond 20 min the kernels became brittle.

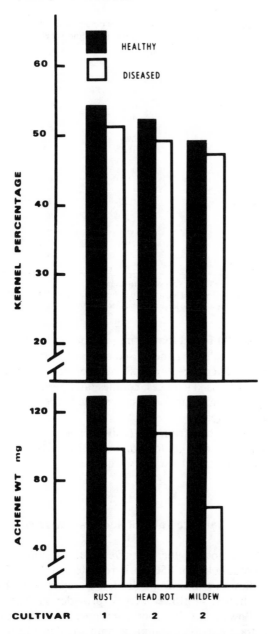

Fig. 2—Comparison of kernel percentages and achene (seed) weights from healthy sunflower plants and plants affected with rust, head rot, and downy mildew. Means of 25 random plants, cultivar 1 = Commander and 2 = Mingren. Adapted from Zimmer and Zimmerman (25).

Table 5—Some physical and chemical characteristics on dry basis (DB) of whole seed, hull, and kernels of selected sunflower cultivars†. Adapted from Earle et al. (10).

			Mingren§	
Structure	Arrowhead	Mingren	Large	Small
Seed				
g/1000 seeds‡	91	102	180	77
Hull, %	44	44	52	43
Moisture, %	7.2	6.8	7.1	8.2
Oil, % (DB)	29.8	31.2	21.0	30.7
Protein, N × 6.25, % (DB)	18.1	15.9	19.0	17.2
Hull				
Moisture, %	9.3	6.9	6.4	7.7
Oil, % (DB)	0.9	0.4	0.5	1.3
Protein, N × 6.12, % (DB)	2.5	1.7	4.1	2.9
Kernel				
g/1000 kernels	51	57	86	44
Moisture, %	9.8	5.8	5.7	4.4
Oil, % (DB)	54.5	54.4	46.7	53.0
Protein, N × 6.25, % (DB)	30.8	26.6	36.4	26.6

† From 1966 crop grown in northern U.S. or southern Canada.
‡ Calculated from weights of kernel and hull to avoid inclusion of empty or insect damaged seed.
§ One sample of Mingren separated into large and small seed fractions.

Talley et al. (24) found that 15 min of roasting at 177 C was optimum as scored by a trained taste panel based on flavor, color, texture, odor, appearance, and product acceptance. Sunflower kernels imparted a definite nut-like flavor when used in cakes, cookies, and pies. Chopping or grinding before baking was sometimes necessary as the kernels were too crisp or the flavor too concentrated.

Some information from extensive chemical analyses of whole sunflower seed, hulls, and kernels conducted by Earle et al. (10) is reported in Table 5. They noted that kernels were especially high but the hull low in oil and protein.

Nutritive Value of Sunflower Kernels

Some information on the nutritive value of sunflower kernels from Agriculture Handbook No. 456 (1) is reported in Table 6 along with the daily dietary allowances for an adult male as reported by Bogert et al. (2). One pound of hulled sunflower kernels provides sufficient quantities for daily dietary allowances of all food components, as specified by Bogert et al. (2) except food energy, calcium, vitamin A, and riboflavin. They indicated that although protein quantities may be adequate, there may be an imbalance of essential amino acids. One should consult Agriculture Handbook No. 456 (1), Bogert et al. (2) or other books on nutrition, and one's doctor before large amounts of sunflower kernels are used in the diet.

Table 6—Nutritive values for household measures and market units of raw unsalted sunflower kernels and recommended daily dietary allowances for the adult male.

Food, approximate measures, units & weights	Values for edible part of foods					
	Water	Food energy	Protein	Fat	Carbo-hydrates	Calcium
	%	cal	g	g	g	mg
In hull (54% kernels)†						
454 g (1⅓ cups kernels)	4.8	1,371	58.8	115.8	48.7	294
1 cup (⅓ cup kernels, 35 g)	4.8	257	11.0	21.7	9.1	55
Hulled‡						
454 g (1 lb)	4.8	2,540	108.9	214.6	90.3	544
1 cup (145 g)	4.8	812	34.8	68.6	28.9	174
Daily dietary allowances§						
Adult male (22–35 years)	--	2,800	65.0	--	--	800

	Phos-phorus	Iron	Sodium	Potas-sium	Vit. A	Thia-min	Ribo-flavin	Niacin
	mg	mg	mg	mg	IU	mg	mg	mg
In hull								
454 g	2,050	17.4	73	2,253	120	4.80	0.56	13.2
1 cup	384	3.3	14	422	30	0.90	0.11	2.5
Hulled								
454 g	3,797	32.2	136	4,173	230	8.89	1.04	24.5
1 cup	1,214	10.3	44	1,334	70	2.84	0.33	7.8
Daily allowances	800	10.0	--	--	5,000	1.40	1.70	18.0

† Measure and weight with hull included, adapted from Adams (1).
‡ Adapted from Adams (1).
§ Adapted from Bogert et al. (2), blank values no daily allowances.

Uses for Sunflower Kernels in Cooking

Millette (18) listed several uses of sunflower kernels such as (a) blending with honey, butter, and salt to make a spread; (b) a nut substitute in candies, cookies, and muffins; (c) sprinkling on syrup on pancakes or waffles; (d) blending in cake frosting; (e) adding to meat, fish, and vegetable dishes; (f) an ice cream topping; (g) a fondue dip; (h) adding to or sprinkling on salads; and (i) a snack with most beverages. Millette (18) tested recipes using sunflower kernels. Recipes are available from several sources including Fisher Nut Company, St. Paul, MN 55114; Dahlgren and Company, Crookston, MN 56716, and Sigco Sun Products, Breckenridge, MN 56520.

Quality of Birdfeed (Petfood) Sunflower

Small sunflower seeds which pass through a 7.1 mm (18/64 in) round-hole sieve generally are used for birdfeed, either pure or in various blends. Blends contain different percentages of nonoilseed sunflower, oilseed sunflower, wheat, oat, corn, proso millet, and/or grain sorghum. Ideally, the sunflower in birdfeed should be (a) black and white in color, bright, and clean; (b) plump with well developed seed and kernel; (c) free of extraneous matter such as sunflower plant parts, weed seed, gravel and soil; (d) of relatively low moisture level to allow storage at a wide range of temperatures without spoilage; and (e) free from insects or insect damaged seeds—in short, an attractive product to persons who feed birds or animal pets.

Weathering of seed may cause the hull to stain and the outside layer to break off. A killing frost prior to maturity causes brown seeds which are low in volume weight, and hence difficult to pack. Dockage is more difficult to remove from the smaller sunflower seed used for birdfeed, which often includes whole and broken kernels.

The moisture percentage of sunflower seeds for birdfeed should be below 7% since they are stored at varying temperatures from 0 C or lower under winter warehouse conditions to room temperature in retail outlets and the homes of people feeding sunflower seed. Sunflower seed should be fumigated with a registered pesticide before packaging because insect activity increases as storage temperature increases.

BYPRODUCTS

Kernel Fragments

Kernel fragmentation occurs in hulling of seeds, especially if the moisture percentage of whole seed is below 10%. These fragments referred to as "chips" and "halves" can be used (a) as baking ingredients, especially if they are pure and well preserved; (b) in blended bird or animal pet feed; or (c) as a source of sunflower oil.

Sunflower Meal

Sunflower meal is a high protein byproduct, especially of oilseed sunflower, which is nutritious and valuable for human food or livestock feed. More information on sunflower meal can be found in Chapter 13.

Sunflower Hulls

The hulling process provides large amounts of hulls as byproducts. Several uses for hulls include: (a) livestock and poultry bedding (7); (b) a roughage component either pulverized or whole in ruminant feeds (8, 9, 12, 13, 16); (c) compressed fireplace logs (21); (d) fertilizer, since the ash is high in phosphorus and potassium (3, 15); (e) the production of furfural, as sunflower hulls contain 14% less pentosans, 2% less cellulose, and 13% more lignin than corn cobs (10, 15); (f) the extraction of reducing sugars (3, 15);

(g) the extraction of waxes (15, 17); and (h) the manufacturing of agglomerated panel boards (3). Additional information on sunflower hulls can be found in Chapter 13.

Chemical Composition of Sunflower Hulls

Cancalon (3) found sunflower hulls contained 5.17% lipids, including wax, hydrocarbons, fatty acid, sterols, and triterpenic alcohols; 4.0% protein which was 13.4% aspartic acid, 10.1% glutamic acid, 9.1% hydroxyproline, 8.0% alanine, and lesser amounts of other amino acids; 50% carbohydrates, principally cellulose and lignin; 25.7% reducing sugars of which 52% was xylose, 27% arabinose, 12% galacturonic acid, and 9% galactose; 8.2% moisture; and 2.1% ash.

Sunflower Hulls as Feed for Ruminants

Finely ground sunflower hulls mix readily with other feed ingredients, absorb liquids such as molasses, and can be pelleted for easy handling according to Kinard (13).

Jordan and Hanke (12) reported on lambs fed diets of sunflower hulls and of alfalfa hay (*Medicago sativa* L.), with other ingredients identical. The lambs on sunflower hulls gained at 94.1% of the rate of those on alfalfa. Lambs readily consumed 0.45 kg (1 lb) per day of pelleted sunflower hulls containing 14% protein within 3 or 4 days after feed adjustment trials started. Dinusson (8) reported lambs digested 48% of the ration dry matter when fed alfalfa but when nonoilseed sunflower hulls were substituted for one half the alfalfa the dry matter digestibility dropped below 30%. Treatment of sunflower hulls with 4 or 8% sodium hydroxide increased digestibility as a lamb roughage but such treatment failed to be of practical benefit.

Dinusson et al. (9) reported that sunflower hull pellets are high in acid detergent fiber and lignin content suggesting a low usable energy content. Therefore, sunflower hull pellets may compose a portion of the roughage component of a ruminant ration consistent with adequate energy intake of the animals for body maintenance, weight gain, or milk production. Heifers fed pelleted sunflower hulls alone died from impacted abomasum, because of the low digestibility and slow rate of passage, according to Dinusson (8). Marx (16) reported dairy heifers died of impaction on a ration of 50% sunflower hulls and 50% corn silage supplemented with urea as the roughage and 1.4 kg (3.0 lb) barley per head daily. Dairy steers remained on the same trial for 90 days with no ill effects with an average daily gain of 0.89 kg compared to 1.07 kg for a corn silage and barley (*Hordeum vulgare* L.) ration.

LITERATURE CITED

1. Adams, C. F. 1975. Nutritive value of American foods. USDA Agric. Handb. 456.
2. Bogert, L. J., G. M. Briggs, and D. H. Calloway. 1973. Nutrition and physical fitness. W. B. Saunders Co., Philadelphia, Pennsylvania.
3. Cancalon, P. 1971. Chemical composition of sunflower seed hulls. J. Am. Oil Chem. Soc. 48:629–632.
4. Carlson, E. C., P. F. Knowles, and J. E. Dillé. 1972. Sunflower varietal resistance to sunflower moth larvae. Calif. Agric. 26:11–13.
5. Christensen, C. M. 1971. Evaluating condition and storability of sunflower seeds. J. Stored Prod. Res. 7:163–169.
6. –––––. 1972. Moisture content of sunflower seeds in relation to invasion by storage fungi. Plant Dis. Rep. 56:173–175.
7. Clandinin, D. R. 1958. Sunflower seed oil meal. p. 557–575. In A. M. Altschul (ed.) Processed plant protein foodstuffs. Academic Press, Inc., N.Y.
8. Dinusson, W. E. 1977. Sunflower by-products as feeds. p. 1–2. In Proc. 2nd Sunflower Forum. Sunflower Association of America, Fargo, N.D.
9. –––––, C. N. Haugse, D. O. Erickson, and R. D. Knutson. 1973. Sunflower hull and corn roughage pellets, triticale and ergot in rations for beef cattle. North Dakota Farm Res. 30(4):35–39.
10. Earle, F. R., C. H. VanEtten, T. F. Clark, and I. A. Wolff. 1968. Compositional data on sunflower seed. J. Am. Oil Chem. Soc. 45:876–879.
11. Jones, B. D. 1972. Methods of aflatoxin analysis. Tropical Products Institute, London. 58 p.
12. Jordan, R. M., and H. E. Hanke. 1969. Self feeding, pelleted sunflower hulls, ensiled corn and feeding of antibiotics for feedlot lambs. Feedstuffs p. 26–28 (22 March).
13. Kinard, D. H. 1975. Feeding value of sunflower meal and hulls. Feedstuffs. p. 26–31 (3 November).
14. Lofgren, J. R. 1976. Seed quality of confectionery sunflowers (Helianthus annus L.). p. 2–3. In Proc. 1st Sunflower Forum. Sunflower Association of America, Fargo, N.D.
15. Mantha, K. S., and V. V. R. Subrahmanyan. 1973. Sunflower seed and oil: a review. J. Oil Tech. Assoc. India. 5:11–17, 26–34.
16. Marx, G. D. 1977. Utilization of sunflower silage, sunflower hulls with poultry litter and sunflower hulls mixed with corn silage for growing dairy animals. J. Dairy Sci. 60 (Supp I):112–113.
17. Meerov, J. S. 1974. Obtaining of waxes from sunflower hull. Maslo-Zhir. Prom. 7:30–32.
18. Millette, R. A. 1974. Seeds from the sunflower. North Dakota State University, Fargo, Cir. HE-120. 3 p.
19. Nagarajan, V., R. V. Bhat, and P. G. Tulpule. 1974. Aflatoxin production in sunflower (Helianthus annuus) seed varieties. Current Sci. 43:603–604.
20. Putt, E. D. 1940. Observations on morphological characters and flowering processes in the sunflower (Helianthus annuus L.). Sci. Agric. 21:167–179.
21. –––––. 1963. Sunflowers. Field Crop Abstr. 16:1–6.
22. Robertson, J. A. 1975. Use of sunflower seed in food products. Crit. Rev. Food Sci. Nutr. 6:201–240.
23. –––––, and J. K. Thomas. 1976. Chemical and microbial changes in dehulled confectionery sunflower kernels during storage under controlled conditions. J. Milk Food Technol. 39:18–23.
24. Talley, L. J., B. J. Brummett, and E. E. Burns. 1972. Sunflower food products. Texas Agric. Exp. Stn. MP1026. 10 p.
25. Zimmer, D. E., and D. C. Zimmerman. 1972. Influence of some diseases on achene and oil quality of sunflower. Crop Sci. 12:859–861.

JAMES R. LOFGREN: B.S., M.S., Ph.D.; Agronomist-Plant Breeder, Dahlgren and Company, Crookston, Minnesota. Sunflower breeder and agronomist; production and simulated hail damage studies on sunflower while with the University of Minnesota; and development and production of nonoilseed and oilseed sunflower hybrids with private industry.

Chapter 15

Future of Sunflower as an Economic Crop in North America and the World

HARRY O. DOTY, JR.

The economic future of sunflower (*Helianthus annuus* L.) in North America and the world must be appraised in comparison to its major oilseed competitors. This appraisal must consider the relative value or desirability of sunflower products and factors affecting their markets.

Two distinct types of sunflower are produced, the oilseed and the nonoilseed. The farmer selects the cultivar for his intended market. Sunflower production in the USA in 1977 was about 90% oilseed and 10% nonoilseed types. Oil is the main product of the oilseed type, with meal an important byproduct. The edible kernel is the main product of the nonoilseed type, and hulls are a minor byproduct.

Oilseed sunflower produces two kinds of oil, one high in polyunsaturates (high in linoleic acid) and another high in monounsaturates (high in oleic acid) depending primarily on the temperature during the growing season. Sunflower oil high in linoleic acid generally is produced in Canada and USA in the northern growing areas or southern areas from late plantings in July and August, while oil high in oleic acid is produced in the South from spring plantings, primarily in April and May. Sunflower seed containing different kinds of oil should be kept separate so that the unique characteristics of each may be used.

Hulls may be removed from oilseed sunflower before oil extraction or seeds may be left intact, depending on the extraction process used. Sunflower oil usually is removed from the seed by screw pressing followed by solvent extraction. The remaining product is sunflower meal. Hulls also may be removed from the meal after the oil has been extracted. Sunflower meal and hulls are used in livestock feed. Sunflower meal can be processed further to produce edible high protein products which may be utilized in

From: Carter, Jack F. (ed.). 1978. *Sunflower Science and Technology.* Agronomy 19. Copyright © 1978 by the American Society of Agronomy, Crop Science Society of America, and Soil Science Society of America, 677 South Segoe Road, Madison, WI 53711 USA.

many food products; however, the process is not now commercially feasible. When sunflower oil is refined, a byproduct, called refiner's foots or soapstock, is produced. This product is utilized in manufacturing livestock feeds, fatty acids, and soap.

Nonoilseed sunflower seeds are roasted and marketed whole, sometimes with a coating of salt, or the seed may be dehulled to produce edible kernels for the "nut" and confectionery markets. Hulls are a byproduct. About 40 to 50% of the nonoilseed production is processed for human food, with the remainder sold for bird and petfood (12, 33). The bird and petfood market uses mostly the small nonoilseed type sunflower but also uses some oilseed sunflower.

Due to quantity of oilseed sunflower produced and the rapid increase in USA oilseed sunflower production in recent years, the future of sunflower as an economic crop in the world is aligned closely with oilseed sunflower. Therefore, this assessment of the future for sunflower relates primarily to oilseed sunflower.

CHANGES IN THE OILSEED INDUSTRY

The oilseed industry, like many others, changes constantly. These changes will continue in the future. Far-reaching changes for the oilseed industry appear more imminent than ever before (14).

Increased Food Production

Increased food production from oilseeds was discussed at the World Food Conference in Rome in the fall of 1976. If international or governmental agencies decided to increase food production to provide adequate food or prevent mass starvation in the world, the question arises as to which oilseeds production should be expanded. Little attention would be paid to byproducts such as cottonseed oil, corn oil, lard, tallow, etc., since the quantity of byproducts available depends upon production of the main product. Some byproducts, however, could make a new and important contribution to our food supply. For example, a change to the production of gossypol-free, glandless cotton (*Gossypium* spp.) would make all cottonseed meal presently used in feed for ruminants available for all feeds and edible protein uses. Under conditions of high demand for food production, areas planted to cotton might be decreased drastically so that food crops could be grown on cotton land and the world would rely more on synthetics to supply its fiber needs.

If more low-cost edible vegetable protein appeared most needed, then soybean (*Glycine max* L. Merr.) probably would be chosen for increased production. Soybean produces a high yield of protein having the best amino acid balance for food among the vegetable proteins. If the decision were made to increase oil production, then sunflower might be chosen for in-

Table 1—Average yield and value per hectare of four oil crops for selected countries, 1971-75. After Boutwell et al. (11).

Crop	Country	Yield		Value†		
		Oil	Meal	Oil	Meal	Total
		——— kg/ha ———		————$/ha———		
Oil palm	W. Malaysia	3,895‡	560	1,576	67	1,643
Soybean	USA	323	1,426	153	247	400
Sunflower	USSR	590	738	341	101	442
Peanut	USA	828	1,090	556	175	731
Peanut	Nigeria	226	274	151	44	195

† Using 1971-75 average prices for Europe.
‡ Includes 3,475 kg palm oil and 420 kg palm kernel oil.

creased production in the temperate zones and oil palm for the tropics (Table 1). Since soybean, sunflower, and oil palm are grown in different locations the decision might be made to expand production of all three crops.

In the USA, oil yield per hectare of oilseed sunflower is higher in some areas than the yield shown in Table 1. Increased oil production in the USA is attributed to increased use of hybrid oilseed sunflower, which has about 25% greater yield per hectare than open-pollinated cultivars grown previously. This yield increase greatly improved the competitive position of sunflower. A great wealth of unexplored genetic material exists for use in improving sunflower. Using modern breeding methods, improved sunflower hybrids are expected to have better disease and insect resistance. Increased sunflower seed production per hectare, as well as increased oil content of the seed, are anticipated. Both result in increased sunflower oil production per hectare and make sunflower a more economically viable crop.

The oil palm produces higher oil yields than any oilseed crop in the world (Table 1). Oil palm trees, however, require 4 years from planting to first oil production. Hence there is a delay in meeting food emergencies because of the inelastic supply of palm oil. Further, palm oil is a saturated oil of lower food quality than sunflower oil.

Oil palm is a tropical crop that grows in a narrow latitude near the Equator in an area receiving about 200 cm of rainfall evenly distributed during the year. Sunflower is more widely adapted than most major economic crops. It is grown over greater latitudes, with less rainfall, and a shorter growing season. It also can be double cropped and farmers can use grain planting and harvesting equipment with slight modifications for sunflower production.

Relative Value of Oil and Meal

The practice of growing cultivars that produce seeds of higher oil concentration, along with more efficient extraction processes, have resulted in higher yields of oil per ton of seed, although the relative value of the meal has increased more than that of the oil (15). This change was caused by increased demand and higher prices for protein meals to feed livestock, the greatly increased supplies of fats and oils at lower prices (resulting from increased world production), and the large displacement of fats and oils by petroleum-based synthetics. The values of the sunflower oil and meal are now about 75 and 25%, respectively. Because sunflower is grown primarily for its oil, this change in value relationships does not favor sunflower compared to other oilseeds, such as soybean, which derived about 60% of its value from the protein meal and 40% from the oil in recent years (22).

Synthetics

Synthetic products made from petroleum have replaced oilseed products in many uses. Most substitution has occurred in the industrial markets, although some synthetics have been approved by the Food and Drug Administration for food and feed. Synthetics now dominate two formerly large markets for fats and oils, the paint and coating and the detergent industries. The glycerin market was another large, fat-related market largely lost to synthetics. Synthetic urea and other nonprotein nitrogen products have displaced large quantities of oilseed meals in livestock feeds and will continue to do so in the future (15).

In recent years people have become more concerned about the possible pollution of our environment by using synthetics. Consequently, these concerns and recent new laws regarding air and water pollution help natural fats and oils and their products compete in some markets. Synthetics may lose some of their current markets in food with the greater emphasis placed on consumer safety. Also, some new and improved fat-based products may regain some old markets and capture some new markets for fats and oils.

Synthetics in the past have made large inroads in industrial fats and oils markets. Recent and pending large increases in petroleum prices may result in natural fats and oils products being more competitive with petroleum-derived synthetics.

Substituting Fats and Oils

Each fat or oil has unique characteristics for specific uses, but there is a high degree of interchangeability among them for many uses. Additional processing with increased cost, however, may be necessary for a specific oil to substitute for another oil, such cost determining whether substitution is economical. Small price differences among competing oils can change the proportions of oils used in the product. Complete substitution of one fat or oil for another is impossible in many products, but usually a lower rate of substitution may be feasible. Blends of several fats and oils sometimes are needed to obtain an acceptable substitute for a specific fat or oil.

Not all fats and oils are suitable for food, because some are toxic. Vegetable oils usually are "refined" before they are suitable for food. Some oils, however, are not suitable for food even after refining. Sunflower oil is a "high grade" food oil, and sells at a higher price than many other fats and oils. Food fats and oils also can be used in animal feeds or industrial products, but usually at a lower price. Tallow is classified as edible or inedible in the USA based on federal regulations regarding the source of the raw material used in its manufacture. Linseed oil usually is used industrially because it oxidizes (producing bad flavor) rapidly, although it has been used in the USA for food when needed.

Whether a fat or oil is of vegetable, animal, or marine origin makes no difference for industrial use or feed and is only of limited consideration for food. Oil origin, however, is important to the kosher and vegetarian trades, where vegetable oils have an advantage over animal fats.

New Processing Methods

The methods used in processing oilseeds for oil and meal have changed greatly over the years. Countries in the world are in different stages of economic development; therefore, advanced methods may be used in some countries, antiquated methods in others. Some innovations have improved the quality of oils, or permitted substitution of one oil for another oil in certain uses. Oil substitution may cause economic changes throughout the oilseed industry and finally result in changes in the proportion of oilseed crops grown by farmers around the world. For example, in 1948, soybean oil composed 35% of the vegetable oils used in the manufacture of margarine in the USA, and manufacturers believed that technical problems would restrict soybean oil composition of margarine to 50%. After processing innovations, however, soybean oil comprised 87% of the vegetable oil in margarine in the USA by 1958.

Innovations still occur today. Recent research has fractionated palm oil, and one fraction is a liquid oil comparable to peanut or olive oil (11). Manufacturers can produce a liquid palm oil with specific desirable characteristics at low temperatures. This process is used in Europe now and may be used in the USA in the near future. Processed palm oil may now compete in the cooking and salad oil market, a major sunflower market area, from which it was excluded previously. These liquid palm oils can be used in frying, margarine, cracker spray, and salad oils.

Future innovations may favor one oil over another oil. Many such innovations benefit lower grade oils as they are improved or "upgraded." Sunflower oil is already one of the highest quality edible oils as extracted without further processing. Most of the processing innovations are designed to improve oils which then compete with sunflower and other high quality natural oils.

Polyunsaturated Fatty Acids

A steady market for premium priced fat products high in polyunsaturation has developed in recent years because of popular concern about high blood cholesterol and heart attacks. Experiments with humans have shown that blood cholesterol levels can be reduced by decreasing total fat intake and substituting polyunsaturated fatty acids for saturated ones (1, 2). Many people eat polyunsaturated fat products because they may have high blood cholesterol levels, they are under doctor's orders, or they choose them as a preventative measure. If it were established scientifically that high cholesterol levels cause heart attacks, rapid expansion in the polyunsaturated food fat market would occur. Then, high linoleic acid sunflower oil and safflower oil with their high degree of unsaturation relative to other food oils (Table 2) would be in great demand.

The potential market in the USA for the saturated vegetable oils, such as unfractionated coconut oil and palm oil, is not as promising as a few years ago, and are restricted or not used by some institutions because of their high saturation. The School Lunch Program of the USDA limits the use of saturated vegetable fats in their engineered food program; e.g., in the filled cheese product of the School Lunch Program, the lipids shall not contain more than 50% of saturated fatty acids (5). Their breakfast cake must contain at least 5% of the total calories as linoleic acid (4).

The U.S. Congress's Senate Select Committee on Nutrition and Human Needs (1977) in its study, Dietary Goals for the United States, recommends that Americans reduce the amount of fat and cholesterol in their diet, eat less saturated fat (both animal fat and hydrogenated vegetable oil), and eat more polyunsaturated fat (35). Currently, polyunsaturated fat accounts for 7% of total caloric intake, a figure that the Committee recommends be increased to 10%. Saturated fat is now 16% and monounsaturated fat 19% of total caloric intake, and the Committee recommends that each of these be reduced to 10%. If Americans put these recommendations into practice, it would expand the demand for the high polyunsaturated sunflower oil.

Linoleic Acid

The linoleic acid in the fat products consumed is important to human health. Linoleic acid is an essential fatty acid for good nutrition and must be obtained from food because it cannot be produced by the human body (13). It is necessary for growth and reproduction and protects against excessive loss of water and damage from radiation (1, 13). It is essential for healthy skin—lack of it can cause dryness and scaling. Linoleic acid also is involved in the metabolism of cholesterol and may aid in the excretion of cholesterol and products resulting from its breakdown (24). Most unsaturated fats are good sources of linoleic acid. High linoleic sunflower oil and safflower oil

Table 2—Fat content and major fatty acid composition of selected fats and oils in decreasing order of linoleic acid content. After Leverton (24) except where noted.

Fats and oils	Total fat	Fatty acids†		
			Unsaturated	
		Saturated‡	Oleic§	Linoleic¶
			%	
Safflower	100	10	13	74
Sunflower, (high linoleic) northern grown, USA	100	11	14	70
Corn	100	13	26	55
Cottonseed	100	23	17	54
Soybean	100	14	25	50
Sesame	100	14	38	42
Sunflower, (high oleic) southern grown, USA (21)	100	9	49	41
Peanut	100	18	47	29
Lard (18)	100	38	46	10
Palm (18)	100	45	40	8
Olive	100	11	76	7
Tallow, beef (18)	100	48	44	2
Coconut	100	80	5	1

† Total is not expected to equal "total fat."
‡ Includes fatty acids with chains from 8 through 18 carbon atoms.
§ Monounsaturated.
¶ Polyunsaturated.

contain 70% or more, corn, cottonseed, and soybean oils contain 50% or more, and high oleic Southern-grown sunflower oil and sesame oil contain 40% or more linoleic acid (Table 2). Other fats and oils such as coconut, palm, and lard contain only 1%, 8%, and 10% linoleic acid, respectively.

OILSEED SUNFLOWER MARKETS

The future of oilseed sunflower may be evaluated by considering supply and demand for the sunflower oil and meal. Oil and meal are obtained from several competing oil-bearing vegetable materials—soybean [*Glycine max* (L.) Merr.], peanut (*Arachis hypogaea* L.), sunflower (*Helianthus annuus* L.), coconut (*Cocos nucifera* L.), oil palm (*Elaeis guineensis* Jacq.), rapeseed (*Brassica napus* L.), cottonseed (*Gossypium* species), flaxseed (*Linum isitatissimum* L.), olive (*Olea europa* L.), safflower (*Carthamus tinctorius* L.), and sesame (*Sesamum indicum* L.). The future of sunflower will be determined by the demand for sunflower oil and its meal byproduct, and its competitive position with other oils and meals.

Food Fat Markets

Fats and Oils

The extracted oil accounts for 75% of the total value of the oilseed sunflower crop and determines the value of sunflower in world markets. This discussion, therefore, will focus on the oil rather than on the meal or other products.

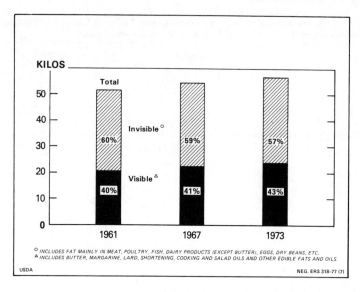

Fig. 1—U.S. Consumption of visible and invisible food fats per capita.

Fats and oils are widely traded international commodities and are highly substitutable one for another. Therefore, prices and trade of fats and oils in one country influence the market for fats and oils in the world. A discussion of some key characteristics of fats and oils and their markets may explain how sunflower oil competes with world oils for the North American and other markets.

Fats and oils can be edible or inedible and are of animal, vegetable, or marine origin. They can be primary products such as sunflower, linseed, and safflower oils or byproducts such as tallow, cottonseed oil, corn oil, and lard. Fats and oils are classified as drying, semidrying, and nondrying, and they can be saturated or unsaturated based on their chemical properties. They are classified as food or industrial, and edible or inedible oils based on their market use.

Trends in Food Fat Consumption in the USA

Total consumption of food fats is comprised of the so-called invisible and visible fats. Invisible fats are consumed as dairy products (other than butter), eggs, animal products (other than lard and edible beef fat), fruits, vegetables, and cereals. Consumption of visible and invisible food fats in the USA increased from about 52 kg/person in 1961 to 57 in 1973—about 0.5 kg/year (Fig. 1). This is based on data computed by the Consumer and Food Economics Research Institute, Science and Education Administration, USDA. Both visible and invisible fat consumption are increasing in the USA. Consumption of the visible kinds of fats, which include butter, lard, margarine, shortening, cooking and salad oils, and other fats and oils is in-

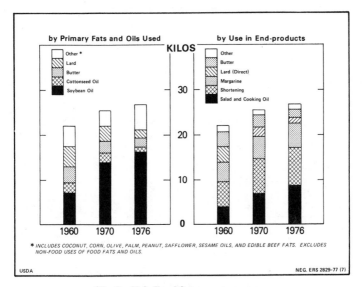

Fig. 2—U.S. Food fat use per person.

creasing more rapidly. Visible fat intake in 1973 accounted for 43% of the total fat intake compared with 40% in 1961. Consequently, invisible fat consumption was responsible for 60% of the total in 1961 and declined to 57% in 1973. The following discussion considers only the visible fat consumption where sunflower competes with other fats and oils for markets.

Consumption of visible food fats and oils in the USA has increased about 20% since 1960, from 22.0 kg/person to 26.7 kg in 1976 (Fig. 2). Prior to 1960 consumption was relatively stable at about 20 kg/person. Increased consumption of food fats and oils results from the changing eating habits in the USA. Consumption of convenience and snack foods has increased sharply. Similar growth has occurred in the fast-food restaurant industry, which uses large amounts of fats and oils in preparing hamburger meat, french fried potatoes, chicken, and fish.

Dramatic changes have occurred in the food fat economy. A shift from solid fats to liquid oils and from animal fats to vegetable oils has occurred. Use of salad and cooking oil per person more than doubled from about 4 kg to 9 kg from 1960 to 1976. The use of shortening and margarine produced mainly from vegetable oil also increased. This trend favors increased sunflower oil use and production.

Edible vegetable oils comprise an increasing share of the growing fats and oils market. The USA food fat market in 1956 was divided nearly equally, 47% being in animal fats, including lard, butter, and edible beef fats, and 53% in vegetable oils. By 1976, it was 18 and 82%, respectively (Fig. 3). Consumption of edible vegetable oil increased during the past two decades from 11 kg to 21 kg/capita. Meanwhile, use of animal fat decreased from about 10 kg to about 5 kg/capita. This trend toward increased use of vegetable oil favors increased markets for the high quality sunflower oil.

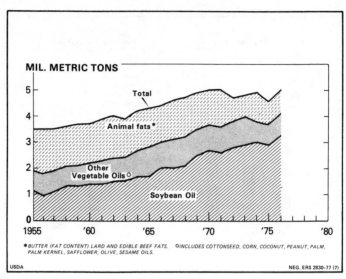

MIL. METRIC TONS

*BUTTER (FAT CONTENT) LARD AND EDIBLE BEEF FATS.   ○INCLUDES COTTONSEED, CORN, COCONUT, PEANUT, PALM,
PALM KERNEL, SAFFLOWER, OLIVE, SESAME OILS.

USDA   NEG. ERS 2830-77 (7)

Fig. 3—U.S. Disappearance of fats and oils in food products.

Vegetable oils now dominate the edible oil and fat markets mainly because of (a) more soybean oil at competitive price levels; (b) increased hydrogenation processing allowing the manufacture of shortenings from 100% vegetable oils; (c) consumer shifts from butter to lower-priced vegetable oil margarines; and (d) trends of diet and cholesterol conscious consumers toward liquid, unsaturated oils and away from solid, saturated fats (23). The edible oil industry has met the increased demand for unsaturated fats by producing more vegetable oil shortening, salad and cooking oils, "soft" margarines, and by using more edible oil in commercial food preparations.

Use of primary fats and oils in USA food fat products has increased 62% in the past two decades, from about 3.3 to over 5.4 million metric tons. Patterns of use of the various fats and oils changed significantly during this period.

Soybean oil has become the dominant food fat in the USA. It comprises about 60% of all fats and oils in food products. In 1950 soybean and cottonseed oil each accounted for about 20% of all food fats utilized, while lard was in first place with about 30% of the market. Use of soybean oil has doubled from the early 1960's to 1976, from 7 to 15 kg/person while use of cottonseed oil dropped from 3 to 1 kg. Use of soybean oil expanded very rapidly because farmers responded with more production to expanding demand for soybean oil and meal. In contrast, cottonseed oil and lard are byproducts and therefore do not respond to demand and price changes. Butter production depends on the milk supply and its use in other dairy products.

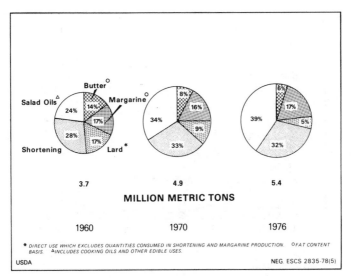

Fig. 4—U.S. Consumption of food fats and oils.

Food Fat Products

Substitution has taken place in the USA among the three major food fat product groups—the butter and margarine spreads, the lard and shortening cooking fats, and salad and cooking oils (Fig. 4).

The salad and cooking oil proportion of the USA food fat market increased from 24% in 1960 to 39% in 1976. The use of shortening increased from 28 to 32% of the total market, while the use of lard decreased from 17 to 5%. Margarine use remained the same at 17%; however, butter declined from 14 to 6% from 1960 to 1976. Vegetable oils used in salad and cooking oil production increased from less than 1 million in 1960 to 2 million metric tons in 1976. At present, salad and cooking oil is the largest market for soybean oil, reaching 1.5 million metric tons in 1976, almost four times that in 1960. During the 1960 to 1976 period, soybean oil's proportion of the salad and cooking oil market increased from 45 to 77%. During the same period the proportion of cottonseed oil dropped from 38 to 9% and corn oil from 13 to 7%, although there was slightly higher usage of corn oil. Smaller quantities of peanut, safflower, and olive oil also are used as salad and cooking oils.

Fats and oils used in the production of shortening in the USA increased 80% from 1 million to 1.8 million metric tons during the period 1960 to 1976. This is the second largest market for soybean oil, and it increased from 0.5 million metric tons in 1960 to 1.1 million metric tons in 1976. Soybean oil increased from 50 to 60% of the market. Use of animal fats in the

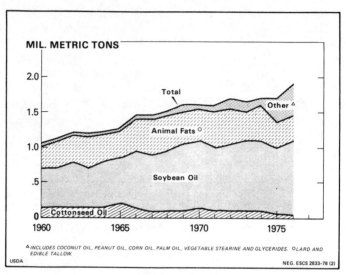

Fig. 5—U.S. Fats and oils used in shortening.

production of shortening increased rapidly in the 1950's. Use in 1976 at 0.35 metric tons, however, is about the same as in 1960; therefore, the share held by animal fat decreased from 33 to 20% during this period. Cottonseed oil also decreased from 16 to 3% during this period. The use of individual fats and oils in shortening varies with price relationships. The price of palm oil was considerably lower than soybean oil in the USA from 1973-76 so its use doubled, with 2.6 million metric tons used for shortening in 1976. Palm oil use in shortening increased from less than 1% in 1960 to over 13% in 1976.

Fats and oils used in margarine production increased from 0.6 million metric tons in 1960 to 9.5 million in 1976. Many consumers in the USA gradually have changed their table spread (for bread, etc.) from butter to margarine. This change is due to the lower price of margarine and the improvement in flavor and other qualities of margarine. The proportion of soybean oil in margarine has been constant at about 80% during this period. Cottonseed oil has declined from 10 to 2%, while corn oil increased from 4 to over 10%. Small amounts of safflower, peanut, coconut, and palm oils, as well as animal fats also were used.

Sunflower Oil Use in Food Fat Products

Edible oils usually sell at a higher price than inedible oils. Sunflower oil is an edible oil, having its main markets in edible products. Although it has been produced commercially in the USA since 1966, the author has not discussed the use of sunflower oil in food fat products earlier because production has been small, and data on various uses have not been reported. Although oilseed sunflower production has been growing rapidly in the USA since 1966, most of the production has been exported as seed. Exporters have estimated that approximately 70% of USA sunflower production in 1975 was exported (6). As a result, relatively small amounts of sunflower have been available for crushing in the USA. Oilseed crushers located in the areas where flax and cotton were formerly more important are looking for other oilseeds to crush. The oilseed mills now have excess crushing capacity, which is a major problem, and such mills, with slight modification, can crush sunflower. More sunflower was produced in 1977 so greater quantities of seed should be available for crushing and export than in previous years.

The Statistical Reporting Service of the USDA has officially estimated 1977 sunflower plantings for the four major producing states (Minnesota, North Dakota, South Dakota, and Texas) at 889,518 ha (2,198,000 A), and production at 1,247,845 metric tons (2,751 million lb) with the oilseed type representing 89% of the total. Value of sunflower production was estimated at about $270 million. Trade estimates place sunflower production for all states at about 17% above the four-state total.

Sunflower Oil Use in Shortening

Sunflower grown in the South from spring plantings in April or May produces oil that is more monounsaturated (high oleic acid) than oil from sunflower grown in the North or late (July and August) plantings in the South (Table 2). This oil is preferred over the more highly unsaturated (high linoleic) sunflower oil in the manufacture of shortening. This Southern-produced oil is of high grade and good stability, making it desirable for manufacturing shortening or cooking oil. It does not have many of the undesirable characteristics of palm, soybean, and several other oils.

The specific food fats used in producing shortening varies annually and is highly dependent upon the price of the finished oil or oil that has been refined, bleached, deodorized, and hydrogenated. Lower quality food fats, such as palm oil, soybean oil, lard, tallow, etc., usually are lower priced and thus more likely to be used than sunflower oil in making shortening (Fig. 5). The animal fats are excluded automatically from consideration in the production of vegetable shortening. Substantial quantities of sunflower oil could be used domestically in the manufacture of shortening in the future if sunflower oil sells at a price competitive with soybean oil.

Sunflower Oil Use in Margarine

Sunflower oil, high in polyunsaturated fatty acids (high linoleic acid oil), is used to manufacture premium margarines. This oil has a high ratio of polyunsaturated to saturated fatty acids. Premium margarine is designed for consumers wishing to reduce their intake of saturated fat. Sunflower oil competes with safflower and corn oil for use in making premium margarine (Fig. 6). Several companies have introduced premium sunflower oil

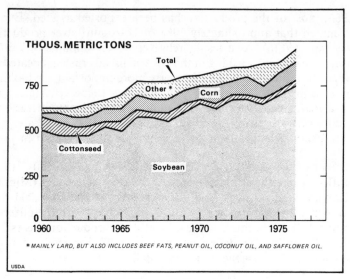

Fig. 6—U.S. Fats and oils used in margarine.

margarine to the USA market. Consumers have accepted the product and it is marketed now over a large area. Factors that would contribute to successful market promotion of margarine made from sunflower oil are the sunflower name, the use of sunflower seed as an edible nut-like food, the high level of polyunsaturated fatty acids in the oil, and the attractive yellow and brown flower.

The higher saturated sunflower oil, produced from seed grown under high temperatures, common in Southern production areas, also can be used to manufacture an excellent margarine, but of lower grade. Cottonseed, soybean, and peanut oils would provide the principal competition for Southern (high oleic) sunflower oil in this U.S. market (Fig. 6).

Sunflower Oil Use in Salad Oil

Sunflower oil has excellent acceptance as a salad oil. Dewaxed, refined sunflower oil results in a salad oil with excellent stability and a high nutritional value (10). The high linoleic acid sunflower oil, with its high polyunsaturate content, can be used to manufacture premium mayonnaise, salad dressing, and salad oil. Since salads are eaten widely by diet-conscious people to control weight and lower cholesterol intake, these highly polyunsaturated salad oil products appeal to them. Safflower and corn oil are the main competitors of sunflower oil in the high quality salad oil market (Fig. 7).

The higher saturated Southern oil also may be used for salad oil. Cottonseed oil was used widely as a salad oil in the USA a few years ago, and it is still the number 2 salad oil. Its use has declined in recent years because of

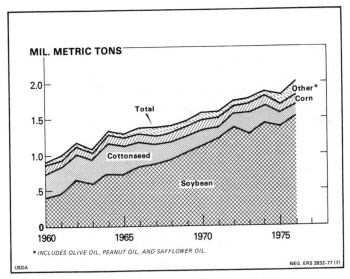

Fig. 7—U.S. Vegetable oils used in salad and cooking oils.

improvements in soybean oil and declining cotton production. Apparently high oleic acid sunflower oil could gain a large segment of the salad oil market formerly held by cottonseed oil.

Sunflower Oil Use in Cooking Oil

Cooking oils in recent years have been the fastest-growing segment of the U.S. fats and oils market. Much of this growth may be attributed to the increasing importance of snack items, such as potato chips, and the rapid emergence of fast-food outlets featuring fried chicken, seafood, and other items. The oil is used several times in fast-food outlets; therefore, a highly stable oil is required. Without special processing, soybean oil may develop off-flavors, after repeated use at high temperatures, because of its unstable linolenic acid component (34).

Sunflower oil gives excellent performance as a cooking oil because it lacks linolenic acid, which catalyzes to polymers when heated. This catalysis causes a thickening and darkening of the oil, and leaves a deposit on the frying vessel. Refined sunflower oil is an ideal frying oil for the rapidly-expanding, snack food industry. Products fried with sunflower oil usually will absorb less of the frying oil than when fried in most other vegetable oils. Also, products fried in sunflower oil are reported to have a longer retail shelf-life than products fried in some other oils (10).

The potato chip industry uses a standard frying oil of 70:30 cotton-seed:corn oil mixture. A taste panel consistently scored potato chips fried in 100% sunflower oil above those fried in the cottonseed-corn oil mixture (16).

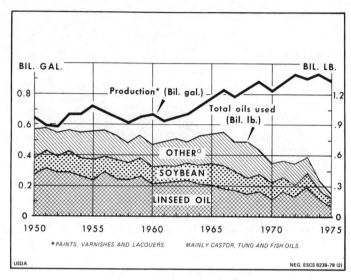

Fig. 8—Surface coatings production and oils utilized.

The Southern sunflower oil is more stable than the Northern oil and probably will be used mostly in cooking oil. Some manufacturers of snack foods already are using Southern sunflower oil for frying, with excellent results. A tremendous market may be available for Southern sunflower oil as a cooking and salad oil in the USA, if it can capture the market formerly held by cottonseed oil only 10 years ago.

A smaller market in cooking oil is available for Northern sunflower oil. Some consumers prefer to fry their foods in a polyunsaturated oil for dietary and health reasons. Safflower and corn oil will compete with Northern sunflower oil for this market.

Industrial Oil Markets

Sunflower oil is a high quality, edible oil commanding a higher price than some of the other fats and oils and, therefore, it is excluded from many uses as an industrial oil. Drying oil products appear to offer the most potential for industrial use of sunflower oil, with this market limited to the highly unsaturated sunflower oil. This type of sunflower oil has good drying-oil properties and, because of its low linolenic acid content, does not yellow with age. Nonyellowing is desirable especially when the oil is used in white or pastel shades of paint. Yellowing is a major problem of linseed oil, the most widely used drying oil, and consequently other natural oils and synthetic materials have been replacing it. This segment of the natural oil paint market is held by safflower, and sunflower would have to compete with it. Since World War II, the protective coating industry has been shifting to synthetic materials, and this change has resulted in a gradual downward trend in the use of natural oils in drying oil products (Fig. 8). In 1976 the 248,000 metric tons of fats and oils used in drying oil products in the USA

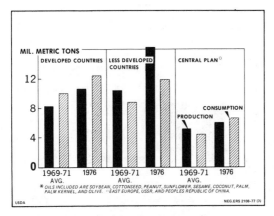

Fig. 9—World edible oil production and apparent consumption.

was about half that used 20 years earlier. Although some sunflower oil may be used in drying oil products, the market is not considered a major potential outlet.

World Trends in Fats and Oils Markets

Trends in World Food Fat Consumption

Food fat consumption varies widely throughout the world, both in the kinds of fat consumed and in the quantity consumed per person. The increased world demand for fats and oils is related closely to population, income level, and price of food fats. World consumption of fats and oils has increased at an average annual rate of 1.2 million metric tons in recent years. "Consumption" is apparent consumption derived from production plus imports less exports. Stocks data for many countries are incomplete or not available.

Consumption patterns in a country or geographic region are influenced greatly by available food fats and oils produced locally. Indigenous production has the advantages of low transportation costs, protective economic policies, and eating habits and consumer preferences for locally produced fats and oils. Only a few different primary fats and oils are used in some areas of the world. In contrast, a wide assortment of food fats and oils are available and used interchangeably by such major importers as Japan and Western European countries.

Oil and fat consumption per person in developed countries is high and overall consumption usually exceeds production. Consumption in developed countries in 1976 was 12.5 million metric tons while production was 10.5 million metric tons (Fig. 9). The developed countries of Western Europe and Japan follow the trend; however, noticeable exceptions are the USA and Canada, which export edible oils (23).

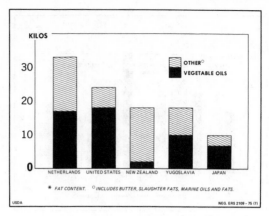

Fig. 10—Oils and fats consumption per person, 1971.

The pattern of fat consumption for less developed countries is directly opposite that of developed countries. Fat consumption per capita is low, and these countries produce more vegetable oils than they need. Consumption in less developed countries of the world in 1976 was 12 million metric tons while production was 16 million metric tons (Fig. 9). Eastern Asia is the leading exporter, followed by Central Africa and Brazil. As the general economic conditions of developing countries improve, so does the demand for fats and oils.

Fats and oils production and consumption are more nearly equal in the "centrally planned" countries. In 1976 fat production was estimated at 6 million metric tons while consumption was estimated at 6.5 million metric tons (Fig. 9). The USSR is usually a net exporter while the Peoples Republic of China and Eastern Europe are net importers.

Patterns of fat usage also vary in different regions of the world. Oil and fat consumption is highest in Western Europe at 30 kg/capita or higher, low in Asia at about 5 kg/capita, and lowest in Africa at about 2 kg/capita (3).

Countries also differ greatly in the quantity of fats and oils consumed per person. In 1971, consumption of fats per person in the Netherlands average 33 kg, of which half were vegetable oils and the remainder were animal slaughter fats, marine oils, and butter (Fig. 10). Vegetable oils composed about 75% of the 24 kg/person of fat consumption in the USA in 1971. Three countries, each having a per capita consumption of fats and oils of about 18 kg in 1971, but consuming different types of fats and oils, were New Zealand, Yugoslavia, and Spain. Vegetable oil consumption was 10, 55, and 89% of fat use in New Zealand, Yugoslavia, and Spain, respectively. Vegetable oils composed about 75% of the 10 kg/person of fat use in Japan in 1971.

These large variations in fats and oils consumption per person illustrate the great potential for expanding fats and oils intake in regions of low usage. Fats and oils consumption, however, may be expanded easier in regions with medium or higher consumption than in those of low consump-

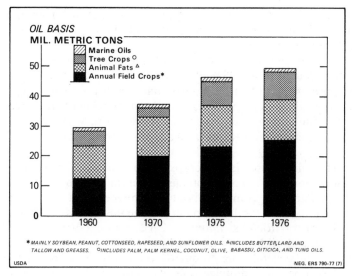

Fig. 11—World production of oils and fats.

tion. These data show regions where vegetable oils are popular, and they might be future markets for sunflower oil.

Expansion in demand for edible oils in recent years has been primarily in the developed countries of Europe, Japan, the Soviet Union, etc. These countries have their own crushing capacity. They import oilseeds so that, by crushing them, they have the edible oil for food and the protein meal for livestock feed or human food. The USA has exported much of its whole sunflower seed to these markets in recent years, but soybeans now are used for most of the oilseed requirements of these countries.

Trends in World Sources of Fats and Oils

World production of fats and oils, including oil equivalent of oilseeds, increased about 60% between 1960 and 1976, from 30 million to 50 million metric tons. Several sources provide significant portions of the world supply of oils and protein meals. These sources are (a) annual field crops, (b) animal fats, (c) tree crops, and (d) marine oils.

The greatest expansion since 1960 has occurred in the annual field crops, which approximately doubled in production to over 25 million metric tons and now compose over 50% of the world total (Fig. 11). Food fat sources from this group include soybean, sunflower, peanut (groundnut), rapeseed, safflower, and sesame. Oil output from this group may expand or contract greatly from year to year, and is the most flexible in the world supply of fats and oils and protein meals. The huge expansion in soybean production is the main reason for the large world increase of fats and oils in recent years. In addition, this category also includes cottonseed, corn, and other oils that are byproducts of processing a crop for another purpose. Flaxseed is also in this category, although its oil is considered inedible in most areas of the world.

The second major source is such animal fats as butter, lard, and tallow, which now account for 28% of the world edible oil output. Lard and tallow are byproducts of the meat industry and butter production depends on the supply of milk and use in other dairy products. Animal fats increased from 11 to 14 million metric tons from 1960 to 1976, with tallow showing the most increase. Meat meal and tankage are byproducts of animal fat production.

The third source is tree crops, from which oil production also is increasing rapidly, from 5 million metric tons in 1960 to 9 million metric tons in 1976. They now compose about 20% of the world oil output. Coconut, palm, palm kernel, and olive oil are edible oils in this group along with industrial castor (*Ricinus communis* L.) and tung (*Aleurites fordii* L.) oils. The perennial oil-bearing trees generally grow in tropical and subtropical climates and generally are productive for several decades. Palm oil production has more than doubled, and coconut oil also has increased greatly during this period.

The fourth source is marine oil, a byproduct of the fish meal industry, which has shown little change in production of about 1 million metric tons annually during the period.

As stated earlier, the world consumption of fats and oils is increasing at a rate of more than 1 million metric tons per year. Consumption patterns result in about 33% of the world output of fats and oils moving in world trade. Between 1960 and 1976 world production increased 60% and exports doubled from 8 to 16 million metric tons. This growth has been limited primarily to edible vegetable oils from annual field crops and perennial tree crops.

World Area and Production of Sunflower

World sunflower prodution in 1977 is forecast at about 12.0 million metric tons, assuming a 6.5 million metric ton crop for USSR (9). This is a remarkable comeback from the 10 million metric tons produced in 1976 and approaches the record crop of 1973 of slightly over 12 million metric tons (Table 3).

By far the leading sunflower producer in 1977 was the USSR, with 54% of the world's production. The USA, with 10% of the total, jumped into second place ahead of Argentina (8%) and Romania (6%). Other leading producers, in order of quantity produced, were South Africa, Yugoslavia, Turkey, Bulgaria, Spain, and Australia (Table 3). There is much land suitable for growing sunflower in the world and farmers in some areas need alternative crops to replace crops whose production is declining due to reduced demand.

Table 3—World area and production of sunflower seeds; 1973-1977. After Anon. (9).

Country	Area 1973	1974	1975	1976	1977	Production 1973	1974	1975	1976	1977
	——— 1,000 ha ———					———1,000 metric tons———				
Argentina	1,338	1,087	1,196	1,411	1,460	800	970	732	1,085	900
Australia	242	151	210	128	223	102	84	113	83	137
Bulgaria	255	262	231	230	230	440	366	406	390	400
Romania	512	604	604	521	514	756	671	728	800	750
South Africa	346	241	239	288	389	233	254	209	255	476
Spain	416	437	623	463	580	293	286	318	270	250
Turkey	481	425	390	418	450	560	420	488	505	400
USA†	301	263	478	378	830	353	272	541	463	1,225
USSR	4,745	4,686	4,492	4,534	4,500	7,385	6,784	4,990	5,277	6,500
Yugoslavia	224	200	194	174	209	433	298	272	319	437
Others	630	536	607	632	639	607	506	602	516	517
World total	9,490	8,892	9,264	9,177	10,024	12,042	10,911	9,399	10,008	11,992

† Unofficial data compiled from trade sources and selected state reports.

The tremendous increase in U.S. sunflower production in 1977 was due partially to farmers shifting almost completely to hybrids from open-pollinated cultivars. The area planted in the USA also increased significantly. Sunflower production in North America may have a long-term expansion potential. Sunflower has been grown primarily on land formerly planted to flax and cotton. Future expansion of sunflower in the USA is likely on land located on the fringe of the present corn, soybean, and cotton growing areas.

Projection of World Production and Consumption of Fats and Oils

World production and consumption of fats and oils that are used primarily as food fats both have been increasing at an average annual rate of about 3% (11, 27). World population is increasing at a rate of about 2%/year and the global per capita consumption rate of fats and oils also is increasing. Production of edible vegetable oils has been increasing at a faster rate of slightly over 4% (27, 34). Since 1955, little difference has occurred in the rate of growth in fat consumption among the developed, less developed, and the centrally planned countries.

High Protein Meal Markets

High Protein Meal Sources, Production, and Substitutes

The several agricultural and marine sources that provide significant portions of the world supply of oil and protein meal were discussed earlier. It must be remembered that high protein meals are mostly byproducts and their availability depends on production of the primary oil product. Only soybean and fish produce meals more valuable than the oil extracted from these various raw materials. For soybean meal, this situation may be re-

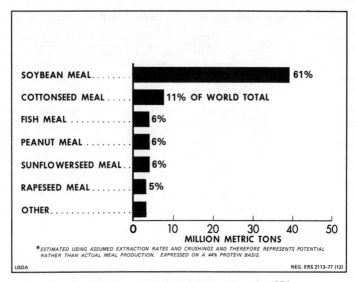

Fig. 12—World production of high protein meals, 1976.

versed when oil prices are extremely high. Sunflower meal, including hulls, accounts for only about 25% of the total value of sunflower.

The world production of the major protein meals, including soybean, cottonseed, peanut, fish, rapeseed, sunflower, copra, linseed, sesame, and palm kernel, totaled 75 million metric tons in 1976 (8). Soybean meal was the leader, with 61% of the total, followed by cottonseed meal with 11% (Fig. 12). Peanut meal, fish meal, rapeseed meal, and sunflower meal each accounted for 5 to 6% of the total. Copra meal and linseed meal were each about 2%, and sesame meal and palm kernel meal about 1% of total production.

In 1976 world exports of the major protein meals amounted to 34 million metric tons or about 45% of world production. Soybean meal comprised 73% of these protein meal exports while sunflower meal accounted for less than 2%. Most of the sunflower meal is used in the country where it is processed.

Production of major high protein meals from soybean, cottonseed, peanut, and linseed meal in the USA was about 21 million metric tons in 1976, or over 27% of the world protein meal production. Exports of 5 million metric tons were about 15% of the world total. Soybean meal composed over 90% of the protein meal produced, cottonseed meal about 6%, and peanut meal and linseed meal each about 1%. Sunflower meal amounted to less than 1% of the protein meal produced in the USA in 1976.

Considerable substitution takes place among the various sources of protein used in livestock feeds, depending on price and availability. This interchangeability is limited by such nutritional factors as differences in protein and fiber content, as well as physical characteristics of protein source materials.

In the future a large portion of the market for protein ingredients in ruminant rations may be filled by urea or other nonprotein nitrogen compounds, but these compounds would not be used in nonruminant rations. Hence overall demand for protein meals is expected to remain stable or increase due to the expanding demand for protein and livestock products in the USA and the rest of the world. This demand translates to increased demand for protein meal to feed livestock as well as for use as vegetable protein in human diets.

Several of the essential amino acids in protein now can be synthesized. Synthetic lysine, methionine, and glycine are being manufactured commercially. These amino acids can be used to supplement the limiting amino acids in meals and reduce the quantity of protein concentrates needed. They also permit the use of amino acid-fortified, cheaper concentrates, such as sunflower meal, and thus tend to broaden the market for sunflower meal.

Sunflower Meal

Sunflower meal is the byproduct obtained when the oil is removed from sunflower seed. Like other oilseed meals, it may differ in composition and nutritive value for the following reasons: the method of oil extraction, the operating procedures of the plant, the composition of the seed, and percentage of hull remaining in the meal. Hulls are high in fiber content and limit the use of sunflower meal even in ruminant rations. Prepress, solvent-extracted sunflower meal from dehulled seed usually contains approximately 44% protein. Sunflower meal containing the hull is about 28% protein. Feed manufacturers and livestock and poultry feeders, when considering meals in formulating feeds, are interested primarily in protein percentage, protein quality, and price. Due to variation in composition caused by factors mentioned earlier, they also will evaluate the fat and fiber content. Sunflower meal has been used in ruminant rations for many years in some countries where it is now a standard feed ingredient. Although sunflower meal is new as a high protein feed ingredient in North America, it has been accepted readily by feed manufacturers and farmers alike.

Sunflower meal is not known to contain any growth-depressing or toxic substances (30), a primary advantage over several other protein sources. Toxic substances found in other protein sources affect their processing and limit their use in some rations. Sunflower meal appears similar to other oilseed meals except for its unique gray-green color. Its mineral and energy content compares favorably with other oilseed meals (20).

Numerous animal feeding studies by private industry and universities have demonstrated that meal from sunflower seed of high oil content is an excellent, high quality, protein ingredient for feeding poultry, swine, and ruminants. Sunflower meal should be used with care in nonruminant rations so that adequate lysine and energy are available from other ingredients in the ration.

Oilseed meals, because they are bulky and costly to ship usually are consumed by livestock near the meal production facility. Processors easily sell all the sunflower meal produced at prices competitive with cottonseed meal.

The demand for sunflower meal depends upon livestock output and price. In the USA, beef and veal production is expected to decline slightly in 1978–80; however, overall production trend will increase slightly. Pork production is expected to increase in the near future, then level off and probably decline slightly. Chicken, turkey, egg, and milk production are all expected to increase modestly. Livestock production elsewhere in the world is expected to increase at a moderate rate. Therefore, the future outlook for increased demand for high protein feed ingredients, including sunflower meal, is good.

Sunflower Hulls

The hull, a byproduct of oil extraction, comprises 22 to 28% of the total weight of oilseed sunflower (30). Sunflower hulls may be removed before or immediately following oil extraction or may remain with the meal. Hulls contain about 50% crude fiber, 4% crude protein, and 2% fat (29).

Sunflower hulls are a coarse roughage, high in fiber but suitable for ruminant rations. Finely ground hulls mix well with other feed ingredients. They add bulk to a concentrated ration, are a carrier for some ingredients, and absorb liquids, such as molasses, that may be added to rations. If hulls are removed prior to oil extraction they sometimes are added back to sunflower meal. Sunflower hulls have sold well to feed manufacturers and livestock feeders at a price comparable to cottonseed hulls.

Sunflower hulls often are used as a fuel to generate steam in sunflower-producing countries, other than the USA (12). In Canada and Turkey the hulls at one time were pressed into cylindrical shapes and sold as fire logs (19). Hulls sometimes are used in the USSR to make furfural and ethyl alcohol (31). Other proposed uses include building or insulation board, litter for livestock or poultry, and packing material.

Edible Sunflower Protein

Several studies have predicted that the present small but growing market for edible vegetable protein will enlarge greatly in the future. Production of edible soybean protein products in the USA is estimated at 408,000 metric tons in 1977 (7). The majority of the European Community countries now produce some kind of edible soybean protein (26). Sunflower protein, marketed like soybean protein as defatted flour, concentrates, and isolates has potential use in a variety of food products. Research has shown that sunflower protein can be incorporated into new food formulas and also used to produce enriched, traditional food products acceptable to native populations (25, 32). Nutritionists report, however, that sunflower protein flour would be more desirable if it contained more of the amino acids, lysine and isoleucine.

The absence of toxins in sunflower protein is advantageous to its use in human foods. Several oilseed protein materials contain toxins that must be removed or rendered harmless before they can be used for human food. Many potential consumers already are familiar with the edible, roasted, sunflower seed, such acquaintance making the introduction of edible sunflower protein easier.

Sunflower protein flour has properties making it more desirable for use in food supplements than the soybean protein flour dominating the edible vegetable protein market today. It has a desirable, bland flavor and is low in sugars that produce flatulence (gases) in the digestive tract. Soybean flour, on the other hand, has a beany flavor which may be objectionable in some uses, and when eaten produces undesirable flatulence. Soybean protein isolate is produced to overcome these disadvantages but it is much more expensive than soybean flour.

The major problem limiting use of sunflower protein in human diets at present is a color change that can occur in the flour. Upon oxidation of chlorogenic acid, a phenolic compound in sunflower flour, the color of the flour changes from white to beige, to green, to brown. This color change may effect its acceptability in food products. As the color changes, the pH also changes from acid to alkaline (17, 28, 32).

WORLD COMPETITION FOR SUNFLOWER

World production of fats and oils (including oil equivalent of oilseeds), was 49.5 million metric tons in 1976 (8). Vegetable oils were 70% of the total, animal fats 28%, and marine oils, 2%. Animal fats were almost equally divided between butter, lard, and tallow. The seven leading oilseeds accounted for 60% of the world production. Soybean was the leader, with 21% of the total. Sunflower, peanut, palm, coconut, rapeseed, and cottonseed each accounted for 6 or 7% of the total.

World exports of fats and oils, including oil equivalent of oilseeds, were 15.8 million metric tons in 1976 (8). Vegetable oils were 78%, animal fats 18%, and marine oils 3% of the total. The seven leading oils accounted for 70% of the world's exports of fats and oils. Soybean alone was 28%, followed by palm with 13%, coconut 11%, and peanuts 6%. Sunflower and rapeseed each had 5% and cottonseed 2% of the total.

Major food fats compete for the present and expanding world fats and oils market. An examination of soybean, oil palm, and peanut, the three leading competitors to sunflower, for domination of the worlds oilseed, oil and meal markets in the future follows.

Soybean

The soybean is the most important single source of edible oil and high protein meal. Soybean provided over a fifth of the world fats and oils production in 1976. Soybean oil and soybean, on an oil basis, compose the largest

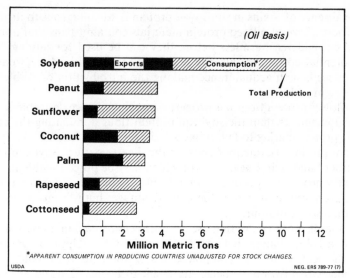

Fig. 13—World production and exports of major vegetable oils, 1976.

volume of oil moving in world trade, accounting for over 25% of the world oil exports in 1976. Nearly 50% of the total production of soybean, on an oil basis, was marketed outside the country of origin (Fig. 13).

World production of soybean has continued upward for the past 25 years and has more than doubled between 1965 and 1977. The USA is the leading producer of soybean, accounting for about 60% of the total. The USA also is largely responsible for this sharp increase in world output, although in recent years production in Brazil has expanded significantly. Brazil recently has become the number 2 producer, slightly ahead of the Peoples Republic of China. These three countries produce 90% of the world's soybeans.

The large increase in soybean production has resulted mainly from an expansion in the areas planted, because soybean yields have increased only slightly in the past two decades. Future expansion of areas planted to soybean in the USA will be more difficult to obtain than in the past. There is great potential for long-term expansion in South America, however, especially Brazil, Argentina, and Paraguay. There is no commercial production of hybrid soybean, and none seems likely in the near future. Increasing yield per hectare probably will continue slowly, resulting from more adaptable cultivars and advances in resistance and control of diseases, insects, and nematodes.

Soybean is important because of its oil and high protein meal. The oil and meal content varies from region to region and even from farm to farm. Also the oil percentage in the USA tends to increase, when soybean is grown in warmer climates, while protein content tends to decrease (22). The meal has represented 60% of the value of soybean in recent years. Soybean meal accounts for about 50% of the world production of oilseed meals and nearly 90% of oilseed meals produced in the USA. It has the best amino acid balance of all the vegetable protein materials. Soybean meal is used widely in feeds for livestock and poultry, especially in feed formulations for poultry and hogs—monogastric animals which cannot utilize large amounts of fiber.

Soybean is also an important source of edible soy protein in the diet of peoples of the Far East. There is increasing interest in edible vegetable protein in the Western world, and soybeans are the best known source. Meat extenders and meat analogs manufactured from soybean meal now are being produced and marketed with some success. The increasing world population needs a more nutritious diet; therefore, there is vast potential for edible protein.

Soybean oil is the "workhorse" of the edible oil products industry. It is used in the manufacture of almost all edible fat products because of many technological improvements made in processing the oil. Soybean oil has quality and saturation in the middle range among edible vegetable oils (Table 2).

Oil Palm

Palm oil, produced from the fruit of the oil palm, is the strongest competitor of other edible oils in the world. Palm trees begin to bear fruit when 4 years old and produce for about 30 years, with few costs except for harvesting. The oil palm, a native of Africa, is found along most of the western coast of Africa. Aggressive planting programs during the 1960's and 1970's, however, have made West Malaysia the largest producer, with Indonesia ranking second. West Malaysia has increased production areas and been the leader in adopting the high oil-yielding, Tenera-cross cultivars. Oil palm, unlike soybean and some of the other oilseeds, produces a relatively small quantity of low quality meal.

World palm oil production has been expanding rapidly from 1.7 to 3.1 million metric tons from 1970 to 1976—over an 80% increase. West Malaysia has provided 50% of the world increase. Indonesia also had a significant increase in total production. In addition, enough area has been planted in recent years to double 1974 production in West Malaysia by 1980. Over 90% of West Malaysian production will be exported, so palm oil should account for an increasingly larger share of total world supplies and exports of edible vegetable oils. Palm oil now ranks second to soybean in international trade.

European countries import and consume most of the palm oil in world trade. Europe is also a major importer of sunflower seed and oil. Thus, expanded world palm oil production will compete with sunflower exports.

Palm oil is used primarily for edible uses—shortening and to a lesser extent for margarine and cooking oil. For quality reasons, palm oil use has not increased significantly in margarine and cooking oil markets. An additional drawback of palm oil is its high saturation. Technological improvements for processing palm oil and its low price, however, may encourage increased use in margarine and cooking oils.

The major competitive advantage of palm oil is its high yield (Table 1), resulting in a low cost of production. No firm figures are available; however, limited information indicates a production cost of $0.25/kg when palm kernel oil sells at $0.44/kg (11). Soybean oil usually sells for more than palm oil because of its higher quality. Sunflower oil sells at a higher price than soybean oil; therefore, an even larger price spread exists between palm and sunflower oils.

Peanut

Peanut usually ranks third in world production behind soybean and sunflower as an oilseed crop. Peanut also is an important oilseed in world trade, usually ranking fourth behind soybean, palm, and coconut. Peanut production and trade are subject to wide variations due to weather conditions. India, People's Republic of China, USA, Senegal, and Sudan are the largest producers. Production in these countries generally is upward, although overall world production has changed little.

Peanut has been produced for many years in the USA with price supports tied to an allotment of about 0.65 million ha since 1960. Production has increased greatly with the same production area. Under this program, U.S. peanuts were produed for the edible market and sold at about twice the price of peanuts sold on the world market for crushing. A new price support program for peanuts is effective in 1978, as required by provisions of the Food and Agriculture Act of 1977. In addition to individual farmer area allotments, each allotment holder will have a poundage quota for the domestic edible market. The higher of two price support levels will be limited to "quota peanuts." For "additional peanuts" produced under a farmers area allotment but in excess of his weight quota, the farmer generally will receive a lower support price. These peanuts may be crushed, exported, and used in the domestic edible market under certain conditions.

If farmers producing peanuts in the USA were not restricted in their plantings, peanut might become a real competitor of sunflower. With present increased yields, some peanut farmers have stated that they can produce peanuts at the world price. The average yield of peanut in the hull in the USA in 1975 was 2,875 kg/ha or nearly triple yields in 1955. The average annual rate of increase during this period was about 5.2%. Continued yield increases are expected. The 1975 average peanut yield for the state of Georgia was 3,693 kg/ha. Individual farmers have reported yields of over 6,725 kg/ha. Areas planted to peanuts in the USA have the potential to expand greatly if permitted by law. Presumably peanut yield outside of the USA also can be increased by use of improved cultivars and new technology employed in the USA.

Peanut oil is used primarily as a cooking and salad oil. It has excellent quality and a flavor preferred by some consumers, making it especially valuable as a frying fat. For this reason, it usually sells at a higher price than some other vegetable oils. It is, however, relatively high in saturated fatty acids (Table 2). Peanuts contain about twice the oil percentage of soybeans. Therefore, peanut meal is not as important to the total value of peanuts as soybean meal is to soybeans, but it is a good high protein meal used in livestock feeds.

SUMMARY

In the future sunflower could become a more important economic crop in North America and the world. Factors supporting increased production of sunflower in the future are: the yield potential and superior agronomic characteristics of hybrid oilseed sunflower; the need for alternative crops in some areas; excessive crushing capacity in areas where flax and cotton were formerly more important; increasing world demand for edible fats and oils; and the health benefits of polyunsaturated oils.

The potential for expansion of sunflower production in the world is good, and much land is suitable for growing sunflower. Sunflower production in North America also may have potential for long-term expansion. Sunflower in the USA has been grown primarily on land formerly planted to flax and cotton. Future expansion of sunflower in the USA is likely on land located on the fringe of the present corn, soybean, and cotton-growing areas.

A great wealth of unexplored genetic material exists for use in improving sunflower. By using modern breeding methods, we expect improved hybrids to have better disease and insect resistance. Increased sunflower seed production per hectare as well as increased oil percentage of the seed are anticipated. Both result in increased sunflower oil production per hectare, and economically they make sunflower a more viable crop. The large-scale switch by U.S. farmers from open-pollinated to higher yielding hybrid sunflower last year greatly improved the competitive position of sunflower.

Sunflower possesses several agronomic characteristics which are advantageous to expanding its production. It is deep-rooted and uses soil moisture efficiently; thus is well-adapted to growing in drier regions than most crops. Sunflower also has one of the shortest growing seasons of the major economic crops in the world, allowing double cropping of sunflower in areas having growing seasons over 200 days. It also can be successfully grown in a wider range of latitudes than most crops. With minor adjustments, equipment used for planting and harvesting of grain can be used for sunflower.

Farmers in some areas need alternative crops to replace crops whose production is declining from reduced demand. Production of flax and cotton in the USA has been declining, and sunflower is proving to be a successful alternative crop. Oilseed crushers located in the areas where flax and cotton were formerly more important are seeking other oilseeds to crush to use their excess crushing capacity. These oilseed mills, with slight equipment modification, can crush sunflower.

With world population increasing and with global per capita consumption of fats and oils also increasing, food fat consumption will continue to rise and require increased production of edible vegetable oils. Oilseed crops most likely to expand production to fill this need for more oil are sunflower, oil palm, and soybean. Since these crops are grown in different locations, some expansion in production may take place for all three crops.

Sunflower oil is a high quality food oil and prospects are excellent for increased use in all three major food fat products—cooking and salad oils, margarine, and shortening. Cooking oil is the fastest growing segment of the food fat market and high oleic acid sunflower oil is a superior cooking oil.

Blood cholesterol levels can be reduced by decreasing fat intake and substituting polyunsaturated fatty acids for saturated ones. High linoleic acid sunflower oil is high in polyunsaturated fatty acids and makes excellent polyunsaturated margarine and salad oils. This oil is finding greater use in these markets. If it were established scientifically that high cholesterol levels cause heart attacks, then rapid expansion in the polyunsaturated food fat market would occur and high linoleic sunflower oil would be in great demand.

Under the present structure and technology, industrial use of sunflower oil is minor and the potential future growth is small.

With an anticipated moderate rate of growth in world livestock and poultry production in the years ahead, any increase in the production of high protein sunflower meal is expected to find a ready market as a feed ingredient. Sunflower hulls are a roughage ingredient, high in fiber, suitable for use in ruminant rations, and may serve as a carrier of molasses and other ingredients or be used to add bulk to a ration. The absence of toxins in sunflower meal favors its use over many oilseed meals as a high protein supplement in human foods if a method could be developed to prevent the flour from changing color.

LITERATURE CITED

1. Anonymous. 1968. Recommended dietary allowances. Seventh Edition. National Academy of Sciences, Washington, D.C. p. 10–13.
2. Anonymous. 1972. Diet and coronary heart disease. The Food and Nutrition Board of the National Academy of Sciences-National Research Council and the Council on Foods and Nutrition of the American Medical Association. NAS-NRC, Washington, D.C.
3. Anonymous. 1973. Food consumption statistics, 1955–71. Organization for Economic Cooperation and Development, Paris, France.
4. Anonymous. 1974. School breakfast and nonfood assistance programs and state administrative expenses. Federal Register. National Archives and Records Service, Washington, D.C. March 27. (39 F.R. 11249).
5. Anonymous. 1974. National school lunch program—cheese alternate products. Federal Register. National Archives and Records Service, Washington, D.C. August 29. (39 F.R. 31514).
6. Anonymous. 1976. Sunflower handbook. Honeymead Products Company (Division of Grain Terminal Association), Minneapolis, Minnesota. p. 10.
7. Anonymous. 1977. Fats and oils situation. Economic Research Service, USDA, FOS-289, Table 9. p. 12. (October).

8. Anonymous. 1977. Foreign agricultural circular-oilseeds and products. Foreign Agriculture Service, USDA. FOP 18-77. p. 11. (September).

9. Anonymous. 1977. Foreign agricultural circular-oilseeds and products. Foreign Agriculture Service, USDA. FOP 23-77. (December).

10. Anderson, L. R. 1970. Potential utilization of sunflower oil in the United States. *In* Proc. 4th Int. Sunflower Conf. (Memphis, Tennessee), National Cottonseed Products Association, Memphis.

11. Boutwell, W., H. Doty, D. Hacklander, and A. Walter. 1976. Analysis of the fats and oils industry to 1980-with implications for palm oil imports. Economic Research Service, USDA. ERS-627.

12. Cobia, David, and David Zimmer. 1975. Sunflowers—production, pests, and marketing. North Dakota State University, Fargo. Ext. Bull. 25.

13. Coons, C. M. 1960. Fats and fatty acids. Food—the yearbook of agriculture 1959. USDA, Washington, D.C.

14. Doty, H. O., Jr. 1975. Decisionmaking in the oilseed processing industry. Economic Research Service, USDA. ERS-598.

15. ————, and J. V. Lawler. 1969. Synthetics and substitutes for oilseed products. Synthetic and substitutes for agricultural products, a compendium. Economic Research Service, USDA. Misc. Pub. 1141. p. 60–81.

16. Evans, C. D., and Roy Shaw. 1968. Flavor and oxidative stability of sunflower oil. *In* Proc. 3rd Int. Sunflower Conf. (Crookston, Minnesota).

17. Gheyasuddin, S., C. M. Cater, and K. F. Mattil. 1972. Preparation of a colorless sunflower protein isolate. The Cotton Gin and Oil Mill Press, p. 12. (May 13).

18. Goddard, V. R., and L. Goodall. 1959. Fatty acids in food fats. ARS, USDA. Home Economics Res. Rep. No. 7.

19. Helgeson, D. L., D. W. Cobia, R. C. Coon, W. C. Hardie, L. W. Schaffner, and D. F. Scott. 1977. The economic feasibility of establishing oil sunflower processing plants in North Dakota. North Dakota Agric. Exp. Stn. Bull. 503.

20. Kinard, D. H. 1975. Feeding value of sunflower meal and hulls. Feedstuffs. p. 26. (Nov. 3).

21. Kinman, M. L., and F. R. Earle. 1964. Agronomic performance and chemical composition of the seed of sunflower hybrids and introduced varieties. Crop Sci. 4:417–420.

22. Kromer, G. W. 1967. Factors affecting soybean oil and meal yields. Economic Research Service, USDA. ERS-338.

23. ————. 1977. World oil sources and trends in consumption. Speech before the Fifth Western Hemisphere Nutrition Congress, Quebec, P. Q., Canada. Economic Research Service, USDA. Unnumbered Rep. (August).

24. Leverton, R. M. 1974. Fats in food and diet. ARS, USDA. Agric. Inf. Bull. 361.

25. MacGregor, D. 1970. Formulation of new sunflower seed products. p. 107. *In* Proc. 4th Int. Sunflower Conf. (Memphis, Tennessee), National Cottonseed Products Association, Memphis.

26. McClure, S., and B. Simons. 1976. Utilization of soy protein in the European Community. Foreign Agriculture Service, USDA. FAS. M-270.

27. Moe, L. E., and M. M. Mohtadi. 1971. World supply and demand prospect for oilseeds and oilseed products in 1980. Foreign Agriculture Economics Rep. 71. Economic Research Service, USDA.

28. Pomenta, J. L., and E. E. Burns. 1970. Factors affecting chlorogenic, quinic, and coffeic acid levels in sunflower kernels. p. 114. *In* Proc. 4th Int. Sunflower Conf. (Memphis, Tennessee), National Cottonseed Products Association, Memphis.

29. Robertson, J. A. 1975. Use of sunflowerseed in food products. Crit. Rev. Food Sci. Nutr. 6(2):201–240.

30. Smith, K. J. 1969. Feeding value of sunflower meal. p. 93. *In* Proc. 18th Cottonseed Processing Clinic. New Orleans, Louisiana.

31. Suslov, V. M. 1968. Economical significance of sunflowers in the U.S.S.R. p. 9. *In* Proc. 3rd Int. Sunflower Conf. (Crookston, Minnesota).

32. Talley, L. D., J. C. Brummett, and E. E. Burns. 1970. Utilization of sunflower in human food products. p. 110. *In* Proc. 4th Int. Sunflower Conf. (Memphis, Tennessee), National Cottonseed Products Association, Memphis.

33. Thomason, F. G. 1974. The U.S. sunflower seed situation. Economic Research Service, USDA. ERS-590.

34. Trotter, W. K., H. O. Doty, Jr., W. D. Givan, and J. V. Lawler. 1973. Potential for oilseed sunflowers in the United States. Economic Research Service, USDA. AER-237.

35. U.S. Congress. Senate Select Committee on Nutrition and Human Needs. 1977. Dietary goals for the United States. 2nd Ed. U.S. Government Printing Office, Washington, D.C. (Sup. Doc. No. Y4.N95:D56/977-2).

HARRY O. DOTY, JR.: B.S. and M.S. in agricultural economics. Agricultural Marketing Research Economist, Fibers and Oils Program Area, Economics, Statistics, and Cooperatives Service, USDA, Washington, D.C. Authored and coauthored a large number of publications and articles dealing mainly with economic research on oilseeds, fats, and oils, oilseed meals, and their products; USDA, 1949 to date. Received, along with coauthors, the USDA Superior Service Award in May 1976 for an economic study of the impact of expanding world palm oil production on the U.S. and world fats and oils economy.

INDEX

oil, 459
Varnish, 394
Vermont, 8
Vernalization, 91
Veronezh, 5
Verticillium albo-atrum, 244, 246, 248
Verticillium dahliae, 245, 246
  wilt, 225, 244, 251
Verticillium wilt, 247, 248, 307, 324, 445
Victoria, state of, 16
Viguiera, 33
Virginia, 2, 3
Virus, 255, 256
Vitamin E, 413
Vitamins, 422, 428
Volunteer plants, 105, 231, 232, 235

Weeds, 246
West Malaysia, 483
West Virginia, 8
Western Australia, 17
Wheat, 242, 246, 388, 443, 453
Whetzelinia, 236
White blister rust, 249
White mold, 237
White rust, 325
Wild bees, 377
Wild birds, 1, 14
Wild sunflower, 297
Wilt, 241, 242, 243, 244, 245, 254
Wilting coefficient, 94
Winter crop, 96
Winterization, 417
Witches broom, 256
World collection, 299

W

Warburg effect, 90
Warriors, 3
Washington State, 8
Water potentials, 119, 121
Water requirement, 93
Water use efficiency, 121
Waxes, 412
Weed competition, 124
Weed control, 102, 105, 124, 375, 392

Y

Yellows, 251
Yellows (also aster yellows), 256
Yield, 390
  components, 308
  compensation, 229
  seed, 10
  stability of, 289, 309
Yugoslavia, 7, 250, 255
  damage by birds in, 266
  losses to birds in, 269

The assistance of J. J. Hammond, Agronomy Department, North Dakota State University, Fargo, in writing the Fortran computer program to alphabetize the Index is gratefully acknowledged. The program is available for similar use elsewhere.